Nicola Gigli
Measure Theory in Non-Smooth Spaces

Nicola Gigli

Measure Theory in Non-Smooth Spaces

Managing Editor: Agnieszka Bednarczyk-Drąg
Series Editor: Gianluca Crippa
Language Editor: Sara Tavares

Open Access
Partial Differential Equations and Measure Theory

DE GRUYTER
OPEN

ISBN 978-3-11-055082-5
e-ISBN (PDF) 978-3-11-055083-2

Library of Congress Cataloging-in-Publication Data
A CIP catalog record for this book has been applied for at the Library of Congress.

Published by De Gruyter Open Ltd, Warsaw/Berlin
Part of Walter de Gruyter GmbH, Berlin/Boston
The book is published with open access at www.degruyter.com.
Cover illustration: © *sakkmesterke*

Managing Editor: Agnieszka Bednarczyk-Drąg
Series Editor: Gianluca Crippa
Language Editor: Sara Tavares

www.degruyteropen.com

Contents

Luigi Ambrosio and Shouhei Honda

New stability results for sequences of metric measure spaces with uniform Ricci bounds from below

1.1 Introduction

In this paper we establish new stability properties for sequences of metric measure spaces $(X, \mathsf{d}_i, \mathfrak{m}_i)$ convergent in the measured Gromov-Hausdorff sense (mGH for short). Even though some results are valid under weaker assumptions, to give a unified treatment of the several topics treated in this paper we confine our discussion to sequences of $RCD(K, \infty)$ metric measure spaces, with $K \in \mathbb{R}$ independent of i. A pointed mGH limit of a sequence of Riemannian manifolds with a uniform lower Ricci curvature bound, called Ricci limit space, gives a typical example of a $RCD(K, \infty)$ metric measure space. This paper provides new results even for such sequences and for the corresponding Ricci limit spaces. Our stability results, relative to spectral properties and Hessians, extend the ones in [33], [34] for compact Ricci limit spaces.

The stability of the curvature-dimension conditions has been treated in the seminal papers [41], [48], while stability of the "Riemannian" condition (i.e. the quadratic character of Cheeger's energy) has been estabilished in [7]. Consider complex objects derived from the metric measure structure like derivations, Lagrangian flows associated to derivations, heat flows, Hessians. It is by now quite clear that treating the stability of such objects is possible by adopting the so-called extrinsic approach (even though we do not exclude other possibilities). This approach assumes that $(X_i, \mathsf{d}_i) = (X, \mathsf{d})$ are independent of i, and that $\mathfrak{m}_i$ weakly converge to $\mathfrak{m}$ in duality with $C_{\mathrm{bs}}(X)$, the space of continuous functions with bounded support. We follow this approach, also because this paper builds upon the recent papers [31] (for stability of heat flows and Mosco convergence of Cheeger's energies) and [14] (for strong convergence of derivations) which use the same one. See also [31, Theorem 3.15] for a detailed comparison between the extrinsic approach and other intrinsic ones, with or without doubling assumptions. In a broader context, see also the recent monograph [47] for detailed analysis of convergence and concentration for metric measure structures.

Before moving to a more precise technical description of the paper's content, we discuss the main applications:

Spectral gap. We discuss joint continuity w.r.t. $\big(p, (X, \mathsf{d}, \mathfrak{m})\big)$ of the p-spectral gaps

$$\big(\lambda_{1,p}(X, \mathsf{d}, \mathfrak{m})\big)^{1/p} \tag{1.1}$$

w.r.t. mGH convergence. Here, for $p \in [1, \infty)$, $\lambda_{1,p}$ is the first positive eigenvalue of the p-Laplacian when $p > 1$, and Cheeger's constant when $p = 1$, see (1.54) for the precise definition in our setting. This extends the analysis of [31] from $p = 2$ to general p and even to the case when p depends on i. See Theorem 1.9.4 and also Theorem 1.9.6, dealing with the case $p_i \to \infty$, with

$$\left(\lambda_{1,\infty}(X, \mathsf{d}, \mathfrak{m})\right)^{1/\infty} := \frac{2}{\operatorname{diam}(\operatorname{supp}\mathfrak{m})}. \tag{1.2}$$

These general continuity properties were conjectured in [33] in the Ricci limit setting, and so we provide an affirmative answer to the conjecture in the more general setting of $RCD(K, \infty)$ spaces. In particular, Theorem 1.9.4 yields that Cheeger's constants are continuous w.r.t. mGH convergence.

The class $RCD^*(K, N)$ of metric measure spaces has been proposed in [28] and deeply investigated in [8], [27] and [12] in the nonsmooth setting. Recall that in the class of smooth weighted n-dimensional Riemannian manifolds $(M^n, \mathsf{d}, e^{-V}\mathrm{vol}_{M^n})$ the $RCD^*(K, N)$ condition, $n \le N$, is equivalent to

$$\mathrm{Ric} + \mathrm{Hess}(V) - \frac{\nabla V \otimes \nabla V}{N - n} \ge KI.$$

Analogously, it is well-known that the condition $RCD(K, \infty)$ for $(M^n, \mathsf{d}, e^{-V}\mathrm{vol}_{M^n})$ is equivalent to $\mathrm{Ric} + \mathrm{Hess}(V) \ge KI$.

By combining the continuity of (1.1) with the compactness property of the class of $RCD^*(K, N)$-spaces w.r.t. the mGH convergence, we also establish a uniform bound

$$C_1 \le \left(\lambda_{1,p}(X, \mathsf{d}, \mathfrak{m})\right)^{1/p} \le C_2, \tag{1.3}$$

where C_i are positive constants depending only on K, $N < \infty$, and two-sided bounds of the diameter, i.e. C_i do *not* depend on p (Proposition 1.11.1).

Suspension theorems. The second application is related to almost spherical suspension theorems of positive Ricci curvature. For simplicity we discuss here only the case when $N \ge 2$ is an integer, but our results (as those in [48], [37], [38], [19]) cover also the case $N \in (1, \infty)$. In [19] Cavalletti-Mondino proved that for any $RCD^*(N-1, N)$-space, the quantity (1.1) is greater than or equal than $\left(\lambda_{1,p}(\mathbf{S}^N, \mathsf{d}, \mathfrak{m}_N)\right)^{1/p}$ for any $p \in [1, \infty)$, where $\mathbf{S}^N$ is the unit sphere in $\mathbb{R}^{N+1}$, d is the standard metric of sectional curvature 1, and $\mathfrak{m}_N$ is the N-dimensional Hausdorff measure. Moreover, equality implies that the metric measure space is isomorphic to a spherical suspension. Under our notation (1.2) as above, this observation is also true when $p = \infty$, which corresponds to the Bonnet-Myers theorem in our setting (see [48] by Sturm). Note that [19] also provides rigidity results as the following one: for a fixed $p \in [1, \infty]$, if $(\lambda_{1,p})^{1/p}$ is close to $\left(\lambda_{1,p}(\mathbf{S}^N, \mathsf{d}, \mathfrak{m}_N)\right)^{1/p}$, then the space is Gromov-Hausdorff close to the spherical suspension of a compact metric space, a so-called *almost* spherical suspension theorem. The converse is known for $p \in \{2, \infty\}$ in [37, 38] by Ketterer and we extend the result to general p; in addition, combining this with the joint spectral continuity result we can remove the p-dependence in the almost spherical suspension theorem, i.e. if $(\lambda_{1,p})^{1/p}$ is close to $(\lambda_{1,p}(\mathbf{S}^N, \mathsf{d}, \mathfrak{m}_N))^{1/p}$ for *some* $p \in [1, \infty]$, then this happens for any other $q \in [1, \infty]$, see Corollary 1.11.6. This seems to be new even for compact n-dimensional Riemannian manifolds endowed with the n-dimensional Hausdorff measure. In particular, by using Petrunin's compatibility result [44] between Alexandrov spaces and curvature-dimension conditions, this extension of the result to general p also holds for all finite-dimensional Alexandrov spaces with curvature bounded below by 1, which is also new.

Stability of Hessians and of Gigli's measure-valued Ricci tensor. The final application deals with stability of Hessians and Ricci tensor with respect to mGH convergence. These notions come from the second order differential calculus on $RCD(K, \infty)$ spaces fully developed by Gigli in [29], starting from ideas from Γ-calculus. For Ricci limit spaces, analogous stability results were estabilished in [34]. In this respect, the main novelty of this paper is the treatment of $RCD(K, \infty)$ spaces, dropping also the dimensionality assumption. The main results are the stability of Hessians, see Corollary 1.10.4 and Corollary 1.10.3, and a kind of localized stability of the measure-valued Ricci tensor. In connection with the latter, specifically, we prove in Theorem 1.10.5 that local lower bounds of the form

$$\mathbf{Ric}(\nabla f, \nabla f) \geq \zeta |\nabla f|^2 \mathfrak{m},$$

with $\zeta \in C(X)$ bounded from below, are stable under mGH convergence. This way, also nonconstant bounds from below on the Ricci tensor can be proved to be stable (see also [39] for stability results in the same spirit, obtained from a localization of the Lagrangian definition of curvature/dimension bounds). On the other hand, since our approach is extrinsic, this result becomes of interest from the intrinsic point of view only when ζ's depending on the metric structure, as $\varphi \circ \mathsf{d}$, are considered. See also Remark 1.10.7 for an analogous stability property of the $BE(K, N)$ condition with K and N dependent on $x \in X$.

We believe that these stability results and the tools developed in this paper could be the basis for the analysis of the stability of the other calculus tools and concepts developed in [29], as exterior and covariant derivatives, Hodge laplacian, etc. However, we will not pursue this point of view in this paper.

Organization of the paper. In Section 1.2 we introduce the main measure-theoretic preliminaries. In Section 1.3 we discuss convergence of functions f_i in different measure spaces relative to $\mathfrak{m}_i$; here the main new ingredient is a notion of L^{p_i} convergence which also covers the case when the exponents p_i converge to $p \in [1, \infty)$. We discuss the case of strong convergence, and of weak convergence when $p > 1$. Section 1.4 recalls the main terminology and the main known facts about $RCD(K, \infty)$ spaces and the regularizing properties of the heat flow h_t. Less standard facts proved in this section are: the formula provided in Proposition 1.4.5 for $u \mapsto \int_X |\nabla u| \, d\mathfrak{m}$ (somehow reminiscent of the duality tangent/cotangent bundle at the basis of [29]), of particular interest for the proof of lower semicontinuity properties, and the weak isoperimetric property of Proposition 1.4.7.

In Section 1.5 we enter the core of the paper, somehow "localizing" the Mosco convergence result of Cheeger's energies of [31]. The main result is Theorem 1.5.7 where we prove, among other things, that the measures $|\nabla f_i|_i^2 \mathfrak{m}_i$ weakly converge to $|\nabla f|^2 \mathfrak{m}$ whenever f_i strongly converge to f in $H^{1,2}$ (i.e., f_i L^2-strongly converge to f and the Cheeger energies of f_i converge to the Cheeger energy of f). To prove this, the main difficulty is the localization of the lim inf inequality of [31]; we obtain it using the recent results in [14], for families of derivations with convergent L^2 norms (in this case, gradient derivations, see Theorem 1.5.6 in this paper). Section 1.6 covers the stability properties of BV functions, the main result is that $f \in BV(X, \mathsf{d}, \mathfrak{m}_i)$ whenever $f_i \in BV(X, \mathsf{d}, \mathfrak{m}_i)$ L^1-strongly converge to f, with $L = \liminf_i |\mathrm{D}f_i|(X) < \infty$. In addition, $|\mathrm{D}f|(X) \le L$. The proof of these stability properties strongly relies on the results of Section 1.5 and, notwithstanding the well-estabilished Eulerian-Lagrangian duality for Sobolev and BV spaces (see [3] for the latter spaces) it seems harder to get from the Lagrangian point of view.

Section 1.7 covers compactness results for BV and $H^{1,p}$, also in the case when p depends on i. In the proof of these facts we use the (local) strong L^2 compactness properties for sequences bounded $H^{1,2}$ proved in [31]; the generalization from the exponent 2 to higher exponents is quite simple, while the treatment of smaller powers and the improvement from L^p_{loc} to L^p convergence (essential for our results in Section 1.9) requires the existence of uniform isoperimetric profiles. We review the state of the art on this topic in Theorem 1.7.2. In Section 1.8 we prove Γ-convergence of the p_i-Cheeger energies $\mathsf{Ch}^i_{p_i}$ relative to $(X, \mathsf{d}, \mathfrak{m}_i)$ (set equal to the total variation functional $f \mapsto |\mathrm{D}f|(X)$ in BV when $p = 1$), namely

$$\liminf_{i\to\infty} \mathsf{Ch}^i_{p_i}(f_i) \ge \mathsf{Ch}_p(f)$$

whenever f_i L^{p_i}-strongly converge to f, and the existence of a sequence f_i with this property satisfying $\limsup_i \mathsf{Ch}^i_{p_i}(f_i) \le \mathsf{Ch}_p(f)$. The only difference with the case $p = 2$ considered in [31] is that, in general, we are not able to achieve the lim inf inequality with L^{p_i}-weakly convergent sequences, unless a uniform isoperimetric assumption on the spaces grants relative compactness w.r.t. strong L^{p_i} convergence. Under this assumption, Mosco and Γ-convergence coincide.

Finally, Section 1.9, Section 1.10 and Section 1.11 cover the above mentioned stability results for p-eigenvalues and eigenfunctions (using Section 1.7 and Section 1.8), for Hessians and Ricci tensors (using Section 1.5), and the dimensional results relative to the suspension theorems (using Section 1.9).

Acknowledgement. The first author acknowledges helpful conversations on the subject of this paper with Fabio Cavalletti, Andrea Mondino and Giuseppe Savaré. The second author acknowledges the support of the JSPS Program for Advancing Strategic International Networks to Accelerate the Circulation of Talented Researchers, the

Grant-in-Aid for Young Scientists (B) 16K17585 and the warm hospitality of SNS.
The authors warmly thank the referee for the detailed reading of the paper and for the constructive comments.

1.2 Notation and basic setting

Metric concepts. In a metric space (X, d), we denote by $B_r(x)$ and $\overline{B}_r(x)$ the open and closed balls respectively, by $\mathrm{C_{bs}}(X)$ the space of bounded continuous functions with bounded support, by $\mathrm{Lip_{bs}}(X) \subset \mathrm{C_{bs}}(X)$ the subspace of Lipschitz functions. We use the notation $\mathrm{C_b}(X)$ and $\mathrm{Lip_b}(X)$ for bounded continuous and bounded Lipschitz functions respectively.

For $f : X \to \mathbb{R}$ we denote by $\mathrm{Lip}(f) \in [0, \infty]$ the Lipschitz constant and by $\mathrm{lip}(f)$ the slope, namely

$$\mathrm{lip}(f)(x) := \limsup_{y \to x} \frac{|f(y) - f(x)|}{\mathrm{d}(y, x)}. \tag{1.4}$$

We also define the asymptotic Lipschitz constant by

$$\mathrm{Lip}_a f(x) = \inf_{r>0} \mathrm{Lip}\big(f|_{B_r(x)}\big) = \lim_{r \to 0^+} \mathrm{Lip}\big(f|_{B_r(x)}\big), \tag{1.5}$$

which is upper semicontinuous.

The metric algebra $\mathcal{A}_{\mathrm{bs}}$. We associate to any separable metric space (X, d) the smallest $\mathcal{A} \subset \mathrm{Lip_b}(X)$ containing

$$\min\{\mathrm{d}(\cdot, x), k\} \quad \text{with } k \in \mathbb{Q} \cap [0, \infty],\ x \in D \text{ and } D \subset X \text{ countable and dense} \tag{1.6}$$

which is a vector space over $\mathbb{Q}$ and is stable under products and lattice operations. It is a countable set and it depends only on the choice of the set D (but this dependence will not be emphasized in our notation, since the metric space will mostly be fixed). We shall work with the subalgebra $\mathcal{A}_{\mathrm{bs}}$ of functions with bounded support.

Measure-theoretic notation. The Borel σ-algebra of a metric space (X, d) is denoted $\mathcal{B}(X)$. The Borel signed measures with finite total variation are denoted by $\mathcal{M}(X)$, while we use the notation $\mathcal{M}^+(X)$, $\mathcal{M}^+_{\mathrm{loc}}(X)$, $\mathcal{P}(X)$ for nonnegative finite Borel measures, Borel measures which are finite on bounded sets and Borel probability measures.

We use the standard notation $L^p(X, \mathfrak{m})$, $L^p_{\mathrm{loc}}(X, \mathfrak{m})$ for the L^p spaces when $\mathfrak{m}$ is nonnegative ($p = 0$ is included and denotes the class of $\mathfrak{m}$-measurable functions). Notice that, in this context where no local compactness assumption is made, L^p_{loc} means p-integrability on bounded subsets.

Given metric spaces (X, d_X) and (Y, d_Y) and a Borel map $f : X \to Y$, we denote by $f_\#$ the induced push-forward operator, mapping $\mathcal{P}(X)$ to $\mathcal{P}(Y)$, $\mathcal{M}^+(X)$ to $\mathcal{M}^+(Y)$ and, if the preimage of bounded sets is bounded, $\mathcal{M}^+_{\mathrm{loc}}(X)$ to $\mathcal{M}^+_{\mathrm{loc}}(Y)$. Notice that, for all $\mu \in \mathcal{M}^+(X)$, $f_\#\mu$ is also well defined if f is μ-measurable.

Convergence of measures. We say that $\mathfrak{m}_n \in \mathcal{M}_{\mathrm{loc}}(X)$ weakly converge to $\mathfrak{m} \in \mathcal{M}_{\mathrm{loc}}(X)$ if $\int_X v \, \mathrm{d}\mathfrak{m}_n \to \int_X v \, \mathrm{d}\mathfrak{m}$ as $n \to \infty$ for all $v \in C_{\mathrm{bs}}(X)$. When all the measures $\mathfrak{m}_n$ as well as $\mathfrak{m}$ are probability measures, this is equivalent to requiring that $\int_X v \, \mathrm{d}\mathfrak{m}_n \to \int_X v \, \mathrm{d}\mathfrak{m}$ as $n \to \infty$ for all $v \in C_{\mathrm{b}}(X)$. We shall also use the following well-known proposition.

Proposition 1.2.1. *If $\mathfrak{m}_n$ weakly converge to $\mathfrak{m}$ in $\mathcal{M}^+_{\mathrm{loc}}(X)$, and if $\limsup_{i\to\infty} \int_X \Theta \, \mathrm{d}\mathfrak{m}_i$ is finite for some Borel $\Theta : X \to (0, \infty]$, then*

$$\lim_{i\to\infty} \int_X v \, \mathrm{d}\mathfrak{m}_i = \int_X v \, \mathrm{d}\mathfrak{m} \tag{1.7}$$

for all $v : X \to \mathbb{R}$ continuous with $\lim_{\mathrm{d}(x,\bar{x})\to\infty} |v|(x)/\Theta(x) = 0$ for some (and thus all) $\bar{x} \in X$. If $\Theta : X \to [0, \infty)$ is continuous and

$$\limsup_{n\to\infty} \int_X \Theta \, \mathrm{d}\mathfrak{m}_n \le \int_X \Theta \, \mathrm{d}\mathfrak{m} < \infty,$$

then (1.7) *holds for all $v : X \to \mathbb{R}$ continuous with $|v| \le C\Theta$ for some constant C.*

Metric measure space. Throughout this paper, a *metric measure space* is a triple $(X, \mathrm{d}, \mathfrak{m})$, where (X, d) is a complete and separable metric space and $\mathfrak{m} \in \mathcal{M}^+_{\mathrm{loc}}(X)$.

As explained in the introduction, in this paper we always consider metric measure spaces according to the previous definition. When a sequence convergent in the measured-Gromov Hausdorff sense is considered, we shall always assume (up to an isometric embedding in a common space) that the sequence has the structure $(X, \mathrm{d}, \mathfrak{m}_i)$ with $\mathfrak{m}_i \in \mathcal{M}^+_{\mathrm{loc}}(X)$ weakly convergent to $\mathfrak{m} \in \mathcal{M}^+_{\mathrm{loc}}(X)$. In particular, this convention forces us to drop the condition $\operatorname{supp} \mathfrak{m} = X$, used in many papers where individual spaces are considered.

1.3 Convergence of functions

In our setting, we are dealing with a sequence $(\mathfrak{m}_i) \subset \mathcal{M}^+_{\mathrm{loc}}(X)$ weakly convergent to $\mathfrak{m} \in \mathcal{M}^+_{\mathrm{loc}}(X)$. Assuming that f_i in suitable Lebesgue spaces relative to $\mathfrak{m}_i$ are given, we discuss in this section suitable notions of weak and strong convergence for f_i. Motivated by the convergence results of Section 1.8 and Section 1.9, we extend the analysis of [31] and [14] to the case when the exponents $p_i \in [1, \infty)$ are allowed to vary, with $p_i \to p \in [1, \infty)$. For weak convergence we only consider the case $p > 1$ (we do not need L^1-weak convergence), while for strong convergence, in connection with the results of Section 1.6, we also consider the case $p = 1$.

Weak convergence. Assume that $p_i \in [1, \infty)$ converge to $p \in (1, \infty)$. We say that $f_i \in L^{p_i}(X, \mathfrak{m}_i)$ L^{p_i}-weakly converge to $f \in L^p(X, \mathfrak{m})$ if $f_i\mathfrak{m}_i$ weakly converge to $f\mathfrak{m}$ in

$\mathcal{M}_{loc}(X)$, with

$$\limsup_{i\to\infty} \|f_i\|_{L^{p_i}(X,\mathfrak{m}_i)} < \infty. \tag{1.8}$$

For $\mathbb{R}^k$-valued maps we understand the convergence componentwise.

It is obvious that L^{p_i}-weak convergence is stable under finite sums. The proof of the following result is very similar to the proof when p and $\mathfrak{m}$ are fixed, and is omitted.

Proposition 1.3.1. *If $f_i \in L^{p_i}(X, \mathfrak{m}_i; \mathbb{R}^k)$ L^{p_i}-weakly converge to $f \in L^p(X, \mathfrak{m}; \mathbb{R}^k)$, then*

$$\|f\|_{L^p(X,\mathfrak{m};\mathbb{R}^k)} \le \liminf_{i\to\infty} \|f_i\|_{L^{p_i}(X,\mathfrak{m}_i;\mathbb{R}^k)}.$$

Moreover, any sequence $f_i \in L^{p_i}(X, \mathfrak{m}_i; \mathbb{R}^k)$ such that (1.8) *holds admits a L^{p_i}-weakly convergent subsequence.*

Strong convergence. We discuss the simpler case $p_i = p$ first. If $p > 1$ we say that $f_i \in L^p(X, \mathfrak{m}_i; \mathbb{R}^k)$ L^p-strongly converge to $f \in L^p(X, \mathfrak{m}; \mathbb{R}^k)$ if, in addition to weak L^p-convergence, one has $\limsup_i \|f_i\|_{L^p(X,\mathfrak{m}_i;\mathbb{R}^k)} \le \|f\|_{L^p(X,\mathfrak{m};\mathbb{R}^k)}$. If $k = p = 1$, we say that $f_i \in L^1(X, \mathfrak{m}_i)$ L^1-strongly converge to $f \in L^1(X, \mathfrak{m})$ if $\sigma \circ f_i$ L^2-strongly converges to $\sigma \circ f$, where $\sigma(z) = \mathrm{sign}(z)\sqrt{|z|}$ is the signed square root.

In the following remark we see that strong convergence can be written in terms of convergence of the probability measures naturally associated to the graphs of f_i; this also holds for vector-valued maps and we will use this fact in the proof of Proposition 1.3.3.

Remark 1.3.2 (Convergence of graphs versus L^p-strong convergence). If $p > 1$ one can use the strict convexity of the map $z \in \mathbb{R}^k \mapsto |z|^p$ to prove that $F_i : X \to \mathbb{R}^k$ L^p-strongly converge to F if and only if $\mu_i = (Id \times F_i)_\# \mathfrak{m}_i$ weakly converge to $\mu = (Id \times F)_\# \mathfrak{m}$ in duality with

$$\mathrm{C_p}(X \times \mathbb{R}^k) := \left\{ \psi \in C(X \times \mathbb{R}^k) : |\psi(x, z)| \le C|z|^p \text{ for some } C \ge 0 \right\} \tag{1.9}$$

(see for instance [5, Section 5.4], [31]).
If $p = k = 1$, we can use the fact that the signed square root is a homeomorphism of $\mathbb{R}$, and the equivalence estabilished in the quadratic case to get the same result.

We recall in the following proposition a few well-known properties of L^p-strong convergence, see also [35], [31] for a more detailed treatment of this topic.

Proposition 1.3.3. *For all $p \in [1, \infty)$ the following properties hold:*
(a) If f_i L^p-strongly converge to f the functions $\varphi \circ f_i$ L^p-strongly converge to $\varphi \circ f$ for all $\varphi \in \mathrm{Lip}(\mathbb{R})$ with $\varphi(0) = 0$.
(b) If f_i, g_i L^p-strongly converge to f, g respectively, then $f_i + g_i$ L^p-strongly converge to $f + g$.

(c) If f_i L^p-strongly (resp. L^p-weakly) converge to f, then ϕf L^p-strongly (resp. L^p-weakly) converge to ϕf for all $\phi \in C_b(X)$ (resp. $\phi \in C_{bs}(X)$).

(d) If f_i L^2-strongly converge to f and g_i L^2-weakly converge to g, then

$$\lim_{i\to\infty} \int_X f_i g_i \, d\mathfrak{m}_i = \int_X fg \, d\mathfrak{m}.$$

If g_i are also L^2-strongly convergent, then $f_i g_i$ are L^1-strongly convergent.

(e) If (g_i) is uniformly bounded in L^∞ and L^1-strongly convergent to g, then

$$\lim_{i\to\infty} \|g_i\|_{L^{p_i}(X,\mathfrak{m}_i)} = \|g\|_{L^p(X,\mathfrak{m})}$$

whenever $p_i \in [1,\infty)$ converge to $p \in [1,\infty)$.

Proof. (a) In the case $p > 1$ this is a simple consequence of Remark 1.3.2, since $\mu_i = (Id \times f_i)_\# \mathfrak{m}_i$ weakly converge to $\mu = (Id \times f)_\# \mathfrak{m}$ in duality with the space $C_p(X \times \mathbb{R})$ in (1.9). Since $\tilde{\psi}(x,z) = \psi(x, \varphi(z))$ belongs to $C_p(X \times \mathbb{R})$ for all $\psi \in C_p(X \times \mathbb{R})$ it follows that $(Id \times \varphi \circ f_i)_\# \mathfrak{m}_i$ weakly converge to $\mu = (Id \times \varphi \circ f)_\# \mathfrak{m}$ in duality with $C_p(X \times \mathbb{R})$, and then Remark 1.3.2 applies again to provide the L^p-strong convergence of $\varphi \circ f_i$ to $\varphi \circ f$.

In the case $p = 1$, since $\sigma(\varphi(f_i)) = \operatorname{sign}(\varphi \circ f_i)\sqrt{|\varphi| \circ f_i}$, from the strong L^2-convergence of $\sqrt{\varphi^\pm \circ f_i}$ to $\sqrt{\varphi^\pm \circ f}$ and the additivity of L^2-strong convergence (proved in the first line of the proof of (b), independently of (a)) we get the result.

(b) The case $p > 1$ is dealt with, for instance, in [35], see Corollary 3.26 and Proposition 3.31 therein. In order to prove additivity for $p = 1$ we can reduce ourselves, thanks to the stability under left composition proved in (a), to the sum of nonnegative functions u_i, v_i. Since $\sqrt{u_i}$ and $\sqrt{v_i}$ are L^2-strongly convergent, using the identity $\sqrt{u_i + v_i} = \sqrt{\sqrt{u_i}^2 + \sqrt{v_i}^2}$ we obtain that also $\sqrt{u_i + v_i}$ is strongly L^2-convergent.

The proof of (c) is a simple consequence of the definitions of L^p-strong convergence, splitting φ and f_i in positive and negative parts to deal also with the case $p = 1$.

The proof of the first part of statement (d) is a simple consequence of

$$\liminf_i \|f_i + t g_i\|_{L^2(X,\mathfrak{m}_i)} \ge \|f + tg\|_{L^2(X,\mathfrak{m})} \qquad \forall t \in \mathbb{R},$$

see also Section 1.8 where a similar argument is used in connection with Mosco convergence. In order to prove L^1-strong convergence when also g_i are L^2-strongly convergent, we can reduce ourselves to the case when f_i and g_i are nonnegative. Then, convergence of the L^2 norms of $\sqrt{f_i g_i}$ follows by the first part of the statement; weak convergence of $\sqrt{f_i g_i}\mathfrak{m}_i$ to $\sqrt{fg}\mathfrak{m}$ follows by Remark 1.3.2, with $k = p = 2$, $F_i = (f_i, g_i)$ and $\psi(z) = \sqrt{|z_1||z_2|}$.

For the proof of (e), let $N = \sup_i \|g_i\|_{L^\infty(X,\mathfrak{m}_i)}$ and notice first that (g_i) is uniformly bounded in L^{p_i}. Hence, the lim inf inequality follows by the L^{p_i}-weak convergence of g_i to g. The proof of the lim sup inequality follows by statement (a) with $\varphi(z) =$

$|z|^p \wedge N^p$, which ensures that $\int_X \varphi(g_i)\,\mathrm{d}\mathfrak{m}_i \to \int_X \varphi(g)\,\mathrm{d}\mathfrak{m} = \|g\|^p_{L^p(X,\mathfrak{m})}$, noticing that $p_i \to p$ implies $\int_X \varphi(g_i)\,\mathrm{d}\mathfrak{m}_i - \int_X |g_i|^{p_i}\,\mathrm{d}\mathfrak{m}_i \to 0$. □

Now we turn to the general case $p_i \to p \in [1,\infty)$. We say that L^{p_i}-strongly converge to f if $f_i \in L^{p_i}(X,\mathfrak{m}_i)$, L^{p_i}-weakly convergent to $f \in L^p(X,\mathfrak{m})$ and if for any $\epsilon > 0$ we can find an additive decomposition $f_i = g_i + h_i$ with

(i) (g_i) uniformly bounded in L^∞, and strongly L^1-convergent;
(ii) $\sup_i \|h_i\|_{L^{p_i}(X,\mathfrak{m}_i)} < \epsilon$.

It is obvious from the definition that also L^{p_i}-strong convergence is stable under finite sums. In the following proposition we show that stability under composition with Lipschitz maps φ holds and that L^{p_i} convergence implies convergence of the L^{p_i} norms.

Proposition 1.3.4 (Properties of L^{p_i}-strong convergence). *The following properties hold:*
(a) If f_i L^{p_i}-strongly converge to f, the functions $\varphi \circ f_i$ L^{p_i}-strongly converge to $\varphi \circ f$ for all $\varphi \in \mathrm{Lip}(\mathbb{R})$ with $\varphi(0) = 0$.
(b) If (f_i) is L^{p_i}-strongly convergent to $f \in L^p(X,\mathfrak{m})$, then

$$\lim_{i\to\infty} \|f_i\|_{L^{p_i}(X,\mathfrak{m}_i)} = \|f\|_{L^p(X,\mathfrak{m})}.$$

Proof. (a) Possibly splitting φ in positive and negative parts we can assume $\varphi \geq 0$. Since φ is a contraction, taking also Proposition 1.3.3(a) into account, it is immediate to check that decompositions $f_i = g_i + h_i$ induce decompositions $\varphi \circ g_i + (\varphi \circ f_i - \varphi \circ g_i)$ of $\varphi \circ f_i$; in addition, if ψ is any L^{p_i}-weak limit point of $(\varphi \circ f_i)$, from the lower semicontinuity of L^{p_i} convergence we get

$$\begin{aligned}
&\|\psi - \varphi \circ g\|_{L^p(X,\mathfrak{m})} \leq \liminf_{i\to\infty} \|\varphi \circ h_i\|_{L^{p_i}(X,\mathfrak{m}_i)} \leq \mathrm{Lip}(\varphi)\epsilon \\
&\|\varphi \circ f - \varphi \circ g\|_{L^p(X,\mathfrak{m})} \leq \mathrm{Lip}(\varphi)\|f - g\|_{L^p(X,\mathfrak{m})} \leq \mathrm{Lip}(\varphi) \liminf_{i\to\infty} \|h_i\|_{L^{p_i}(X,\mathfrak{m}_i)} \\
&\leq \mathrm{Lip}(\varphi)\epsilon,
\end{aligned}$$

where g denotes the L^{p_i}-strong limit of g_i. Since ϵ is arbitrary, we obtain that $\psi = \varphi \circ f$, and this proves the L^{p_i}-strong convergence of f_i to f.

(b) The lim inf inequality follows by weak convergence. If $f_i = g_i + h_i$ is a decomposition as in (i), (ii), and if g is the L^{p_i}-strong limit of g_i, the lim sup inequality is a direct consequence of the inequality $\|f - g\|_{L^p(X,\mathfrak{m})} < \epsilon$ and of

$$\lim_{i\to\infty} \|g_i\|_{L^{p_i}(X,\mathfrak{m}_i)} = \|g\|_{L^p(X,\mathfrak{m})},$$

ensured by Proposition 1.3.3(e). □

1.4 Minimal relaxed slopes, Cheeger energy and $RCD(K,\infty)$ spaces

In this section we recall basic facts about minimal relaxed slopes, Sobolev spaces and heat flow in metric measure spaces $(X, \mathsf{d}, \mathfrak{m})$, see [6] and [28] for a more systematic treatment of this topic. For $p \in (1,\infty)$ the p-th Cheeger energy $\mathsf{Ch}_p : L^p(X,\mathfrak{m}) \to [0,\infty]$ is the convex and $L^p(X,\mathfrak{m})$-lower semicontinuous functional defined as follows:

$$\mathsf{Ch}_p(f) := \inf\left\{ \liminf_{n\to\infty} \frac{1}{p}\int_X \mathrm{Lip}_a^p(f_n)\,d\mathfrak{m} \;:\; f_n \in \mathrm{Lip}_b(X)\cap L^p(X,\mathfrak{m}),\ \|f_n - f\|_p \to 0 \right\}. \tag{1.10}$$

The original definition in [22] involves generalized upper gradients of f_n in place of their asymptotic Lipschitz constant, but many other pseudo gradients (upper gradients, or the slope $\mathrm{lip}(f) \le \mathrm{Lip}_a(f)$, which is a particular upper gradient) can be used and all of them lead to the same definition. Indeed, all these pseudo gradients produce intermediate functionals between the functional in (1.10) and the functional based on the minimal p-weak upper gradient of [46], which are shown to be coincident in [1] (see also the discussion in [6, Remark 5.12]).

The Sobolev spaces $H^{1,p}(X, \mathsf{d}, \mathfrak{m})$ are simply defined as the finiteness domains of Ch_p. When endowed with the norm

$$\|f\|_{H^{1,p}} := \left(\|f\|^p_{L^p(X,\mathfrak{m})} + p\mathsf{Ch}_p(f)\right)^{1/p}$$

these spaces are Banach, and reflexive if (X, d) is doubling (see [1]).

The case $p = 2$ plays an important role in the construction of the differentiable structure, following [29]. For this reason we use the disinguished notation $\mathsf{Ch} = \mathsf{Ch}_2$ and it can be proved that $H^{1,2}(X, \mathsf{d}, \mathfrak{m})$ is Hilbert if Ch is quadratic.

In connection with the definition of Ch, for all $f \in H^{1,2}(X, \mathsf{d}, \mathfrak{m})$ one can consider the collection $RS(f)$ of all functions in $L^2(X,\mathfrak{m})$ larger than a weak $L^2(X,\mathfrak{m})$ limit of $\mathrm{Lip}_a(f_n)$, with $f_n \in \mathrm{Lip}_b(X)$ and $f_n \to f$ in $L^2(X,\mathfrak{m})$. This collection describes a convex, closed and nonempty set, whose element with smallest $L^2(X,\mathfrak{m})$ norm is called minimal relaxed slope and denoted by $|\nabla f|$. We use the not completely appropriate nabla notation, instead of the notation $|\mathrm{D}f|$ of [29], since we will be dealing only with quadratic Ch. Notice also that a similar construction can be applied to Ch_p, and provides a minimal p-relaxed gradient that can indeed depend on p (see [26]). However, either under the doubling and Poincaré assumptions [22], or under curvature assumptions [30] this dependence disappears. In any case, we will only be dealing with the 2-minimal relaxed slope in this paper.

When Ch is quadratic we denote by $\langle\nabla f, \nabla g\rangle$ the canonical symmetric bilinear form from $[H^{1,2}(X, \mathsf{d}, \mathfrak{m})]^2$ to $L^1(X,\mathfrak{m})$ defined by

$$\langle\nabla f, \nabla g\rangle := \lim_{\epsilon\to 0} \frac{|\nabla(f+\epsilon g)|^2 - |\nabla f|^2}{2\epsilon} \tag{1.11}$$

(where the limit is understood in the $L^1(X, \mathfrak{m})$ sense). Notice also that the expression $\langle \nabla f, \nabla g \rangle$ still makes sense $\mathfrak{m}$-a.e. (i.e. up to $\mathfrak{m}$-negligible sets) for any $f, g \in \mathrm{Lip_b}(X)$ (not necessarily in the $H^{1,2}$ space, when $\mathfrak{m}(X) = \infty$), since f, g coincide on bounded sets with functions in the Sobolev class, and gradients satisfy the locality property on open and even on Borel sets.

Because of the minimality property, $|\nabla f|$ provides integral representation to Ch, so that

$$\int_X \langle \nabla f, \nabla g \rangle \, d\mathfrak{m} = \lim_{\epsilon \to 0} \frac{\mathsf{Ch}(f + \epsilon g) - \mathsf{Ch}(f)}{\epsilon}$$

and it is not hard to improve weak to strong convergence.

Theorem 1.4.1. *For all $f \in D(\mathsf{Ch})$ one has*

$$\mathsf{Ch}(f) = \frac{1}{2} \int_X |\nabla f|^2 \, d\mathfrak{m}$$

and there exist $f_n \in \mathrm{Lip_b}(X) \cap L^2(X, \mathfrak{m})$ with $f_n \to f$ in $L^2(X, \mathfrak{m})$ and $\mathrm{Lip}_a(f_n) \to |\nabla f|$ in $L^2(X, \mathfrak{m})$. In particular, if $H^{1,2}(X, \mathsf{d}, \mathfrak{m})$ is reflexive, there exist $f_n \in \mathrm{Lip_b}(X) \cap L^2(X, \mathfrak{m})$ satisfying $f_n \to f$ in $L^2(X, \mathfrak{m})$ and $|\nabla(f_n - f)| \to 0$ in $L^2(X, \mathfrak{m})$.

Most standard calculus rules can be proved, when dealing with minimal relaxed slopes. For the purposes of this paper the most relevant ones are:

Locality on Borel sets. $|\nabla f| = |\nabla g|$ $\mathfrak{m}$-a.e. on $\{f = g\}$ for all $f, g \in H^{1,2}(X, \mathsf{d}, \mathfrak{m})$;

Pointwise minimality. $|\nabla f| \le g$ $\mathfrak{m}$-a.e. for all $g \in RS(f)$;

Degeneracy. $|\nabla f| = 0$ $\mathfrak{m}$-a.e. on $f^{-1}(N)$ for all $f \in H^{1,2}(X, \mathsf{d}, \mathfrak{m})$ and all $\mathcal{L}^1$-negligible $N \in \mathcal{B}(\mathbb{R})$;

Chain rule. $|\nabla(\phi \circ f)| = |\phi'(f)||\nabla f|$ for all $f \in H^{1,2}(X, \mathsf{d}, \mathfrak{m})$ and all $\phi : \mathbb{R} \to \mathbb{R}$ Lipschitz with $\phi(0) = 0$.

Leibniz rule. If $f, g \in H^{1,2}(X, \mathsf{d}, \mathfrak{m})$ and $h \in \mathrm{Lip_b}(X)$, then

$$\langle \nabla f, \nabla(gh) \rangle = h \langle \nabla f, \nabla g \rangle + g \langle \nabla f, \nabla h \rangle \qquad \mathfrak{m}\text{-a.e. in } X.$$

Another object canonically associated to Ch and then to the metric measure structure is the heat flow h_t, defined as the $L^2(X, \mathfrak{m})$ gradient flow of Ch, according to the Brezis-Komura theory of gradient flows, see for instance [17]. This theory provides a continuous contraction semigroup h_t in $L^2(X, \mathfrak{m})$ which, under the growth condition

$$\mathfrak{m}\big(B_r(\bar{x})\big) \le c_1 e^{c_2 r^2} \qquad \forall r > 0, \tag{1.12}$$

extends to a continuous and mass preserving semigroup (still denoted h_t) in all $L^p(X, \mathfrak{m})$ spaces, $1 \le p < \infty$. In addition, h_t preserves upper and lower bounds with constants, namely $f \le C$ $\mathfrak{m}$-a.e. (resp. $f \ge C$ $\mathfrak{m}$-a.e.) implies $h_t f \le C$ $\mathfrak{m}$-a.e. (resp. $h_t f \ge C$ $\mathfrak{m}$-a.e.) for all $t \ge 0$.

We shall use h_t only in the case when Ch is quadratic, as a regularizing operator. We adopt the notation

$$D(\Delta) := \left\{ f \in H^{1,2}(X, \mathsf{d}, \mathfrak{m}) : \Delta f \in L^2(X, \mathfrak{m}) \right\} \tag{1.13}$$

namely $D(\Delta)$ is the class of functions $f \in H^{1,2}(X, \mathsf{d}, \mathfrak{m})$ satisfying $-\int_X vg \,d\mathfrak{m} = \int_X \langle \nabla f, \nabla v \rangle \,d\mathfrak{m}$ for all $v \in H^{1,2}(X, \mathsf{d}, \mathfrak{m})$, for some $g \in L^2(X, \mathfrak{m})$ (and then, since g is uniquely determined, $\Delta f := g$). When Ch is quadratic the semigroup h_t is linear (and this property is equivalent to Ch being quadratic) and it is easily seen that

$$\lim_{t \downarrow 0} h_t f = f \qquad \text{strongly in } H^{1,2} \text{ for all } f \in H^{1,2}(X, \mathsf{d}, \mathfrak{m}).$$

We shall extensively use the typical regularizing properties (independent of curvature assumptions)

$$h_t f \in W^{1,2}(X, \mathsf{d}, \mathfrak{m}) \text{ for all } f \in L^2(X, \mathfrak{m}),\ t > 0 \text{ and } \mathsf{Ch}(h_t f) \le \frac{\|f\|^2_{L^2(X,\mathfrak{m})}}{2t}, \tag{1.14}$$

$$h_t f \in D(\Delta) \text{ for all } f \in L^2(X, \mathfrak{m}),\ t > 0 \text{ and } \|\Delta h_t f\|^2_{L^2(X,\mathfrak{m})} \le \frac{\|f\|^2_{L^2(X,\mathfrak{m})}}{t^2}, \tag{1.15}$$

as well as the commutation rule $h_t \circ \Delta = \Delta \circ h_t$, $t > 0$.

Finally, we describe the class of $RCD(K, \infty)$ metric measure spaces of [7], where thanks to the lower bounds on Ricci curvature even stronger properties of h_t can be proved.

Definition 1.4.2 ($CD(K, \infty)$ and $RCD(K, \infty)$ spaces). *We say that a metric measure space $(X, \mathsf{d}, \mathfrak{m})$ satisfying the growth bound (1.12) (for some constants c_1, c_2 and some $\bar{x} \in X$) is a $RCD(K, \infty)$ metric measure space, with $K \in \mathbb{R}$, if:*

(a) the Relative Entropy Functional $\mathrm{Ent}(\mu) : \mathcal{P}_2(X) \to \mathbb{R} \cup \{\infty\}$ *given by*

$$\mathrm{Ent}(\mu) := \begin{cases} \int_X \rho \log \rho \,d\mathfrak{m} & \text{if } \mu = \rho\mathfrak{m} \ll \mathfrak{m}; \\ \infty & \text{otherwise} \end{cases} \tag{1.16}$$

where

$$\mathcal{P}_2(X) := \left\{ \mu \in \mathcal{P}(X) : \int_X \mathsf{d}^2(\bar{x}, x) \,d\mathfrak{m}(x) < \infty \right\},$$

is K-convex along Wasserstein geodesics in $\mathcal{P}_2(X)$, namely

$$\mathrm{Ent}(\mu_t) \le (1-t)\mathrm{Ent}(\mu_0) + t\mathrm{Ent}(\mu_1) - \frac{K}{2} t(1-t) W_2^2(\mu_0, \mu_1)$$

for all $\mu_0, \mu_1 \in D(\mathrm{Ent}) := \{\mu : \mathrm{Ent}(\mu) < \infty\}$, for some constant speed geodesic μ_t from μ_0 to μ_1 (so, this condition forces $D(\mathrm{Ent}), W_2)$ to be geodesic). This condition corresponds to the $CD(K, \infty)$ condition of [41], [48].

(b) Ch *is quadratic. This is the axiom added to the Lott-Sturm-Villani theory in [7].*

Remark 1.4.3 (On the growth condition (1.12)). *Notice that* (1.12) *is needed to give a meaning to the integral in* (1.16), *as it ensures the integrability of the negative part of* $\rho \log \rho$. *On the other hand, adopting a suitable convention on the meaning to be given to* Ent *in these cases of indeterminacy (so that the* $CD(K,\infty)$ *condition makes sense), it has been proved in [48] that* (1.12) *can be deduced from the* $CD(K,\infty)$ *condition, and that the constants* c_i *can be estimated in terms of* K *and of the measure of two concentric balls centered at* $\bar{x} \in \operatorname{supp} \mathfrak{m}$.

It is not hard to prove that the support of any $\mathrm{RCD}(K,\infty)$ (or even $CD(K,\infty)$ space) is length, namely the infimum of the length of the absolutely continuous curves connecting any two points $x, y \in \operatorname{supp} \mathfrak{m}$ is $\mathrm{d}(x,y)$. See [7] (dealing with finite reference measures), [9] (for the σ-finite case) and [8] for various characterizations of the class of $RCD(K,\infty)$ spaces. We quote here a few results, which essentially derive from the identification of h_t as the gradient flow of Ent w.r.t. the Wasserstein distance and the contractivity properties w.r.t. that distance.

It is proved in [7] that the formula

$$h_t g(x) := \int_X g(y) d\tilde{h}_t \delta_x(y) \qquad x \in X,\ t \geq 0$$

where $\tilde{h}_t$ is the dual K-contractive semigroup acting on $\mathcal{P}_2(X)$, provides a pointwise version of the semigroup on $L^2 \cap L^\infty(X,\mathfrak{m})$ with better continuity properties. This result is recalled in the next proposition. In the formula

$$\tilde{h}_t \mu := \int \tilde{h}_t \delta_x \,\mathrm{d}\mu(x)$$

provides a canonical extension of $\tilde{h}_t$ to the whole of $\mathcal{P}(X)$, used in Proposition 1.6.3.

Proposition 1.4.4 (Regularizing properties of h_t). *Let* $(X,\mathrm{d},\mathfrak{m})$ *be a* $RCD(K,\infty)$ *metric measure space. Then, any* $f \in H^{1,2}(X,\mathrm{d},\mathfrak{m})$ *with* $|\nabla f| \in L^\infty(X,\mathfrak{m})$ *has a Lipschitz representative* $\tilde{f}$, *with* $\mathrm{Lip}(\tilde{f}) = \||\nabla f|\|_{L^\infty(X,\mathfrak{m})}$ *and the following properties hold for all* $t > 0$:

(a) *if* $f \in L^2 \cap L^\infty(X,\mathfrak{m})$ *one has* $h_t f \in \mathrm{Lip}_b(X) \cap H^{1,2}(X,\mathrm{d},\mathfrak{m})$ *with*

$$|\nabla h_t f| = \mathrm{lip}(h_t f) \quad \mathfrak{m}\text{-a.e.}, \qquad \mathrm{Lip}(h_t f) \leq \frac{1}{\sqrt{2 \mathrm{I}_{2K}(t)}} \|f\|_{L^\infty(X,\mathfrak{m})}; \tag{1.17}$$

(b) *for all* $f \in H^{1,2}(X,\mathrm{d},\mathfrak{m})$ *with* $|\nabla f| \in L^\infty(X,\mathfrak{m})$ *the Bakry-Émery condition holds in the form*

$$\mathrm{Lip}_a(h_t f, x) \leq e^{-Kt} h_t |\nabla f|(x) \qquad \forall x \in X; \tag{1.18}$$

(c) *if* $\mu \in \mathcal{P}_2(X)$, *then* $\tilde{h}_t \mu = f_t \mathfrak{m}$, *with*

$$\int_X f_t \log f_t \,\mathrm{d}\mathfrak{m} \leq \frac{1}{2\mathrm{I}_{2K}(t)} \left(r^2 + \int_X \mathrm{d}^2(x,\bar{x}) \,\mathrm{d}\mu(x) \right) - \log\left(\mathfrak{m}(B_r(\bar{x}))\right)$$

for all $\bar{x} \in X$ and $r > 0$.

Proof. (a) is proved in [7, 8], (b) in [45]. The inequality (c) follows by Wang's log-Harnack inequality, see [8, Theorem 4.8] for a proof in the $RCD(K, \infty)$ context. □

In $RCD(K, \infty)$ spaces we have a useful formula to represent the functional $\int_X |\nabla f| \, d\mathfrak{m}$.

Proposition 1.4.5. *For all $f \in H^{1,2}(X, \mathsf{d}, \mathfrak{m})$ one has that $|\nabla f|$ is the essential supremum of the family $\langle \nabla f, \nabla v \rangle$ as v runs in the family of 1-Lipschitz functions in $H^{1,2}(X, \mathsf{d}, \mathfrak{m})$. Moreover, for all $g : X \to [0, \infty)$ lower semicontinuous, one has*

$$\int_X |\nabla f| g \, d\mathfrak{m} = \sup \sum_k \int_X \langle \nabla f, \nabla v_k \rangle w_k \, d\mathfrak{m} \tag{1.19}$$

where the supremum runs among all finite collections of 1-Lipschitz functions $v_k \in H^{1,2}(X, \mathsf{d}, \mathfrak{m})$ and all $w_k \in C_{\mathrm{bs}}(X)$ with $\sum_k |w_k| \leq g$.

Proof. The proof of the representation of $|\nabla f|$ as essential supremum has been achieved in [15, Lemma 9.2]. We sketch the argument: denoting by M the essential supremum in the statement, one has obviously the inequalities $M \leq |\nabla f|$ $\mathfrak{m}$-a.e. and $|\langle \nabla f, \nabla v \rangle| \leq M \,\mathrm{Lip}(v)$ $\mathfrak{m}$-a.e. for all $v \in H^{1,2}(X, \mathsf{d}, \mathfrak{m})$ Lipschitz and bounded. By localization, this last inequality is improved to $|\langle \nabla f, \nabla v \rangle| \leq M \mathrm{Lip}_a(v)$ $\mathfrak{m}$-a.e. for all $v \in H^{1,2}(X, \mathsf{d}, \mathfrak{m})$ Lipschitz and bounded and then a density argument provides the inequality $|\langle \nabla f, \nabla v \rangle| \leq M |\nabla v|$ for all $v \in H^{1,2}(X, \mathsf{d}, \mathfrak{m})$ Lipschitz and bounded, which leads to $|\nabla f| \leq M$ choosing $v = f$.

In order to prove (1.19) we remark that the representation of $|\nabla f|$ as essential supremum yields

$$\int_X g |\nabla f| \, d\mathfrak{m} = \sup \sum_k c_k \int_{B_k} \langle \nabla f, \nabla v_k \rangle \, d\mathfrak{m}$$

where the supremum runs among all finite Borel partitions B_k of X, constants $c_k \leq \inf_{B_k} g$ and all choices of bounded 1-Lipschitz functions $v_k \in H^{1,2}(X, \mathsf{d}, \mathfrak{m})$. By inner regularity, the supremum is unchanged if we replace the Borel partitions by finite families of pairwise disjoint compact sets K_k. In turn, these families can be approximated by functions $w_k \in C_{\mathrm{bs}}(X)$ with $\sum_k |w_k| \leq g$. □

Now we recall three useful functional inequalities available in $RCD(K, \infty)$ spaces.

Proposition 1.4.6. *If $(X, \mathsf{d}, \mathfrak{m})$ is a $RCD(K, \infty)$ metric measure space, for all $f \in \mathrm{Lip}_{\mathrm{bs}}(X)$ one has*

$$\int_X |h_t f - f| \, d\mathfrak{m} \leq c(t, K) \int_X |\nabla f| \, d\mathfrak{m} \tag{1.20}$$

with $c(t, K) \sim \sqrt{t}$ as $t \downarrow 0$.

Proof. Fix $g \in L^\infty(X, \mathfrak{m})$ with $\|g\|_{L^\infty(X,\mathfrak{m})} \le 1$ and let us estimate the derivative of $t \mapsto \int_X g h_t f \,\mathrm{d}\mathfrak{m}$:

$$\begin{aligned}\left|\int_X g\Delta h_t f \,\mathrm{d}\mathfrak{m}\right| &= \left|\int_X g h_{t/2}\Delta h_{t/2} f \,\mathrm{d}\mathfrak{m}\right| = \left|\int_X h_{t/2} g\Delta h_{t/2} f \,\mathrm{d}\mathfrak{m}\right| \\ &= \left|\int_X \langle \nabla h_{t/2} g, \nabla h_{t/2} f\rangle \,\mathrm{d}\mathfrak{m}\right| \le \frac{1}{\sqrt{2\mathsf{I}_{2K}(t/2)}} \int_X |\nabla h_{t/2} f| \,\mathrm{d}\mathfrak{m} \\ &\le \frac{e^{-Kt/2}}{\sqrt{2\mathsf{I}_{2K}(t/2)}} \int_X |\nabla f| \,\mathrm{d}\mathfrak{m}.\end{aligned}$$

By integration, and then taking the supremum w.r.t. g, we get (1.20). □

When the space has finite diameter and $K \le 0$ we will also use, as a replacement of the isoperimetric inequality (presently known in the $RCD(K, \infty)$ setting only when $K > 0$), the following inequality, which is an easy consequence of Proposition 1.4.4(c).

Proposition 1.4.7. *If $(X, \mathsf{d}, \mathfrak{m})$ is a $RCD(K, \infty)$ metric measure space with $\mathfrak{m}(X) = 1$, and if $D = \operatorname{supp} \mathfrak{m}$ is finite, for all $\epsilon > 0$ we can find $M = M(\epsilon, D, K) \ge 1$ such that*

$$\int_{\{f \ge M \int_X f \,\mathrm{d}\mathfrak{m}\}} f \,\mathrm{d}\mathfrak{m} \le \epsilon\left(\int_X f \,\mathrm{d}\mathfrak{m} + \int_X |\nabla f| \,\mathrm{d}\mathfrak{m}\right).$$

for all $f \in \mathrm{Lip}_b(X)$ nonnegative.

Proof. The standard entropy inequality

$$\int_A g \,\mathrm{d}\mathfrak{m} \log\left(\frac{1}{\mathfrak{m}(A)} \int_A g \,\mathrm{d}\mathfrak{m}\right) \le \int_A g \log g \,\mathrm{d}\mathfrak{m} \le \int_X g \log g \,\mathrm{d}\mathfrak{m} + \frac{1}{e}\mathfrak{m}(X \setminus A)$$

provides a modulus of continuity ω_E, depending only on $E \ge 0$, such that g nonnegative and $\int_X g \log g \,\mathrm{d}\mathfrak{m} \le E$ imply $\int_A g \,\mathrm{d}\mathfrak{m} \le \omega_E(\mathfrak{m}(A))$.

Assume first $\int_X f \,\mathrm{d}\mathfrak{m} = 1$ and let $M > 0$. For all $t > 0$ we apply Proposition 1.4.6 and Proposition 1.4.4(c) with $r = D$ to get

$$\begin{aligned}\int_{\{f \ge M\}} f \,\mathrm{d}\mathfrak{m} &\le \int_{\{f \ge M\}} h_t f \,\mathrm{d}\mathfrak{m} + \int_X |h_t f - f| \,\mathrm{d}\mathfrak{m} \\ &\le \omega_{E_t}\left(\frac{1}{M}\right) + c(K, t) \int_X |\nabla f| \,\mathrm{d}\mathfrak{m}\end{aligned} \tag{1.21}$$

with

$$E_t = \frac{D^2}{\mathsf{I}_{2K}(t)} \ge \int_X h_t f \log h_t f \,\mathrm{d}\mathfrak{m}.$$

By a scaling argument, the inequality (1.21) implies

$$\int_{\{f\geq M\int_X f\,\mathrm{dm}\}} f\,\mathrm{dm} \leq \omega_{E_t}\Big(\frac{1}{M}\Big)\int_X f\,\mathrm{dm} + c(K,t)\int_X |\nabla f|\,\mathrm{dm} \qquad \forall t,\ M>0.$$

Then, given $\epsilon > 0$ we choose first $t > 0$ sufficiently small such that $c(t, K) < \epsilon$ and then M sufficiently large to reach our conclusion □

Finally, we close this section by reminding higher order properties, strongly inspired by Bakry's calculus, which played a fundamental role in the recent developments of the theory.

Proposition 1.4.8. *Let $(X, \mathrm{d}, \mathfrak{m})$ be a $RCD(K, \infty)$ space. Then*

$$\||\nabla f|\|_{L^4(X,\mathfrak{m})} \leq c\|f\|_\infty\|(\Delta - K^- I)f\|_{L^2(X,\mathfrak{m})} \tag{1.22}$$

for all $f \in L^\infty(X, \mathfrak{m}) \cap D(\Delta)$, and

$$\|\nabla|\nabla g|^2\|^2_{L^2(X,\mathfrak{m})} \leq -\int_X \big(2K|\nabla g|^4 + 2|\nabla g|^2\langle\nabla g, \nabla\Delta g\rangle\big)\,\mathrm{dm} \tag{1.23}$$

for all $g \in H^{1,2}(X, \mathrm{d}, \mathfrak{m}) \cap \mathrm{Lip}_b(X) \cap D(\Delta)$ with $\Delta g \in H^{1,2}(X, \mathrm{d}, \mathfrak{m})$.

Proof. See [11, Theorem 3.1] for (1.22), [45, Section 3] for (1.23). □

1.5 Local convergence of gradients under Mosco convergence

The main goal of this section is to localize the Mosco convergence result of [31], proving convergence results for $\langle\nabla u_i, \nabla v_i\rangle_i$ to $\langle\nabla u, \nabla v\rangle$ when u_i are strongly convergent in $H^{1,2}$ to u, and v_i are weakly convergent in $H^{1,2}$ to v. When both sequences are strongly convergent, we obtain at least the weak convergence as measures. Our main tools are the Theorem 1.5.4 borrowed from [31] and the convergence results of [14] in the more general context of derivations (see Theorem 1.5.6).

Definition 1.5.1 (Mosco convergence). *We say that the Cheeger energies $\mathsf{Ch}^i := \mathsf{Ch}_{\mathfrak{m}_i}$ Mosco converge to Ch if both the following conditions hold:*

(a) (Weak-lim inf). *For every $f_i \in L^2(X, \mathfrak{m}_i)$ L^2-weakly converging to $f \in L^2(X, \mathfrak{m})$, one has*

$$\mathsf{Ch}(f) \leq \liminf_{i\to\infty} \mathsf{Ch}^i(f_i).$$

(b) (Strong-lim sup). *For every $f \in L^2(X, \mathfrak{m})$ there exist $f_i \in L^2(X, \mathfrak{m}_i)$, L^2-strongly converging to f with*

$$\mathsf{Ch}(f) = \lim_{i\to\infty} \mathsf{Ch}^i(f_i). \tag{1.24}$$

One of the main results of [31] is that Mosco convergence holds if $(X, \mathsf{d}, \mathfrak{m}_i)$ are $RCD(K, \infty)$ spaces with

$$\mathfrak{m}_i(B_r(\bar{x})) \le c_1 e^{c_2 r^2} \qquad \forall r > 0, \ \forall i \tag{1.25}$$

for some $\bar{x} \in X$ and c_1, $c_2 > 0$. This result holds even in the larger class of $CD(K, \infty)$ spaces and the uniform growth condition (1.25), that we prefer to emphasize, is actually a consequence of the local weak convergence of $\mathfrak{m}_i$ to $\mathfrak{m}$ and of the uniform lower bound on Ricci curvature (see Remark 1.4.3).

Following [31], we define weak and strong convergence in the Sobolev space $H^{1,2}$ in a natural way, and with a variable reference measure.

Definition 1.5.2 (Convergence in the Sobolev spaces). *We say that $f_i \in H^{1,2}(X, \mathsf{d}, \mathfrak{m}_i)$ are weakly convergent in $H^{1,2}$ to $f \in H^{1,2}(X, \mathsf{d}, \mathfrak{m})$ if f_i are L^2-weakly convergent to f and $\sup_i \mathsf{Ch}^i(f_i)$ is finite. Strong convergence in $H^{1,2}$ is defined by requiring L^2-strong convergence of the functions, and that $\mathsf{Ch}(f) = \lim_i \mathsf{Ch}^i(f_i)$.*

Notice that the sequence $f_i = h$, with $h \in \mathrm{Lip}_{bs}(X)$ fixed, need not be strongly convergent in $H^{1,2}$, as the following simple example taken from [14] shows. The reason is that this sequence should not be considered as a constant one since the supports of $\mathfrak{m}_i$ can well be pairwise disjoint.

Example 1.5.3. *Take $X = \mathbb{R}^2$ endowed with the Euclidean distance, $f(x_1, x_2) = x_2$ and let*

$$\mathfrak{m}_i = i\mathcal{L}^2 \llcorner ([0,1] \times [0, \tfrac{1}{i}]), \qquad \mathfrak{m} = \mathcal{H}^1 \llcorner [0,1] \times \{0\}.$$

Then, it is easily seen that $|\nabla f|_i = 1$ while $|\nabla f| = 0$.

It is immediate to check that weak convergence in $H^{1,2}$ is stable under finite sums; it follows from (1.26) below that the same holds for strong convergence in $H^{1,2}$. Also, Theorem 1.7.4 below (borrowed from [31]) yields that weakly convergent sequences are also L^2_{loc}-strongly convergent, and provides conditions under which this can be improved to L^2-strong convergence.

Theorem 1.5.4 (Mosco convergence under uniform Ricci bounds). *If $(X, \mathsf{d}, \mathfrak{m}_i)$ are $RCD(K, \infty)$ spaces satisfying (1.25), then Ch^i Mosco converge to Ch. In addition*

$$\lim_{i \to \infty} \int_X \langle \nabla v_i, \nabla w_i \rangle_i \, d\mathfrak{m}_i = \int_X \langle \nabla v, \nabla w \rangle \, d\mathfrak{m}, \tag{1.26}$$

whenever (v_i) strongly converge in $H^{1,2}$ to v and (u_i) weakly converge in $H^{1,2}$ to u and the heat flows h^i relative to $(X, \mathsf{d}, \mathfrak{m}_i)$ converge to the heat flow h relative to $(X, \mathsf{d}, \mathfrak{m})$ in the following sense:

$$\forall t \ge 0,\ h^i_t f_i \ L^2\text{-strongly converge to } h_t f \text{ whenever } f_i \ L^2\text{-strongly converge to } f. \tag{1.27}$$

Proof. See [31, Theorem 6.8] for the Mosco convergence and [31, Theorem 6.11] for the L^2-strong convergence of $h_t^i f_i$ to $h_t f$. The proof of (1.26) is elementary: since $v_i + tw_i$ weakly converge in $H^{1,2}$ to $v + tw$ for all $t > 0$, by Mosco convergence we have

$$\begin{aligned} & \mathsf{Ch}(v) + 2t \int_X \langle \nabla v, \nabla w \rangle \, d\mathfrak{m} + t^2 \mathsf{Ch}(w) \\ = \; & \mathsf{Ch}(v + tw) \le \liminf_{i \to \infty} \mathsf{Ch}^i(v_i + tw_i) \\ = \; & \liminf_{i \to \infty} \mathsf{Ch}^i(v_i) + 2t \int_X \langle \nabla v_i, \nabla w_i \rangle_i \, d\mathfrak{m}_i + t^2 \mathsf{Ch}^i(g_i) \\ \le \; & \mathsf{Ch}(v) + 2t \liminf_{i \to \infty} \int_X \langle \nabla v_i, \nabla w_i \rangle_i \, d\mathfrak{m} + t^2 \limsup_{i \to \infty} \mathsf{Ch}^i(w_i). \end{aligned}$$

Since $\sup_i \mathsf{Ch}^i(w_i)$ is finite, we may let $t \downarrow 0$ to deduce the lim inf inequality; replacing w by $-w$ gives (1.26). □

In the following corollary we prove standard consequences of the Mosco convergence of Theorem 1.5.4, which refine (1.27) (see also [31, Corollary 6.10] for a discrete counterpart of this result, involving the resolvents).

Corollary 1.5.5. *Under the same assumptions of Theorem 1.5.4, one has*
(a) if $f_i \in H^{1,2}(X, \mathsf{d}, \mathfrak{m}_i)$, $f_i \in D(\Delta_i)$ L^2-strongly converge to f and $\Delta_i f_i$ is uniformly bounded in L^2, then $f \in D(\Delta)$, $\Delta_i f_i$ L^2 weakly converge to Δf and f_i strongly converge in $H^{1,2}$ to f;
(b) for all $t > 0$, $h_t^i f_i$ strongly converge in $H^{1,2}$ to $h_t f$ whenever f_i L^2-strongly converge to f.

Proof. (a) Using the integration by parts formula we see that f_i is weakly convergent in $H^{1,2}$. Let $\chi \in H^{1,2}(X, \mathsf{d}, \mathfrak{m})$ and let $\chi_i \in H^{1,2}(X, \mathsf{d}, \mathfrak{m}_i)$ be strongly convergent to χ in $H^{1,2}$. Let g be a L^2-weak limit point of $\Delta_i f_i$ as $i \to \infty$, so that (1.26) gives (along a subsequence, that for simplicity we do not denote explicitly)

$$\int_X g\chi \, d\mathfrak{m} = \lim_{i \to \infty} \int_X \chi_i \Delta_i f_i \, d\mathfrak{m}_i = -\lim_{i \to \infty} \int_X \langle \nabla \chi_i, \nabla f_i \rangle_i \, d\mathfrak{m}_i = -\int_X \langle \nabla \chi, \nabla f \rangle \, d\mathfrak{m}.$$

This proves $f \in D(\Delta)$ and $g = \Delta f$, so that compactness implies $\Delta_i f_i$ L^2-weakly converge to Δf. We can take the limit in the integration by parts formula $\int_X |\nabla f_i|_i^2 \, d\mathfrak{m}_i = -\int_X f_i \Delta_i f_i \, d\mathfrak{m}_i$ to prove the strong $H^{1,2}$ convergence of f_i to f.

Now, we can prove (b). From (1.15) we know that $\Delta_i h_t^i f_i$ is bounded in L^2 for all $t > 0$, hence (a) provides the strong convergence in $H^{1,2}$ of $h_t^i f_i$ to $h_t f$. □

In order to localize the previous results (see in particular (1.26)) we shall use the next theorem, proved in [14, Theorem 5.3]. It shows that any sequence (f_i) strongly convergent in $H^{1,2}$ to f induces gradient derivations which are strongly converging to the gradient derivation of the limit function, using as class of test functions the family $h_{\mathbb{Q}^+}\mathcal{A}_{\mathrm{bs}}$ defined below

$$h_{\mathbb{Q}_+}\mathcal{A}_{\mathrm{bs}} := \{h_s f : f \in \mathcal{A}_{\mathrm{bs}},, \ s \in \mathbb{Q}_+\} \subset \mathrm{Lip}_b(X). \tag{1.28}$$

Notice that $h_{\mathbb{Q}_+}\mathcal{A}_{\mathrm{bs}}$ depends only on the limit metric measure structure, and it is dense in $H^{1,2}(X, \mathsf{d}, \mathfrak{m})$, see [14, Theorem B.1]. Notice also that, since $\operatorname{supp} \mathfrak{m}$ can well be a strict subset of X, the $\mathrm{Lip}_b(X)$ extension of $f \in h_{\mathbb{Q}_+}\mathcal{A}_{\mathrm{bs}}$ is not necessarily unique, and therefore $\langle \nabla v, \nabla f\rangle_i$ might depend on this extension when $v \in H^{1,2}(X, \mathsf{d}, \mathfrak{m}_i)$ (while $\langle \nabla v, \nabla f\rangle$ does not for $v \in H^{1,2}(X, \mathsf{d}, \mathfrak{m})$). Nevertheless, the following convergence theorem is independent of the extension.

Theorem 1.5.6 (Strong convergence of gradients). *Assume that $(X, \mathsf{d}, \mathfrak{m})$ is a $RCD(K, \infty)$ metric measure space, that Ch^i are quadratic and that Mosco converge to Ch. Let $v_i \in H^{1,2}(X, \mathsf{d}, \mathfrak{m}_i)$ be strongly convergent in $H^{1,2}$ to $v \in H^{1,2}(X, \mathsf{d}, \mathfrak{m})$. Then, for all $f \in h_{\mathbb{Q}^+}\mathcal{A}_{\mathrm{bs}}$, $\langle \nabla v_i, \nabla f\rangle_i$ L^2-strongly converge to $\langle \nabla v, \nabla f\rangle$.*

Theorem 1.5.7 (Continuity of the gradient operators). *Assume that $(X, \mathsf{d}, \mathfrak{m}_i)$ are $RCD(K, \infty)$ metric measure spaces, let $v \in H^{1,2}(X, \mathsf{d}, \mathfrak{m})$ and let $v_i \in H^{1,2}(X, \mathsf{d}, \mathfrak{m}_i)$ be strongly convergent in $H^{1,2}$ to v. Then:*
(a) the following tightness on bounded sets holds:

$$\lim_{R\to\infty} \limsup_{i\to\infty} \int_{X\setminus B_R(\bar{x})} |\nabla v_i|_i^2 \, d\mathfrak{m}_i = 0. \tag{1.29}$$

(b) If w_i weakly converge to w in $H^{1,2}$ the measures $\langle \nabla v_i, \nabla w_i\rangle_i \mathfrak{m}_i$ weakly converge in duality with $h_{\mathbb{Q}_+}\mathcal{A}_{\mathrm{bs}}$ to $\langle \nabla v, \nabla w\rangle \mathfrak{m}$, and if $\langle \nabla v_i, \nabla w_i\rangle_i$ is bounded in L^p for some $p \in (1, \infty)$ also weakly in L^p.
(c) If w_i strongly converge to w in $H^{1,2}$ then $\langle \nabla v_i, \nabla w_i\rangle_i$ L^1-strongly converge to $\langle \nabla v, \nabla w\rangle$.

Proof. (a) In order to prove (1.29) we choose $\chi_R : X \to [0, 1]$ $1/R$-Lipschitz with $\chi_R \equiv 0$ on $B_R(\bar{x})$, $\chi_R \equiv 1$ on $X \setminus B_{2R}(\bar{x})$ and notice that the Leibniz rule gives

$$\int_X |\nabla v_i|_i^2 \chi_R \, d\mathfrak{m}_i = \int_X \langle \nabla v_i, \nabla (v_i \chi_R)\rangle_i \, d\mathfrak{m}_i - \int_X \langle \nabla v_i, \nabla \chi_R\rangle v_i \, d\mathfrak{m}_i$$

so that we can use (1.26) to get

$$\limsup_{i\to\infty} \int_X |\nabla v_i|_i^2 \chi_R \, d\mathfrak{m}_i \le \int_X \langle \nabla v, \nabla (v\chi_R)\rangle \, d\mathfrak{m} + \frac{1}{R}\Big(\int_X |\nabla v|^2 \, d\mathfrak{m}\Big)^{1/2} \|v\|_{L^2(X,\mathfrak{m})}.$$

Using the Leibniz rule once more we get

$$\limsup_{i\to\infty} \int_X |\nabla v_i|_i^2 \chi_R \, d\mathfrak{m}_i \le \int_X |\nabla v|^2 \chi_R \, d\mathfrak{m} + \frac{2}{R}\Big(\int_X |\nabla v|^2 \, d\mathfrak{m}\Big)^{1/2} \|v\|_{L^2(X,\mathfrak{m})}$$

which gives (1.29).

Let us now prove (b). Let $f \in h_{\mathbb{Q}_+}\mathcal{A}_{\mathrm{bs}}$. Using the Leibniz rule we can write

$$\int_X \langle \nabla v_i, \nabla w_i \rangle_i f \, \mathrm{d}\mathfrak{m}_i = -\int_X \langle \nabla v_i, \nabla f \rangle_i w_i \, \mathrm{d}\mathfrak{m}_i + \int_X \langle \nabla v_i, \nabla (w_i f) \rangle_i \, \mathrm{d}\mathfrak{m}_i$$

and use (1.26) together with the L^2-strong convergence of $\langle \nabla v_i, \nabla f \rangle_i$ to $\langle \nabla v, \nabla f \rangle$, ensured by Theorem 1.5.6, to conclude the weak convergence in duality with $h_{\mathbb{Q}_+}\mathcal{A}_{\mathrm{bs}}$ of $\langle \nabla v_i, \nabla w_i \rangle_i \mathfrak{m}_i$. Assuming in addition that $\langle \nabla v_i, \nabla w_i \rangle_i$ satisfy a uniform L^p bound for some $p > 1$, let $\xi \in L^p(X, \mathfrak{m})$ be the L^p-weak limit of a subsequence (not relabelled for simplicity of notation). Then, (1.29) gives

$$\limsup_{i \to \infty} \left| \int_X \langle \nabla v_i, \nabla w_i \rangle_i \varphi \psi_R \, \mathrm{d}\mathfrak{m}_i - \int_X \langle \nabla v_i, \nabla w_i \rangle_i \varphi \, \mathrm{d}\mathfrak{m}_i \right| = o(R)$$

with $\varphi \in h_{\mathbb{Q}_+}\mathcal{A}_{\mathrm{bs}}$ and $\psi_R = 1 - \chi_R \in \mathrm{Lip}_{\mathrm{bs}}(X)$ with χ_R chosen as in the proof of (a). Hence, we take to the limit as $i \to \infty$ to get

$$\left| \int_X \xi \varphi \psi_R \, \mathrm{d}\mathfrak{m} - \int_X \langle \nabla v, \nabla w \rangle \varphi \, \mathrm{d}\mathfrak{m} \right| = o(R).$$

Since $h_{\mathbb{Q}_+}\mathcal{A}_{\mathrm{bs}}$ is dense in $L^q(X, \mathfrak{m})$, with q dual exponent of p, we can pass to the limit as $R \to \infty$ and use the arbitrariness of φ to obtain that $\xi = \langle \nabla v, \nabla w \rangle$.

In order to prove (c), by polarization and the linearity of L^1-strong convergence it is not restrictive to assume $v_i = w_i$. It is then sufficient to apply (1.30) of Lemma 1.5.8 below (whose proof uses only (a), (b) of this proposition) to obtain the inequality $\liminf_i \int_A |\nabla f_i|_i \, \mathrm{d}\mathfrak{m}_i \geq \int_A |\nabla f| \, \mathrm{d}\mathfrak{m}$ on any open set $A \subset X$. Assume that $\xi \in L^2(X, \mathfrak{m})$ is a L^2-weak limit point of $|\nabla f_i|_i$; from the lim inf inequality we get $\int_A \xi \, \mathrm{d}\mathfrak{m} \geq \int_A |\nabla f| \, \mathrm{d}\mathfrak{m}$ for any open set A with $\mathfrak{m}(\partial A) = 0$. A standard approximation then gives $\xi \geq |\nabla f|$ $\mathfrak{m}$-a.e. in X. Since the $H^{1,2}$ strong convergence gives

$$\limsup_{i \to \infty} \int_X |\nabla f_i|_i^2 \, \mathrm{d}\mathfrak{m}_i \leq \int_X |\nabla f|^2 \, \mathrm{d}\mathfrak{m} \leq \int_X \xi^2 \, \mathrm{d}\mathfrak{m},$$

we obtain the L^2-strong convergence of $|\nabla f_i|_i$. Combinig the inequality above with $\liminf_i \| |\nabla f_i|_i \|_{L^2(X, \mathfrak{m}_i)} \geq \|\xi\|_{L^2(X, \mathfrak{m})}$ we obtain that $\xi = |\nabla f|$. □

Lemma 1.5.8. *If $f_i \in H^{1,2}(X, \mathrm{d}, \mathfrak{m}_i)$ weakly converge in $H^{1,2}$ to f, then*

$$\liminf_{i \to \infty} \int_X g |\nabla f_i|_i \, \mathrm{d}\mathfrak{m}_i \geq \int_X g |\nabla f| \, \mathrm{d}\mathfrak{m} \tag{1.30}$$

for any lower semicontinuous $g : X \to [0, \infty]$ and then

$$\liminf_{i \to \infty} \int_A |\nabla f_i|_i^2 \, \mathrm{d}\mathfrak{m}_i \geq \int_A |\nabla f|^2 \, \mathrm{d}\mathfrak{m} \tag{1.31}$$

for any open set $A \subset X$.

Proof. Since truncation preserves L^2_{loc}-strong convergence and uniform L^2 bounds, in the proof of (1.30) we can assume with no loss of generality that f_i are uniformly bounded. Since any lower semicontinuous function is the monotone limit of a sequence of Lipschitz functions with bounded support, we also assume $g \in \mathrm{Lip_{bs}}(X)$. Also, taking into account the inequality $|\nabla h^i_t f_i|_i \le e^{-Kt} h^i_t |\nabla f|_i$, we can estimate

$$\begin{aligned}\liminf_{i\to\infty} \int_X g|\nabla f_i|_i \,\mathrm{d}\mathfrak{m}_i &\ge \liminf_{i\to\infty} \int_X h^i_t g|\nabla f_i|_i \,\mathrm{d}\mathfrak{m}_i - \limsup_{i\to\infty} \int_X |h^i_t g - g||\nabla f_i|_i \,\mathrm{d}\mathfrak{m}_i \\ &\ge e^{Kt} \liminf_{i\to\infty} \int_X g|\nabla h^i_t f_i|_i \,\mathrm{d}\mathfrak{m}_i - C \limsup_{i\to\infty} \|h^i_t g - g\|_{L^2(X,\mathfrak{m}_i)},\end{aligned}$$

with $C = \sup_i (2\mathrm{Ch}^i(f_i))^{1/2}$. Since (1.20) gives

$$\lim_{t\to 0} \limsup_{i\to\infty} \int_X |h^i_t g - g|^2 \,\mathrm{d}\mathfrak{m}_i = 0,$$

this means that as soon as we have the lim inf inequality for $h^i_t f_i$, $h_t f$ for all $t > 0$, we have it for f_i, f.

Hence, possibly replacing f_i by $h^i_t f_i$ we see thanks to (1.17) that we can assume with no loss of generality that f_i are uniformly Lipschitz. Under this assumption, we first prove (1.30) in the case when $g = \chi_A$ is the characteristic function of an open set $A \subset X$, we fix finitely many $v_k \in H^{1,2}(X, \mathrm{d}, \mathfrak{m})$ with $\mathrm{Lip}(v_k) \le 1$, as well as finitely many $w_k \in C_{bs}(X)$ with $\operatorname{supp} w_k \subset A$ and $\sum_k |w_k| \le 1$. Let us also fix $v_{k,i}$ strongly convergent in $H^{1,2}$ to v_k. Now, notice that

$$\lim_{i\to\infty} \int_X \langle \nabla f_i, \nabla v_{k,i}\rangle_i w_k \,\mathrm{d}\mathfrak{m}_i = \int_X \langle \nabla f, \nabla v_k\rangle w_k \,\mathrm{d}\mathfrak{m} \qquad \forall k. \tag{1.32}$$

Indeed, (1.32) follows at once from the weak L^2 convergence of $\langle \nabla f_i, \nabla v_{k,i}\rangle_i$ to $\langle \nabla f, \nabla v_k\rangle$ provided by Theorem 1.5.7(b). Adding w.r.t. k, since $\mathrm{Lip}(v_{k,i}) \le 1$ and $\sum_k |w_k| \le \chi_A$, from (1.19) with $g \equiv \chi_A$ we get (1.30).

For general g we use the formula

$$\int g h \,\mathrm{d}\mu = \int_0^\infty \int_{\{g>t\}} h \,\mathrm{d}\mu \,\mathrm{d}t$$

(with $\mu = \mathfrak{m}_i$ and $\mu = \mathfrak{m}$) and Fatou's lemma.

The proof of (1.31) is a direct consequence of (1.30), of the superadditivity of the lim inf operator, and of the elementary identity

$$\int_A u^2 \,\mathrm{d}\mathfrak{m} = \sup \left\{ \sum_k \mathfrak{m}(A_k)^{-1} \left(\int_{A_k} |u| \,\mathrm{d}\mathfrak{m} \right)^2 \right\},$$

where the supremum runs among the finite disjoint families of open subsets A_k of A with $\mathfrak{m}(A_k) > 0$, of (1.30) and of the superadditivity of the lim inf operator. □

1.6 *BV* functions and their stability

In this section we first recall basic facts about BV functions in metric measure spaces. The most important result of this section, estabilished in Theorem 8.1.1, is the extension of a well-known fact, namely the stability of BV functions under L^1-strong convergence, to the case when even the family of spaces is variable.

Definition 1.6.1 (The class $BV(X, \mathsf{d}, \mathfrak{m})$ and $|\mathrm{D}f|(X)$). *We say that $f \in L^1(X, \mathfrak{m})$ belongs to $BV(X, \mathsf{d}, \mathfrak{m})$ if there exist functions $f_n \in L^1(X, \mathfrak{m}) \cap \mathrm{Lip_b}(X)$ convergent to f in $L^1(X, \mathfrak{m})$ with*

$$L := \liminf_{n\to\infty} \int_X \mathrm{lip}(f_n)\,\mathrm{d}\mathfrak{m} < \infty, \tag{1.33}$$

where $\mathrm{lip}(g)$ denotes the local Lipschitz constant of g, see (1.4). If $f \in BV(X, \mathsf{d}, \mathfrak{m})$, the optimal L in (1.33) (i.e. the inf of $\liminf$) is called total variation of f and denoted by $|\mathrm{D}f|(X)$. By convention, we put $|\mathrm{D}f|(X) = \infty$ if $f \in L^1 \setminus BV(X, \mathsf{d}, \mathfrak{m})$.

It is immediate to check from the definition of total variation that $\varphi \circ f \in BV(X, \mathsf{d}, \mathfrak{m})$ for all $f \in BV(X, \mathsf{d}, \mathfrak{m})$ and all $\varphi : \mathbb{R} \to \mathbb{R}$ 1-Lipschitz with $\varphi(0) = 0$, with

$$|\mathrm{D}(\varphi \circ f)|(X) \le |\mathrm{D}f|(X). \tag{1.34}$$

In addition, the very definition of $|\mathrm{D}f|(X)$ provides the lower semicontinuity property

$$|\mathrm{D}f|(X) \le \liminf_{n\to\infty} |\mathrm{D}f_n|(X) \qquad \text{whenever } f_n \to f \text{ in } L^1(X, \mathsf{d}, \mathfrak{m}).$$

Still using the lower semicontinuity, arguing as in [43], one can prove the coarea formula

$$|\mathrm{D}f|(X) = \int_0^\infty |\mathrm{D}\chi_{\{f>t\}}|(X)\,\mathrm{d}t \qquad \forall f \in L^1(X, \mathfrak{m}),\ f \ge 0. \tag{1.35}$$

In the following proposition, whose proof was suggested to the first author by S. Di Marino, we provide a useful equivalent representation of $|\mathrm{D}f|(X)$.

Proposition 1.6.2. *For all $f \in L^1(X, \mathfrak{m})$ one has*

$$|\mathrm{D}f|(X) = \inf \liminf_{n\to\infty} \int_X \mathrm{Lip}_a(f_n)\,\mathrm{d}\mathfrak{m},$$

where the infimum runs amont all $f_n \in \mathrm{Lip_{bs}}(X)$ convergent to f in $L^1(X, \mathfrak{m})$.

Proof. By a diagonal argument it is sufficient, for any $f \in \mathrm{Lip_b}(X)$ with $\mathrm{lip}(f) \in L^1(X, \mathfrak{m})$, to find $f_n \in \mathrm{Lip_{bs}}(X)$ convergent to f in $L^1(X, \mathfrak{m})$ with $\mathrm{Lip}_a(f_n) \to g$ in $L^1(X, \mathfrak{m})$ and $g \le \mathrm{lip}(f)$ $\mathfrak{m}$-a.e. in X. By a further diagonal argument, it is sufficient to find f_n

when $f \in \mathrm{Lip}_{\mathrm{bs}}(X)$. Under this assumption, we know by Theorem 1.4.1 that there exist $f_n \in \mathrm{Lip}_{\mathrm{b}}(X)$ satisfying $f_n \to f$ in $L^2(X, \mathfrak{m})$ with $\mathrm{Lip}_a(f_n) \to |\nabla f|$ in $L^2(X, \mathfrak{m})$. Since f has bounded support, also f_n can be taken with equibounded support, hence both convergences occur in $L^1(X, \mathfrak{m})$. Since $|\nabla f| \le \mathrm{lip}(f)$ $\mathfrak{m}$-a.e., we are done. □

In the following proposition we list more properties of BV functions in $RCD(K, \infty)$ spaces.

Proposition 1.6.3. *Let $(X, \mathsf{d}, \mathfrak{m})$ be a $RCD(K, \infty)$ space. Then, the following properties hold:*

(a) if $f \in \mathrm{Lip}_{\mathrm{b}}(X) \cap L^1(X, \mathfrak{m}) \cap H^{1,2}(X, \mathsf{d}, \mathfrak{m})$ one has

$$|\mathrm{D}f|(X) = \int_X |\nabla f| \, \mathrm{d}\mathfrak{m}; \tag{1.36}$$

(b) if $f \in BV(X, \mathsf{d}, \mathfrak{m})$ one has

$$|\mathrm{D}h_t f|(X) \le e^{-Kt} |\mathrm{D}f|(X); \tag{1.37}$$

(c) for all $f \in BV(X, \mathsf{d}, \mathfrak{m})$ one has

$$\int_X |P_t f - f| \, \mathrm{d}\mathfrak{m} \le c(t, K) |\mathrm{D}f|(X) \tag{1.38}$$

with $c(t, K) \sim \sqrt{t}$ as $t \downarrow 0$.

Proof. (a) Let $f \in \mathrm{Lip}_{\mathrm{b}}(X) \cap L^1(X, \mathfrak{m}) \cap H^{1,2}(X, \mathsf{d}, \mathfrak{m})$ and apply (1.18) and the inequality $\mathrm{lip}(g) \le \mathrm{Lip}_a(g)$ to get

$$|\mathrm{D}h_t f|(X) \le \int_X |\nabla h_t f| \, \mathrm{d}\mathfrak{m} \le e^{-Kt} \int_X |\nabla f| \, \mathrm{d}\mathfrak{m}.$$

Letting $t \downarrow 0$ provides the inequality $\le$ in (a). In order to prove the converse inequality we have to bound from below the number L in (1.33) along all sequences $(f_n) \subset \mathrm{Lip}_{\mathrm{b}}(X)$ convergent to f in $L^1(X, \mathfrak{m})$. It is not restrictive to assume that the lim inf is a finite limit and also, since f is bounded, that f_n are uniformly bounded. The finiteness of $\int_X |\nabla f_n| \, \mathrm{d}\mathfrak{m}$ gives immediately $f_n \in H^{1,2}(X, \mathsf{d}, \mathfrak{m})$. In addition, for all $t > 0$ it is easily seen that $h_t f_n$ weakly converge to $h_t f$ in $H^{1,2}(X, \mathsf{d}, \mathfrak{m})$, hence the convexity of

$$g \mapsto \int_X |\nabla g| \, \mathrm{d}\mathfrak{m} \qquad g \in H^{1,2}(X, \mathsf{d}, \mathfrak{m})$$

and Mazur's lemma give

$$L \ge e^{Kt} \liminf_{n \to \infty} \int_X |\nabla (h_t f_n)| \, \mathrm{d}\mathfrak{m} \ge e^{Kt} \int_X |\nabla h_t f| \, \mathrm{d}\mathfrak{m}.$$

We can use the lower semicontinuity of the total variation to get the inequality ≥ in (a).

The proof of (b) in the case of bounded functions uses (1.18) as in the proof of (a) and it is omitted. The general case can be recovered by a truncation argument.

The proof of (c) is an immediate consequence of (1.20) and the definition of BV. □

The following theorem provides the stability of the BV property under mGH-convergence. It will be generalized in Theorem 1.8.1, but we prefer to give a direct proof in the BV case, while the proof of Theorem 1.8.1 will focus more on the Sobolev case.

Theorem 1.6.4 (Stability of the BV property under mGH convergence). *Let* $(X, \mathsf{d}, \mathfrak{m}_i)$ *be* $RCD(K, \infty)$ *spaces satisfying* (1.25). *If* $f_i \in BV(X, \mathsf{d}, \mathfrak{m}_i)$ L^1*-strongly converge to* f *with* $\sup_i |Df_i|_i(X) < \infty$, *then* $f \in BV(X, \mathsf{d}, \mathfrak{m})$ *and*

$$|Df|(X) \le \liminf_{i\to\infty} |Df_i|_i(X). \tag{1.39}$$

Proof. In the proof it is not restrictive to assume that the functions f_i are uniformly bounded. Indeed, since the truncated functions $f_i^N := N \wedge f_i \vee -N$ L^1-converge to $f^N := N \wedge f \vee -N$, if we knew that $f_N \in BV(X, \mathsf{d}, \mathfrak{m})$, with

$$|Df^N|(X) \le \liminf_{i\to\infty} |Df_i^N|_i(X),$$

then we could apply (1.34) to f_i^N and use the lower semicontinuity of the total variation to obtain (1.39).

After this reduction to uniformly bounded sequences, let us fix $t > 0$ and consider the functions $h_t^i f_i$, which are uniformly bounded, uniformly Lipschitz (thanks to (1.17)), in $H^{1,2}(X, \mathsf{d}, \mathfrak{m}_i)$ and converge to $h_t f \in H^{1,2}(X, \mathsf{d}, \mathfrak{m})$. If we were able to prove

$$|Dh_t f|(X) \le \liminf_{i\to\infty} |Dh_t^i f_i|_i(X) \tag{1.40}$$

then we could use (1.37) to obtain

$$|Dh_t f|(X) \le e^{-Kt} \liminf_{i\to\infty} |Df_i|_i(X)$$

and we could use once more the lower semicontinuity of the total variation to conclude our argument.

Thanks to these preliminary remarks, in the proof of the proposition it is not restrictive to assume that f_i are equibounded and equi-Lipschitz, with $f_i \in H^{1,2}(X, \mathsf{d}, \mathfrak{m}_i)$, $f \in H^{1,2}(X, \mathsf{d}, \mathfrak{m})$. Assuming also with no loss of generality that the $\liminf$ in (1.39) is a finite limit, we have that f_i are equibounded in $H^{1,2}$, so that they converge weakly to f in $H^{1,2}$. Hence, thanks to the representation (1.36) of the total variation on Lipschitz functions, we need to prove that

$$\int_X |\nabla f| \, d\mathfrak{m} \le \liminf_{i\to\infty} \int_X |\nabla f_i|_i \, d\mathfrak{m}_i. \tag{1.41}$$

This is a consequence of Lemma 1.5.8 with $g \equiv 1$. □

1.7 Compactness in $H^{1,p}$ and in BV

In this section, building upon the basic compactness result in $H^{1,2}$ of [31], we provide new compactness results. In order to state them in global form (i.e. moving from L^p_{loc}-strong to L^p-strong convergence) and in order to reach exponents p smaller than 2, suitable uniform isoperimetric estimates along the sequence of spaces will be needed.

Definition 1.7.1 (Isoperimetric profile). *Assume $\mathfrak{m}(X) = 1$. We say that $\omega : (0,\infty) \to (0, 1/2]$ is an isoperimetric profile for $(X, \mathrm{d}, \mathfrak{m})$ if for all $\epsilon > 0$ one has the implication*

$$\mathfrak{m}(A) \le \omega(\epsilon) \qquad \Rightarrow \qquad \mathfrak{m}(A) \le \epsilon|\mathrm{D}\chi_A|(X) \tag{1.42}$$

for any Borel set $A \subset X$.

A stronger formulation is

$$\mathfrak{m}(A) \le \Phi(|\mathrm{D}\chi_A|(X)) \quad \text{whenever } \mathfrak{m}(A) \le 1/2$$

for some $\Phi : [0,\infty] \to [0,1]$ nondecreasing with $\Phi(0) = 0$ and $\Phi(u) = o(u)$ as $u \downarrow 0$, but the formulation (1.42), which involves only the control of sets with sufficiently small measure, is more adapted to our needs.

If $(X, \mathrm{d}, \mathfrak{m})$ has ω as isoperimetric profile, one has the following property: for any $\epsilon > 0$ and any $t \in \mathbb{R}$ such that $\mathfrak{m}(\{f > t\}) \le \omega(\epsilon)$, one has

$$\int_{\{f\ge t\}} (f-t)^p \,\mathrm{d}\mathfrak{m} \le p^p \epsilon^p \int_X \mathrm{lip}^p(f)\,\mathrm{d}\mathfrak{m}. \tag{1.43}$$

In order to prove (1.43) it is sufficent to apply (1.35) to get

$$\int_{\{g\ge 0\}} g \,\mathrm{d}\mathfrak{m} \le \epsilon \int_X \mathrm{lip}(g)\,\mathrm{d}\mathfrak{m} \qquad \text{whenever } \mathfrak{m}(\{g > 0\}) \le \omega(\epsilon).$$

By applying this to $g = [(f-t)^+]^p$, with the Hölder inequality we reach our conclusion. By the definition of Ch_p we also get

$$\int_{\{f\ge t\}} (f-t)^p \,\mathrm{d}\mathfrak{m} \le p^{p+1}\epsilon^p \mathsf{Ch}_p(f) \quad \forall f \in H^{1,p}(X,\mathrm{d},\mathfrak{m}) \text{ with } \mathfrak{m}(\{f > t\}) \le \omega(\epsilon). \tag{1.44}$$

The following theorem provides classes of spaces for which the existence of an isoperimetric profile is known. Notice that $RCD(K,N)$ spaces with $K > 0$ and $N < \infty$ have always finite diameter.

Theorem 1.7.2 (Isoperimetric profiles). *The class of spaces $(X, \mathrm{d}, \mathfrak{m})$ with $\mathfrak{m}(X) = 1$ having an isoperimetric profile includes:*

(a) $RCD(K,\infty)$ spaces with $K > 0$;

(b) *$RCD(K,\infty)$ spaces with finite diameter.*

Proof. Statement (a) follows from Bobkov's inequality that, when particularized to characteristic functions, gives $\sqrt{K}\mathcal{I}(\mathfrak{m}(A)) \le |D\chi_A|(X)$, where $\mathcal{I}$ is the Gaussian isoperimetric function. The proof given in [16, Theorem 8.5.3] can be adapted without great difficulties to the context of $RCD(K,\infty)$ metric measure spaces (notice that the setting of Markov triples of [16], with a Γ-invariant algebra of functions, does not seem to apply to $RCD(K,\infty)$ spaces), see [13] for a proof.

Statement (b) is a direct consequence of Proposition 1.4.7 and of the definition of BV which, choosing $f = \chi_A$, grant the inequality

$$\mathfrak{m}(A) \le \epsilon\big(\mathfrak{m}(A) + |D\chi_A|(X)\big)$$

as soon as $M(\epsilon, D, K)\mathfrak{m}(A) \le 1$. □

Remark 1.7.3 (Sharp isoperimetric profiles). See also [18] for comparison results and for a description of the sharp isoperimetric profile in the case when $N < \infty$, in the much more general class of $CD(K,N)$ spaces (assuming finiteness of the diameter when $K \le 0$).

The following compactness theorem is one of the main results of [31], see Theorem 6.3 therein. We adapted the statement to our needs, adding also a compactness in $L^2_{\rm loc}$ independent of the equi-tightness condition (1.46). We say that a sequence (f_i) $L^2_{\rm loc}$-strongly converges to f if $f_i\varphi$ L^2-strongly converges to $f\varphi$ for all $\varphi \in C_{\rm bs}(X)$.

Theorem 1.7.4. *Assume that $(X, \mathsf{d}, \mathfrak{m}_i)$ are $RCD(K,\infty)$ spaces and $f_i \in H^{1,2}(X, \mathsf{d}, \mathfrak{m}_i)$ satisfy*

$$\sup_i \int_X |f_i|^2 \, d\mathfrak{m}_i + \mathsf{Ch}^i(f_i) < \infty \tag{1.45}$$

and (for some and thus all $\bar{x} \in X$)

$$\lim_{R\to\infty} \limsup_{i\to\infty} \int_{X\setminus B_R(\bar{x})} |f_i|^2 \, d\mathfrak{m}_i = 0. \tag{1.46}$$

Then (f_i) has a L^2-strongly convergent subsequence to $f \in H^{1,2}(X, \mathsf{d}, \mathfrak{m})$. In general, if only (1.45) holds, (f_i) has a subsequence $L^2_{\rm loc}$-strongly convergent to $f \in H^{1,2}(X, \mathsf{d}, \mathfrak{m})$.

Proof. The first part, as we said, is [31, Theorem 6.3]. For the second part, having fixed $\bar{x} \in X$, it is sufficient to apply the first part to the sequences $f_i\chi_R$, where $\chi_R \in \mathrm{Lip}(X, [0,1])$ with $\chi_R \equiv 1$ on $B_R(\bar{x})$ and $\chi_R \equiv 0$ on $X \setminus B_{R+1}(\bar{x})$, and then to apply a standard diagonal argument. □

Under suitable finiteness assumptions, coupled with the existence of a common isoperimetric profile, we can extend this result to L^{p_i} compactness, assuming Sobolev or BV bounds, as follows.

Proposition 1.7.5. *Assume that* $(X, \mathsf{d}, \mathfrak{m}_i)$, $(X, \mathsf{d}, \mathfrak{m})$ *are* $RCD(K, \infty)$ *spaces satisfying* $\mathfrak{m}_i(X) = 1$, $\mathfrak{m}(X) = 1$ *and with a common isoperimetric profile.*

Assuming that $p_i > 1$ *converge to* p *in* $[1, \infty)$ *and that* $f_i \in H^{1,p_i}(X, \mathsf{d}, \mathfrak{m}_i)$ *satisfy*

$$\sup_i \int_X |f_i|^{p_i}\, d\mathfrak{m}_i + \mathsf{Ch}^i_{p_i}(f_i) < \infty,$$

the family (f_i) *has a* $L^{p_{i(j)}}$*-strongly convergent subsequence* $(f_{i(j)})$*. Analogously, if* $p_i = 1$ *and*

$$\sup_i \int_X |f_i|\, d\mathfrak{m}_i + |Df_i|_i(X) < \infty,$$

then the family (f_i) *has a* L^1*-strongly convergent subsequence* $(f_{i(j)})$*.*

Proof. By L^{p_i}-weak compactness we can assume that the weak limit $f \in L^p(X, \mathfrak{m})$ exists.

The case $p_i = 2$ for infinitely many i is already covered by Theorem 1.7.4. Indeed, the condition (1.46) is automatically satisfied under the isoperimetric assumption, splitting

$$\int_{X \setminus B_R(\bar{x})} |f_i|^2\, d\mathfrak{m}_i \le \int_{\{|f_i| \ge M\}} |f_i|^2\, d\mathfrak{m}_i + M^2 \mathfrak{m}_i(X \setminus B_R(\bar{x}))$$

and using (1.44) with $p = 2$, letting first $R \to \infty$ and then $M \uparrow \infty$.

Hence, we need only to consider the cases $p_i > 2$ for i large enough and $p_i < 2$ for i large enough.

In the case when $p_i > 2$ for i large enough the proof is simpler since for any $\delta > 0$ we can write $f_i = g_i + h_i$ with $\|h_i\|_{L^{p_i}(X,\mathfrak{m}_i)} < \delta$, $\|g_i\|_{L^\infty(X,\mathfrak{m}_i)}$ equibounded and $\sup_i \mathsf{Ch}^i_{p_i}(g_i) < \infty$. Since $2\mathsf{Ch}^i_2(g_i) \le \left(p_i \mathsf{Ch}^i_{p_i}(g_i)\right)^{2/p_i}$, it follows that $\mathsf{Ch}^i_2(g_i)$ is bounded as well. Hence, by what we already proved in the case $p = 2$ we can find a subsequence $g_{i(j)}$ L^2-strongly convergent and then (since (g_i) are equibounded) L^{p_i}-strongly convergent. The decomposition $f_i = g_i + h_i$ can be achieved using (1.44) with $p = p_i$, which gives

$$\lim_{M\to\infty} \sup_i \int_{\{|f_i| > M\}} (|f_i| - M)^{p_i}\, d\mathfrak{m}_i = 0.$$

This is due to the fact that Markov's inequality and the uniform L^1 bound on f_i give

$$\lim_{M\to\infty} \sup_i \mathfrak{m}_i(\{|f_i| > M\}) = 0.$$

Hence, we can first choose $\epsilon > 0$ sufficiently small, in such a way that

$$\sup_i p_i^{p_i+1} \epsilon^{p_i} \mathsf{Ch}^i_{p_i}(f_i) < \delta$$

and then M in such a way that $\sup_i \mathfrak{m}_i(\{|f_i| \ge M\}) \le \omega(\epsilon)$, setting

$$g_i = (f_i \vee -M) \wedge M.$$

In the case $p_i < 2$ for i large enough the decomposition $f_i = g_i + h_i$ can still be achieved using (1.44) (with $\epsilon \sup_i |\mathrm{D}f_i|(X) < \delta$ in the case $p_i = 1$). Since $p_i < 2$, this time we need one more regularization step to achieve the compactness of g_i. More precisely, we write $g_i = (g_i - h_t^i g_i) + h_t^i g_i$; since $h_t^i g_i$ are uniformly Lipschitz we obtain that $\sup_i \mathsf{Ch}_2(h_t^i g_i)$ is uniformly bounded; hence, we can extract a L^2-strongly convergent (and also L^{p_i}-strongly convergent) subsequence. It remains to prove that

$$\lim_{t\downarrow 0} \limsup_{i\to\infty} \int_X |g_i - h_t^i g_i|^{p_i} \, \mathrm{d}\mathfrak{m}_i = 0. \tag{1.47}$$

This is an immediate consequence of (1.38) and the uniform boundedness of (g_i). □

1.8 Mosco convergence of p-Cheeger energies

The definition of Mosco convergence can be immediately adapted to the case when the exponent p is different from 2 and even i-dependent. Adopting the convention $\mathsf{Ch}_1(f) = |\mathrm{D}f|(X)$ to include also the case $p = 1$, if $p_i \in [1, \infty)$ converge to $p \in [1, \infty)$ we say that the p_i-Cheeger energies $\mathsf{Ch}_{p_i}^i$ relative to $(X, \mathsf{d}, \mathfrak{m}_i)$ Mosco converge to Ch_p, the p-Cheeger energy relative to $(X, \mathsf{d}, \mathfrak{m})$, if:

(a) (*Weak*-lim inf). For every $f_i \in L^{p_i}(X, \mathfrak{m}_i)$ L^{p_i}-weakly converging to $f \in L^p(X, \mathfrak{m})$, one has

$$\mathsf{Ch}_p(f) \le \liminf_{i\to\infty} \mathsf{Ch}_{p_i}^i(f_i).$$

(b) (*Strong*-lim sup). For every $f \in L^p(X, \mathfrak{m})$ there exist $f_i \in L^{p_i}(X, \mathfrak{m}_i)$ L^{p_i}-strongly converging to f with

$$\mathsf{Ch}_p(f) = \lim_{i\to\infty} \mathsf{Ch}_{p_i}^i(f_i). \tag{1.48}$$

We speak instead of Γ-convergence if the same notions of convergence occur in (a) and (b), namely the lim inf inequality is only required along L^{p_i}-strongly convergent sequences. Obviously Mosco convergence implies Γ-convergence and we have provided in Proposition 1.7.5 a compactness result that allows to improve, under the assumptions on $(X, \mathsf{d}, \mathfrak{m}_i)$ stated in the proposition, Γ to Mosco convergence.

Theorem 1.8.1. *Let* $(X, \mathsf{d}, \mathfrak{m}_i)$ *be* $RCD(K, \infty)$ *spaces satisfying* (1.25) *and let* $(p_i) \subset [1, \infty)$ *be convergent to* $p \in [1, \infty)$. *Then* $\mathsf{Ch}_{p_i}^i$ Γ*-converge to* Ch_p. *Under the assumption of Proposition 1.7.5 one has Mosco convergence.*

Proof. lim inf *inequality,* $p > 1$. Possibly replacing f_i by their L^{p_i} approximations involved in the definition of Ch_{p_i}, we need only to prove the weaker inequality

$$p\mathsf{Ch}_p(f) \le \liminf_{i\to\infty} \int_X \mathrm{Lip}_a^{p_i}(f_i) \, \mathrm{d}\mathfrak{m}_i. \tag{1.49}$$

Assume first that f_i are uniformly bounded in $H^{1,2}$ and equi-Lipschitz. Then, Lemma 1.5.8 and the inequality $|\nabla f|_i \le \mathrm{lip}(f)$ give

$$\int_X g|\nabla f|\,\mathrm{d}\mathfrak{m} \le \liminf_{i\to\infty} \int_X g|\nabla f_i|_i\,\mathrm{d}\mathfrak{m}_i \le \liminf_{i\to\infty} \int_X g\mathrm{lip}(f_i)\,\mathrm{d}\mathfrak{m}_i$$

for any g lower semicontinuous and nonnegative. This, in combination with the elementary duality identity

$$\frac{1}{p}\int_X |\nabla f|^p\,\mathrm{d}\mathfrak{m} = \sup\left\{\int_X g|\nabla f|\,\mathrm{d}\mathfrak{m} - \frac{1}{q}\int_X g^q\,\mathrm{d}\mathfrak{m} : g \in C_{\mathrm{bs}}(X),\ g \ge 0\right\} \tag{1.50}$$

with q dual exponent of p (applied also to the spaces $(X, \mathsf{d}, \mathfrak{m}_i)$ with $p = p_i$), provides the inequality

$$\int_X |\nabla f|^p\,\mathrm{d}\mathfrak{m} \le \liminf_{i\to\infty} \int_X \mathrm{Lip}_a^{p_i}(f_i)\,\mathrm{d}\mathfrak{m}_i. \tag{1.51}$$

In order to remove the additional assumptions on f_i we now consider the intermediate case when f_i are uniformly bounded in L^∞ and in L^2. Let us fix $t > 0$ and consider the functions $h_t^i f_i$, which are uniformly bounded, uniformly Lipschitz (thanks to (1.17)), in $H^{1,2}(X, \mathsf{d}, \mathfrak{m}_i)$ and weakly converge in $H^{1,2}$ to $h_t f \in H^{1,2}(X, \mathsf{d}, \mathfrak{m})$ by Theorem 1.5.4. Then we can use (1.17), (1.18) and (1.51) with $h_t^i f_i$ to get

$$e^{Kpt}\int_X \mathrm{Lip}_a^p(h_{2t}f)\,\mathrm{d}\mathfrak{m} \le \int_X |\nabla h_t f|^p\,\mathrm{d}\mathfrak{m} \le e^{-Kpt}\liminf_{i\to\infty}\int_X \mathrm{Lip}_a^{p_i}(f_i)\,\mathrm{d}\mathfrak{m}_i.$$

Letting $t \downarrow 0$ then provides (1.49).

We consider the general case f_i; possibly splitting in positive and negative parts, we assume $f_i \ge 0$. We consider the truncation 1-Lipschitz functions (notice that the quadratic regularization near the origin is necessary in the case $p \ge 2$, to get L^2 integrability)

$$\varphi_N(t) := \begin{cases} \frac{N}{2}z^2 & \text{if } 0 \le z \le \frac{1}{N}; \\ -\frac{1}{2N} + z & \text{if } \frac{1}{N} \le z \le N; \\ -\frac{1}{2N} + N & \text{if } N \le z \end{cases}$$

and $f_i^N := \varphi_N \circ f_i$. Since f_i^N L^{p_i}-strongly converge to $f^N := \varphi_N \circ f$, we obtain

$$\mathrm{Ch}_p(f^N) \le \liminf_{i\to\infty} \mathrm{Ch}_{p_i}^i(f_i^N) \le \liminf_{i\to\infty} \mathrm{Ch}_{p_i}^i(f_i).$$

By letting $N \to \infty$ we reach our conclusion.

lim inf *inequality, $p = 1$.* The proof is analogous, in the case when the f_i are uniformly bounded it is sufficient to prove (1.49) for the regularized functions $h_t^i f_i$, $h_t f$, without using the duality formula (1.50). The uniform boundedness assumption on f_i can be removed as in the case $p > 1$, with the simpler truncations $\varphi_N(z) = \min\{N, x\}$.

lim sup *inequality.* For $p > 1$, let us consider $f \in H^{1,p}(X, \mathsf{d}, \mathfrak{m})$ and $f^N \in \mathrm{Lip_{bs}}(X)$ with $\mathrm{Lip}_a(f^N) \to |\nabla f|$ in $L^p(X, \mathfrak{m})$. For any N one has, by the upper semicontinuity of the asymptotic Lipschitz constant,

$$\limsup_{i\to\infty} p_i \mathsf{Ch}^i_{p_i}(f^N) \le \limsup_{i\to\infty} \int_X \mathrm{Lip}_a^{p_i}(f^N)\,\mathrm{d}\mathfrak{m}_i \le \int_X \mathrm{Lip}_a^p(f^N)\,\mathrm{d}\mathfrak{m}.$$

Since f^N L^{p_i} converge to f^N, by a diagonal argument, we can then define $f_i = f^{N(i)}$ with $N(i) \to \infty$ as $i \to \infty$ in such a way that f_i L^{p_i} converge to f and $\limsup_i \mathsf{Ch}^i_{p_i}(f_i) \le \mathsf{Ch}_p(f)$. For $p = 1$ the proof is similar and uses Proposition 1.6.2. □

1.9 *p*-spectral gap

Throughout this section we assume that $\mathfrak{m}(X) = 1$ when a single space is considered and, when a sequence is considered, also $\mathfrak{m}_i(X) = 1$. For any $p \in [1, \infty)$ and any $f \in L^p(X, \mathfrak{m})$ we put

$$c_p(f) := \left(\inf_{a\in\mathbb{R}} \int_X |f - a|^p\,\mathrm{d}\mathfrak{m} \right)^{1/p}. \tag{1.52}$$

We also recall that for any $f \in L^1(X, \mathfrak{m})$ there exists a *median of* f, i.e. a real number m such that

$$\mathfrak{m}(\{f > m\}) \le \frac{1}{2} \qquad \text{and} \qquad \mathfrak{m}(\{f < m\}) \le \frac{1}{2}.$$

In the following remark we recall a few well-known facts about the minimization problem (1.52) (see also [53, Lemma 2.2], [21]).

Remark 1.9.1. *For* $p \in (1, \infty)$, *thanks to the strict convexity of* $z \mapsto |z|^p$ *there is a unique minimizer* a *in* (1.52), *and it is characterized by*

$$\int_X |f - a|^{p-2}(f - a)\,\mathrm{d}\mathfrak{m} = 0.$$

It is also well known that, when $p = 1$, *medians are minimizers in* (1.52), *the converse seems to be less well known, so let us provide a simple proof. Assume that* a *is a minimizer and assume by contradiction that* $\mathfrak{m}(\{f > a\}) > 1/2$ *(if* $\mathfrak{m}(\{f < a\}) > 1/2$ *the argument is similar). We can then find* $\delta > 0$ *such that* $\mathfrak{m}(\{f > a + \delta\}) > 1/2$ *and a simple computation gives*

$$\begin{aligned}\int_X |f - (a + \delta)|\,\mathrm{d}\mathfrak{m} - \int_X |f - a|\,\mathrm{d}\mathfrak{m} &= \delta\big(\mathfrak{m}(\{f < a + \delta\}) - \mathfrak{m}(\{f \ge a + \delta\})\big) \\ &\quad - 2\int_{\{a<f<a+\delta\}} (f - a)\,\mathrm{d}\mathfrak{m} < 0,\end{aligned}$$

contradicting the minimality of a.

In particular, for any $p \in [1, \infty)$ there exists a minimizer of (1.52), and it will be denoted by $m_p(f)$; by convention, it will be any median of f when $p = 1$. Analogously, when we say that $m_{p_i}(f_i)$ converge to $m_p(f)$ we understand this convergence in the set-theoretic sense when $p = 1$ (i.e. limit points of $m_{p_i}(f_i)$ are medians).

Lemma 1.9.2. *Let p_i converge to p in $[1, \infty)$ and let $f_i \in L^{p_i}(X, \mathfrak{m}_i)$ be an L^{p_i}-strongly convergent sequence to $f \in L^p(X, \mathfrak{m})$. Then*

$$\lim_{i\to\infty} m_{p_i}(f_i) = m_p(f) \qquad \text{and} \qquad \lim_{i\to\infty} c_{p_i}(f_i) = c_p(f).$$

Proof. Since

$$\limsup_{i\to\infty} c_{p_i}(f_i) \le \lim_{i\to\infty} \left(\int_X |f_i - b|^{p_i} \, d\mathfrak{m}_i \right)^{1/p_i} = \left(\int_X |f - b|^p \, d\mathfrak{m} \right)^{1/p} \qquad \forall b \in \mathbb{R},$$

taking the infimum w.r.t. b gives the upper semicontinuity of $c_{p_i}(f_i)$.

On the other hand, since it is easily seen that $|m_{p_i}(f_i)| \le 2\|f_i\|_{L^{p_i}(X,\mathfrak{m}_i)}$, the family $m_{p_i}(f_i)$ has limit points as $i \to \infty$, and if $m_{p_{i(k)}}(f_{i(k)}) \to a$ as $k \to \infty$ one has

$$\begin{aligned} \liminf_{k\to\infty} c_{p_{i(k)}}(f_{i(k)}) &= \liminf_{k\to\infty} \left(\int_X |f_i - m_{p_{i(k)}}(f_{i(k)})|^{p_i} \, d\mathfrak{m}_i \right)^{1/p_i} \\ &= \left(\int_X |f - a|^p \, d\mathfrak{m} \right)^{1/p} \ge c_p(f). \end{aligned} \tag{1.53}$$

If we apply this to limit points of subsequences $i(k)$ on which the $\liminf_k c_{p_{i(k)}}(f_{i(k)})$ is achieved, this gives that $c_{p_i}(f_i) \to c_p(f)$. In addition, the inequality (1.53) gives that any limit point of $m_{p_i}(f_i)$ is a minimizer. □

Now, for $p \in [1, \infty)$ let

$$\lambda_{1,p}(X, \mathsf{d}, \mathfrak{m}) := \inf_f \frac{1}{c_p^p(f)} \int_X \mathrm{Lip}_a^p(f) \, d\mathfrak{m}, \tag{1.54}$$

where the infimum runs among all nonconstant Lipschitz functions f on X. By the very definition of Ch_p, the infimum above does not change if we minimize $p\mathsf{Ch}_p(f)/c_p^p(f)$ in the class of nonconstant functions $f \in H^{1,p}(X, \mathsf{d}, \mathfrak{m})$. Furthermore, whenever a minimizer exists, we may normalize it in such a way that $c_p(f) = \|f\|_{L^p(X,\mathfrak{m})} = 1$ (i.e. the infimum in (1.52) is attained at $a = m_p(f) = 0$).

For $p \in (1, \infty)$, Remark 1.9.1 and the definition of Ch_p gives other characterizations of $\lambda_{1,p}(X)$:

$$\lambda_{1,p}(X, \mathsf{d}, \mathfrak{m}) = \inf \left\{ \int_X \mathrm{Lip}_a^p(f) \, d\mathfrak{m} : f \in \mathrm{Lip}(X, \mathsf{d}), \int_X |f|^p \, d\mathfrak{m} = 1, \int_X |f|^{p-2} f \, d\mathfrak{m} = 0 \right\}$$

$$= \inf\left\{\int_X \mathrm{lip}^p(f)\,\mathrm{dm} : f \in \mathrm{Lip}(X,\mathrm{d}),\ \int_X |f|^p\,\mathrm{dm} = 1,\ \int_X |f|^{p-2} f\,\mathrm{dm} = 0\right\}$$

$$= \inf\left\{p\mathrm{Ch}_p(f) : f \in H^{1,p}(X,\mathrm{d},\mathfrak{m}),\ \int_X |f|^p\,\mathrm{dm} = 1,\ \int_X |f|^{p-2} f\,\mathrm{dm} = 0\right\}. \tag{1.55}$$

Remark 1.9.3. *If $\mathfrak{m}(X) = 1$, let us define the Cheeger constant $h(X, \mathrm{d}, \mathfrak{m})$ of $(X, \mathrm{d}, \mathfrak{m})$ by*

$$h(X,\mathrm{d},\mathfrak{m}) := \inf_A \frac{M^-(A)}{\mathfrak{m}(A)},$$

where the infimum runs among all Borel subsets A of X with $0 < \mathfrak{m}(A) \le 1/2$, and $M^-(A)$ is the lower Minkowski content of A, namely (here $I_r(A)$ is the open r-neighbourhood of A)

$$M^-(A) := \liminf_{r\to 0^+} \frac{\mathfrak{m}\left(I_r(A)\right) - \mathfrak{m}(A)}{r}.$$

Then, in [4] it has been proved that

$$h(X,\mathrm{d},\mathfrak{m}) = \inf_A \frac{|\mathrm{D}\chi_A|(X)}{\mathfrak{m}(A)},$$

where as before the infimum runs among all Borel subsets A of X with $0 < \mathfrak{m}(A) \le \mathfrak{m}(X)/2$ (the same result holds if we use the upper Minkowski content in the definition of h). On the other hand, by applying Lemma 1.9.2 with $\mathfrak{m}_i = \mathfrak{m}$, from Proposition 1.6.2 we get

$$\lambda_{1,1}(X,\mathrm{d},\mathfrak{m}) = \inf\left\{\frac{|\mathrm{D}f|(X)}{c_1(f)} : f \in BV(X,\mathrm{d},\mathfrak{m}),\ f \not\equiv \text{constant}\right\}. \tag{1.56}$$

Since $c_1(\chi_A) = \mathfrak{m}(A)$ for $\mathfrak{m}(A) \le 1/2$, the coarea formula for BV maps shows that the Cheeger constant h coincides also with the quantities in (1.56).

In the following theorem we prove a generalized continuity property (1.57) of the first eigenvalue, allowing also the exponents $p_i \to p \in [1,\infty)$ to depend on i. As the proof shows, this property holds even in the extreme case when $\operatorname{diam}\operatorname{supp}(\mathfrak{m}) = 0$, with the convention

$$(\lambda_{1,p}(X,\mathrm{d},\mathfrak{m}))^{1/p} := \infty \qquad \text{if } \operatorname{diam}\operatorname{supp}(\mathfrak{m}) = 0.$$

Note that (1.57) in the case when $\operatorname{diam}\operatorname{supp}(\mathfrak{m}) = 0$ will be used in the proof of Corollary 1.11.6.

Theorem 1.9.4. *Assume that $(X, \mathrm{d}, \mathfrak{m}_i)$, $(X, \mathrm{d}, \mathfrak{m})$ are $RCD(K,\infty)$ spaces satisfying $\mathfrak{m}_i(X) = 1$, $\mathfrak{m}(X) = 1$ with a common isoperimetric profile (for instance either $K > 0$ or uniformly bounded diameters of* $\operatorname{supp}\mathfrak{m}_i$*). If p_i converge to p in $[1,\infty)$, then*

$$\lim_{i\to\infty} \lambda_{1,p_i}(X,\mathrm{d},\mathfrak{m}_i) = \lambda_{1,p}(X,\mathrm{d},\mathfrak{m}). \tag{1.57}$$

In particular the Cheeger constants are continuous w.r.t. the measured Gromov-Hausdorff convergence.

Proof. For any $f \in H^{1,p}(X, \mathsf{d}, \mathfrak{m})$ with $c_p(f) = \|f\|_p = 1$, by Theorem 1.8.1, there exists a sequence $f_i \in H^{1,p_i}(X, \mathsf{d}, \mathfrak{m}_i)$ L^{p_i}-strongly converging to f with $\limsup_i \mathsf{Ch}^i_{p_i}(f_i) \le \mathsf{Ch}_p(f)$. Applying Lemma 1.9.2 yields

$$\limsup_{i\to\infty} \lambda_{1,p_i}(X, \mathsf{d}, \mathfrak{m}_i) \le \limsup_{i\to\infty} \frac{p_i \mathsf{Ch}^i_{p_i}(f_i)}{\left(c_{p_i}(f_i)\right)^{p_i}} \le \mathsf{Ch}_p(f).$$

Taking the infimum w.r.t. f gives the upper semicontinuity of $\lambda_{1,p_i}(X, \mathsf{d}, \mathfrak{m}_i)$.

In order to prove the lower semicontinuity, we can assume with no loss of generality that $\lambda_{1,p_i}(X, \mathsf{d}, \mathfrak{m}_i)$ is a bounded convergent sequence. For any $i \ge 1$ take $f_i \in H^{1,p_i}(X, \mathsf{d}, \mathfrak{m}_i)$ with

$$\left|\lambda_{1,p_i}(X, \mathsf{d}, \mathfrak{m}_i) - p_i \mathsf{Ch}^i_{p_i}(f_i)\right| < \frac{1}{i} \quad \text{and} \quad c_{p_i}(f_i) = \int_X |f_i|^{p_i}\, d\mathfrak{m}_i = 1.$$

By Proposition 1.7.5, without loss of generality we can assume that the L^{p_i}-strong limit $f \in L^p(X, \mathfrak{m})$ of f_i exists. Thus, Theorem 8.1.1 gives $\mathsf{Ch}_p(f) \le \liminf_i \mathsf{Ch}^i_{p_i}(f_i)$. As a consequence, since Lemma 1.9.2 gives $c_p(f) = \|f\|_{L^p(X,\mathfrak{m})} = 1$, we have

$$\liminf_{i\to\infty} \lambda_{1,p_i}(X, \mathsf{d}, \mathfrak{m}_i) = \liminf_{i\to\infty} p_i \mathsf{Ch}^i_{p_i}(f_i) \ge p\mathsf{Ch}_p(f) \ge \lambda_{1,p}(X, \mathsf{d}, \mathfrak{m}).$$

□

For $p \in (1, \infty)$ and $\Omega \subset X$ Borel, let us denote

$$\Lambda_p(\Omega, \mathsf{d}, \mathfrak{m}) := \left\{ f \in H^{1,p}(X, \mathsf{d}, \mathfrak{m}) : \int_\Omega |f|^p\, d\mathfrak{m} = 1,\ f = 0 \ \mathfrak{m}\text{-a.e. on } X \setminus \Omega \right\}.$$

Accordingly, we define $\lambda^D_{1,p}(\Omega, \mathsf{d}, \mathfrak{m})$ as the infimum of the p-energy with Dirichlet conditions

$$\lambda^D_{1,p}(\Omega, \mathsf{d}, \mathfrak{m}) := \inf\left\{p\mathsf{Ch}_p(f) : f \in \Lambda_p(\Omega, \mathsf{d}, \mathfrak{m})\right\}. \tag{1.58}$$

Lemma 1.9.5. *Let $p \in (1, \infty)$.*

(1) For any Borel subsets Ω_1, Ω_2 of X with $\mathfrak{m}(\Omega_1 \cap \Omega_2) = 0$, we have

$$\lambda_{1,p}(X, \mathsf{d}, \mathfrak{m}) \le \max\left\{\lambda^D_{1,p}(\Omega_1, \mathsf{d}, \mathfrak{m}), \lambda^D_{1,p}(\Omega_2, \mathsf{d}, \mathfrak{m})\right\}. \tag{1.59}$$

(2) If $p \in [2, \infty)$ and $f \in H^{1,p}(X, \mathsf{d}, \mathfrak{m})$ is a minimizer of the right hand side of (1.54) *with $m_p(f) = 0$, then*

$$\int_X \langle\nabla f, \nabla g\rangle |\nabla f|^{p-2}\, d\mathfrak{m} = \lambda_{1,p}(X, \mathsf{d}, \mathfrak{m}) \int_X |f|^{p-2} f g\, d\mathfrak{m} \tag{1.60}$$

for any $g \in H^{1,p}(X,\mathsf{d},\mathfrak{m})$. In particular, choosing $g = f^{\pm}$ gives

$$\lambda_{1,p}(X,\mathsf{d},\mathfrak{m}) = p\mathsf{Ch}_p(f^{\pm})\left(\int_X |f^{\pm}|^p\,\mathrm{d}\mathfrak{m}\right)^{-1}. \tag{1.61}$$

Proof. We first prove (1.59). Take $f_i \in H^{1,p}(X,\mathsf{d},\mathfrak{m})$ with $\int_{\Omega_i} |f_i|^p\,\mathrm{d}\mathfrak{m} = 1$ and $f_i = 0$ $\mathfrak{m}$-a.e. on $X \setminus \Omega_i$. Then, choosing thanks to a continuity argument $\alpha \in \mathbb{R}$ such that $\int_X |f_1 + \alpha f_2|^{p-2}(f_1 + \alpha f_2)\,\mathrm{d}\mathfrak{m} = 0$, we get

$$\begin{aligned}(1+|\alpha|^p)\lambda_{1,p}(X,\mathsf{d},\mathfrak{m}) &= \lambda_{1,p}(X,\mathsf{d},\mathfrak{m})\left(\int_{\Omega_1} |f_1|^p\,\mathrm{d}\mathfrak{m} + \int_{\Omega_2} |\alpha f_2|^p\,\mathrm{d}\mathfrak{m}\right)\\ &= \lambda_{1,p}(X,\mathsf{d},\mathfrak{m})\int_X |f_1 + \alpha f_2|^p\,\mathrm{d}\mathfrak{m} \\ &\le p\mathsf{Ch}_p(f_1+\alpha f_2) = p\mathsf{Ch}_p(f_1) + p|\alpha|^p\mathsf{Ch}_p(f_2).\end{aligned} \tag{1.62}$$

By taking the infimum w.r.t. f_1 and f_2 we obtain (1.59).

Next we prove (1.60). Let

$$F(s,t) := \int_X |f + sg - t|^{p-2}(f + sg - t)\,\mathrm{d}\mathfrak{m}.$$

Then, it is easy to check that

$$F_s(s,t) = (p-1)\int_X g|f + sg - t|^{p-2}\,\mathrm{d}\mathfrak{m}$$

and that

$$F_t(s,t) = (1-p)\int_X |f + sg - t|^{p-2}\,\mathrm{d}\mathfrak{m}.$$

The implicit function theorem yields that $s \mapsto m_p(f+sg)$ is differentiable at $s = 0$.

Recall that according to [30], we can represent $p\mathsf{Ch}_p(f)$ as $\int_X |\nabla f|^p\,\mathrm{d}\mathfrak{m}$, where $|\nabla f|$ is the 2-minimal relaxed slope (as always, in this paper). Then, the direct calculation of the left hand side of

$$\frac{\mathrm{d}}{\mathrm{d}s}\left(\frac{p\mathsf{Ch}_p(f+sg)}{\|(f+sg) - m_p(f+sg)\|^p_{L^p(X,\mathfrak{m})}}\right)\Bigg|_{s=0} = 0$$

with the differentiability of $m_p(f+sg)$ at $s = 0$ proves (1.60). □

In the following stability result we need the extra assumption

$$\begin{gathered}\limsup_{i\to\infty} \|f_i\|_{L^{p_i}(X,\mathfrak{m}_i)} \le \|f\|_{L^\infty(X,\mathfrak{m})}\\ \text{whenever } p_i \to \infty,\ \sup_i \|f_i\|_{L^{p_i}(X,\mathfrak{m}_i)} + \Big(\int_X |\nabla f_i|^{p_i}\,\mathrm{d}\mathfrak{m}_i\Big)^{1/p_i} < \infty \\ \text{and } f_i \text{ strongly } L^p\text{-converge to } f \text{ for some (and thus all) } p \in (1,\infty).\end{gathered} \tag{1.63}$$

This is a kind of extension of Theorem 1.9.4 to the case $p = \infty$. We believe that it should be possible to avoid this assumption, possibly making an additional hypothesis on the decay rate of the common isoperimetric profile. Nevertheless, this assumption is harmless for the applications of Theorem 1.9.6 below in Section 1.11. Indeed, in the setting of Section 1.11, as soon as $p_i > N$ the functions f_i and f are equibounded and equi-Hölder on $\operatorname{supp} \mathfrak{m}_i$, $\operatorname{supp} \mathfrak{m}$ respectively; denoting by f_i, f suitable equibounded and equi-Hölder extensions of f_i, f to the whole of X, the Hausdorff convergence of $\operatorname{supp} \mathfrak{m}_i$ to $\operatorname{supp} \mathfrak{m}$ and the weak convergence of $f_i\mathfrak{m}_i$ to $f\mathfrak{m}$ easily imply the uniform convergence of f_i to f on $\operatorname{supp} \mathfrak{m}$, so that

$$\limsup_{i\to\infty} \|f_i\|_{L^{p_i}(X,\mathfrak{m}_i)} \le \limsup_{i\to\infty} \|f_i\|_{L^{p_i}(X,\mathfrak{m})} \le \limsup_{i\to\infty} \|f\|_{L^{p_i}(X,\mathfrak{m})} \le \|f\|_{L^\infty(X,\mathfrak{m})}.$$

Theorem 1.9.6. *Let $(X, \mathsf{d}, \mathfrak{m}_i)$, $(X, \mathsf{d}, \mathfrak{m})$ be RCD(K, ∞) metric measure spaces with $\mathfrak{m}_i(X) = 1$, $\mathfrak{m}(X) = 1$ and a common isoperimetric profile (e.g. either $K > 0$ or equibounded diameters of* $\operatorname{supp} \mathfrak{m}_i$*). If $p_i \in [1, \infty)$ diverge to ∞ and* (1.63) *holds, one has*

$$\lim_{i\to\infty} \left(\lambda_{1,p_i}(X, \mathsf{d}, \mathfrak{m}_i)\right)^{1/p_i} = \frac{2}{\operatorname{diam} \operatorname{supp}(\mathfrak{m})}. \tag{1.64}$$

Proof. Let $x_1, x_2 \in \operatorname{supp} \mathfrak{m}$; thanks to the weak convergence of $\mathfrak{m}_i$ to $\mathfrak{m}$ we can find $x_{j,i}$ convergent to x_j as $i \to \infty$, $j = 1, 2$. Let $r = \mathsf{d}(x_1, x_2)$, $r_i = \mathsf{d}(x_{1,i}, x_{2,i})$ and let us define nonnegative Lipschitz functions $\delta_{j,i} \in \operatorname{Lip}(X, \mathsf{d})$ by

$$\delta_{j,i}(x) := \max\left\{\frac{r_i}{2} - \mathsf{d}(x_{j,i}, x), 0\right\},$$

uniformly convergent as $i \to \infty$ to

$$\delta_j(x) := \max\left\{\frac{r}{2} - \mathsf{d}(x_j, x), 0\right\}.$$

Then, since $\{B_{r_i/2}(x_{j,i})\}_{j=1,2}$ are nonempty disjoint subsets of X, and since $\delta_{j,i}$ are 1-Lipschitz, for any $p \in (1, \infty)$, (1.59) and the Hölder inequality give that

$$\begin{aligned}
\left(\lambda_{1,p_i}(X, \mathsf{d}, \mathfrak{m}_i)\right)^{1/p_i} &\le \max_{j=1,2}\left\{\left(\lambda^D_{1,p_i}\left(B_{r_i/2}(x_{j,i})\right)\right)^{1/p_i}\right\}\\
&\le \max_{j=1,2}\left\{\left(\frac{1}{\mathfrak{m}_i\left(B_{r_i/2}(x_{j,i})\right)}\int_{B_{r_i/2}(x_{j,i})} \delta_{j,i}^{p_i}\, d\mathfrak{m}_i\right)^{-1/p_i}\right\}\\
&\le \max_{j=1,2}\left\{\left(\frac{1}{\mathfrak{m}_i\left(B_{r_i/2}(x_{j,i})\right)}\int_{B_{r_i/2}(x_{j,i})} \delta_{j,i}^{p}\, d\mathfrak{m}_i\right)^{-1/p}\right\}
\end{aligned}$$

for all sufficiently large i. Thus by letting $i \to \infty$ we have

$$\limsup_{i\to\infty} \left(\lambda_{1,p_i}(X, \mathsf{d}, \mathfrak{m}_i)\right)^{1/p_i} \le \max_{j=1,2}\left\{\left(\frac{1}{\mathfrak{m}\left(B_{r/2}(x_j)\right)}\int_{B_{r/2}(x_j)} \delta_j^{p}\, d\mathfrak{m}\right)^{-1/p}\right\}.$$

Letting $p \to \infty$ yields

$$\limsup_{i\to\infty} \left(\lambda_{1,p_i}(X,\mathsf{d},\mathfrak{m}_i)\right)^{1/p_i} \le \max_{j=1,2}\{\|\delta_j\|^{-1}_{L^\infty(X,\mathfrak{m})}\} = \frac{2}{r} = \frac{2}{\mathsf{d}(x_1,x_2)}.$$

By minimizing w.r.t. x_1 and x_2 we get the lim sup inequality in (1.64).

Next we check the lim inf inequality in (1.64). We can assume with no loss of generality that the limit $\lim_i \left(\lambda_{1,p_i}(X,\mathsf{d},\mathfrak{m}_i)\right)^{1/p_i}$ exists and is finite. For any i such that $p_i > 2$ take a minimizer $f_i \in H^{1,p_i}(X,\mathsf{d},\mathfrak{m}_i)$ of the right hand side of (1.55) (whose existence is granted by Proposition 1.7.5). Set $\tilde{f}_i := f_i^\pm/\|f_i^\pm\|_{L^{p_i}(X,\mathfrak{m}_i)}$ and $\tilde{f}_i := \tilde{f}_i^+ - \tilde{f}_i^-$. Since Lemma 1.9.5 yields

$$\lambda_{1,p_i}(X,\mathsf{d},\mathfrak{m}_i) = p_i \mathsf{Ch}^i_{p_i}(\tilde{f}_i^\pm),$$

by the compactness property provided by Theorem 1.8.1 we can also assume that $\tilde{f}_i^+$ L^p-strongly converge for all $p > 1$ to a nonnegative $g \in \bigcap_{p>1} H^{1,p}(X,\mathsf{d},\mathfrak{m})$, and that $\tilde{f}_i^-$ L^p-strongly converge for all $p > 1$ to a nonnegative $h \in \bigcap_{p>1} H^{1,p}(X,\mathsf{d},\mathfrak{m})$, so that $\tilde{f}_i$ strongly L^p-converge for all $p > 1$ to $f = g - h$. For $p > 1$ fixed, taking the limit as $i \to \infty$ in the equality

$$\|\tilde{f}_i^+\|^p_{L^p(X,\mathfrak{m}_i)} + \|\tilde{f}_i^+\|^p_{L^p(X,\mathfrak{m}_i)} = \|\tilde{f}_i\|^p_{L^p(X,\mathfrak{m}_i)}$$

we obtain that $g = f^+$ and $h = f^-$. We now claim that both f^+ and f^- have unit L^∞ norm. The proof of the upper bound is a simple consequence of the inequalities $\|\tilde{f}_i^\pm\|_{L^p(X,\mathfrak{m}_i)} \le \|\tilde{f}_i^\pm\|_{L^{p_i}(X,\mathfrak{m}_i)} = 1$ for $p_i \ge p$, by letting first $i \to \infty$ and then $p \to \infty$, while the proof of the lower bound is a direct consequence of (1.63).

Theorem 1.8.1 and the inequality (actually, as we already remarked, equality holds under our curvature assumption, see [30]) between p-minimal relaxed slope and 2-minimal relaxed slope $|\nabla f|$ give

$$\|\nabla f^\pm\|_{L^p(X,\mathfrak{m})} \le \left(p\mathsf{Ch}_p(f^\pm)\right)^{1/p} \le \liminf_{i\to\infty}\left(p_i \mathsf{Ch}^i_p(\tilde{f}_i^\pm)\right)^{1/p_i}$$

for any $p \ge 2$, thus letting $p \to \infty$ gives

$$\||\nabla f^\pm|\|_{L^\infty(X,\mathfrak{m})} \le \lim_{i\to\infty}\left(\lambda_{1,p_i}(X,\mathsf{d},\mathfrak{m}_i)\right)^{1/p_i}.$$

Therefore $f^\pm$ have Lipschitz representatives, still denoted by $f^\pm$, with Lipschitz constants at most the right hand side above. The relatively open subsets $\Omega^\pm := \{f^\pm > 0\} \cap \operatorname{supp}\mathfrak{m}$ of $\operatorname{supp}\mathfrak{m}$ are disjoint and nonempty. Let

$$r(\Omega^\pm) := \sup_{x\in\Omega^\pm}\left(\inf_{y\in\partial\Omega^\pm\cap\operatorname{supp}(\mathfrak{m})} \mathsf{d}(x,y)\right).$$

Using the inequality $r(\Omega^+) + r(\Omega^-) \le \operatorname{diam}(\operatorname{supp}\mathfrak{m})$, ensured by the length property+ of $(\operatorname{supp}\mathfrak{m},\mathsf{d})$, we get

$$\frac{2}{\operatorname{diam}\operatorname{supp}(\mathfrak{m})} \le \max\left\{\frac{1}{r(\Omega^+)}, \frac{1}{r(\Omega^-)}\right\}. \tag{1.65}$$

For $\delta \in (0,1)$, take points $x^\pm \in \Omega^\pm$ with $f^\pm(x^\pm) \geq 1-\delta$, and take points $y^\pm \in \partial\Omega^\pm \cap \operatorname{supp}\mathfrak{m}$; since $f^\pm(y^\pm) = 0$, we have

$$1-\delta \leq |f^\pm(x^\pm) - f^\pm(y^\pm)| \leq \operatorname{Lip}(f^\pm)\mathsf{d}(x^\pm, y^\pm),$$

so that $\|f^\pm\|_{L^\infty(X,\mathfrak{m})} = 1$ and the arbitrariness of $y^\pm$ give

$$1 \leq \operatorname{Lip}(f^\pm)r(\Omega^\pm) \leq \liminf_{i\to\infty} \left(\lambda_{1,p_i}(X,\mathsf{d},\mathfrak{m}_i)\right)^{1/p_i} \cdot r(\Omega^\pm).$$

Thus

$$\max\left\{\frac{1}{r(\Omega^+)}, \frac{1}{r(\Omega^-)}\right\} \leq \liminf_{i\to\infty} \left(\lambda_{1,p_i}(X,\mathsf{d},\mathfrak{m}_i)\right)^{1/p_i} \tag{1.66}$$

and (1.65) and (1.66) yield the lim inf inequality in (1.64). □

1.10 Stability of Hessians and Ricci tensor

Recall that derivations, according to [29] (the definitions being inspired by [52]), are linear functionals $\boldsymbol{b} : H^{1,2}(X,\mathsf{d},\mathfrak{m}) \to L^0(X,\mathfrak{m})$ satisfying the quantitative locality property

$$|\boldsymbol{b}(u)| \leq h|\nabla u| \qquad \mathfrak{m}\text{-a.e. in } X, \text{ for all } u \in H^{1,2}(X,\mathsf{d},\mathfrak{m})$$

for some $h \in L^0(X,\mathfrak{m})$. The minimal h, up to $\mathfrak{m}$-negligible sets, is denoted $|\boldsymbol{b}|$. The simplest example of derivation is the gradient derivation $\boldsymbol{b}_v(u) := \langle\nabla v, \nabla u\rangle$ induced by $v \in H^{1,2}(X,\mathsf{d},\mathfrak{m})$, which satisfies $|\boldsymbol{b}_v| = |\nabla v|$ $\mathfrak{m}$-a.e. in X. By a nice duality argument, it has also been proved in [29, Section 2.3.1] that the $L^\infty(X,\mathfrak{m})$-module generated by gradient derivations is dense in the class of L^2 derivations. In the language of [29], L^2-derivations correspond to L^2-sections of the tangent bundle $T(X,\mathsf{d},\mathfrak{m})$ viewed as dual of the L^2-sections of cotangent bundle $T^*(X,\mathsf{d},\mathfrak{m})$ (the latter built starting from differentials of Sobolev functions), see [29, Section 2.3] for more details.

Even though higher order tensors will not play a big role in this paper, except for the Hessians, let us describe the basic ingredients of the theory developed for this purpose in [29]. In a metric measure space $(X,\mathsf{d},\mathfrak{m})$, for $p \in [1,\infty]$ let $L^p(T^r_s(X,\mathsf{d},\mathfrak{m}))$ denote the space of L^p-tensor fields of type (r,s) on $(X,\mathsf{d},\mathfrak{m})$, defined as in [29]. A tensor field of type (r,s) is a $L^\infty(X,\mathfrak{m})$-multilinear map

$$T : \bigotimes_{k=1}^{r} T(X,\mathsf{d},\mathfrak{m}) \otimes \bigotimes_{k=r+1}^{r+s} T^*(X,\mathsf{d},\mathfrak{m}) \to L^0(X,\mathfrak{m})$$

satisfying, for some $g \in L^0(X,\mathfrak{m})$ a continuity property

$$|T(u\otimes v)| \leq g|u\otimes v|_{HS} \qquad \mathfrak{m}\text{-a.e. in } X.$$

w.r.t. a suitable Hilbert-Schmidt norm on the tensor products. The minimal (up to $\mathfrak{m}$-negligible sets) g is denoted $|T|$ and L^p tensor fields correspond to tensor fields satisying $|T| \in L^p(X,\mathfrak{m})$.

In particular derivations correspond to (0, 1)-tensor fields. We recall the following facts and definitions:
(1) any choice of $g^0, \ldots, g^{r+s} \in W^{1,2}(X, \mathsf{d}, \mathfrak{m})$ induces a product tensor field T acting as follows

$$\langle T, \bigotimes_{k=1}^{r} \nabla f^k \otimes \bigotimes_{k=r+1}^{r+s} df^k \rangle = g_0 \prod_{k=1}^{r} \boldsymbol{b}_{f^k}(g^k) \cdot \prod_{k=r+1}^{r+s} \boldsymbol{b}_{g^k}(f^k)$$

and denoted $g^0 \bigotimes_1^r dg^k \otimes \bigotimes_{r+1}^{r+s} \nabla g^k$. Since derivations correspond to (0, 1)-tensor fields, we recover in particular the concept of gradient derivations.
(2) Denoting, as in [45], [29] (recall that $D(\Delta)$ is defined as in (1.13))

$$\mathrm{Test}F(X, \mathsf{d}, \mathfrak{m}) := \left\{ f \in \mathrm{Lip}_{\mathrm{b}}(X) \cap D(\Delta) : \Delta f \in H^{1,2}(X, \mathsf{d}, \mathfrak{m}) \right\},$$

the space of finite combinations of tensor products

$$ST_s^r(X, \mathsf{d}, \mathfrak{m}) := \left\{ \sum_{j=1}^{N} g^{j,0} \bigotimes_{k=1}^{r} dg^{j,k} \otimes \bigotimes_{k=r+1}^{r+s} \nabla g^{j,k} : N \geq 1,\ g^{j,i} \in \mathrm{Test}F(X, \mathsf{d}, \mathfrak{m}) \right\}$$

is dense in $L^p(T_s^r(X, \mathsf{d}, \mathfrak{m}))$ for $p \in [1, \infty)$. This is due to the fact that the very definition of tensor product involves a completion procedure of the class of finite sums of elementary products. Notice also that h_t maps $\mathrm{Lip}_{\mathrm{b}}(X)$ into $\mathrm{Test}F(X, \mathsf{d}, \mathfrak{m})$ for all $t > 0$.
(3) If $(X, \mathsf{d}, \mathfrak{m})$ is a $RCD(K, \infty)$ space, the space $W^{2,2}(X, \mathsf{d}, \mathfrak{m})$ is defined in [29] to be the space of all functions $f \in H^{1,2}(X, \mathsf{d}, \mathfrak{m})$ such that

$$\begin{aligned} 2\int_X \varphi \mathrm{Hess}(f)(dg \otimes dh) &= -\int_X \langle \nabla f, \nabla g \rangle \,\mathrm{div}(\varphi \nabla h)\, d\mathfrak{m} - \int_X \langle \nabla f, \nabla h \rangle \,\mathrm{div}(\varphi \nabla g)\, d\mathfrak{m} \\ &\quad - \int_X \varphi \big\langle \nabla f, \nabla \langle \nabla g, \nabla h \rangle \big\rangle \, d\mathfrak{m} \end{aligned} \tag{1.67}$$

for $\varphi, f, g \in \mathrm{Test}F(X, \mathsf{d}, \mathfrak{m})$, with $\mathrm{Hess}(f)$ a (0, 2) tensor field in L^2. This is a Hilbert space when endowed with the norm

$$\|f\|_{W^{2,2}(X,\mathsf{d},\mathfrak{m})} := \left(\|f\|^2_{H^{1,2}(X,\mathsf{d},\mathfrak{m})} + \||\mathrm{Hess}(f)|\|^2_{L^2(X,\mathfrak{m})} \right)^{1/2}.$$

It has been proved in [29, Corollary 3.3.9] that $H^{1,2}(X, \mathsf{d}, \mathfrak{m}) \cap D(\Delta) \subset W^{2,2}(X, \mathsf{d}, \mathfrak{m})$, with

$$\int_X |\mathrm{Hess}(f)|^2 \, d\mathfrak{m} \leq \int_X (\Delta f)^2 + K^- |\nabla f|^2 \, d\mathfrak{m}. \tag{1.68}$$

Notice that (1.67) makes sense because of (1.23); on the other hand, as soon as $f \in W^{2,2}(X, \mathsf{d}, \mathfrak{m})$, by approximation the formula extends from $\varphi \in \mathrm{Test}F(X, \mathsf{d}, \mathfrak{m})$ to $\varphi \in \mathrm{Lip}_{\mathrm{b}}(X)$. In particular, in our convergence results we shall use the choice $\varphi \in h_{\mathbb{Q}_+}\mathcal{A}_{\mathrm{bs}}$, where h is the semigroup relative to the limit metric measure structure. Also, arguing as in [29, Theorem 3.3.2(iv)], we immediately obtain that, given $f \in H^{1,2}(X, \mathrm{d}, \mathfrak{m})$, $f \in W^{2,2}(X, \mathrm{d}, \mathfrak{m})$ if and only if there is $h \in L^2(X, \mathfrak{m})$ satisfying

$$\Big|\sum_k \Big(- \int_X \langle \nabla f, \nabla g_k\rangle \operatorname{div}(\varphi_k \psi_k \nabla h_k)\,\mathrm{d}\mathfrak{m} - \int_X \langle \nabla f, \nabla h_k\rangle \operatorname{div}(\varphi_k \psi_k \nabla g_k)\,\mathrm{d}\mathfrak{m} - \int_X \varphi_k \psi_k \langle \nabla f, \nabla \langle \nabla g_k, \nabla h_k\rangle\rangle \Big)\Big| \le \int_X h \Big|\sum_k \varphi_k \psi_k \nabla g_k \otimes \nabla h_k\Big|\,\mathrm{d}\mathfrak{m} \tag{1.69}$$

for any finite collection of $\varphi_k, \psi_k \in h_{\mathbb{Q}_+}\mathcal{A}_{\mathrm{bs}}$, $g_k, h_k \in \mathrm{Test}F(X, \mathrm{d}, \mathfrak{m})$. In addition, the smallest h up to $\mathfrak{m}$-negligible sets is precisely $|\mathrm{Hess}(f)|$.

We shall also use the simplified notation $\mathrm{Hess}(f)(g, h)$.

Remark 1.10.1. *If we have finitely many $g_k, h_k \in H^{1,2}(X, \mathrm{d}, \mathfrak{m})$ and $g_i^k, h_i^k \in H^{1,2}(X, \mathrm{d}, \mathfrak{m}_i)$ are strongly convergent to g_k, h_k in $H^{1,2}$ and uniformly Lipschitz, then*

$$\Big|\sum_k \varphi_k \nabla g_i^k \otimes \nabla h_i^k\Big|_i \quad L^2\text{-strongly converge to} \quad \Big|\sum_k \varphi_k \nabla g^k \otimes \nabla h^k\Big| \tag{1.70}$$

for any choice of $\varphi_k \in C_b(X)$. Indeed, we can use the identity

$$\Big|\sum_k \varphi_k \nabla g_i^k \otimes \nabla h_i^k\Big|_i^2 = \sum_{k,l} \varphi_k \varphi_l \langle \nabla g_i^k, \nabla g_i^l\rangle_i \langle \nabla h_i^k, \nabla h_i^l\rangle_i$$

and Theorem 1.5.7(c) which provides the L^1-strong convergence of $\langle \nabla g_i^k, \nabla g_i^l\rangle_i$ to $\langle \nabla g^k, \nabla g^l\rangle$; since these gradients are equibounded we can use Proposition 1.3.3(a) to improve the convergence to L^2 (actually any L^p, $p < \infty$) convergence, so that the products L^1-strongly converge.

Let us consider the regularization of h_t

$$h_\rho f := \int_0^\infty \rho(s) h_s f\,\mathrm{d}s, \tag{1.71}$$

with $\rho \in C_c^\infty((0, \infty))$ convolution kernel and, when necessary, let us define h_ρ^i in an analogous way. Since

$$\Delta h_\rho f = - \int_0^\infty \rho'(s) h_s f\,\mathrm{d}s \quad \text{if } f \in L^2(X, \mathfrak{m}), \quad \Delta h_\rho f = \int_0^\infty \rho(s) h_s \Delta f\,\mathrm{d}s \quad \text{if } f \in D(\Delta), \tag{1.72}$$

it is immediately seen that h_ρ maps $L^2(X, \mathfrak{m})$ into $\mathrm{Test}F(X, \mathrm{d}, \mathfrak{m})$ and retains many properties of h, namely

$$\sup|h_\rho f| \le \sup|f|, \qquad \mathrm{Lip}(h_\rho f) \le e^{K^- \tau}\,\mathrm{Lip}(f), \tag{1.73}$$

(with $\tau = \sup \operatorname{supp} \rho$) if f is bounded and/or Lipschitz, and

$$\int_X |\nabla h_\rho f|^2 \, d\mathfrak{m} \le \int_X |\nabla f|^2 \, d\mathfrak{m} \qquad \text{if } f \in H^{1,2}(X, \mathsf{d}, \mathfrak{m}), \tag{1.74}$$

$$\int_X |\Delta h_\rho f|^2 \, d\mathfrak{m} \le \int_X |\Delta f|^2 \, d\mathfrak{m} \qquad \text{if } f \in D(\Delta). \tag{1.75}$$

Then, we define

$$\begin{aligned} \mathrm{Test}_\star F(X, \mathsf{d}, \mathfrak{m}) \quad := \quad & \left\{ h_\rho \left(L^2 \cap L^\infty(X, \mathfrak{m})\right) : \rho \in C_c^\infty((0, \infty)) \text{ convolution kernel} \right\} \\ & \subset \mathrm{Test}F(X, \mathsf{d}, \mathfrak{m}). \end{aligned}$$

By letting $\rho \to \delta_0$ it is immediately seen from (1.73), (1.74), (1.75) that the class $\mathrm{Test}_\star F(X, \mathsf{d}, \mathfrak{m})$ is dense in $\mathrm{Test}F(X, \mathsf{d}, \mathfrak{m})$, namely for any $f \in \mathrm{Test}F(X, \mathsf{d}, \mathfrak{m})$ there exist $f_n \in \mathrm{Test}_\star F(X, \mathsf{d}, \mathfrak{m})$ strongly convergent in $H^{1,2}$ to f, with $\sup |f_n| \le \sup |f|$, $\mathrm{Lip}(f_n) \le \mathrm{Lip} f$, and $\Delta f_n \to \Delta f$ strongly in $H^{1,2}$.

In the next proposition we show a canonical approximation of test functions in the class $\mathrm{Test}F(X, \mathsf{d}, \mathfrak{m})$ by test functions for the approximating metric measure structures. Notice that we do not know if condition (b) can be improved, getting strong $H^{1,2}$ convergence of $|\nabla f_i|_i^2$.

Proposition 1.10.2. *Let $f \in \mathrm{Test}F(X, \mathsf{d}, \mathfrak{m})$. Then there exist $f_i \in \mathrm{Test}_\star F(X, \mathsf{d}, \mathfrak{m}_i)$ with $\|f_i\|_{L^\infty(X, \mathfrak{m}_i)} \le \|f\|_{L^\infty(X, \mathfrak{m})}$ and $\sup_i \mathrm{Lip}(f_i) < \infty$, such that f_i and $\Delta_i f_i$ strongly converge to f and Δf in $H^{1,2}$, respectively. Moreover, these properties yield:*
(a) $|\nabla f_i|_i^2$ L^1-strongly and L^2_{loc}-strongly converge to $|\nabla f|^2$;
(b) $|\nabla f_i|_i^2$ weakly converge to $|\nabla f|^2$ in $H^{1,2}$.

Proof. Let us assume first that $f = h_\rho g$ for some $g \in L^2 \cap L^\infty(X, \mathfrak{m})$ and some convolution kernel ρ. We define f_i as $h^i_\rho g_i$, with g_i L^2-strongly convergent to g, with $\|g_i\|_{L^\infty(X, \mathfrak{m}_i)} \le \|g\|_{L^\infty(X, \mathfrak{m})}$. It is clear from the construction that $\|f_i\|_{L^\infty(X, \mathfrak{m}_i)} \le \|f\|_{L^\infty(X, \mathfrak{m})}$ and that $\sup_i \mathrm{Lip}(f_i) < \infty$. From (1.14) and (1.15), together with the first formula in (1.72) (applied to h^i_ρ), we obtain that both f_i and $\Delta_i f_i$ are bounded in $H^{1,2}$, and their strong convergence is a direct consequence of Corollary 1.5.5(b) and of (1.72) again.

The weak convergence in $H^{1,2}$ of $|\nabla f_i|_i^2$ to $|\nabla f|^2$ follows by the apriori estimates (1.22) and (1.23), that ensure the uniform bounds in $H^{1,2}$, and by Theorem 1.5.7(c) that identifies the L^1-strong limit (and therefore the weak $H^{1,2}$ limit) as $|\nabla f|^2$. Theorem 1.7.4 provides the relative compactness in L^2_{loc} of $|\nabla f_i|_i^2$ and then proves L^2_{loc}-convergence of $|\nabla f_i|_i^2$ to $|\nabla f|^2$ as well.

When $f \in \mathrm{Test}F(X, \mathsf{d}, \mathfrak{m})$ we apply the previous approximation procedure to $h_\rho f$ and then we make a diagonal argument, letting $\rho \to \delta_0$, noticing that the first identity in (1.72) grants the strong convergence in $H^{1,2}$ of $\Delta_i h^i_\rho f_i$ to $\Delta h_\rho f$, while the second identity in (1.72) grants

$$\|\Delta h_\rho f\|_{L^2(X, \mathfrak{m})} \le \|\Delta f\|_{L^2(X, \mathfrak{m})}, \qquad \|\nabla \Delta h_\rho f\|_{L^2(X, \mathfrak{m})} \le \nabla \Delta f\|_{L^2(X, \mathfrak{m})}.$$

□

Theorem 1.10.3 (Stability of $W^{2,2}$ regularity and weak convergence of Hessians). *Let $f_i \in W^{2,2}(X, \mathsf{d}, \mathfrak{m}_i)$ with $\sup_i \|f_i\|_{W^{2,2}(X,\mathsf{d},\mathfrak{m}_i)} < \infty$, and assume that f_i strongly converge in $H^{1,2}$ to $f \in H^{1,2}(X, \mathsf{d}, \mathfrak{m})$.*
Then $f \in W^{2,2}(X, \mathsf{d}, \mathfrak{m})$ and $\mathrm{Hess}_i(f_i)$ L^2-weakly converge to $\mathrm{Hess}(f)$ in the following sense: whenever $g_i \in H^{1,2}(X, \mathsf{d}, \mathfrak{m}_i)$ are uniformly Lipschitz and strongly converge in $H^{1,2}$ to $g \in H^{1,2}(X, \mathsf{d}, \mathfrak{m})$,

$$\mathrm{Hess}_i(f_i)(g_i, g_i) \ L^2\text{-weakly converge to } \mathrm{Hess}(f)(g, g).$$

In addition, $|\mathrm{Hess}(f)| \le H$ $\mathfrak{m}$-a.e. for any L^2-weak limit point H of $|\mathrm{Hess}_i(f_i)|$, and in particular

$$\int_X |\mathrm{Hess}(f)|^2 \, d\mathfrak{m} \le \liminf_{i\to\infty} \int_X |\mathrm{Hess}_i(f_i)|^2 \, d\mathfrak{m}_i. \tag{1.76}$$

Proof. Let $g \in \mathrm{Test}F(X, \mathsf{d}, \mathfrak{m})$ and let H be a L^2-weak limit point of $|\mathrm{Hess}_i(f_i)|$. Let (g_i) be provided by Proposition 1.10.2. We will first prove convergence of the Hessians under this stronger convergence assumption on g_i.

In order to identify the L^2-weak limit of $\mathrm{Hess}(f_i)(g_i, g_i)$ we want to take the limit as $i \to \infty$ in the expression

$$-2 \int_X \langle \nabla f_i, \nabla g_i \rangle_i \, \mathrm{div}(\varphi \nabla g_i) \, d\mathfrak{m}_i - \int_X \varphi \langle \nabla f_i, \nabla |\nabla g_i|_i^2 \rangle_i \, d\mathfrak{m}_i$$

with $\varphi \in h_{\mathbb{Q}_+}\mathcal{A}_{\mathrm{bs}}$. Let us analyze the first term. Since $\mathrm{div}(\varphi \nabla g_i) = \varphi \Delta_i g_i + \langle \nabla g_i, \nabla \varphi \rangle$, this term L^2-strongly converges to $\mathrm{div}(\varphi \nabla g) = \varphi \Delta g + \langle \nabla g, \nabla \varphi \rangle$. On the other hand, by Theorem 1.5.7(b), the term $\langle \nabla f_i, \nabla g_i \rangle_i$ L^2-weakly converges to $\langle \nabla f, \nabla g \rangle$. This proves the convergence of the first term.

Let us analyze the second term. Since Proposition 1.10.2(b) shows that $|\nabla g_i|_i^2$ weakly converge in $H^{1,2}$ to $|\nabla g|^2$, we can apply Theorem 1.5.7(b) again to obtain the convergence of $\int_X \varphi \langle \nabla f_i, \nabla |\nabla g_i|_i^2 \rangle_i \, d\mathfrak{m}_i$ to $\int_X \varphi \langle \nabla f, \nabla |\nabla g|^2 \rangle \, d\mathfrak{m}$.

This completes the proof under the additional assumption on g_i. In the general case it is sufficient to apply the already proved convergence result to $h_\rho^i g_i$, with ρ convolution kernel with support in $(0, \infty)$, noticing the uniform Lipschitz bound on g_i yields

$$\begin{aligned} &\int_X |\mathrm{Hess}(f_i)(g_i, g_i) - \mathrm{Hess}(f_i)(h_\rho^i g_i, h_\rho^i g_i)| \, d\mathfrak{m}_i \\ &\quad \le \int_X |\mathrm{Hess}(f_i)| |\nabla g_i \otimes \nabla g_i - \nabla h_\rho^i g_i \otimes \nabla h_\rho^i g_i| \, d\mathfrak{m}_i \\ &\quad \le C \int_X |\mathrm{Hess}(f_i)| |\nabla g_i - \nabla h_\rho^i g_i|_i \, d\mathfrak{m}_i \end{aligned}$$

and that the strong $H^{1,2}$ convergence of $h^i_\rho g_i$ to $h_\rho g$ yields

$$\lim_{\rho\to\delta_0}\limsup_{i\to\infty}\int_X |\nabla g_i - \nabla h^i_\rho g_i|_i^2\, \mathrm{dm}_i = 0.$$

The inequality $|\mathrm{Hess}(f)| \le H$ can be proved as follows. We start from the observation that, by bilinearity,

$$-\int_X \langle\nabla f_i, \nabla g_i\rangle_i \operatorname{div}(\varphi\psi\nabla h_i)\, \mathrm{dm}_i - \int_X \langle\nabla f_i, \nabla h_i\rangle_i \operatorname{div}(\varphi\psi\nabla g_i)\, \mathrm{dm}_i -$$
$$+\int_X \varphi\psi\big\langle\nabla f_i, \nabla\langle\nabla g_i, \nabla h_i\rangle_i\big\rangle_i\, \mathrm{dm}_i$$

converges to

$$-\int_X \langle\nabla f, \nabla g\rangle \operatorname{div}(\varphi\psi\nabla h)\, \mathrm{dm} - \int_X \langle\nabla f, \nabla h\rangle \operatorname{div}(\varphi\psi\nabla g)\, \mathrm{dm} - \int_X \varphi\psi\big\langle\nabla f, \nabla\langle\nabla g, \nabla h\rangle\big\rangle\, \mathrm{dm}$$

for any $\varphi, \psi \in h_{\mathbb{Q}_+}\mathcal{A}_{\mathrm{bs}}$ whenever $g_i, h_i \in \mathrm{Test}F(X, \mathsf{d}, \mathrm{m}_i)$ are uniformly Lipschitz and strongly converge in $H^{1,2}$ to $g, h \in \mathrm{Test}F(X, \mathsf{d}, \mathrm{m})$ respectively. This, taking also Remark 1.10.1 into account, enables to take the limit in (1.69) written for f_i, to get

$$\Big|\sum_k\Big(-\int_X \langle\nabla f, \nabla g_k\rangle \operatorname{div}(\varphi_k\psi_k\nabla h_k)\, \mathrm{dm} - \int_X \langle\nabla f, \nabla h_k\rangle \operatorname{div}(\varphi_k\psi_k\nabla g_k)\, \mathrm{dm}$$
$$- \int_X \varphi_k\psi_k\big\langle\nabla f, \nabla\langle\nabla g_k, \nabla h_k\rangle\big\rangle\Big)\Big| \le \int_X H\Big|\sum_k \varphi_k\psi_k\nabla g_k\otimes\nabla h_k\Big|\, \mathrm{dm}$$

for any finite collection of $\varphi_k, \psi_k \in h_{\mathbb{Q}_+}\mathcal{A}_{\mathrm{bs}}, g_k, h_k \in \mathrm{Test}F(X, \mathsf{d}, \mathrm{m})$. This proves that $|\mathrm{Hess}(f)| \le H$ m-a.e. in X. □

In the next corollary we use the bounds on laplacians of f_i to obtain at the same time strong convergence in $H^{1,2}$ and the uniform bound in $W^{2,2}$, so that the conclusions of Theorem 1.10.3 apply.

Corollary 1.10.4 (Weak stability of Hessians under Laplacian bounds). *Let $f_i \in D(\Delta_i)$ with*

$$\sup_i\big(\|f_i\|_{L^2(X,\mathrm{m}_i)} + \|\Delta_i f_i\|_{L^2(X,\mathrm{m}_i)}\big) < \infty$$

and assume that f_i L^2-strongly converge to f. Then $f \in D(\Delta)$ and
(i) f_i strongly converge to f in $H^{1,2}$;
(ii) $\Delta_i f_i$ L^2-weakly converge to Δf;
(iii) the Hessians of f_i are weakly convergent to the Hessian of f as in Theorem 1.10.3.

Proof. Statements (i) and (ii) follows by Corollary 1.5.5(a), while statement (iii) is a consequence of Theorem 1.10.3 and of (1.68). □

In the final part of his work [29], motivated also by the measure-valued Γ_2 operator introduced in [45], Gigli introduced a weak Ricci tensor **Ric**. It is a sort of measure-valued $(0,2)$-tensor, whose action on gradients of functions $f \in \mathrm{Test}F(X,\mathsf{d},\mathfrak{m})$ is given by

$$\mathbf{Ric}(\nabla f, \nabla f) := \Delta \frac{1}{2}|\nabla f|^2 - |\mathrm{Hess}(f)|^2\mathfrak{m} - \langle \nabla f, \nabla \Delta f\rangle \mathfrak{m}, \tag{1.77}$$

where the potentially singular part w.r.t. $\mathfrak{m}$ comes from the distributional laplacian Δ. The measure defined in (1.77) is bounded from below by $K|\nabla f|^2\mathfrak{m}$ and it is a capacitary measure, namely it vanishes on sets with null capacity (w.r.t. the Dirichlet form associated to Ch); hence, its duality with functions in $H^{1,2}(X,\mathsf{d},\mathfrak{m})$ is well defined.

Actually, **Ric** can be defined as a bilinear form on a larger class $H^{1,2}_H(T(X,\mathsf{d},\mathfrak{m}))$ of vector fields, weakly differentiable in a suitable sense, which includes gradient vector fields of functions in $\mathrm{Test}F(X,\mathsf{d},\mathfrak{m})$; on the other hand, using the linearity property of Proposition 3.6.9 in [29], as well as the continuity property (3.6.13) of Theorem 3.6.7, one can prove that (1.79) holds if $\mathbf{Ric}(v,v) \geq \zeta|v|^2$ for all $v \in H^{1,2}_H(T(X,\mathsf{d},\mathfrak{m}))$. For this reason we confine ourselves to the smaller class of vector fields.

Using the tools developed so far we are able to prove a kind of upper semicontinuity, in the measure-valued sense, for **Ric** under measured Gromov-Hausdorff convergence.

Theorem 1.10.5 (Upper semicontinuity of Ricci curvature). *Assume that $(X,\mathsf{d},\mathfrak{m}_i)$ are $RCD(K_i,\infty)$ spaces satisfying*

$$\mathbf{Ric}_i(\nabla f, \nabla f) \geq \zeta|\nabla f|_i^2 \qquad \forall f \in \mathrm{Test}F(X,\mathsf{d},\mathfrak{m}_i) \tag{1.78}$$

for some $\zeta \in C(X)$ with ζ^- bounded. Then

$$\mathbf{Ric}(\nabla f, \nabla f) \geq \zeta|\nabla f|^2 \qquad \forall f \in \mathrm{Test}F(X,\mathsf{d},\mathfrak{m}). \tag{1.79}$$

Proof. Setting $K = \sup \zeta^-$, from (1.78) and from the characterization of $RCD(K,\infty)$ spaces based on Bochner's inequality in [8] we obtain that $(X,\mathsf{d},\mathfrak{m}_i)$ are $RCD(K,\infty)$ spaces. By a truncation argument, it is not restrictive to assume that $\zeta \in C_b(X)$. Assume that $f \in \mathrm{Test}F(X,\mathsf{d},\mathfrak{m})$ and let $f_i \in \mathrm{Test}F(X,\mathsf{d},\mathfrak{m}_i)$ be strongly convergent in $H^{1,2}$ to f, with $\sup_i(\sup_X |f_i| + \mathrm{Lip}(f_i)) < \infty$, $\Delta_i f_i$ strongly convergent to Δf in $H^{1,2}$ and $|\nabla f_i|_i^2$ weakly convergent in $H^{1,2}$ to $|\nabla f|^2$. "The existence of a sequence (f_i) with these properties is granted by Proposition 1.10.2.

We want to take the limit as $i \to \infty$ in the integral formulation

$$-\frac{1}{2}\int_X \langle \nabla \varphi_i, \nabla |\nabla f_i|_i^2\rangle_i \,d\mathfrak{m}_i - \int_X \varphi_i |\mathrm{Hess}(f_i)|_i^2 \,d\mathfrak{m}_i - \int_X \varphi_i \langle \nabla f_i, \nabla \Delta_i f_i\rangle_i \,d\mathfrak{m}_i$$
$$\geq \int_X \zeta \varphi_i |\nabla f_i|_i^2 \,d\mathfrak{m}_i \tag{1.80}$$

of (1.78), with $\varphi_i \in H^{1,2}(X,\mathsf{d},\mathfrak{m}_i)$ bounded and nonnegative, thus getting the integral formulation of (1.79). To this aim, for $\varphi \in H^{1,2}(X,\mathsf{d},\mathfrak{m})$, let φ_i be uniformly bounded, nonnegative and strongly convergent in $H^{1,2}$ to φ. First of all, since $|\nabla f_i|_i^2$ L^1-strongly converge to $|\nabla f|^2$, the right hand sides converge to $\int_X \zeta\varphi|\nabla f|^2\,\mathrm{d}\mathfrak{m}$. Also the convergence of the third term in the left hand side to $\int_X \varphi\langle\nabla f, \nabla\Delta f\rangle\,\mathrm{d}\mathfrak{m}$ is ensured by Theorem 1.5.7(b). To handle the first term, we just use (1.26). Finally, in connection with the Hessians, possibly extracting a subsequence we obtain a L^2-weak limit point H of $|\mathrm{Hess}_i(f_i)|$, with $H \geq |\mathrm{Hess}(f)|$ $\mathfrak{m}$-a.e. in X.

Summing up, taking the limit as $i \to \infty$ in (1.80) one obtains the inequality

$$-\frac{1}{2}\int_X \langle\nabla\varphi, \nabla|\nabla f|^2\rangle\,\mathrm{d}\mathfrak{m} - \int_X \varphi H^2\,\mathrm{d}\mathfrak{m} - \int_X \varphi\langle\nabla f, \nabla\Delta f\rangle\,\mathrm{d}\mathfrak{m} \geq \int_X \zeta\varphi|\nabla f|^2\,\mathrm{d}\mathfrak{m}.$$

Using the inequality $H \geq |\mathrm{Hess}(f)|$ $\mathfrak{m}$-a.e. in X we conclude the proof. □

Remark 1.10.6. *For any $r \in (0,1)$, it is easy to construct a sequence (g_i^r) of Riemannian metrics on $\mathbf{S}^2$ with sectional curvature bounded below by 1 such that $(\mathbf{S}^2, g_i^r) \to [0,\pi]\times_{\sin}\mathbf{S}^1(r)$ in the Gromov-Hausdorff sense, where $\mathbf{S}^1(r) := \{x \in \mathbf{R}^2; |x| = r\}$ (the limit space is an Alexandrov space of curvature ≥ 1). Note that $[0,\pi]\times_{\sin}\mathbf{S}^1(r) \to [0,\pi]$ as $r \to 0$ in the Gromov-Hausdorff sense, and that*

$$\frac{\mathcal{H}^2(B_s(x_0))}{\mathcal{H}^2([0,\pi]\times_{\sin}\mathbf{S}^1(r))} = \frac{\mathcal{H}^2(B_s(x_\pi))}{\mathcal{H}^2([0,\pi]\times_{\sin}\mathbf{S}^1(r))} = \frac{1}{2}\int_0^s \sin t\,\mathrm{d}t$$

for any $s \in [0,\pi]$, where $x_0 = (0,\star)$ and $x_\pi = (\pi,\star)$. Thus, by a diagonal argument, there exist Riemannian metrics (g_i) on $\mathbf{S}^2$ (in fact $g_i := g_i^{r_i}$ for some $r_i \to 0$) with sectional curvature bounded below by 1 such that $(\mathbf{S}^2, g_i, \mathcal{H}^2/\mathcal{H}^2(\mathbf{S}^2)) \to ([0,\pi], g, \upsilon)$ in the measured Gromov-Hausdorff sense, where g is the Euclidean metric and υ is the Borel probability measure on $[0,\pi]$ defined by

$$\upsilon([r,s]) = \frac{1}{2}\int_r^s \sin t\,\mathrm{d}t$$

for any $r, s \in [0,\pi]$ with $r \leq s$. Let us consider eigenfunctions $f_i \in C^\infty(\mathbf{S}^2)$ of the first positive eigenvalues of Δ_i with $\|f_i\|_{L^2(\mathbf{S}^2,\mathfrak{m}_i)} = 1$, where $\mathfrak{m}_i = \mathcal{H}^2/\mathcal{H}^2(\mathbf{S}^2)$ w.r.t. g_i. Then, by [24] we can assume with no loss of generality that f_i strongly converge to f in $H^{1,2}$, with f eigenfunction of the first positive eigenvalue of Δ. It is known that $\Delta f = 2f$ and that $\lim_i \||\mathrm{Hess}_i(f_i) + f_i g_i|\|_{L^2(X,\mathfrak{m}_i)} = 0$. Moreover we can prove that $f(t) = 3\cos t$. Note that these observations correspond to the Bonnet-Mayers theorem and the rigidity on singular spaces. See [23, 24] for the proofs.

In particular $\lim_i \||\mathrm{Hess}_i(f_i)|\|_{L^2(\mathbf{S}^2,\mathfrak{m}_i)} = 2\lim_i \|f_i\|_{L^2(\mathbf{S}^2,\mathfrak{m}_i)} = 2$. On the other hand, it was proven in [35] that g_i L^2-weakly converge to g on $[0,\pi]$. Thus $\mathrm{Hess}(f) + fg = 0$ in L^2. In particular $\||\mathrm{Hess}(f)|\|_{L^2([0,\pi],\upsilon)} = \|f\|_{L^2([0,\pi],\upsilon)} = 1$. Thus these facts give

$$\lim_{i\to\infty} Ric_i(\nabla f_i, \nabla f_i)(\mathbf{S}^2, g_i, \mathfrak{m}_i) < Ric(\nabla f, \nabla f)([0,\pi], g, \upsilon),$$

i.e. the Ricci curvatures are strictly increasing even in the case when f_i, $|\nabla f_i|_i^2$, $\Delta_i f_i$ *are uniformly bounded, and strongly converge to* f, $|\nabla f|^2$, Δf *in* $H^{1,2}$, *respectively. In this respect, Theorem 1.10.5 might be sharp. Moreover this example also tells us that, in general, the condition that* $\Delta_i f_i$ L^2*-strongly converge to* Δf *does not imply that* $|\mathrm{Hess}_i(f_i)|$ L^2*-strongly converge to* $|\mathrm{Hess}(f)|$.

Remark 1.10.7. *With a very similar argument one can prove stability of the* $BE(K,N)$ *condition*

$$\frac{1}{2}\Delta|\nabla f|^2 \geq \langle \nabla f, \nabla \Delta f\rangle + \frac{(\Delta f)^2}{N} + K|\nabla f|^2,$$

with $K : X \to (-\infty, +\infty]$ *lower semicontinuous and bounded from below,* $N : X \to (0,\infty]$ *upper semicontinuous. Notice that the strategy of passing to an integral formulation, adopted in [8, Theorem 5.8], seems to work only when* K *and* N *are constant.*

1.11 Dimensional stability results

In this section only we state results that depend on the assumption $N < \infty$. We recall that the definition of $RCD^*(K,N)$ space has been proposed in [28] and deeply investigated and characterized in various ways in [27] (via the so-called Entropy power functional, a dimensional modification of Shannon's logarithmic entropy) and in [12] (via nonlinear diffusion semigroups induced by Rényi's N-entropy), see also [8] in connection with the stability point of view. Starting from $RCD(K,\infty)$, the conditions $RCD^*(K,N)$ amount to the following reinforcement of Bochner's inequality

$$\Delta\frac{1}{2}|\nabla f|^2 \geq \frac{1}{N}(\Delta f)^2\mathfrak{m} + \langle \nabla f, \nabla\Delta f\rangle\mathfrak{m} + K|\nabla f|^2\mathfrak{m} \tag{1.81}$$

in the class $\mathrm{Test}F(X,\mathsf{d},\mathfrak{m})$.

Proposition 1.11.1. *There exist positive and finite constants* $C_i(\alpha, N)$, $i = 1,2$, *such that for any* $RCD^*(K,N)$*-space* $(Y,\mathsf{d},\mathfrak{m})$ *with* $\mathrm{supp}\,\mathfrak{m} = Y$, $\mathfrak{m}(Y) = 1$ *and finite diameter one has*

$$0 < C_1(K(\mathrm{diam}\,Y)^2, N) \leq \mathrm{diam}\,Y \left(\lambda_{1,p}(Y,\mathsf{d},\mathfrak{m})\right)^{1/p} \leq C_2(K(\mathrm{diam}\,Y)^2, N) < \infty \tag{1.82}$$

for any $p \in [1,\infty]$.

Proof. Since the rescaled metric measure space

$$\left(Y, (\mathrm{diam}\,Y)^{-1}\mathsf{d}, \mathfrak{m}\right)$$

is an $RCD^*(K(\mathrm{diam}\,Y)^2, N)$-space, and

$$\lambda_{1,p}\left(Y, (\mathrm{diam}\,Y)^{-1}\mathsf{d}, \mathfrak{m}\right) = (\mathrm{diam}\,Y)^p\,\lambda_{1,p}(Y,\mathsf{d},\mathfrak{m}),$$

it suffices to check (1.82) under diam $Y = 1$.

Let $\mathcal{M}(K, N)$ be the set of all isometry classes of $RCD^*(K, N)$ spaces $(Y, \mathsf{d}, \mathfrak{m})$ satisfying supp $\mathfrak{m} = Y$, diam $Y = 1$ and $\mathfrak{m}(Y) = 1$. It is known that this set is sequentially compact w.r.t. the measured Gromov-Hausdorff convergence by [8, 27]. We consider the function F on $\mathcal{M}(K, N) \times [1, \infty]$ defined by

$$F\left((Y, \mathsf{d}, \mathfrak{m}), p\right) := \left(\lambda_{1,p}(Y, \mathsf{d}, \mathfrak{m})\right)^{1/p}.$$

Hence, Theorem 1.9.4 and Theorem 1.9.6 yield that F is continuous. In particular the maximum and the minimum exist. Moreover, by the definition of $RCD^*(K, N)$ space depend only on the parameters N and K. This shows (1.82). □

Remark 1.11.2. *The finiteness of N in Proposition 1.11.1 is essential, i.e. the estimate* $C_1(KR^2) \le \operatorname{diam} Y \left(\lambda_{1,p}(Y, \mathsf{d}, \mathfrak{m})\right)^{1/p} \le C_2(KR^2)$ *does not hold for* $RCD(K, \infty)$*-spaces. Indeed, the standard n-dimensional unit sphere with the standard probability measure* $(\mathbf{S}^n, \mathsf{d}_n, \mathfrak{m}_n)$ *satisfies*

$$\lim_{n\to\infty} \lambda_{1,2}(\mathbf{S}^n, \mathsf{d}_n, \mathfrak{m}_n) = \infty.$$

For any $N \in (1, \infty)$ and any $p \in [1, \infty]$ let us denote $(\lambda_{1,p}^N)^{1/p}$ the infimum of $(\lambda_{1,p})^{1/p}$ in the set $\mathcal{M}(N)$ of all isometry classes of $RCD^*(N-1, N)$ probability spaces. For $p = 2$ the sharp Poincaré inequality for $CD^*(N-1, N)$-spaces given in [48] by Sturm yields $(\lambda_{1,2}^N)^{1/2} = N^{1/2}$ which coincides with $(\lambda_{1,2}(\mathbf{S}^N, \mathsf{d}, \mathfrak{m}_N))^{1/2}$ if N is an integer. The Bonnet-Meyers theorem for $CD^*(N-1, N)$-spaces given in [48] by Sturm gives $(\lambda_{1,\infty}^N)^{1/\infty} = 2/\pi$ which also coincides with $(\lambda_{1,\infty}(\mathbf{S}^N, \mathsf{d}, \mathfrak{m}_N))^{1/\infty}$ if N is an integer.

The following rigidity theorem is proven by Ketterer in [37, 38].

Theorem 1.11.3. *For any* $p \in \{2, \infty\}$, *any* $N \in (1, \infty)$, *and any* $RCD^*(N-1, N)$*-space* $(Y, \mathsf{d}, \mathfrak{m})$ *with* supp $\mathfrak{m} = Y$, *the equality*

$$\left(\lambda_{1,p}(Y, \mathsf{d}, \mathfrak{m})\right)^{1/p} = \left(\lambda_{1,p}^N\right)^{1/p}$$

holds if and only if $(Y, \mathsf{d}, \mathfrak{m})$ *is isometric to the spherical suspension of an* $RCD^*(N-2, N-1)$*-space.*

Furthermore for any $p \in \{2, \infty\}$, *any* $N \in (1, \infty)$, *and any* $\epsilon > 0$ *there exists* $\delta := \delta(p, N, \epsilon) > 0$ *such that if an* $RCD^*(N-1, N)$*-space* $(Y, \mathsf{d}, \mathfrak{m})$ *satisfies* supp $\mathfrak{m} = Y$ *and*

$$\left|\left(\lambda_{1,p}(Y, \mathsf{d}, \mathfrak{m})\right)^{1/p} - \left(\lambda_{1,p}^N\right)^{1/p}\right| < \delta,$$

then

$$\left|\left(\lambda_{1,q}(Y, \mathsf{d}, \mathfrak{m})\right)^{1/q} - \left(\lambda_{1,q}^N\right)^{1/q}\right| < \epsilon,$$

for any $q \in \{2, \infty\}$ *and there exists an* $RCD^*(N-2, N-1)$*-space* (Z, ρ, ν) *such that*

$$d_{GH}\left((Y, \mathsf{d}, \mathfrak{m}), ([0, \pi] \times_{\sin}^{N-1} (Z, \rho, \nu))\right) < \epsilon.$$

The following theorem is proven by Cavalletti-Mondino in [18, 19].

Theorem 1.11.4. *We have the following.*

(i) *For any $p \in [1, \infty)$ and $N \in \mathbf{N}_{\geq 2}$, we have $(\lambda_{1,p}^N)^{1/p} = (\lambda_{1,p}(\mathbf{S}^N, \mathsf{d}_N, \mathfrak{m}_N))^{1/p}$.*

(ii) *For any $p \in [1, \infty)$, any $N \in (1, \infty)$ and any $RCD^*(N-1, N)$-space $(Y, \mathsf{d}, \mathfrak{m})$ with* $\operatorname{supp} \mathfrak{m} = Y$, *if the equality*

$$\left(\lambda_{1,p}(Y, \mathsf{d}, \mathfrak{m})\right)^{1/p} = \left(\lambda_{1,p}^N\right)^{1/p}.$$

holds, then $(Y, \mathsf{d}, \mathfrak{m})$ is isometric to the spherical suspension of an $RCD^*(N-2, N-1)$-space.

Furthermore for any $p \in [1, \infty)$, any $N \in (1, \infty)$, and any $\epsilon > 0$ there exists $\delta := \delta(p, N, \epsilon) > 0$ such that if an $RCD^*(N-1, N)$-space $(Y, \mathsf{d}, \mathfrak{m})$ satisfies $\operatorname{supp} \mathfrak{m} = Y$ *and*

$$\left|\left(\lambda_{1,p}(Y, \mathsf{d}, \mathfrak{m})\right)^{1/p} - \left(\lambda_{1,p}^N\right)^{1/p}\right| < \delta,$$

then $\left|\operatorname{diam}(Y, \mathsf{d}) - \pi\right| < \epsilon$.

We now give a model metric measure space whose $(\lambda_{1,p})^{1/p}$ attains $(\lambda_{1,p}^N)^{1/p}$ for general N.

Proposition 1.11.5. *For any $N \in (1, \infty)$, let $([0, \pi], \mathsf{d}, \upsilon_N)$ with d equal to the Euclidean distance and*

$$\upsilon_N(A) := \frac{1}{\int_0^\pi \sin^{N-1} t \, dt} \int_A \sin^{N-1} t \, dt.$$

Then $([0, \pi], \mathsf{d}, \upsilon_N)$ is an $RCD^*(N-1, N)$-space with

$$\left(\lambda_{1,p}([0, \pi], \mathsf{d}, \upsilon_N)\right)^{1/p} = (\lambda_{1,p}^N)^{1/p} \qquad \forall p \in [1, \infty].$$

Proof. By [19, Theorem 1.4], for any $p \in [1, \infty]$, $(\lambda_{1,p}^N)^{1/p}$ coincides with the infimum in the smaller class

$$\inf \left\{\lambda_{1,p}([0, \pi], \mathsf{d}, \mathfrak{m}); ([0, \pi], \mathsf{d}, \mathfrak{m}) \in \mathcal{M}(N)\right\}. \tag{1.83}$$

By Theorem 1.9.4 and the sequencial compactness of $\mathcal{M}(N)$, there exists a Borel probability measure $\mathfrak{m}^p$ on $[0, \pi]$ such that $(\lambda_{1,p}([0, \pi], \mathsf{d}, \mathfrak{m}^p))^{1/p} = (\lambda_{1,p}^N)^{1/p}$. Then the maximal diameter theorem and p-Obata theorem for general $N \in (1, \infty)$ yield $\mathfrak{m}^p = \upsilon_N$. This completes the proof. □

As a corollary of Theorem 1.9.4 and Theorem 1.9.6, we have a generalization of Theorem 1.11.3 and Theorem 1.11.4 as follows. It is worth pointing out that this is new even in the class of smooth metric measure spaces, and shows that the parameter δ in Theorem 1.11.4 can be chosen independently of p:

Corollary 1.11.6. *For any $N \in (1, \infty)$ and any $\epsilon > 0$ there exists $\delta := \delta(N, \epsilon) > 0$ such that if an $RCD^*(N-1, N)$ space $(X, \mathsf{d}, \mathfrak{m})$ satisfies* $\operatorname{supp} \mathfrak{m} = X$, $\mathfrak{m}(X) = 1$ *and*

$$\left| \left(\lambda_{1,p}(X, \mathsf{d}, \mathfrak{m})\right)^{1/p} - \left(\lambda_{1,p}^N\right)^{1/p} \right| < \delta$$

for some $p \in [1, \infty]$, then

$$\left| \left(\lambda_{1,q}(X, \mathsf{d}, \mathfrak{m})\right)^{1/q} - \left(\lambda_{1,q}^N\right)^{1/q} \right| < \epsilon$$

for all $q \in [1, \infty]$.

Proof. We first prove that if an $RCD^*(N-1, N)$-space $(Y, \mathsf{d}, \mathfrak{m})$ satisfies $\operatorname{supp} \mathfrak{m} = Y$ and $\operatorname{diam}(Y, \mathsf{d}) = \pi$, then $(\lambda_{1,p}(Y, \mathsf{d}, \mathfrak{m}))^{1/p} = (\lambda_{1,p}^N)^{1/p}$ for any $p \in [1, \infty)$.

By Theorem 1.11.3, there exists an $RCD^*(N-2, N-1)$-space (Z, ρ, ν) such that $(Y, \mathsf{d}, \mathfrak{m})$ is isometric to $([0, \pi] \times_{\sin}^{N-1} (Z, \rho, \nu))$ and, from now on, we make this identification. Note that for any $f \in L^1([0, \pi], \upsilon_N)$ the function $f_0(y) := f(t)$ for $y = (t, z)$ is in $L^1(Y, \mathsf{d}, \mathfrak{m})$, and satisfies $c_p(f_0) = c_p(f)$, $\|f_0\|_{L^p} = \|f\|_{L^p}$ for any $f \in L^p([0, \pi], \upsilon_N)$. In addition

$$\int_Y f_0 \, \mathrm{d}\mathfrak{m} = \int_0^\pi f \, \mathrm{d}\upsilon_N. \tag{1.84}$$

Let $g \in \operatorname{Lip}([0, \pi], \mathsf{d})$ with $c_p(g) = \|g\|_{L^p} = 1$ (w.r.t. υ_N). Using the agreement of minimal relaxed slope with local Lipschitz constant in metric measure spaces satisfying the doubling and $(1, p)$-Poincaré condition (first proved in [22], see also [1]), it is easy to check that $|\nabla g_0|(t, z) = |\nabla g|(t)$ for any $t \in (0, \pi)$, any $z \in Z$. Applying (1.84) for $f = |\nabla g|^p$ yields

$$\lambda_{1,p}(Y, \mathsf{d}, \mathfrak{m}) \le \int_Y |\nabla g_0|^p \, \mathrm{d}\mathfrak{m} = \int_0^\pi |\nabla g|^p \, \mathrm{d}\upsilon_N.$$

Taking the infimum for g with Proposition 1.11.5 yields

$$(\lambda_{1,p}(Y, \mathsf{d}, \mathfrak{m}))^{1/p} = (\lambda_{1,p}([0, \pi], \mathsf{d}, \upsilon_N))^{1/p} = (\lambda_{1,p}^N)^{1/p}$$

because $c_p(g_0) = \|g_0\|_{L^p} = 1$.

We are now in a position to finish the proof of Corollary 1.11.6. The proof is done by contradiction via a standard compactness argument. Assume that the assertion is false. Then there exist $\epsilon > 0$, $p_i \in [1, \infty]$, $q_i \in [1, \infty]$ and $RCD^*(N-1, N)$-spaces $(X_i, \mathsf{d}_i, \mathfrak{m}_i)$ with $\operatorname{supp} \mathfrak{m}_i = X_i$ and $\mathfrak{m}_i(X_i) = 1$ such that

$$\lim_{i \to \infty} \left| \left(\lambda_{1,p_i}(X_i, \mathsf{d}_i, \mathfrak{m}_i)\right)^{1/p_i} - \left(\lambda_{1,p_i}^N\right)^{1/p_i} \right| = 0$$

and

$$\left| \left(\lambda_{1,q_i}(X_i, \mathsf{d}_i, \mathfrak{m}_i)\right)^{1/q_i} - \left(\lambda_{1,q_i}^N\right)^{1/q_i} \right| \ge \epsilon.$$

By the sequential compactness of $\mathcal{M}(N)$, without loss of generality we can assume (after embedding isometrically (X_i, d_i) into a common metric space (X, d)), that $X_i = X$, $\mathsf{d}_i = \mathsf{d}$ and that the measured Gromov-Hausdorff limit $(X, \mathsf{d}, \mathfrak{m})$ of the spaces $(X, \mathsf{d}, \mathfrak{m}_i)$ exists, and is an $RCD^*(N-1, N)$-space. We assume also that the limits $p, q \in [1, \infty]$ of p_i, q_i exist. Then Theorem 1.9.4 and Theorem 1.9.6 yield that

$$\left(\lambda_{1,p}(X, \mathsf{d}, \mathfrak{m})\right)^{1/p} = \left(\lambda_{1,p}^N\right)^{1/p}$$

and that

$$\left(\lambda_{1,q}(X, \mathsf{d}, \mathfrak{m})\right)^{1/q} \neq \left(\lambda_{1,q}^N\right)^{1/q}.$$

This contradicts Theorem 1.11.4 with the argument above. □

Bibliography

[1] L. Ambrosio, M. Colombo, S. Di Marino: Sobolev spaces in metric measure spaces: reflexivity and lower semicontinuity of slope. Advanced Studies in Pure Mathematics, 67 (2015), 1–58.
[2] L. Ambrosio, G. Crippa: Continuity equations and ODE flows with non-smooth velocity. Proc. Roy. Soc. Edinburgh Sect. A 144 (2014), 1191–1244.
[3] L. Ambrosio, S. DiMarino: Equivalent definitions of *BV* space and of total variation on metric measure spaces. Journal of Functional Analysis, 266 (2014), 4150–4188.
[4] L. Ambrosio, S. DiMarino, N. Gigli: Perimeter as relaxed Minkowski content in metric measure spaces. ArXiv preprint, 1603.08412.
[5] L. Ambrosio, N. Gigli, G. Savaré: Gradient flows in metric spaces and in the space of probability measures, Lectures in Mathematics, ETH Zürich, Birkhäuser (2008).
[6] L. Ambrosio, N. Gigli, G. Savaré: Calculus and heat flow in metric measure spaces and applications to spaces with Ricci bounds from below, Inventiones Mathematicae, 195 (2014), 289–391.
[7] L. Ambrosio, N. Gigli, G. Savaré: Metric measure spaces with Riemannian Ricci curvature bounded from below, Duke Math. J. 163 (2014), 1405–1490.
[8] L. Ambrosio, N. Gigli, G. Savaré: Bakry-Émery curvature-dimension condition and Riemannian Ricci curvature bounds, Annals of Probability, 43 (2015), 339–404.
[9] L. Ambrosio, N. Gigli, A. Mondino, T. Rajala: Riemannian Ricci curvature lower bounds in metric measure spaces with σ-finite measure. Transactions of the AMS, 367 (2015), 4661–4701.
[10] L. Ambrosio, B. Kirchheim: Currents in metric spaces, Acta Math. 185 (2000), 1–80.
[11] L. Ambrosio, A. Mondino, G. Savaré: On the Bakry-Émery condition, the gradient estimates and the Local-to-Global property of $RCD^*(K, N)$ metric measure spaces. Journal of Geometric Analysis, 26 (2016), 24–56.
[12] L. Ambrosio, A. Mondino, G. Savaré: Nonlinear diffusion equations and curvature conditions in metric measure spaces. ArXiv preprint 1509.07273.
[13] L. Ambrosio, A. Mondino: Gaussian-type isoperimetric inequalities in $RCD(K, \infty)$ probability spaces for positive K. Atti Accad. Naz. Lincei Rend. Lincei Mat. Appl. **27** (2016), 497–514.
[14] L. Ambrosio, F. Stra, D. Trevisan: Weak and strong convergence of derivations and stability of flows with respect to MGH convergence. ArXiv preprint 1603.05561.
[15] L. Ambrosio, D. Trevisan: Well posedness of Lagrangian flows and continuity equations in metric measure spaces, Analysis and PDE 7 (2014), 1179–1234.

[16] D. Bakry, I. Gentil, M. Ledoux: Analysis and Geometry of Markov Diffusion operators. Grundlehren der mathematisches Wissenschaften 348, Springer, 2014
[17] H. Brezis: Opérateurs maximaux monotones et semi-groupes de contractions dans les espaces de Hilbert. North-Holland Publishing Co., 1973.
[18] F. Cavalletti, A. Mondino: Sharp and rigid isoperimetric inequalities in metric-measure spaces with lower Ricci curvature bounds, ArXiv preprint 1502.06465.
[19] F. Cavalletti, A. Mondino: Sharp geometric and functional inequalities in metric measure spaces with lower Ricci curvature bounds, ArXiv preprint 1505.02061.
[20] F. Cavalletti, A. Mondino: Isoperimetric inequalities for finite perimeter sets in metric-measure spaces with lower Ricci curvature bounds. In progress.
[21] I. Chavel: Isoperimetric inequalities, Cambridge Tracts in Math. 145, Cambridge Univ. Press, Cambridge, U.K. (2001).
[22] J. Cheeger: Differentiability of Lipschitz functions on metric measure spaces. Geom. Funct. Anal., 9 (1999), 428–517.
[23] J. Cheeger, T. H. Colding: Lower bounds on Ricci curvature and the almost rigidity of warped products, Ann. of Math., 144 (1996), 189–237.
[24] J. Cheeger, T. H. Colding: On the structure of spaces with Ricci curvature bounded below, III, J. Differential Geom. 54 (2000), 37–74.
[25] J. Cheeger, B. Kleiner, A. Schioppa: Infinitesimal structure of differentiability spaces, and metric differentiation. ArXiv preprint 1503.07348.
[26] S. Di Marino, S., G. Speight: The p-weak gradient depends on p, Proc. Amer. Math. Soc. 143 (2015), 5239–5252.
[27] M. Erbar, K. Kuwada, K.-T. Sturm: On the equivalence of the entropic curvature-dimension condition and Bochner's inequality on metric measure spaces, Invent. Math., 201 (2015), 993–1071.
[28] N. Gigli: On the differential structure of metric measure spaces and applications. Mem. Am. Math. Soc., 236 (2015), no. 1113.
[29] N. Gigli: Nonsmooth differential geometry – An approach tailored for spaces with Ricci curvature bounded from below. ArXiv preprint 1407.0809. To appear on Mem. Am. Math. Soc.
[30] N. Gigli, B. Han: Independence on p of weak upper gradients on RCD spaces. ArXiv preprint 1407.7350.
[31] N. Gigli, A. Mondino, G. Savaré: Convergence of pointed non-compact metric measure spaces and stability of Ricci curvature bounds and heat flows. Proceedings of the London Mathematical Society, 111 (2015), 1071–1129.
[32] M. Gromov: Metric structures for Riemannian and non-Riemannian spaces. Modern Birkhäuser Classics, Birkhäuser, Boston, MA, english ed., 2007.
[33] S. Honda: Cheeger constant, p-Laplacian and Gromov-Hausdorff convergence, arXiv:1310.0304v3.
[34] S. Honda: Elliptic PDEs on compact Ricci limit spaces and applications. arXiv:1410.3296v5, To appear in Memoirs of the AMS.
[35] S. Honda: Ricci curvature and L^p-convergence. J. Reine Angew Math., 705 (2015), 85–154.
[36] S. Keith: A differentiable structure for metric measure spaces. Adv. Math., 183 (2004), 271–315.
[37] C. Ketterer: Cones over metric measure spaces and the maximal diameter theorem, J. Math. Pures Appl. (9) 103 (2015), 1228–1275.
[38] C. Ketterer: Obata's rigidity theorem for metric measure spaces, Anal. Geom. Metr. Spaces, 3 (2015), 278–295.
[39] C. Ketterer: On the geometry of metric measure spaces with variable curvature bounds. ArXiv preprint, 1506.03279.

[40] S. Lisini: Characterization of absolutely continuous curves in Wasserstein spaces. Calc. Var. Partial Differential Equations, 28 (2007), 85–120.
[41] J. Lott, C. Villani: Ricci curvature for metric-measure spaces via optimal transport, Ann. of Math., 169 (2009), 903–991.
[42] E. Milman: On the role of convexity in isoperimetry, spectral gap and concentration, Invent. Math., 177 (2009), 1–43.
[43] M. Miranda: Functions of bounded variation on "good" metric spaces. J. Math. Pures Appl., 82 (2003), 975–1004.
[44] A. Petrunin: Alexandrov meets Lott-Villani-Sturm, Münster J. Math., 4 (2011), 53–64.
[45] G. Savaré: Self-improvement of the Bakry-Émery condition and Wasserstein contraction of the heat flow in $RCD(K,\infty)$ metric measure spaces. Discrete Contin. Dyn. Syst., 34 (2014), 1641–1661.
[46] N. Shanmugalingam: Newtonian spaces: an extension of Sobolev spaces to metric measure spaces, Rev. Mat. Iberoamericana, 16 (2000), 243–279.
[47] T. Shioya: Metric measure geometry – Gromov's theory of convergence and concentration of metrics and measures. IRMA Lectures in Mathematics and Theoretical Physics of the European Mathematical Society, Vol. 25, (2016).
[48] K.-T. Sturm: On the geometry of metric measure spaces, I and II. Acta Math. 196 (2006), 65–131 and 133–177.
[49] M. Valadier: Young measures. Springer, 1990.
[50] L. Veron: Some Existence and Uniqueness Results for Solution of Some Quasilinear Elliptic Equations on Compact Riemannian Manifolds, Colloquia Mathematica Societatis János Bolyai, 62, P.D.E., Budapest (1991), 317–352.
[51] C. Villani, Optimal transport. Old and new, vol. 338 of Grundlehren der Mathematischen Wissenschaften, Springer-Verlag, Berlin, 2009.
[52] N. Weaver: Lipschitz algebras and derivations. II. Exterior differentiation, J. Funct. Anal., 178 (2000), 64–112.
[53] J.-Y. Wu, E.-M. Wang, and Y. Zheng: First eigenvalue of the p-Laplace operator along the Ricci flow, Ann. Global Anal. Geom. 38 (2010), 27–55.

Vladimir I. Bogachev

Surface measures in infinite-dimensional spaces

2.1 Introduction

There are three main constructions of surface measures in $\mathbb{R}^n$ with the standard Lebesgue measure. The most general one is based on the Hausdorff measure H^{n-1} of dimension $n-1$, which is a special case of the p-dimensional Hausdorff measure H^p with $0 \le p \le n$. Its role of a "surface measure" is explained by the following two factors: one, typical sets of finite positive H^{n-1}-measure are surfaces of dimension $n-1$, and two, this measure coincides with other natural candidates in cases where comparisons are possible (in particular, it coincides with the usual Lebesgue measure on hyperplanes). For reasonable sets, this surface measure can be obtained as a limit of normalized volumes of metric ε-neighborhoods of these sets. There is a much older construction that is closer to the intuitive understanding of what a surface measure must be: it is a natural measure on a regular surface $S \subset \mathbb{R}^n$, say, on the graph of a smooth function f on $\mathbb{R}^{n-1}$. This means that this surface measure arises as a limit of flat measures on small pieces of tangent hyperplanes approximating the given surface. Finally, one more construction deals with surfaces that are level sets of regular functions and defines the surface measure of the set $\{F = t\}$ as a certain limit of suitably normalized volumes of "neighborhoods" $\{t - \varepsilon \le F \le t + \varepsilon\}$. Locally, if $\nabla F \neq 0$ on the level set, this construction coincides with the previous one; moreover, all the three constructions coincide in this case. However, in general, a set of finite positive H^{n-1}-measure need not be located on a surface (neither a graph nor a level set); certainly, a level set need not be a graph even locally.

When discussing surface measures in infinite-dimensional spaces, it is customary to recall that there are no exact analogs of Lebesgue measures in infinite-dimensions. This is indeed but not a major problem: there are exact infinite-dimensional analogs of other important measures on $\mathbb{R}^n$, for example, Gaussian, and the local theory of surface measures associated with the standard Gaussian measure on $\mathbb{R}^n$ does not differ much from the classical construction. Apparently, the principal difficulty in constructing surface measures in infinite dimensions is that such measures are related to some intrinsic geometry of the measure but not of the space. In other words, it seems that in many cases there is no natural canonical geometry on the space determining surface measures. For example, we shall see below that the countable power $\mathbb{R}^\infty$ of the

Vladimir I. Bogachev: Department of Mechanics and Mathematics, Moscow State University, 119991 Moscow, Russia, St.-Tikhon's Orthodox Humanitarian University, Moscow, Russia, and National Research University Higher School of Economics, Moscow, Russia

real line equipped with the countable power of the standard Gaussian measure carries surface measures associated with Sobolev smooth functions; however, these surface measures have nothing to do with the standard metric on the space $\mathbb{R}^\infty$ (making it a Polish space); moreover, the measure can be restricted to many (continuum of incomparable) weighted Hilbert spaces in $\mathbb{R}^\infty$ of full measure and the surface measures will be unchanged.

Surface measures on general spaces have become a popular subject of study in recent years due to development of the Malliavin calculus, geometric measure theory in metric measure spaces, and infinite-dimensional stochastic analysis, see [1], [2], [3], [4], [5], [6], [7], [8], [13], [15], [20], [21], [22], [24], [25], [26], [29], [37], [43], [44], [62] and [63], where one can find discussions of diverse problems explicitly or implicitly connected with surface measures in infinite dimensions. In the Gaussian case, surface measures related to Gaussian volume measures by versions of the Gauss–Ostrogradskii formula were considered in the 1960–70s by a number of authors, see, e.g., [69], [66], [67], [39], and [47]. Actually, Skorohod [66], [67] considered surface measures for more general quasi-invariant measures. A rich theory of surface measures on infinite-dimensional spaces equipped with differentiable measures was worked out by A.V. Uglanov in the 1970–80s and presented in his book [75] (see also [73], [74], [77], and [32]). In the same years, an approach to surface measures for Gaussian volume measures was developed in the framework of the Malliavin calculus which provided efficient tools for the study of induced measures. For this approach, see [1], [49], [9], [10], and [11]; far reaching generalizations to the case of differentiable measures were obtained in [56], [57], [58], [59], and [60]. A close construction for configuration spaces was presented in [31]. Hausdorff measures associated with Gaussian measures were studied in [36] (see also [35]); more references for the Gaussian case can be found in [11] and [13].

The goal of this survey is to discuss several approaches to surface measures in infinite dimensions with a particular emphasis on the construction from the recent paper [18] that follows Malliavin's idea, but applies to nonlinear spaces and requires less regularity of the function F generating the surface. This surface measure on $F^{-1}(y)$ is a weak limit of the measures $r^{-1}I_{\{y<F<y+r\}} \cdot \theta_F \cdot \mu$ as $r \to 0$, where θ_F is some weight function (for a sufficiently regular surface, one can think of θ_F as the derivative of F along the "normal to the surface"). In the Gaussian case this construction applies to one-fold Malliavin differentiable functions with gradients having divergences. In the nondegenerate case, these surface measures are equivalent to the standard ones. However, this approach leads to much shorter and simpler proofs; in particular, we shall see that the existence of surface measures is proved in a few lines. We also mention some open problems related to infinite-dimensional surface measures. In particular, surface measures on zero sets of polynomials have not been sufficiently studied and this is an interesting direction of research. There are interesting connections between surface measures and Sobolev and BV functions on infinite-dimensional spaces, connections which have been intensively studied in the last decade, see [2], [3], [6], [7],

[8], [20], [21], [24], [48], and [51]. Finally, surface measures are important for the study of boundary value and variational problems in infinite dimensions, see [28] and [76].

The main construction discussed below is related to the concept of conditional measure (recalled in the next section). It makes sense in great generality while surface measures are usually defined in a more special situation, where one can consider suitable neighborhoods of the "surfaces" $\{F = y\}$ and obtain a reasonable limit after appropriate scaling. For example, the usual surface measure in $\mathbb{R}^d$ arises as a limit of the ratio of the volume of the ε-neighborhood of the surface and 2ε, as $\varepsilon \to 0$. The discussed construction of a surface measure σ^y on the level set $F^{-1}(y)$ is this: we introduce a certain weight function θ_F and set

$$\int f(x)\,\sigma^y(dx) := \lim_{r\to 0}\frac{1}{r}\int_{\{y<F<y+r\}} f(x)\theta_F(x)\,\mu(dx)$$

for a suitable class of functions f (say, bounded Lipschitzian). Under our assumptions the surface measure will be actually a weak limit of the measures

$$r^{-1}I_{\{y<F<y+r\}}\theta_F \cdot \mu.$$

Unlike the case of conditional measures, such constructions require certain constraints on measures and functions in question. In the case of a Gaussian measure μ on a locally convex space X this construction applies to a function F in the second Sobolev class $W^{2,2}(\mu)$ and we take $\theta_F = |D_H F|^2$ (or $\theta_F = |D_H F|$ in a modified construction), where $D_H F$ is the Sobolev gradient of F along the Cameron–Martin space H of the measure μ (or to a function F in the first Sobolev class $W^{1,1}(\mu)$ if $D_H F/|D_H F|_H$ has divergence). The weight function θ_F can be later dismissed provided it is sufficiently nondegenerate; its purpose is to allow degenerate F and lower the required order of differentiability of F. This approach can be also of interest for the study of surface measures on metric measure spaces (see [27], [40], [46], and [71]).

Why is it not enough to deal with conditional measures that exist in much greater generality? The reason is essentially the same as in the finite-dimensional case: the Gauss–Ostrogradskii–Stokes formula and integration by parts. This explains at once why certain smoothness restrictions on the volume measure and the function generating level sets are needed. Another reason is that conditional measures μ^y depend not only on the level sets $F^{-1}(y)$, but also on the image-measure $\mu \circ F^{-1}$ (though, for the induced measures with positive densities this dependence reduces to a constant factor for each fixed y). The presented construction shares this property, but allows a modification that does not.

The paper is organized as follows. Section 2 contains notation and terminology. In Section 3 we first discuss Gaussian surface measures in the finite-dimensional case and then explain how the same method works in infinite dimensions. Section 4 is devoted to the main construction in an abstract setting and its relation to conditional measures. Section 5 provides an additional step needed to obtain surface measures on every level set of a given function (not merely on almost every), which involves a brief discussion of capacities. In Section 6 we consider examples, in particular, returning to Gaussian measures. Finally, surface measures of surfaces of higher codimension are discussed in Section 7.

I am grateful to H. Airault, L. Ambrosio, G. Da Prato, N. Gigli, A. Lunardi, I.I. Malofeev, M. Röckner, and G.I. Zelenov for useful discussions. This research was sup-ported by the Russian Foundation for Basic Research Grant 17-01-00662 at Moscow State University.

2.2 Notation and terminology

Let $C_b^\infty(\mathbb{R}^d)$ be the class of all bounded infinitely differentiable functions on $\mathbb{R}^d$ with bounded derivatives, and let $C_0^\infty(\mathbb{R}^d)$ be its subclass consisting of functions with compact support.

Let X be a completely regular topological space with its Borel σ-field $\mathcal{B}$. Let μ be a bounded nonnegative Radon measure on $\mathcal{B}$, i.e., for every Borel set B and every $\varepsilon > 0$, there is a compact set $K_\varepsilon \subset B$ such that $\mu(B\backslash K_\varepsilon) < \varepsilon$ (see [12] for a discussion of such measures). Below we also use signed Radon measures, i.e., Borel measures m such that $|m|$ is a Radon measure, where $|m| = m^+ + m^-$ is the usual total variation of m.

Functions measurable with respect to the Lebesgue completion of the measure μ are called μ-measurable; such a function can be defined μ-almost everywhere μ-a.e., i.e., outside of a set of measure zero.

Given a a measure μ, the measure with density θ with respect to μ is denoted by $\theta \cdot \mu$, i.e.,

$$\theta \cdot \mu(B) = \int_B \theta \, d\mu.$$

In some assertions we shall assume that μ is concentrated on a countable union of metrizable compact sets; this is always the case if the space X is Souslin or metrizable, or if μ is Gaussian. The main definition does not use this assumption and actually applies to general probability spaces; however, the assumption becomes important in order to compare surface measures with conditional measures, and to ensure that our surface measures are indeed concentrated on the level sets.

Given a measurable function $F\colon X \to \mathbb{R}$ or a measurable mapping $F\colon X \to Y$ with values in a topological space Y, we can take the image-measure $\mu \circ F^{-1}$ defined by the formula

$$\mu \circ F^{-1}(B) := \mu(F^{-1}(B))$$

on the Borel σ-field in $\mathbb{R}$ or Y, respectively.

We recall that a Radon probability measure γ on a locally convex space X is called Gaussian if its one-dimensional images $\gamma \circ l^{-1}$ for all $l \in X^*$ are Gaussian measures on the real line, that is, are either given by densities $(2\pi\sigma)^{-1/2} \exp(-(t-a)^2/(2\sigma))$ with

$\sigma > 0$ or are Dirac point measures. If all these measures are symmetric, then γ is called symmetric or centered; the latter is equivalent to the identity $\gamma(B) = \gamma(-B)$ for all Borel sets B.

The most important examples of Gaussian measures are the standard Gaussian measure γ_n on $\mathbb{R}^n$ with density $(2\pi)^{-n/2}\exp(-|x|^2/2)$, the standard Gaussian measure γ on $\mathbb{R}^\infty$ that is the countable power of the standard Gaussian measure γ_1 on $\mathbb{R}$ (which is defined on the countable power $\mathbb{R}^\infty$ of the real line, i.e., on the space of all real sequences), and the Wiener measure on $C[0,1]$ or $L^2[0,1]$, which can be defined as the image of γ under the mapping

$$(x_n) \mapsto \sum_{n=1}^{\infty} x_n \int_0^t e_n(s)\,ds,$$

where $\{e_n\}$ is an orthonormal basis in $L^2[0,1]$. Certainly, such a definition needs a justification (it turns out that this series converges in $L^2[0,1]$ and even in $C[0,1]$ for almost all $x = (x_n)$). Other equivalent definitions of the Wiener measure can be found in [11] (see also [13] and [14]).

Let γ be a centered Radon Gaussian measure on X. The Cameron–Martin space H of γ consists of all vectors h with finite norm

$$|h|_H := \sup\{l(h):\ l \in X^*,\ \|l\|_{L^2(\gamma)} \le 1\}.$$

It is known that H with this norm is a separable Hilbert space compactly embedded into X; the corresponding inner product is denoted by $(\cdot,\cdot)_H$. A typical example: if γ is the standard Gaussian measure on $\mathbb{R}^\infty$, then $H = l^2$ with the usual Hilbert norm.

For every $h \in H$, there is a measurable linear functional $\widehat{h}$, belonging to the closure of X^* in $L^2(\gamma)$, such that

$$l(h) = \int_X l(x)\widehat{h}(x)\,\gamma(dx) \quad \forall\, l \in X^*.$$

The inner product in H can be defined by the formula

$$(h,k)_H = \int_X \widehat{h}\widehat{k}\,d\gamma.$$

An important role of the Cameron–Martin space is that it is precisely the set of all vectors h such that the shifted measure γ_h defined by $\gamma_h(B) = \gamma(B-h)$ is equivalent to γ; the corresponding Radon–Nikodym derivative is given by the Cameron–Martin formula

$$\exp(\widehat{h} - |h|_H^2/2). \tag{2.1}$$

If a Radon probability measure μ is concentrated on a countable union of metrizable compact sets, then, for any μ-measurable mapping F with values in a complete separable metric space Y, we can find the so-called conditional measures μ^y on X such that the function $y \mapsto \mu^y(B)$ is μ-measurable for each $B \in \mathcal{B}$, μ^y is concentrated on $F^{-1}(y)$ for every y (or $\mu \circ F^{-1}$-a.e. y) and μ is the integral of μ^y against $\mu \circ F^{-1}$, which is written as

$$\mu = \mu^y \cdot \mu \circ F^{-1}(dy),$$

in the sense that

$$\int_X f(x)\,\mu(dx) = \int_Y \int_X f(x)\,\mu^y(dx)\,\mu \circ F^{-1}(dy)$$

for every bounded Borel function f on X; the integral exists due to the assumption of measurability for μ^y, see [12, Chapter 10] or [13, Chapter 1] for details. Actually, conditional measures in a weaker sense exist under more general assumptions about μ and F.

2.3 Surface measures in the Gaussian case

We first consider a smooth function F on $\mathbb{R}^n$ equipped with the standard Gaussian measure γ_n. Suppose that $\nabla F(x) \neq 0$. In this case, the level sets $\{F = t\}$ are smooth surfaces that locally look like graphs of smooth functions. They can be equipped with usual surface (Lebesgue) measures and then Gaussian surface measures can be introduced by simply multiplying these Lebesgue surface measures by the standard Gaussian density. However, we are interested in globally finite Gaussian surface measures. To this end, we shall assume that the function $|\nabla F(x)|^{-1}$ belongs to certain $L^p(\gamma_n)$; the required value of p will become clear soon.

The first step is to verify that the distribution function

$$t \mapsto \gamma_n(x\colon\ F(x) < t)$$

is continuously differentiable. To show this we employ the Malliavin calculus. Let us consider the gradient vector field

$$v(x) = \nabla F(x)$$

and the corresponding differentiation

$$\partial_v g = (\nabla g, v).$$

Let ϱ_n be the standard Gaussian density. For every function $\phi \in C_b^\infty(\mathbb{R})$ we have

$$\int_{\mathbb{R}} \phi'(t)\,\gamma_n \circ F^{-1}(dt) = \int_{\mathbb{R}^n} \phi'(F(x))\,\gamma_n(dx)$$

$$= \int_{\mathbb{R}^n} \phi'(F(x))\partial_v F(x) \frac{1}{\partial_v F(x)} \gamma_n(dx) = \int_{\mathbb{R}^n} \partial_v(\phi \circ F) \frac{1}{\partial_v F} d\gamma_n. \quad (2.2)$$

Integrating by parts by means of the formula

$$\int_{\mathbb{R}^n} \partial_v f \psi \, d\gamma_n = -\int_{\mathbb{R}^n} f \partial_v \psi \, d\gamma_n \int_{\mathbb{R}^n} f\psi[\operatorname{div} v + (v, \nabla \varrho_n/\varrho_n)] \, d\gamma_n,$$

we represent the right-hand side as

$$-\int_{\mathbb{R}^n} \phi \circ F \Big[\partial_v \Big(\frac{1}{\partial_v F}\Big) + \frac{1}{\partial_v F}(\operatorname{div} v + (v, \nabla \varrho_n/\varrho_n))\Big] \, d\gamma_n.$$

We have

$$\partial_v F = |\nabla F|^2, \quad \partial_v(|\nabla F|^2) = 2(D^2 F \cdot \nabla F, \nabla F),$$

$$\partial_v \Big(\frac{1}{\partial_v F}\Big) = -\frac{\partial_v |\nabla F|^2}{|\partial_v F|^2} = -\frac{2(D^2 F \cdot \nabla F, \nabla F)}{|\nabla F|^4},$$

where

$$\operatorname{div} v + (v, \nabla \varrho_n/\varrho_n) = LF$$

and L is the Ornstein–Uhlenbeck operator defined by

$$Lf(x) = \Delta f(x) - (\nabla f(x), x) = \sum_{i=1}^{n} [\partial_{x_i}^2 f(x) - x_i \partial_{x_i} f(x)].$$

Therefore, letting

$$g = \frac{2(D^2 F \cdot \nabla F, \nabla F)}{|\nabla F|^4} + \frac{LF}{|\nabla F|^2},$$

we obtain

$$\int_{\mathbb{R}} \phi'(t) \, \gamma_n \circ F^{-1}(dt) = -\int_{\mathbb{R}} \phi(t) \, \eta(dt),$$

where η is the image under F of the measure with density g with respect to γ_n, provided that

$$\int_{\mathbb{R}^n} |g| \, d\gamma_n < \infty.$$

This means that the generalized derivative of $\gamma_n \circ F^{-1}$ is the measure η. Therefore, $\gamma_n \circ F^{-1}$ is an absolutely continuous measure and its density is $\eta((-\infty, t))$. Moreover, the measure η is also absolutely continuous, since it is obviously absolutely continuous with respect to $\gamma_n \circ F^{-1}$. Hence $\gamma_n \circ F^{-1}$ has a continuous density ϱ_1 and

$$\|\varrho_1\| \le \|g\|_{L^1(\gamma_n)}.$$

The integrability of the function g with respect to the standard Gaussian measure is ensured by the integrability with respect to γ_n of the functions

$$\|D^2 F(x)\|/|\nabla F(x)|^2 \quad \text{and} \quad |LF(x)|/|\nabla F(x)|^2,$$

where $\|D^2F(x)\|$ is the operator norm of the second derivative (one can also use the Hilbert–Schmidt operator). Therefore, if the function $|\nabla F(x)|^{-2-\varepsilon}$ is γ_n-integrable for some $\varepsilon > 0$ then, by Hölder's inequality, it suffices to have the γ_n-integrability of $\|D^2F(x)\|^{1+2/\varepsilon}$, because it is equivalent to the γ_n-integrability of $|LF(x)|^{1+2/\varepsilon}$, so that applying Hölder's inequality to the second function above we see that both functions will be integrable.

Under these assumptions, the function $\gamma_n(F < t)$ is continuously differentiable and we can assign the value

$$\varrho_1(t) := \frac{d}{dt}\gamma_n(F < t)$$

to the surface $S_t := F^{-1}(t)$. However, it is still not a surface measure but just its value on S_t.

Our next step is to observe that we can define a measure in a similar manner: this measure will be the weak limit of the measures

$$\mu_r(B) := (2r)^{-1}\gamma_n(B \cap \{t - r < F < t + r\})$$

as $r \to 0+$. Let us recall that a sequence of Borel measures μ_j on $\mathbb{R}^n$ converges weakly to a Borel measure μ if

$$\int_{\mathbb{R}^n} f\,d\mu = \lim_{j\to\infty} \int_{\mathbb{R}^n} f\,d\mu_j$$

for every bounded continuous function f. Moreover, for nonnegative measures it suffices if this limit exists for all bounded Lipschitz functions f (see, e.g., [13, Chapter 8]); furthermore, it is enough if it exists for all nonnegative bounded Lipschitz functions f.

It remains to observe that our construction also works if we replace the initial measure γ_n by $\psi\gamma_n$, where ψ is a nonnegative bounded Lipschitz function. Indeed, in our previous calculations we replace ϱ_n with $\psi\varrho_n$ noting that $\nabla\psi(x)$ exists almost everywhere and is bounded. Hence, we obtain the new functions

$$\partial_v\Big(\frac{\psi}{\partial_v F}\Big) = \psi\partial_v\Big(\frac{1}{\partial_v F}\Big) + \frac{(v, \nabla\psi)}{|\nabla F|^2},$$

$$g = \frac{2\psi(D^2F\cdot\nabla F, \nabla F)}{|\nabla F|^4} + \frac{\psi LF}{|\nabla F|^2} + \frac{(\nabla F, \nabla\psi)}{|\nabla F|^2}.$$

This function is integrable under the previous assumptions because ψ and $|\nabla\psi|$ are bounded. Thus, the measures μ_r converge weakly to a bounded Borel measure σ^t (when t is fixed) that we can take for a surface measure. However, this is not a true geometric surface measure; because it depends on the function F and not on only on the level set S_t. In particular, this is not the limit of normalized measures of S_t metric neighborhoods. To obtain a true geometric surface measure, we can consider the surface measure

$$\sigma_0^t := |\nabla F|\cdot\sigma^t,$$

which exists at least locally and is finite if, for example, F satisfies the conditions used above (integrability of certain rations with the second derivative and LF) to ensure that σ^t is finite. Moreover, these conditions can be relaxed: we need

$$\|D^2F\|/|\nabla F| + |LF|/|\nabla F| \in L^1(\gamma_n),$$

and for example, it suffices that $\|D^2F\| \in L^2(\gamma_n)$ and $1/|\nabla F| \in L^2(\gamma_n)$. Actually, the same reasoning as above applies to the measure $|\nabla F| \cdot \gamma_n$ in place of γ_n, so that in (2.2) we have to multiply and divide by $|\nabla F|$ in place of $|\nabla F|^2$. This leads to a weaker integrability condition on $1/|\nabla F|$.

Let us show that σ_t^0 is locally the weak limit of the measures $(2r)^{-1}\gamma_n(\cdot \cap S_t^r)$, where S_t^r is the metric r-neighborhood of S_t. This means that for every Lipschitz function $\phi \geq 0$ with bounded support we have

$$\int_{S_t} \phi \, d\sigma_0^t = \lim_{r\to 0}(2r)^{-1} \int_{S_t^r} \phi \, d\gamma_n. \tag{2.3}$$

We can assume that $t = 0$. Let us fix $\varepsilon > 0$. Taking a smooth partition of unity, we can assume that the support of ϕ is contained in a ball K centered at x_0 so small that $1-\varepsilon \leq |\nabla F(x)|/|\nabla F(x_0)| \leq 1+\varepsilon$ on K. We can also assume, changing coordinates, that $\partial_{x_n}F(x_0) = |\nabla F(x_0)|$, $1 - \varepsilon \leq \partial_{x_n}F(x)/\partial_{x_n}F(x_0) \leq 1 + \varepsilon$ on K and $S_0 \cap K$ is the graph of a smooth function of variables $x_1, \dots, x_{n-1}$. For $r > 0$ small enough, the integral of ϕ against σ_0^0 is $(2r)^{-1}|\nabla F(x_0)|\phi \cdot \mu(|F| < r)$ up to a factor $q \in (1 - \varepsilon, 1 + \varepsilon)$. The metric r-neighborhood of S_0 in K is contained in the intersection of K with the set $\{|F| \leq (1+\varepsilon)|\nabla F(x_0)|r\}$. Therefore, for $r > 0$ small enough we have

$$(2r)^{-1} \int_{S_0^r} \phi \, d\gamma_n \leq (1+2\varepsilon)|\nabla F(x_0)| \int_K \phi \, d\sigma^0 \leq (1+2\varepsilon)(1+\varepsilon) \int_K \phi \, d\sigma_0^0.$$

On the other hand, the set $\{|F| < r\} \cap K$ is contained in S_0^{qr} with $q = (1-\varepsilon)^{-1}/|\nabla F(x_0)|$, since if $x \in S_0 \cap K$ and $F(x + se_n) > r$, then $s > r(1-\varepsilon)^{-1}|\nabla F(x_0)|^{-1}$ by the equality $F(x + se_n) - F(x) = s\partial_{x_n}F(x + \theta e_n)$. Therefore,

$$(2r)^{-1}|\nabla F(x_0)| \int_{|F|<r} \phi \, d\gamma_n \leq (2r)^{-1}|\nabla F(x_0)| \int_{S_0^{qr}} \phi \, d\gamma_n,$$

which yields that

$$|\nabla F(x_0)| \int_{S_0} \phi \, d\sigma^0 \leq (1-\varepsilon)^{-1} \lim_{r\to 0}(2r)^{-1} \int_{S_0^r} \phi \, d\gamma_n.$$

These bounds yield (2.3).

We now consider a slightly modified construction an advantage of which is exclusion of the non-degeneracy assumption about ∇F. To this end, from the very beginning we replace the original measure γ_n by the measure

$$\nu := \partial_v F \cdot \gamma_n,$$

assuming that $\partial_v F \in L^1(\gamma_n)$; however, the latter assumption can be dismissed if we agree to deal with local surface measures. This trick enables us to write

$$\int_{\mathbb{R}} \phi'(t)\, \nu \circ F^{-1}(dt) = \int_{\mathbb{R}^n} \phi'(F(x))\partial_v F(x)\, \gamma_n(dx)$$

$$= \int_{\mathbb{R}^n} \partial_v(\phi \circ F)\, d\gamma_n = -\int_{\mathbb{R}^n} \phi \circ F \partial_v \varrho_n \, dx.$$

This shows that the generalized derivative of $\nu \circ F^{-1}$ is the image under F of the measure with density $-(\nabla F(x), x)\varrho_n(x)$ which is finite when $|\nabla F| \in L^2(\gamma_n)$ and absolutely continuous with respect to ν. Repeating the construction above we arrive at different surface measures σ_1^t without any non-degeneracy conditions on ∇F. Again, the obtained surface measures are not "geometric". To return to usual surface measures we have to assume that $\nabla F \neq 0$ and in that case we can consider (at least locally) the surface measures $|\nabla F|^{-1} \cdot \sigma_1^t$. The measures σ^t and σ_1^t are related by the equality $\sigma_1^t = |\nabla F|^2 \cdot \sigma^t$.

Example 2.3.1. Suppose that F is a polynomial of $\mathbb{R}^n$ such that $\nabla F(x) \neq 0$. It is known (see, e.g., [72]) there are numbers $c > 0$ and $\alpha > 0$ such that

$$|\nabla F(x)|^2 \geq c(1 + |x|^2)^{-\alpha};$$

hence the function $|\nabla F|^{-2}$ belongs to all $L^p(\gamma_n)$, $p < \infty$. In this case both constructions apply and yield globally finite surface measures on all surfaces $F^{-1}(t)$.

The described second construction has advantages also in the infinite-dimensional case because it does not involve division by $\partial_v F$. We shall now discuss it still in the Gaussian case but in infinite dimensions, and then present it in full generality.

Now let γ be a centered Radon Gaussian measure on a locally convex space X. Without loss of generality, one can assume this is the standard Gaussian measure on $\mathbb{R}^\infty$ or its restriction to a weighted Hilbert space of sequences $x = (x_n)$ with finite norm

$$\Big(\sum_{n=1}^{\infty} c_n x_n^2\Big)^{1/2}, \quad c_n > 0, \ \sum_{n=1}^{\infty} c_n < \infty.$$

The latter condition ensures this space has measure 1.

Let H be the Cameron–Martin space of γ, i.e., the usual Hilbert space l^2 for the standard Gaussian measure on $\mathbb{R}^\infty$.

Let $\mathcal{FC}$ be the class of all functions of the form

$$f(x) = f_0(l_1(x), \dots, l_n(x)), \quad f_0 \in C_b^\infty(\mathbb{R}^n), \ l_i \in X^*,$$

which for $\mathbb{R}^\infty$ is just the union of all classes $C_b^\infty(\mathbb{R}^n)$. Functions in this class are called smooth cylindrical.

Using the Radon–Nikodym density expression of the shifted measure (see (2.1)) we obtain the equality

$$\int_X t^{-1}[f(x+th)-f(x)]\,\gamma(dx) = \int_X t^{-1}\Big[\exp\big(t\widehat{h}(x)-t^2|h|_H^2/2\big)-1\Big]f(x)\,\gamma(dx)$$

for all $f \in \mathcal{FC}$; hence, it follows by letting $t \to 0$ that

$$\int_X \partial_h f(x)\,\gamma(dx) = \int_X f(x)\widehat{h}(x)\,\gamma(dx), \tag{2.4}$$

where

$$\partial_h f(x) := \lim_{t\to 0} t^{-1}(f(x+th)-f(x)).$$

This simple formula is the basis for our construction.

The Sobolev class $W^{p,1}(\gamma)$, $p \in [1,+\infty)$, is defined as the completion of the class $\mathcal{FC}$ with respect to the Sobolev norm

$$\|f\|_{p,1} = \|f\|_{L^p(\gamma)} + \|D_H f\|_{L^p(\gamma)} = \|f\|_{L^p(\gamma)} + \Big(\int_X |D_H f(x)|_H^p\,\gamma(dx)\Big)^{1/p},$$

where the gradient $D_H f(x) \in H$ (which now plays the role of $\nabla f(x)$) is defined by

$$(D_H f(x), h)_H = \partial_h f(x).$$

If $\{e_n\}$ is an orthogonal basis in H the vector $D_H f(x)$ has coordinates $\partial_{e_n} f(x)$. For the standard Gaussian measure on $\mathbb{R}^\infty$ functions of class $\mathcal{FC}$ are just smooth functions with bounded derivatives in finitely many variables, and $D_H f(x) = \nabla f(x)$.

One defines similarly the Sobolev classes $W^{p,1}(\gamma, E)$ of mappings with values in a separable Hilbert space E; in this case, $D_H f(x)$ is an operator between H and E, and the Hilbert–Schmidt norm $\|\cdot\|_{HS}$ is used to define the Sobolev norm. This means that in place of $|D_H f(x)|_H$ in the previous formula we use the quantity

$$\|D_H f(x)\|_{HS} = \Big(\sum_{n=1}^{\infty} |\partial_{e_n} f(x)|_E^2\Big)^{1/2}.$$

As a result of completion, every Sobolev function $f \in W^{p,1}(\gamma)$ obtains a gradient $D_H f$, an L^p-mapping with values in H. On account of (2.4) it satisfies the integration by parts formula

$$\int_X \psi(x)(D_H f(x), h)_H\,\gamma(dx) = -\int_X f(x)[\partial_h\psi(x) - \psi(x)\widehat{h}(x)]\,\gamma(dx)$$

for all $\psi \in \mathcal{FC}$. Actually, this equality extends to $\psi \in W^{q,1}(\gamma)$, $q = p/(p-1)$. By using this directional integration by parts formula, one can show that γ is differentiable along vector fields $v \in W^{p,1}(\gamma, H)$ in the following sense: there is a function $\beta_v \in L^p(\gamma)$ such that for all functions $f \in W^{p',1}(\gamma)$ with $p' = /(p-1)$ one has

$$\int_X (D_H f(x), v(x))_H \,\gamma(dx) = -\int_X \beta_v(x) f(x)\,\gamma(dx). \tag{2.5}$$

In this case for $v(x) = \sum_{n=1}^{\infty} v_n(x) e_n$ we have

$$\beta_v(x) = \sum_{n=1}^{\infty} (\partial_{e_n} v(x) - v_n(x)\widehat{e}_n(x)),$$

where the series converges in $L^p(\gamma)$. The function β_v is called the logarithmic derivative, or divergence of v with respect to γ. If $v(x) = h \in H$ is constant then $\beta_v = \widehat{h}$. Moreover, we can go a step further:

$$\int_X \psi(x)(D_H f(x), v(x))_H \,\gamma(dx) = -\int_X \psi(x)\beta_v(x) f(x)\,\gamma(dx) - \int_X f(x)(D_H\psi(x), v(x))_H \,\gamma(dx), \tag{2.6}$$

where $f, \psi \in W^{2p',1}(\gamma)$ so that $f\psi \in W^{p',1}(\gamma)$ and (2.5) can be applied to $f\psi$.

It should be noted that (2.5) can hold (with some function β_v) for a vector field v not belonging to a Sobolev class; for example, there are irregular vector fields on the plane with zero divergence in the sense of distributions. If (2.5) holds for all smooth cylindrical functions f then γ is called differentiable along the vector field v.

Inductively one defines higher Sobolev classes $W^{p,k}(\gamma, E)$ with derivatives up to order k; actually, we only need $k = 1, 2$. For example, the class $W^{p,2}(\gamma)$ consists of all functions $f \in W^{p,1}(\gamma)$ such that $D_H f \in W^{p,1}(\gamma, H)$. Therefore, the measure γ is differentiable (in the sense explained above) along the gradient field $v = D_H F$ once $F \in W^{p,2}(\gamma)$. In this case

$$\beta_v = LF,$$

where L is the Ornstein–Uhlenbeck operator; for $\mathbb{R}^\infty$ it is defined as the closure of the operator

$$Lf(x) = \sum_i [\partial_{x_i}^2 f(x) - x_i \partial_{x_i} f(x)]$$

on smooth cylindrical functions. However, belonging to the second Sobolev class is not necessary for the existence of divergence of $D_H F$. This happens already in the finite-dimensional case.

Now, given a function $F \in W^{p,2}(\gamma)$, where γ is the standard Gaussian measure on $\mathbb{R}^\infty$, with some $p > 1$ such that

$$\frac{1}{|D_H F|^2} \in L^{p'}(\gamma),$$

we consider the exact analogs of the two constructions of surface measures on the level sets $S_t = F^{-1}(t)$ considered above for $\mathbb{R}^n$. In the first construction we have the following exact analog of (2.2):

$$\int_{\mathbb{R}} \phi'(t)\,\gamma\circ F^{-1}(dt) = \int_X \partial_v(\phi\circ F)\frac{1}{\partial_v F}\,d\gamma.$$

By using (2.6) and the equality

$$\partial_v (D_H F(x), D_H F(x))_H = 2(D_H^2 F(x)\cdot D_H F(x), D_H F(x))_H,$$

where $D_H^2 F(x)\cdot D_H F(x)$ is the action of the operator $D_H^2 F(x)$ on the vector $D_H F(x)$, we write the right-hand side as

$$-\int_X \phi(F)\Big[\frac{\beta_v}{\partial_v F} - \frac{\partial_v^2 F}{|\partial_v F|^2}\Big]\,d\gamma = -\int_X \phi(F)\Big[\frac{LF}{|D_H F|_H^2} - \frac{2(D_H^2 F\cdot D_H F, D_H F)_H}{|D_H F|_H^4}\Big]\,d\gamma.$$

The integral on the right exists since we have $LF, \|D_H^2 F\|_{HS} \in L^p(\gamma)$ by the assumption that $F\in W^{p,2}(\gamma)$; hence, $|LF|/|D_H F|_H^2, \|D_H^2 F\|_{HS}/|D_H F|_H^2 \in L^1(\gamma)$ by Hölder's inequality. Next we show that similar equalities hold if we replace the measure γ by $\psi\cdot\gamma$, where ψ is a bounded function that is Lipschitz on $\mathbb{R}^\infty$ with respect to the metric

$$d(x,y) = \sum_{n=1}^{\infty} 2^{-n}\max(|x_n - y_n|, 1).$$

To this end, we observe that such a function is Lipschitz along H, i.e.,

$$|\psi(x+h) - \psi(x)| \le L|h|_H \quad \text{for all } x\in\mathbb{R}^\infty \text{ and } h\in H = l^2.$$

Indeed, $\max(|h_n|, 1) \le |h|_H$ for all n, hence $d(x+h, x) = d(h,0) \le |h|_H$. It is known (see [11, Section 5.11]) that this yields the inclusion $\psi\in W^{s,1}(\gamma)$ for all $s\in[1,+\infty)$ and $|D_H\psi|_H \le L$. Hence, in the calculations above we have to replace $1/\partial_v F$ by $\psi/\partial_v F$ so that in place of $\partial_v(1/\partial_v F)$ we have $\partial_v\psi/\partial_v F + \psi\partial_v(1/\partial_v F)$. By the equality

$$\partial_v\psi/\partial_v F = (D_H\psi, D_H F)_H/|D_H F|_H$$

and boundedness of $D_H\psi$, this yields finite integrals in analogous calculations. Therefore, as in the finite-dimensional case, the measures

$$B \mapsto (2r)^{-1}\gamma(B\cap\{t-r<F<t+r\})$$

converge weakly to finite measures σ^t as $r\to 0+$.

Finally, in place of sets $\{t-r<F<t+r\}$ we could deal with sets $\{t<F<t+r\}$ or $\{t\le F<t+r\}$ and divide by r in place of $2r$ in the appropriate places; this will be done in the next section for the sake of some minor technical simplifications.

The measures σ_0^t can again be defined by the equality

$$\sigma_0^t = |D_H F| \cdot \sigma^t$$

provided that $|D_H F|$ is σ^t-integrable. This condition automatically fulfilled if $|D_H F|$ is bounded or under the assumptions $F \in W^{2,p}(\gamma)$ and $1/|D_H F|^2 \in L^{p'}(\gamma)$ used above. Moreover, as in the finite-dimensional case, even weaker assumptions are sufficient: it is enough to have

$$F \in W^{2,2}(\gamma), \quad 1/|D_H F| \in L^2(\gamma).$$

This is verified by the same method with the measure $|D_H F| \cdot \gamma$ in place of γ.

The second construction with the measure $|D_H F|^2 \cdot \gamma$ in place of γ is completely analogous; we obtain finite measures σ_1^t as weak limits of the measures

$$B \mapsto (2r)^{-1}(|D_H F|^2 \cdot \gamma)(B \cap \{t - r < F < t + r\}), \quad r \to 0+.$$

We could also in this case deal with measures of the sets $\{t < F < t+r\}$ or $\{t \le F < t+r\}$ divided by r.

However, now we have a problem that arises also in $\mathbb{R}^n$ if we do not assume the continuity of F (which does not follow from the membership in $W^{p,2}(\mathbb{R}^n)$ for large n). Namely, we cannot assert that σ^t or σ_1^t is concentrated on S_t. We shall solve this problem in the next section in a general setting. To do so we compare our surface measures with conditional measures and show that they are concentrated on the level sets S_t for almost all t (with respect to the image measure). Moreover, in Section 5 we involve Sobolev capacities to construct surface measures concentrated on the corresponding surfaces for all t.

It is worth noting that an exact analog of the finite-dimensional situation considered above arises under the following two conditions: (i) we restrict our measure γ to a weighted Hilbert space Y of sequences $x = (x_n)$ with finite norm $\|x\|_Y$ defined by $\|x\|_Y^2 = \sum_{n=1}^{\infty} c_n x_n^2$, where $c_n > 0$ and $\sum_{n=1}^{\infty} c_n < \infty$ (in this case $\gamma(Y) = 1$), (ii) F is least twice continuously Fréchet differentiable on Y. In that case, the measures σ^t and σ_1^t will be concentrated on S_t for each t. However, in the definition of an analog of the measure σ_0^t we now have two non-equivalent options: we can take the measures $|D_H F|_H \cdot \sigma^t$ or the measures $\|\nabla F\|_Y \cdot \sigma^t$. The former corresponds to r-neighborhoods of S_t with respect to the norm of H, i.e., to the sets $S_t + rU_H$, where U_H is the unit ball of H; while the latter corresponds to the norm of Y. The relation to Y is, on the one hand, natural, since γ is concentrated on Y (but not on H); however, on the other hand, there is no preference in our choice of Y (there are too many suitable spaces).

Consider continuously Fréchet differentiable function F on Y with $\nabla F \neq 0$. We can define surface measures locally by considering small neighborhoods U in which S_t looks like the graph of a continuously differentiable function G on a closed hyperplane Y_0 in Y. This is always possible by the implicit function theorem; moreover, it is possible to choose such a hyperplane Y_0 in such a way that it is orthogonal in Y to a vector h from the Cameron–Martin space H. This simplifies the previous construction, since we can take a constant vector field: $v(x) = h$. To minimize changes in the discussed construction, it is convenient to replace γ by a measure of the form $g \cdot \gamma$, where $g \geq 0$ is a Lipschitz function with support in a ball U_0 of radius r_0 such that the ball with the twice larger radius belongs to the neighborhood U. In addition, functions ψ with support in U will be taken. We can assume that $\partial_h F \geq c > 0$ in U, so the function $1/\partial_h F$ is bounded in U. Hence, both constructions are applicable in this case. In principle, if F is continuously Fréchet differentiable and $\nabla F \neq 0$, we can introduce global surface measures as possibly σ-finite measures by summing local surface measures.

Example 2.3.2. Let us consider the case where $F = \widehat{h}$ is a measurable linear functional. Typically, it has no continuous version. For example, in the case of the standard Gaussian measure on $\mathbb{R}^\infty$ only finite linear combinations of coordinate functions are continuous. Series $\sum_{n=1}^\infty c_n x_n$ with infinitely many nonzero coefficients c_n have no continuous versions; the stochastic integral

$$\int_0^1 \psi(t)\, dx(t)$$

on the Wiener space, where $\psi \in L^2[0,1]$, has a continuous version precisely when ψ has a version of bounded variation. We can assume that $|h|_H = 1$. Then $|D_H \widehat{h}|_H = |h|_H = 1$ and all the three surface measures σ^t, σ_0^t and σ_1^t coincide. They can be calculated by using the connection with conditional measures that are known (see the next section), but this can be also done directly. Let us recall that any Radon measure is uniquely determined by its Fourier transform, i.e., the integrals of the functions $\exp(il)$ for $l \in X^*$. According to our construction, the integral of $\exp(il)$ against σ^t is the limit of the expressions

$$(2r)^{-1} \int_{\{t-r<\widehat{h}<t+r\}} \exp(il)\, d\gamma.$$

We can write $l = c\widehat{h} + \xi$, where $c \in \mathbb{R}$, $\xi = \widehat{u}$ for some $u \in H$ such that ξ and $\widehat{h}$ are orthogonal in $L^2(\gamma)$ (equivalently, $(h,u)_H = 0$). We shall use a proper linear version of ξ (which exists); in this case it is known that $\xi(h) = (u,h)_H = 0$. The orthogonal measurable linear functionals ξ and $\widehat{h}$ are independent Gaussian random variables, hence

$$\begin{aligned}(2r)^{-1} \int_{\{t-r<\widehat{h}<t+r\}} \exp(il)\, d\gamma &= (2r)^{-1} \int_{\{t-r<\widehat{h}<t+r\}} \exp(ic\widehat{h})\, d\gamma \int_X \exp(i\xi)\, d\gamma \\ &= (2r)^{-1}(2\pi)^{-1/2} \int_{t-r}^{t+r} \exp(ics)\exp(-s^2/2)\, ds \int_X \exp(i\xi)\, d\gamma,\end{aligned}$$

which tends to

$$(2\pi)^{-1/2}\exp(ict - t^2/2)\exp(-\|\xi\|_2^2/2)$$

as $r \to 0$. If $t = 0$, then we see that σ^0 coincides up to the factor $(2\pi)^{-1/2}$ with the Gaussian measure η that is the image of γ under the linear mapping $Px = x - \widehat{h}(x)h$. Indeed, the Fourier transform of $\gamma\circ P^{-1}$ at the functional l represented as above equals

$$\int_X \exp\big(il(x - \widehat{h}(x)h)\big)\,\gamma(dx) = \int_X \exp\big(i(c\widehat{h} - \xi)(x - \widehat{h}(x)h)\big)\,\gamma(dx)$$
$$= \int_X \exp\big(i\xi(x)\big)\,\gamma(dx) = \exp(-\|\xi\|_2^2/2),$$

since $\xi(h) = (u, h)_H = 0$. Therefore, the measure σ^t is the shift of the measure σ^0 by the vector th.

We conclude this section by considering an analog of "Gaussian" Hausdorff measures associated with the geometry of the Cameron–Martin space as proposed by Feyel and de La Pradelle in [36]. This construction begins from $\mathbb{R}^n$. Recall (see [12], [33], [34], [79]) that for every $m \in (0, n]$ the classical Hausdorff measure H_m is generated by means of the outer measure H_m^δ defined for each set A by

$$H_m^\delta(A) = \inf\sum_{j=1}^{\infty} C_m(\operatorname{diam} A_j)^m, \quad C_m = 2^{-m}\Gamma(1/2)^m/\Gamma(1+m/2),$$

where inf is taken over all sequences of closed sets A_j of diameter at most δ with union containing A. The values $H_m^\delta(A)$ increase as $\delta \to 0+$ and have a limit (possibly, infinite) denoted by $H_m(A)$. If we apply this method with balls in place of arbitrary closed sets the result will be the spherical Hausdorff measure $\mathcal{S}_m$ that is larger than H_m (on sufficiently regular sets they coincide). Through this approach, the following "Gaussian spherical Hausdorff" measures θ_k on $\mathbb{R}^n$ were introduced:

$$\theta_k(B) = \varrho_n \cdot \mathcal{S}_{n-k}(B),$$

where $\mathcal{S}_{n-k}(B)$ is the limit as $\delta > 0+$ of the infimum of $\sum_{j=1}^\infty \lambda_{n-k}(B_i)$ over all covers of B by closed balls of radius at most δ and $\lambda_{n-k}(B_i)$ is the $(n-k)$-dimensional volume of B_i measured as the $(n-k)$-dimensional volume of the section of B_i by a subspace of dimension $n-k$ passing through the center of B_i (which equals $const(n-k)r_i^{n-k}$, where r_i is the radius of B_i). The number k in this notation refers to "codimension".

The next step is to fix k and take an n-dimensional subspace H_n in H. The orthogonal projection $P_n\colon\ H \to H_n$ admits a measurable linear extension $\widehat{P}_n \to X \to H_n$: if $e_1, \dots, e_n$ is an orthonormal basis in H_n, then

$$\widehat{P}_n x := \widehat{e}_1(x)e_1 + \cdots \widehat{e}_n(x)e_n.$$

Let γ_n be the image of γ under the measurable linear operator $I - \widehat{P}_n$. For every Borel (or Souslin) set A in X set

$$\eta_k^{H_n}(A) := \int_X \theta_k(A_x)\,\gamma_n(dx), \quad A_x = \{y \in H_n\colon\ x + y \in A\}.$$

The section A_x is Borel in H_n if A is Borel (and is Souslin for Souslin A). In addition, the function $x \mapsto \theta_k(A_x)$ is measurable with respect to all Borel measures. Finally, the Gaussian Hausdorff measure η_k of codimension k on X is defined as follows: $\eta_k(A)$ is the supremum of $\eta_k^{H_n}(A)$ over all n-dimensional subspaces in H with $n \geq k$.

These Gaussian Hausdorff measures are related to the initial Gaussian measure γ by the following formula established in [36] and presented here for simplicity in the case $k = 1$: if $f \in W^{p,2}(\gamma)$ for all $p \in [1, +\infty)$ and f is continuous (actually, it suffices that f be quasi-continuous with respect to the Sobolev capacity corresponding to the class $W^{p,1}(\gamma)$, see the next section), then

$$\int_A |D_H f|\, d\gamma = \int_{\mathbb{R}} \eta_1(A \cap \{|D_H f| > 0\} \cap f^{-1}(t))\, dt.$$

If $|D_H f| > 0$, then we integrate $\eta_1(A \cap f^{-1}(t))$ on the right.

Similarly to the finite-dimensional case, an obvious advantage of Gaussian Hausdorff measures is that their construction is absolutely independent of any particular representations of sets. The other side of this universality is that it is rather difficult to calculate surface measures of given sets; this happens already in $\mathbb{R}^n$. Actually, the last formula can help in such calculations: if $|D_H f| > 0$ and $A = B \cap \{f < s\}$, where B is a Borel set and $s \in \mathbb{R}$, then we have

$$\int_{\{f<s\}\cap B} |D_H f|\, d\gamma = \int_{-\infty}^{s} \eta_1(B \cap f^{-1}(t))\, dt.$$

Letting $\eta_{1,t}(B) = \eta_1(B \cap f^{-1}(t))$, for bounded Borel functions ϕ we obtain

$$\int_{\{f<s\}} \phi |D_H f|\, d\gamma = \int_{-\infty}^{s} \int_X \phi\, d\eta_{1,t}\, dt.$$

Therefore, $\eta_1(B \cap f^{-1}(t))$ can be obtained as the derivative of the distribution function $I_B |D_H f| \cdot \gamma(f < t)$; this identifies η_1 on $f^{-1}(t)$ with the surface measure σ_0^t considered above.

It is not clear whether every surface of the form $S = f^{-1}(0)$, where f is a continuous polynomial on a Hilbert space with a Gaussian measure γ such that $\nabla f(x) \neq 0$, has a finite surface measure; certainly, locally all these approaches give nice surface measures. The problem is that it is not known whether the function $|\nabla f(x)|^{-p}$ is integrable (it is even unknown whether there is $p > 0$ for which it is integrable). Unlike the finite-dimensional case, there is no lower bound of the form $Q(x) \geq c(1+|x|^2)^{-\alpha}$ for any continuous polynomial $Q > 0$. It is also worth noting that zero sets of continuous polynomials on a Hilbert space are more complicated sets than in $\mathbb{R}^n$. For example, the cardinality of the set of disjoint connected components of $f^{-1}(0)$ can be continuum and the class of orthogonal projections of such sets coincides with the class of all Souslin sets (see [11, Exercise 6.11.19]). A survey of results on distributions of polynomials is given in [16].

2.4 Surface measures for differentiable measures

Here we described an abstract approach to surface measures suggested in [18], where the proofs of some technical assertions can be found. For the reader's convenience we include justifications of the most important steps.

Let $\mu \geq 0$ be a fixed Radon measure on a completely regular space X and let $\mathcal{B}$ be the Borel σ-algebra of X. Let $\mathcal{F}$ be a class of bounded $\mathcal{B}$-measurable real functions. Recall that a class of functions *separates measures* if two measures coincide whenever they assign equal integrals to all functions in this class. We assume throughout that $\mathcal{F}$ satisfies the following conditions:

(F1) $\mathcal{F}$ is a linear space separating Radon measures on X, and $\phi(f) \in \mathcal{F}$ for all $f \in \mathcal{F}$ and all $\phi \in C_b^\infty(\mathbb{R})$.

For example, if X is a metric space the class of all bounded Lipschitzian functions on X satisfies conditions (F1). We shall see in Section 4 that this class is indeed convenient for many applications.

Another example: given some class $\mathcal{F}_0$ of $\mathcal{B}$-measurable functions, we take for $\mathcal{F}$ the class of all compositions $\phi(f_1, \dots, f_n)$, where $f_i \in \mathcal{F}_0$ and $\phi \in C_b^\infty(\mathbb{R}^n)$. This class is a linear space and is stable under compositions with C_b^∞-functions; certainly, we still need the additional condition that it must separate measures (which trivially holds if $\mathcal{F}_0$ is separating). See also the modification of (F1) for multidimensional mappings considered in Section 7.

It follows from (F1) that $1 \in \mathcal{F}$ and that $fg \in \mathcal{F}$ for all $f, g \in \mathcal{F}$. Indeed, $f^2 \in \mathcal{F}$ for all $f \in \mathcal{F}$, because we can take for ϕ a function in $C_b^\infty(\mathbb{R})$ that coincides with x^2 on the bounded range of f, so it remains to use the equality $2fg = (f+g)^2 - f^2 - g^2$.

Definition 2.4.1. *A vector field on X (or an $\mathcal{F}$-vector field if we need to indicate its relation to $\mathcal{F}$) is a linear mapping (differentiation)*

$$v\colon \mathcal{F} \to L^1(\mu), \quad f \mapsto \partial_v f,$$

such that

$$\partial_v(\phi \circ f) = \phi'(f)\partial_v f \quad \mu\text{-a.e.} \tag{2.7}$$

for all $f \in \mathcal{F}$ and $\phi \in C_b^\infty(\mathbb{R})$.

Similarly we can define more general vector fields for which functions $\partial_v f$ belong to the space $L^0(\mu)$ of μ-measurable functions.

Applying (2.7) to $f, g, f+g$ and ϕ such that $\phi(t) = t^2$ on a sufficiently large interval we obtain the Leibniz rule

$$\partial_v(fg) = f\partial_v g + g\partial_v f \text{ a.e. } \quad \forall f, g \in \mathcal{F}. \tag{2.8}$$

It is worth noting that $\partial_v 1 = 0$ because we can take $\phi = 1$ in (2.7) or, alternatively, we can take $f = g = 1$ in (2.8).

Below a fixed vector field v will play a role of a normal field on level sets (and in some cases one can use indeed unit normal fields).

Definition 2.4.2. *The measure μ is called Skorohod differentiable along v (with respect to $\mathcal{F}$) if there is a Radon measure $d_v\mu$ on $\mathcal{B}$, called the Skorohod derivative of μ along v, such that*

$$\int_X \partial_v f(x)\,\mu(dx) = -\int_X f(x)\,d_v\mu(dx) \quad \forall f \in \mathcal{F}. \tag{2.9}$$

We say that μ is Fomin differentiable along v if $d_v\mu \ll \mu$; in that case the Radon–Nikodym density of $d_v\mu$ with respect to μ is denoted by β_v and is called the logarithmic derivative of μ along v or divergence of v with respect to μ.

For example, let μ be a measure on $\mathbb{R}^d$ with a smooth density ϱ, $\mathcal{F}$ be the class of all bounded Lipschitz functions or the classes $C_b^\infty(\mathbb{R}^d)$, and v be a nonzero constant vector (so that $\partial_v f$ is the usual partial derivative). Then $d_v\mu$ is given by density $\partial_v\varrho$ and $\beta_v = (\partial_v\varrho)/\varrho$, which explains the terminology. If $v = 1$ on the real line the usual Lebesgue measure λ on $[0, 1]$ regarded as a measure on $\mathbb{R}$ is Skorohod differentiable, and $d_1\lambda = \delta_0 - \delta_1$ is the difference of two Dirac measures and Fomin differentiable.

In the case of $\mathbb{R}^d$ with $\mathcal{F}$ as above, a measure μ is Skorohod differentiable along all constant vectors precisely when it has a density ϱ belonging to the class $BV(\mathbb{R}^d)$ of functions of bounded variation, that is, functions in $L^1(\mathbb{R}^d)$ whose first order partial derivatives in the sense of distributions are bounded measures (see [33], [79]); in that case, for a constant vector v, the measure $d_v\mu$ is the partial derivative of μ along v in the sense of distributions. The measure μ is Fomin differentiable along all constant vectors in $\mathbb{R}^d$ precisely when it has a density in the Sobolev class $W^{1,1}(\mathbb{R}^d)$ of integrable functions possessing integrable first order partial derivatives in the sense of distributions. If $\mu = \varrho\,dx$ and $\varrho \in W^{1,1}(\mathbb{R}^d)$, then for any constant vector v one also has $\beta_v = \partial_v\varrho/\varrho$, where we set $\beta_v = 0$ on the set of zeros of ϱ. If a vector field v on $\mathbb{R}^d$ is not constant, but is bounded and Lipschitzian, then

$$\beta_v = \operatorname{div} v + \partial_v\varrho/\varrho.$$

This is also true for vector fields belonging to appropriate Sobolev classes. For a survey of the theory of differentiable measures, see [13].

If μ is a centered Gaussian measure with the Cameron–Martin space H the measure μ is Fomin differentiable along the constant vector field h and $\beta_h = -\widehat{h}$, which is a trivial corollary of the Cameron–Martin formula (2.1).

The original definition of Fomin dealt with constant vector fields on linear spaces. Differentiability of measures along non-constant vector fields was already considered in the 1980–1990s (sometimes implicitly) in the Malliavin calculus and its modifications (see [70], [30], [68], [9], and [31]); close constructions arise in relation to the so-called "carré du champ" operators (see [23]). Obviously, to be differentiable depends on $\mathcal{F}$. However, in reasonable situations differentiability with respect to small classes (separating measures) often yields differentiability with respect to larger classes. For example, in the case of $v(x) = 1$ on the real line, differentiability with respect to C_b^∞ yields differentiability with respect to the class of bounded Lipschitz functions.

Observe that $d_v\mu(X) = 0$, which follows from (2.9) applied to $f = 1$, so $d_v\mu$ is necessarily a signed measure.

We need an extension of ∂_v to functions outside of $\mathcal{F}$.

Definition 2.4.3. *Let μ be Skorohod differentiable along a vector field v. We say that a $\mathcal{B}$-measurable function Ψ belongs to $\mathfrak{D}_v$ if $\Psi \in L^1(\mu) \cap L^1(d_v\mu)$ and there is a sequence of functions $f_n \in \mathcal{F}$ converging to Ψ in $L^1(\mu)$ and in $L^1(d_v\mu)$ such that the functions $\partial_v f_n$ converge in $L^1(\mu)$ to some function w and the functions $f_n\partial_v g$ converge in $L^1(\mu)$ for each $g \in \mathcal{F}$. Then we set $\partial_v\Psi := w$.*

Note that by convergence of $\{f_n\}$ in $L^1(\mu)$ the sequence $\{f_n\partial_v g\}$ converges in $L^1(\mu)$ precisely when it is uniformly μ-integrable. This condition holds if $\{f_n\}$ converges to Ψ in $L^p(\mu)$ for some $p > 1$ and all functions $\partial_v g$ for $g \in \mathcal{F}$ belong to $L^q(\mu)$, $q = p/(p-1)$.

This definition is somewhat technical because we want to make it sufficiently broad. Similar technicalities already arise on the real line if one wants to integrate by parts unbounded functions with respect to measures with densities of bounded variation (but not absolutely continuous) such that the derivatives can be singular. In the case of a Fomin differentiable measure μ with $\beta_v \in L^2(\mu)$, it would be natural to say that a function $\Psi \in L^2(\mu)$ has a Sobolev derivative $\partial_v\Psi \in L^2(\mu)$ if there is a sequence of functions f_n converging to f in $L^2(\mu)$ such that $\{\partial_v f_n\}$ also converges in $L^2(\mu)$. In that case, the limiting function for $\{\partial_v f_n\}$ would satisfy the integration by parts formula. The definition above follows a similar idea under weaker integrability conditions.

The function w (if it exists) is uniquely defined. Indeed, for each $g \in \mathcal{F}$ we have

$$\begin{aligned}\int_X g(x)w(x)\,\mu(dx) &= \lim_{n\to\infty}\int_X g(x)\partial_v f_n(x)\,\mu(dx)\\ &= \lim_{n\to\infty}\int_X [\partial_v(gf_n)(x) - f_n(x)\partial_v g(x)]\,\mu(dx)\end{aligned}$$

$$= -\lim_{n\to\infty} \int_X (gf_n)(x)\, d_v\mu(dx) - \int_X \Psi\partial_v g\, \mu(dx)$$

$$= -\int_X (g\Psi)(x)\, d_v\mu(dx) - \int_X \Psi(x)\partial_v g(x)\, \mu(dx).$$

Therefore, the integral of gw is determined for each $g \in \mathcal{F}$, which uniquely determines w according to Condition (F1).

Remark 2.4.4. If f is bounded a sequence $\{f_n\}$ with the properties mentioned in the definition can be replaced by a uniformly bounded sequence; the technical condition of uniform integrability will therefore be fulfilled automatically. Indeed, taking $\zeta \in C_b^\infty(\mathbb{R})$ such that $\zeta(t) = t$ on an interval containing the range of f, we obtain a new sequence $g_n = \zeta(f_n)$ that is uniformly bounded and converges to f in $L^1(\mu)$ and $L^1(d_v\mu)$. In addition, the functions $\partial_v g_n = \zeta'(f_n)\partial_v f_n$ converge to $\partial_v f$ in $L^1(\mu)$.

Let μ be Skorohod differentiable along v. We shall assume that $F\colon X \to \mathbb{R}$ is a $\mathcal{B}$-measurable function such that

(F2) $\psi(F) \in \mathfrak{D}_v$ for each function $\psi \in C_0^\infty(\mathbb{R})$ and there is a $\mathcal{B}$-measurable function $\partial_v F$ such that $\partial_v(\psi \circ F) = \psi'(F)\partial_v F$ a.e. for each $\psi \in C_0^\infty(\mathbb{R})$. Moreover,

$$\partial_v F \geq 0, \quad \partial_v F \in L^1(\mu).$$

Set

$$\nu := (\partial_v F) \cdot \mu, \quad \eta := d_v\mu \circ F^{-1}. \tag{2.10}$$

The measure ν is finite and nonnegative (it can be zero). The conditional measures on the level sets $F^{-1}(y)$ generated by the measure ν will be denoted by ν^y (in the case where μ is concentrated on a countable union of metrizable compact sets).

We now introduce our surface measures σ^y; it might be reasonable to use the symbol σ_ν^y to emphasize dependence on ν, but we omit this indication for notational simplicity. The definition employs only the differentiability of the distribution functions

$$\Phi_f(y) := \int_{\{F<y\}} f(x)\, \nu(dx)$$

at a given point. In this respect, no topological structures are needed. However, for deriving further properties of our surface measures we shall need some additional assumptions of topological nature. We set

$$\varrho_f(y) = \Phi_f'(y) = \lim_{h\to 0} \frac{\Phi_f(y+h) - \Phi_f(y)}{h} \tag{2.11}$$

if a finite limit exists.

Definition 2.4.5. *Let $y \in \mathbb{R}$. Suppose that Φ_f is differentiable at y for each $f \in \mathcal{F}$ and there is a Radon measure σ^y on $\mathcal{B}$ such that*

$$\int_X f(x)\,\sigma^y(dx) = \varrho_f(y) \quad \forall f \in \mathcal{F}. \tag{2.12}$$

Then σ^y is called the surface measure associated with the level set $F^{-1}(y)$.

Remark 2.4.6. Note that we do not require that the surface measure be concentrated at the level set $F^{-1}(y)$, but under broad assumptions (see the next theorem) it is indeed a measure on $F^{-1}(y)$. In this respect, the situation is similar with conditional measures.

However, if F is continuous (which we do not assume) and, for every $z \in X \backslash F^{-1}(y)$ and every neighborhood U of z, there is a nonnegative function $f \in \mathcal{F}$ with support in U and positive in a neighborhood of z (which is fulfilled if $\mathcal{F}$ contains all bounded Lipschitz functions on a metric space) then σ^y is automatically concentrated on $F^{-1}(y)$. This is readily seen from (2.12) because we can take U such that $|F(x)-y| > |F(z)-y|/2$ for all $x \in U$; by (2.11) this yields $\varrho_f(y) = 0$. Hence, the integral of f against σ^y vanishes, so that z does not belong to the topological support of σ^y.

A similar reasoning shows that σ^y is automatically concentrated on $F^{-1}(y)$ provided that F satisfies $\phi(F) \in \mathcal{F}$ for all $\phi \in C_b^\infty(\mathbb{R})$. In this case, considering suitable functions $f = \phi(F)$, we obtain that the sets $\{F > y + 1/n\}$ and $\{F < y - 1/n\}$ have σ^y-measure zero for all $n \in \mathbb{N}$. The latter condition with compositions is fulfilled if in the Gaussian case (see Section 3 and Section 5) we take for $\mathcal{F}$ the class of all bounded functions in the Sobolev space $W^{2,2}(\mu)$ and $F \in W^{2,2}(\mu)$.

We shall see that the hypothesis of differentiability of Φ_f is fulfilled if μ is Fomin differentiable along v and F satisfies (F2). In typical cases, the assumptions of this definition are ensured by the following condition: the measures $\nu_r = r^{-1}I_{\{y<F<y+r\}} \cdot \nu$ converge weakly as $r \to 0$, which in turn is ensured by their uniform tightness and convergence of the integrals against ν_r of bounded functions from a measure separating class. Exactly this will be implemented below.

In the case of measures on locally convex spaces differentiable along constant vectors, this construction is close to the ones described in [9], [10], [11] and later developed in [58]; however, in our case it requires only one-fold differentiability of F. In [58] the membership of F in the second Sobolev class is required and in [29], in the Gaussian case, also the second derivative is used (the function F is in the first Sobolev class, but its appropriately scaled Malliavin gradient must be also in the first Sobolev class).

The following result from [18] gives broad sufficient conditions for the existence of surface measures and describes their connections with conditional measures.

Theorem 2.4.7. *Let μ be Fomin differentiable along v with the logarithmic derivative β_v. Suppose that* (F1) *holds, a function $F\colon X \to \mathbb{R}$ satisfies* (F2) *and $\mu \circ F^{-1}$ has no atoms* *(i.e., $\mu(F^{-1}(y)) = 0$ for all y). Assume also that at least one of the following conditions holds:*

(i) *X is a complete metric space and $\mathfrak{F}$ contains all bounded Lipschitzian functions;*

(ii) *the measure μ has compact support;*

(iii) *there exists a nonnegative function $W \in \mathfrak{D}_v$ such that $W\beta_v$, $W\partial_v F \in L^1(\mu)$ and the sets $\{W \le R\}$ are compact for all $R \ge 0$.*

Then, for each $y \in \mathbb{R}$, the Radon surface measure σ^y associated with $F^{-1}(y)$ exists.

In addition, if μ is concentrated on a countable union of metrizable compact sets, then, for $\nu \circ F^{-1}$-a.e. y, where $\nu = (\partial_v F) \cdot \mu$, the surface measure σ^y is concentrated on $F^{-1}(y)$ and we have the equality

$$\sigma^y = \varrho_1(y) \cdot \nu^y,$$

where ϱ_1 is the density of $\nu \circ F^{-1}$ and $\{\nu^y\}$ is the system of conditional measures for ν.

It follows that σ^y is absolutely continuous also with respect to the conditional measure μ^y for μ.

The proof will be given below after a number of auxiliary results. However, we can say right now that in all these cases the measure σ^y will be obtained as the limit of the measures $r^{-1}I_{\{y<F<y+r\}} \cdot \nu$ in the week topology; in cases (i) and (ii) this will be an immediate corollary of our assumptions and in case (iii) some little extra work will be needed.

The main point is that, under the assumptions of the theorem, for every $f \in \mathfrak{F}$ the function $\Phi_f(y)$ is continuously differentiable. This can be explained immediately in the case $F \in \mathfrak{F}$. The function $\Phi_f(y)$ is the distribution function of the bounded measure

$$m_f := (f \cdot \nu) \circ F^{-1}$$

on the real line. Therefore, it suffices to show that this measure has a continuous density ϱ_f with respect to Lebesgue measure. This will be done if we show that the derivative of m_f in the sense of generalized functions is a bounded measure η_f without points of nonzero measure. Using the standard reasoning in the Malliavin calculus, we now show that

$$\eta_f = (f \cdot d_v\mu + \partial_v f \cdot \mu) \circ F^{-1}$$

is the generalized derivative of m_f. Let $\psi \in C_0^\infty(\mathbb{R})$. We have

$$\begin{aligned}\int \psi'(t)\, m_f(dt) &= \int \psi'(t)\,(f \cdot \nu) \circ F^{-1}(dt) = \int_X \partial_v(\psi(F))(x) f(x)\,\mu(dx)\\ &= \int_X \partial_v(f\psi(F))(x)\,\mu(dx) - \int_X \psi(F(x))\partial_v f(x)\,\mu(dx)\\ &= -\int_X \psi(F(x))f(x)\,d_v\mu(dx) - \int_X \psi(F(x))\partial_v f(x)\,\mu(dx) = -\int \psi(t)\,\eta_f(dt). \qquad (2.13)\end{aligned}$$

Finally, η_f has no points of nonzero measure if this is true for $\mu \circ F^{-1}$ and $d_v\mu \ll \mu$ (the latter is true in the case of Fomin differentiability). The proof below is similar, we just need to extend (2.13) to more general functions F satisfying (F2).

Remark 2.4.8. It is worth noting that under our assumptions (F1) and (F2) the measure $v \circ F^{-1}$ is absolutely continuous (see below), hence $\mu \circ F^{-1}$ is also absolutely continuous provided that $\partial_v F > 0$ μ-a.e. (then μ and v are equivalent). In particular, in the latter case $\mu \circ F^{-1}$ has no atoms. In general, of course, this is not true, since F can be constant on a positive measure set.

Remark 2.4.9. It will be clear from the proof that if we are interested only in surface measures on $F^{-1}(y)$ for y in some interval I it suffices to have that $\mu(F^{-1}(y)) = 0$ only for $y \in I$. In addition, in many cases the construction can be "localized" by replacing μ with the measure $f \cdot \mu$, where $f \in \mathcal{F}$ has an appropriate support. In this way, one can make assumptions about also local v, say, replacing v by $f \cdot v$ (note that the surface measures in [11] are constructed locally).

The following classical concepts and facts are crucial for the proof of Theorem 2.4.7.

A sequence of Radon measures μ_n converges weakly to a Radon measure μ if, for each bounded continuous function f, we have

$$\int_X f(x)\,\mu(dx) = \lim_{n\to\infty} \int_X f(x)\,\mu_n(dx).$$

By Aleksandrov's theorem, weak convergence of a sequence of Radon probability measures to a Radon probability measure μ is equivalent to the relation $\mu(W) \le \liminf_{n\to\infty} \mu_n(W)$ for every open set W (see [12, Section 8.2]).

By LeCam's theorem (see [12, Corollary 8.6.3]), ofr complete metric spaces, if a sequence of nonnegative Radon measures μ_n is such that the integrals of each bounded Lipschitzian function with respect to these measures converge then this sequence converges weakly to some Radon measure.

Finally, it follows from Prohorov's theorem (see [12, Section 8.6]) that if a sequence of Radon measures v_n on X is uniformly bounded in variation and uniformly tight, i.e., for every $\varepsilon > 0$ there is a compact set K_ε such that $|\mu_n|(X\backslash K_\varepsilon) < \varepsilon$ for all n, and there is a class of bounded Borel functions on X separating Radon measures such that the integrals of f against μ_n converge for each f in this class, then the measures μ_n converge weakly to some Radon measure μ on X (Prohorov's theorem ensures the existence of a Radon measure μ that is a limit point of $\{\mu_n\}$ in the weak topology and the second condition says that this limit point is unique, hence the sequence converges weakly to it).

Note that in order to have the uniform tightness of nonnegative Radon measures v_n it suffices to have a nonnegative Borel function W on X such that the sets $\{W \le R\}$ are compact for all $R \ge 0$ and the integrals of W with respect to the measures v_n are uniformly bounded by some number C. In that case, by the Chebyshev inequality we have

$$\nu_n(X\backslash\{W \le R\}) \le CR^{-1}.$$

Lemma 2.4.10. *Let μ be Skorohod differentiable along v. Then*

(i) *for every $f \in \mathfrak{D}_v$, the measure $(\partial_v f \cdot \mu) \circ f^{-1}$ is absolutely continuous and its distributional derivative is $(f \cdot d_v\mu) \circ f^{-1}$, hence for all t we have*

$$\int_{\{f<t\}} \partial_v f(x)\,\mu(dx) = \int_{-\infty}^{t} \int_{\{f<s\}} f(x)\, d_v\mu(dx)\, ds. \tag{2.14}$$

(ii) *$\phi(f) \in \mathfrak{D}_v$ for each Lipschitzian function ϕ and each $f \in \mathfrak{F}$.*

In addition, $\phi(f) \in \mathfrak{D}_v$ for any continuously differentiable function ϕ with a bounded derivative and any $f \in \mathfrak{D}_v$. In both cases, $\partial_v(\phi \circ f) = \phi'(f)\partial_v f$ μ-a.e.

For a proof, see [18].

Lemma 2.4.11. *We have $\psi(F) \in \mathfrak{D}_v$ for each bounded Lipschitzian function ψ on the real line. In addition, $\partial_v(\psi(F)) = \psi'(F)\partial_v F$.*

The proof of this lemma is easy and can be also found in [18].

Corollary 2.4.12. *Under assumptions* (F1) *and* (F2) *we have*

$$(\nu \circ F^{-1})' = \eta = d_v\mu \circ F^{-1}$$

in the sense of distributions, where ν and η are defined by (2.10). *Hence the measure $\nu \circ F^{-1}$ has a density ϱ_1 of bounded variation, moreover,*

$$\varrho_1(t) = \eta((-\infty, t)) = d_v\mu(x\colon\ F(x) < t).$$

If $\mu \circ F^{-1}$ has no atoms and μ is Fomin differentiable along v, then $|d_v\mu|(\{F = t\}) = 0$ for every t, hence this density is continuous.

Moreover, for every $f \in \mathfrak{F}$ we have

$$((f \cdot \nu) \circ F^{-1})' = (f \cdot d_v\mu) \circ F^{-1} + (\partial_v f \cdot \mu) \circ F^{-1} \tag{2.15}$$

and

$$\|((f \cdot \nu) \circ F^{-1})'\| \le \|f \cdot d_v\mu + \partial_v f \cdot \mu\|.$$

If $\mu \circ F^{-1}$ has no atoms then the measure $(f \cdot \nu) \circ F^{-1}$ has a continuous density ϱ_f of bounded variation and

$$|\varrho_f(y)| \le \|f \cdot d_v\mu + \partial_v f \cdot \mu\| \le \|d_v\mu\| \cdot \|f\|_\infty + \|\partial_v f\|_{L^1(\mu)}.$$

Finally, if $d_v\mu = \beta_v \cdot \mu$, where $\beta_v \in L^q(\mu)$, $q = p/(p-1)$, then

$$|\varrho_f(y)| \le \|\beta_v\|_{L^q(\mu)}\|f\|_{L^p(\mu)} + \|\partial_v f\|_{L^1(\mu)}. \tag{2.16}$$

Proof. Extending (2.13), we find that

$$\int \psi'(t)\,(f\cdot v)\circ F^{-1}(dt) = \int_X \partial_v(\psi\circ F)(x)f(x)\,\mu(dx)$$
$$= -\int_X \psi(F(x))f(x)\,d_v\mu(dx) - \int_X \psi(F(x))\partial_v f(x)\,\mu(dx),$$

which gives (2.15), hence all our assertions follow. □

Remark 2.4.13. It should be noted that the "non-normalized" surface measures introduced above are still not true "surface measures"; they depend not only on the level sets $F^{-1}(y)$ but also on the whole function F. Obviously, the whole measure depends also on our choice of the vector field v. However, there is some scaling invariance of the construction: for example, if we replace F by kF with some number $k > 0$, the set $F^{-1}(0)$ does not change and our measure σ^0 respects this. The sets $\{0 < kF < r\}$ are the old sets $\{0 < F < r/k\}$, so, when evaluating the derivative of the distribution function at zero, we have the factor k coming from $\partial_v(kF)$ and obtain the same quantity.

Nevertheless, if v also depends on F, e.g., if we take for v a suitable gradient of F without normalization, then we loose this invariance. This is a certain disadvantage of our definition which will be partially overcome below (by passing to surface measures normalized by weights or by taking normalized vector fields). One should bear in mind that, even dealing with very nice functions F on infinite-dimensional spaces, the known constructions do not really define surface measures on individual level sets $F^{-1}(y)$. As it happens with usual nice surfaces in $\mathbb{R}^d$, it is still necessary that each fixed surface be included in a special family of level sets. An important exception is the Gaussian Hausdorff measure mentioned in Section 3; as we have seen this is a non-σ-finite measure on all Borel sets and it is not easy to calculate its value on individual surfaces. Another exception is a surface determined by a nondegenerate Fréchet differentiable function on a Banach space (but typical infinitely Sobolev differentiable functions on infinite-dimensional spaces are not even continuous). On the other hand, by using weight functions one can obtain "geometric surface measures" on the basis of our surface measures for a reasonable individual surface.

Proof of Theorem 2.4.7. We can assume that $y = 0$. We know from Corollary 2.4.12 that for every $f \in \mathcal{F}$ the distribution function of the measure $(f\cdot v)\circ F^{-1}$ is differentiable at zero and its derivative is $\varrho_f(0)$. Clearly,

$$\varrho_f(0) = \lim_{n\to\infty} n\int_{B_n} f(x)\,v(dx) = \lim_{n\to\infty}\int_X f(x)\,v_n(dx),$$

where

$$B_n = \{0 < F < n^{-1}\} \quad\text{and}\quad v_n := nI_{B_n}\cdot v.$$

Certainly, in place of n^{-1} we can take numbers $h_n > 0$ decreasing to zero (then the factor n is replaced by h_n^{-1}). The nonnegative measures ν_n are uniformly bounded since their values on the whole space X converge to $\varrho_1(0)$.

If either (i), (ii) or (iii) is fulfilled, it follows that there is a bounded nonnegative Radon measure σ^0 on X such that

$$\int_X f(x)\,\sigma^0(dx) = \lim_{n\to\infty} \int_X f(x)\,\nu_n(dx).$$

Indeed, in case (i) we apply Le Cam's theorem. Note that the measures ν_n are Radon and are concentrated on a common separable subspace. In case (ii) we obviously have the uniform tightness of the measures ν_n, which gives a Radon limit, as explained above.

The same is also true in case (iii) because the integrals

$$\int_X W(x)\,\nu_n(dx) = n\int_{B_n} W(x)\,\nu(dx)$$

are uniformly bounded. This follows by the same reasoning that proves the existence of $\varrho_1(0)$, just in place of μ we take $W\cdot\mu$; our assumptions in (iii) are such that this works. This completes the proof of Theorem 2.4.7.

Let us compare the constructed measures σ^y with the conditional measures ν^y, assuming that μ is concentrated on a countable union of metrizable compact sets. It follows from the definition of $\varrho_f(y)$ that

$$\int_{-\infty}^{+\infty}\int_X f(x)\,\sigma^y(dx)\,dy = \int_X f(x)\,\nu(dx) = \int_{\mathbb{R}}\int_X f(x)\,\nu^y(dx)\,\nu\circ F^{-1}(dy).$$

The integral on the left can be written as

$$\int_{-\infty}^{+\infty}\int_X f(x)\frac{1}{\varrho_1(y)}\,\sigma^y(dx)\,\varrho_1(y)dy = \int_{\mathbb{R}}\int_X f(x)\frac{1}{\varrho_1(y)}\,\sigma^y(dx)\,\nu\circ F^{-1}(dy).$$

Hence the measure $\sigma^y/\varrho_1(y)$ coincides with the conditional measure ν^y for $\nu\circ F^{-1}$-a.e. y due to our assumption that $\mathcal{F}$ separates measures on $\mathcal{B}$, and the essential uniqueness of conditional measures. □

Remark 2.4.14. (i) The assumption that μ is concentrated on metrizable compact sets has not been used for the proof of existence of σ^y; it is only needed if we wish to compare surface measures with conditional measures and localize σ^y on $F^{-1}(y)$.

(ii) In place of the sets $\{y < F < y+h\}$ we could deal with the sets $\{y-h < F < y+h\}$, but then the factor h^{-1} must be replaced by $(2h)^{-1}$.

(iii) It follows from our construction that in cases (i) – (iii) the mapping $y \mapsto \sigma^y$ is continuous provided that the space of probability measures is equipped with the weak topology. Indeed, according to (2.12), whenever $y_j \to y$, for each $f \in \mathcal{F}$ the integral of f against σ^{y_j} converges to the integral of f against σ^y; by the respective assumption this yields weak convergence.

(iv) Although our construction of surface measures is topological in the sense that it involves weak convergence of measures and the latter depends on the initial topology, the resulting measures possess certain topological invariance. Obviously, they do not change if we continuously embed our space into a larger space. Moreover, in many practical situation they do not change even if we consider them on a smaller topological space of full measure continuously embedded into the original space in such a way that the restriction of our measure μ to this smaller space is Radon on it.

(v) Surface measures are closely related to the classes BV of functions of bounded variation. In $\mathbb{R}^n$ the functions of this class are precisely those functions $f \in L^1(\mathbb{R}^n)$ for which the derivatives $\partial_{x_i} f$ in the sense of distributions are measures of bounded variation. The indicator function I_V of a domain belongs to BV precisely when the boundary ∂V of V has finite perimeter. For the infinite-dimensional case, see [2]–[8], [20], [21], [22], [24], [25], [43], [44], [48], [51], [62], and [63]. An interesting question is to determine when level sets of a convex function (for example, a seminorm) have finite surface measures. This question is not trivial even for Gaussian measures and becomes especially challenging for convex (or logarithmically concave) measures, i.e., probability measures whose finite-dimensional projections are measures given by densities of the form e^{-V} with a convex function V with respect to Lebesgue measures on affine subspaces. One of the problems with such measures in infinite dimensions is that it is not known whether they always have vectors of Skorohod differentiability. Returning to the Gaussian case, recall that if γ is a centered Gaussian measure and A is a Borel set of measure $\gamma(A) \geq 1/2$ and $\Phi(a) = \gamma(A)$, where Φ is the standard Gaussian distribution function, then the following isoperimetric inequality holds (see, e.g., [11]):

$$\gamma(A + tU_H) \geq \Phi(a + t) \quad \forall t \geq 0,$$

where U_H is the closed unit ball in the Cameron–Martin space H. Therefore,

$$\frac{\gamma(A + tU_H) - \gamma(A)}{t} \geq \frac{\Phi(a + t) - \Phi(t)}{t}.$$

Hence letting $t \to 0$ we conclude that the liminf of the left-hand side is at least $\Phi'(a) = \varrho(a)$, which is the surface measure of the half-space of γ-measure a. Therefore, half-spaces possess minimal surface measures among sets of γ-measure a.

2.5 Fine versions of surface measures controlled by capacities

The surface measures constructed above on the basis of the Malliavin calculus have the property that they are defined on "almost all" level sets similarly to conditional measures. In this section we give some additional conditions under which there is a more canonical version of σ^y sitting on $F^{-1}(y)$ for each y. Here we assume that μ is Fomin differentiable along v (also some higher integrability of β_v will be assumed). Note that the measures σ^y do not change if we take a different version of F, but the sets $F^{-1}(y)$ can change. We recall that, as noted above, there is no problem if F is continuous and for every point z in the complement of $F^{-1}(y)$ and every neighborhood U of z there is a nonnegative continuous function of class $\mathcal{F}$ positive at z and having support in U.

Another concept coming along with surface measures is capacity (see [11] or [13]). Suppose that $\mathcal{F}$ is equipped with a norm $\|\cdot\|_{\mathcal{F}}$ such that convergence in this norm yields convergence in $L^1(\mu)$. In practical situations, this will be often the norm of a suitable Sobolev space $W^{p,1}(\mu)$, but so far no Sobolev spaces are needed. This norm generates a capacity: for every open set $U \subset X$ we define its capacity associated with $\mathcal{F}$ by the formula

$$C_{\mathcal{F}}(U) = \inf\{\|f\|_{\mathcal{F}} : f \in \mathcal{F}, f \geq 0, f \geq 1 \ \mu\text{-a.e. on } U\}.$$

For any set $B \subset X$ let

$$C_{\mathcal{F}}(B) = \inf\{C_{\mathcal{F}}(U) : U \supset B \text{ is open}\}.$$

Typically, capacities of the sort we consider are tight (see [61], [59], and [60]), i.e., for each $\varepsilon > 0$ there is a compact set K_ε such that $C_{\mathcal{F}}(X\backslash K_\varepsilon) < \varepsilon$. However, we do not assume this property.

Recall that a function f is called $C_{\mathcal{F}}$-quasi-continuous if, for each n, there is a closed set A_n such that $C_{\mathcal{F}}(X\backslash A_n) < 1/n$ and $f|_{A_n}$ is continuous.

It is known that each function $f \in \mathcal{F}$ has a $C_{\mathcal{F}}$-quasi-continuous version (see [13, Section 8.13]), provided that the norm $\|\cdot\|_{\mathcal{F}}$ is strictly convex. This is the case with the L^p-norm with $p \in (1, +\infty)$, and, more generally, with any norm of the form $\|f\|_{\mathcal{F}} = \|T^{-1}f\|_{L^p(m)}$, where m is a probability measure and T is a bounded injective linear operator from $L^p(m)$ to $L^1(\mu)$; in particular, the latter case covers most of Sobolev classes such as $W^{p,1}(\gamma)$ for a Gaussian measure γ (see Section 3). However, in place of such assumptions we simply assume in addition to (F1) and (F2) that

(F3) *F has a quasi-continuous version.*

We now fix a quasi-continuous version of F; the results below refer to this version.

Lemma 2.5.1. *Let μ be Fomin differentiable along v with respect to $\mathcal{F}$ and let* (F1), (F2) *and* (F3) *hold. Suppose that there is $p > 1$ such that*

$$\|f\|_{L^p(\mu)} + \|\partial_v f\|_{L^1(\mu)} \leq \|f\|_{\mathcal{F}}, \quad f \in \mathcal{F}. \tag{2.17}$$

Assume also that $\beta_v \in L^{p/(p-1)}(\mu)$. Then, for every open set $W \subset X$ and any $r > 0$, the measure $\nu = (\partial_v F) \cdot \mu$ satisfies the estimate

$$\nu(W \cap \{y < F < y + r\}) \leq rC(\mu)C_{\mathcal{F}}(W), \quad C(\mu) = 1 + \|\beta_v\|_{L^q(\mu)}, \ q = \frac{p}{p-1}. \tag{2.18}$$

Proof. Let $f \in \mathcal{F}$, $f \geq 0$ and $f \geq 1$ μ-a.e. on W. Then $f \geq 1$ ν-a.e. on W, hence on account of (2.16) and (2.17) we obtain

$$\nu(W \cap \{y < F < y + r\}) \leq \int\limits_{W \cap \{y<F<y+r\}} f(x)\,\nu(dx)$$
$$\leq \int\limits_{y<F<y+r} f(x)\,\nu(dx) \leq rC(\mu)(\|f\|_{L^p(\mu)} + \|\partial_\nu f\|_{L^1(\mu)}) \leq rC(\mu)\|f\|_{\mathcal{F}},$$

which yields the announced estimate by taking inf in f. □

Certainly, we can always equip $\mathcal{F}$ with the norm given by the left-hand side of (2.17). Moreover, this norm is strictly convex (since so is the L^p-norm) and convergence in this norm obviously yields convergence in $L^p(\mu)$, hence in $L^1(\mu)$. However, in concrete examples there might be other natural norms on $\mathcal{F}$ not related to ν, e.g., certain Sobolev norms. Quasi-continuous versions of F depend on our choice of $C_{\mathcal{F}}$, hence on our choice of a norm on $\mathcal{F}$.

Theorem 2.5.2. *Suppose that in Theorem 2.4.7 we have $\beta_\nu \in L^{p/(p-1)}(\mu)$ for some $p > 1$ and that* (2.17) *and* (F3) *hold (which can be ensured by taking the norm on $\mathcal{F}$ defined by the left-hand side of* (2.17)*). Then each σ^y is concentrated on the set $F^{-1}(y)$ and vanishes on all sets of $C_{\mathcal{F}}$-capacity zero.*

Proof. Let us show that $\sigma^y(X\backslash F^{-1}(y)) = 0$. We can assume again that $y = 0$. It suffices to show that σ^0 vanishes on each set $U := \{|F| > \delta\}$, where $\delta > 0$. By assumption, for each n, there is a closed set A_n such that $C_{\mathcal{F}}(X\backslash A_n) < 1/n$ and $F|_{A_n}$ is continuous. The sets

$$U_n = U \cap (X\backslash A_n)$$

are open because $\{|F| \leq \delta\} \cap A_n$ is closed by the continuity of the restriction $F|_{A_n}$. We have $U \subset \bigcap_{n=1}^{\infty} U_n$. Let $k > 1/\delta$. Then $\nu_k(U) = 0$, where, as above, $\nu_k = kI_{\{0<F<k^{-1}\}} \cdot \nu$. By the lemma we have

$$\nu_k(U_n) = \nu_k(X\backslash A_n) \leq C(\mu)n^{-1},$$

hence $\sigma^0(U_n) \leq C(\mu)n^{-1}$, which yields that $\sigma^0(U) = 0$. Note that we could not derive this directly from the equality $\nu_k(U) = 0$, because U need not be open.

We now prove that $\sigma^y(B) = 0$ for every set $B \in \mathcal{B}$ of zero $C_{\mathcal{F}}$-capacity. Again it suffices to consider the case $y = 0$. Let $\varepsilon > 0$. By definition, there is an open set W containing B such that $C_{\mathcal{F}}(W) < \varepsilon$. Therefore, there is a function $f \in \mathcal{F} \geq 0$ such that $f \geq 1$ μ-a.e. on W and $\|f\|_{\mathcal{F}} < \varepsilon$. It follows from the lemma that $|\nu_n(W)| \leq \varepsilon C(\mu)$, which yields that $|\sigma^0(W)| \leq \varepsilon C(\mu)$. Letting $\varepsilon \to 0$ we arrive at the desired conclusion. □

In the framework described above there is no natural way of normalizing our surface measures. One way of making the construction more invariant is this: assuming that there is some intrinsic norm $|v|_H$ (as is the case for Gaussian measures when we use the norm of the Cameron–Martin space H) and $|v(x)|_H > 0$ μ-a.e., one could use the unit field $v/|v|_H$, which leads to the weight $\partial_v F/|v|_H$ in place of $\partial_v F$. However, to avoid possible problems with differentiation along this new field, we just assume that the measure $|v|_H^{-1}\cdot\nu$ is finite and take new measures

$$\sigma_0^y := |v|_H^{-1}\cdot\nu^y$$

on the same level sets $F^{-1}(y)$. These measures are finite for $\nu\circ F^{-1}$-a.e. y, hence also for $\mu\circ F^{-1}$-a.e. y. Finally, if no weight is used, we arrive at surface measures that coincide with conditional measures.

Remark 2.5.3. Note that for any bounded continuous function g the measure $g\cdot\sigma^y$ is naturally defined for every y (not for almost every y). It is readily seen from Theorem 2.5.2 that the same is true for any bounded quasi-continuous function g. In order to assign finite integrals with respect to all surface measures σ^y to some unbounded functions g one can use the following trick: use surface measures generated by the new measure $g\cdot\mu$. Obviously, this requires some additional assumptions about g because our approach is based on positive differentiable measures. However, it works if there is $p > 1$ such that $\partial_v f \in L^p(\mu)$ for all $f\in\mathcal{F}$, $\beta_v\in L^p(\mu)$, $\partial_v F\in L^p(\mu)$, $g\geq 0$ belongs to $\mathfrak{D}_v\cap L^{p'}(\mu)$, $\partial_v g\in L^{p'}(\mu)$, where $p' = p/(p-1)$. Using this trick separately for g^+ and g^-, one can extend this to certain functions of variable sign. One can show that if in this situation g is bounded continuous this procedure yields the usual products $g\cdot\sigma^y$.

It is also worth noting that if we apply the same construction to the original measure μ in place of ν in order to integrate by parts in the equality with $\psi'(F)$ we must artificially add the factor $\partial_v F$ to obtain the expression $\partial_v(\psi(F))$. The effect is that we must impose the assumption of differentiability not on μ, but on the measure $(\partial_v F)^{-1}\cdot\mu$. In principle, this is quite possible but requires some extra assumptions.

In the considered situation we have the following version of the Gauss–Ostrogradskii–Stokes formula with our non-normalized surface measure. Set

$$V_r = F^{-1}(-\infty, r),\quad S_r = F^{-1}(r).$$

Theorem 2.5.4. *Let u be another vector field along which μ is differentiable, satisfying the same hypotheses as v. Then*

$$\int_{V_r}\beta_u(x)\,\mu(dx) = -\int_{S_r}\frac{\partial_u F(x)}{\partial_v F(x)}\,\sigma^r(dx),$$

provided that either the function $\xi := \partial_u F/\partial_v F$ is bounded quasi-continuous or ξ^+ and ξ^- satisfy the additional conditions mentioned in the previous remark.

Proof. Let $\psi_h(s) = 1$ if $s\leq r$, $\psi_h(s) = 0$ if $s\geq r+h$, $\psi_h(s) = C - s/h$ if $r < s < r+h$, $C = 1 + r/h$. Then $\psi_h'(s) = -1/h$ in the interval $(r, r+h)$ and $\psi_h' = 0$ outside the closure of this interval. We have $\partial_u(\psi_h \circ F) = -h^{-1}\partial_u F$ on the set $\{r < F < r+h\}$ and

$$\int_X \psi_h(F(x))\beta_u(x)\,\mu(dx) = -\int_X \partial_u(\psi_h \circ F)(x)\,\mu(dx) = h^{-1}\int_{r<F<r+h} \partial_u F(x)\,\mu(dx).$$

As $h \to 0$, the left-hand side of this identity tends to the integral of β_u over V_r and the right-hand side tends to the surface integral of the function $\partial_u F/\partial_v F$ against the surface measure σ^r. The latter holds if either ξ is bounded quasi-continuous or there are surface measures associated with the new measures $\xi^+ \cdot \mu$ and $\xi^- \cdot \mu$. □

If $u = v$, then we have

$$\int_{V_r} \beta_v(x)\,\mu(dx) = -\int_{S_r} \sigma^r(dx),$$

which gives the total mass of the surface.

2.6 Examples and comments

It is clear from the comments above that all our assumptions are rather general except for the requirement of differentiability of μ along a suitable vector field. Actually, also vector fields of differentiability can be found for quite general measures (see [13, Chapter 11]). The only serious restriction arises if we wish to find this field in a such a way that $\partial_v F$ is not very degenerate in order to connect our surface measures with more traditional surface measures as explained above. For this reason we include $\partial_v F$ in our measure. In particular, identically zero v fits our construction pretty well and produces zero surface measures. In order to avoid such meaningless situations we now consider some examples where for a given F one can find a suitable v with $\partial_v F > 0$ μ-a.e.

Example 2.6.1. Suppose that X is a Banach space, μ is Fomin differentiable along a nonzero constant vector v, and F is a continuous function on X differentiable along v such that $\partial_v F$ is continuous and $c_1 \le \partial_v F \le c_2$ for some positive numbers c_1 and c_2. Then there exist surface measures σ^y on the level sets $F^{-1}(y)$. In addition, one can use equivalent "traditional" surface measures $|\partial_v F|^{-2} \cdot \sigma^y$. One can also use a local version of this construction multiplying μ by a Lipschitzian bump function with a small support in a neighborhood of a point x_0 where $\partial_v F(x_0) > 0$.

In this situation we can apply both (i) and (iii) in Theorem 2.4.7. Applicability of (i) follows from the fact that any Lipschitzian function on X is μ-a.e. differentiable along v, which in turn follows from the one-dimensional case and the existence of differentiable conditional measures on the straight lines $x + \mathbb{R}v$ (see [13, Chapter 3]). Case (iii) applies here if we take for $\mathcal{F}$ the same class of bounded Lipschitzian functions or the class $\mathcal{FC}$ of smooth cylindrical functions. Finally, for W we can take a function of the form $W(x) = w(\|x\|)$, where w is an unbounded Lipschitzian sufficiently slowly increasing function on $\mathbb{R}$ such that W is μ-integrable (one can always find such a function).

An obvious disadvantage of constant vector fields is that they give less chances to obtain positive $\partial_v F$. For example, if X is a Hilbert space and F is Gâteaux differentiable, then it would be optimal in this respect to take $v = \nabla F$, which gives $\partial_v F(x) = \|\nabla F(x)\|^2$. However, even for very nice functions F there might be no natural measures differentiable along ∇F. For example, if we take $F(x) = (x, x)$ and want to define surface measures on the spheres, we have to ensure differentiability of μ along the field $v(x) = x$. However, say, Gaussian measures on infinite-dimensional spaces are not differentiable along this field (see the example below). For this reason, one has to consider vector fields with values in suitable analogs of the Cameron–Martin space.

Example 2.6.2. Let γ be a centered Radon Gaussian measure on a locally convex space X such that its Cameron–Martin space H is infinite-dimensional. Then γ is not Fomin differentiable along the vector field $v(x) = x$. Indeed, it suffices to prove this for the standard Gaussian measure γ on $\mathbb{R}^\infty$. Suppose that $\beta_v \in L^1(\gamma)$ is the divergence of v. Then for every smooth cylindrical function f in variables $x_1, \dots, x_n$ we have

$$\int_X \sum_{i=1}^n x_i \partial_{x_i} f(x)\,\gamma(dx) = -\int_X f(x)\beta_v(x)\,\gamma(dx).$$

The left-hand side equals

$$-\int_X f(x) \sum_{i=1}^n (1 - x_i^2)\,\gamma(dx)$$

by the integration by parts formula. Therefore, the function $S_n(x) = \sum_{i=1}^n (1 - x_i^2)$ equals the conditional expectation of the function β_v with respect to the σ-field generated by $x_1, \dots, x_n$. We recall that the conditional expectation of an integrable function ξ with respect to a smaller σ-algebra $\mathcal{A}$ is an integrable function $\mathrm{E}^{\mathcal{A}}\xi$ that is measurable with respect to $\mathcal{A}$ and satisfies the identity

$$\int_X \eta\xi\, d\gamma = \int_X \eta \mathrm{E}^{\mathcal{A}}\xi\, d\gamma$$

for all bounded functions η measurable with respect to $\mathcal{A}$. By the martingale convergence theorem (see, e.g., [12, Chapter 10]) the sequence of functions S_n converges in $L^1(\gamma)$, but the functions $1 - x_i^2$ are second order polynomials, hence the series of $1 - x_i^2$ converges in all $L^p(\gamma)$ (see [11, Chapter 5]). However, there is no convergence in $L^2(\gamma)$, because these functions are mutually orthogonal in $L^2(\gamma)$ and have equal norms.

Example 2.6.3. Let us return to Section 3 and consider the case of a centered Gaussian measure μ with the Cameron–Martin space H and F belonging to the Sobolev class $W^{2,2}(\mu)$. In this case the vector field $v = D_H F \in W^{2,1}(\mu, H)$ has divergence $\beta_v = LF \in L^2(\mu)$ and $\partial_v F = |D_H F|_H^2$ is in $L^1(\mu)$.

If $F \in W^{p,2}(\mu)$ with some $p \in (1, 2)$, then we can use the vector field

$$v = D_H F/|D_H F|_H^2,$$

for which $\partial_v F = 1$ (and we obtain surface measures from [29]), or the vector field

$$v = D_H F/|D_H F|_H,$$

so that v is a unit normal on the surface (but with respect to the Cameron–Martin norm) and $\partial_v F = |D_H F|_H$; now in both cases it becomes necessary to require that v must have divergence. In the first case it suffices to have $\|D_H^2 F\|_{HS}/|D_H F|_H^2 \in L^p(\mu)$ and in the second case it suffices to have $\|D_H^2 F\|_{HS}/|D_H F|_H \in L^p(\mu)$ (the corresponding surface measures will be different). It is also possible to use a less constructive (but weaker) assumption that $F \in W^{1,1}(\mu)$ and one of these two vector fields has divergence $\beta_v \in L^1(\mu)$.

If X is a Banach space, then we can take for $\mathcal{F}$ the class of all bounded Lipschitzian functions. Conditions (F1)–(F3) are readily verified in this case. Actually, the case of a general locally convex space with a Radon Gaussian measure reduces to this one by the Tsirelson linear isomorphism theorem (see [11, Chapter 3]). If $\beta_v \in L^p(\mu)$, then we can take $\mathcal{F} = \mathcal{FC}$ (smooth cylindrical functions) and apply case (iii) in Theorem 2.4.7. If X is sequentially complete, then for W we can take the Minkowski functional p_Q of an absolutely convex compact set Q of positive measure. It is known that $p_Q \in W^{r,1}(\mu)$ for all $r \geq 1$.

Actually, if one is interested only in a Gaussian measure μ on a Banach space, then the construction of surface measures along these lines becomes straightforward as explained in Section 3.

In particular, if $D_H F \neq 0$ a.e., our construction can be compared with surface measures considered in [1], [49], [11], [13], and [29] (note that the latter work develops a construction based on distribution functions related to the measure μ itself, so that it leads to surface measures that are not invariant under scaling of F, as discussed above). In order to come to usual surface measures one should either deal with a unit field $D_H F/|D_H F|_H$ or deal with $v = D_H F$, and then multiply the obtained surface measures by $1/|D_H F|_H$.

We emphasize that for better surfaces (existing individually such as level sets of continuously Fréchet differentiable functions with nondegenerate derivatives) there is no need to involve variable vector fields $D_H F$: it becomes much simpler to define surface measures locally by using only constant vector fields of differentiability of μ as in Example 2.6.1. In that case no second derivatives of F appear at all and in this way we recover the existence results of [75] (even under weaker assumptions).

In place of a Gaussian measure μ, it is possible to consider a Radon probability measure μ on a locally convex space X that is Fomin differentiable along a continuously embedded dense Hilbert space H. A simple example is the countable power of a probability measure on $\mathbb{R}$ with a smooth density with compact support; then one can take $X = \mathbb{R}^\infty$ and $H = l^2$ is a natural choice (the same measure can be also considered on a weighted Hilbert space of sequences (x_n) such that $\sum_{n=1}^\infty \alpha_n x_n^2 < \infty$, where $\alpha_n > 0$ and $\sum_{n=1}^\infty \alpha_n < \infty$). Then one can also define Sobolev classes. This situation (studied in [58], [59], and [60]) has been the most general considered so far in the linear case. In [29], similar results have been reproved in the Gaussian case.

Remark 2.6.4. (i) It is worth noting that the case of a Fréchet space reduces to that of a separable reflexive Banach space, since every Radon measure on a Fréchet space is concentrated on a compactly embedded separable reflexive space (see [12, Theorem 7.12.4]). For many measures on Banach spaces (Gaussian, differentiable), the class of bounded Lipschitzian functions is a suitable candidate for $\mathcal{F}$, since such functions are almost everywhere differentiable with respect to such measures.

(ii) The class $\mathcal{FC}$ of smooth cylindrical functions and the larger class of bounded Lipschitzian cylindrical functions satisfy condition (F1), but in general they do not have the property of the whole class of bounded Lipschitzian functions that convergence of integrals of such functions with respect to a sequence of probability measures ensures weak convergence of these measures (say, this is not true for infinite-dimensional Hilbert spaces, although is true for $\mathbb{R}^\infty$). This is why we considered cases (ii) and (iii) in Theorem 2.4.7. As already noted, it is possible to define surface measures locally in a suitable sense (for example, on compact sets) by replacing μ by $\zeta\cdot\mu$, where $\zeta \geq 0$ is a bump function whose support gives the desired localization. For example, in the Gaussian case or in the case of a differentiable measure on a Banach space, it is always possible to choose ζ in a such a way that its support will be compact and will contain a given compact set, and the measure $\zeta\cdot\mu$ will remain Fomin differentiable along the same directions as μ. This approach can give local surface measures in more general situations where there are no global surface measures. A possible way of gluing these local surface measures is based on establishing their uniform tightness.

If X is equipped with a suitable tangent structure enabling us to consider v not as a differentiation, but as a true vector field possessing the corresponding norm $|v(x)|$, one might try to use fields of unit length; again the question of their choice arises.

The choice $v = D_H F$ in the Gaussian case mentioned above is connected with another natural object related to Gaussian Hausdorff measures mentioned at the end of Section 3: H-neighborhoods of sets. Given a Borel set B, we take the set $B^r = B + rU_H$, where U_H is the unit ball in the Cameron–Martin space H. The set B^r in general is much smaller than the usual metric r-neighborhood of B. Then, for certain "surfaces" B, the surface measure of B can be obtained as a limit of $\mu(B^r)/r$ as $r \to 0$. However, a precise definition the surface measure of B is more involved.

Among various restrictions on μ and F imposed above, certainly the most stringent one is the existence of vector fields of differentiability for μ. For example, in many cases, given a measure μ on a metric space, one can take for $\mathcal{F}$ the space of bounded Lipschitzian functions; in many cases, such functions possess appropriate gradients μ-almost everywhere, so if F is locally Lipschitzian, then the only problem is to find suitable differentiability fields for the measure. It is not always possible to build such fields from constant vector fields (this happens already for distributions of diffusion processes with non-constant diffusion coefficients, see [13, Chapter 4]). It would be interesting to study vector fields of differentiability of measures in the framework of metric measure spaces.

Remark 2.6.5. It would be interesting to study a possible analog of the Radon transform associated with surface measures in the spirit of the construction developed in [41], [42], [50], [17] for conditional measures on hyperplanes. Recall that the classical Radon transform reconstructs a function on the plane by its integrals over all straight lines (with Lebesgue measure). A natural infinite-dimensional analog (considered in the cited papers) is this: given a Radon probability measure μ, to get some information about a function with given integrals with respect to conditional measures μ^{L+y} on the set $L + y$ for all possible hyperplanes L.

2.7 Surface measures of higher codimension

In this section we consider surface measures on surfaces of higher codimension. Note that Gaussian Hausdorff measures (see Section 3) are defined in a unified way for arbitrary codimension; other approaches exploited the fact that F was a real function. Nevertheless, the construction developed in the previous sections also works in the case of surfaces of higher codimension, but requires a bit more regularity of the mapping

$$F = (F_1, \dots, F_d)\colon X \to \mathbb{R}^d$$

on the level sets of which we wish to define surface measures. We recall that conditional measures are not sensitive at all to this change, they exist even for mappings with values in quite general infinite-dimensional spaces.

Now we need d vector fields $v_1, \dots, v_d$ along which the measure μ is differentiable. However, in the multidimensional case it is reasonable to modify our conditions on $\mathcal{F}$ as follows: $\mathcal{F}$ is a linear space separating Radon measures on X such that

$$\phi(f_1, \dots, f_n) \in \mathcal{F} \quad \forall f_1, \dots, f_n \in \mathcal{F}$$

for all functions $\phi \in C_b^\infty(\mathbb{R}^n)$ with arbitrary n and

$$\partial_{v_i}(\phi(f_1, \dots, f_n)) = \sum_{j=1}^{n} \partial_{x_j}\phi(f_1, \dots, f_n)\partial_{v_i} f_j.$$

In order to modify condition (F2) we shall suppose that $\psi(F) \in \mathfrak{D}_{v_i}$ for all functions $\psi \in C_0^\infty(\mathbb{R}^d)$ and each $i = 1, \ldots, d$. This enables us to define functions $\partial_{v_i} F_j$ as we have done in the one-dimensional case in (F2).

In place of $\partial_v F$ we now take the determinant Δ_F of the so-called Malliavin matrix

$$(\sigma_{ij})_{i,j\le d}, \quad \sigma_{ij} := \partial_{v_i} F_j.$$

Let $M^{ij}(x) = (-1)^{i+j} m^{ji}$, where $m^{ji}(x)$ is the minor in the Malliavin matrix corresponding to the element $\sigma_{ji}(x)$. Thus, M^{ij} is the transposed matrix of cofactors of the Malliavin matrix. If the matrix $(\sigma_{ij}(x))_{i,j\le d}$ is invertible, the inverse matrix will be denoted by $(\gamma^{ij}(x))_{i,j\le d}$. In that case

$$\gamma^{ij} = M^{ij}/\Delta_F.$$

Therefore,

$$\sum_{j\le d} M^{ij}(x)\sigma_{jk}(x) = \Delta_F(x)\delta_{ik}, \tag{2.19}$$

where δ_{ik} is Kronecker's symbol. Indeed, this is true for invertible matrices, but remains valid for any matrix by approximation by invertible matrices.

In the Gaussian case considered above we take $v_i = D_H F_i$, so that

$$\sigma_{ij} = (D_H F_i, D_H F_j)_H$$

and the matrix $(\sigma_{ij})_{i,j\le d}$ is nonnegative definite.

In place of (F2) we suppose that $\Delta_F \ge 0$ and $\Delta_F \in L^1(\mu)$. Set

$$\nu = \Delta_F \cdot \mu.$$

Let

$$U_r = \{x \in \mathbb{R}^d\colon\ |x| < r\} \quad \text{and} \quad W_r = \{|F| < r\}.$$

Let $BV(U_r)$ be the space of functions of bounded variation on U_r and let $W^{p,1}(U_r)$ be the Sobolev class of functions belonging to $L^p(U_r)$ along with their generalized first order partial derivatives.

Theorem 2.7.1. (i) *Suppose that $fM^{ij} \in \mathfrak{D}_{v_j}$ for all $i, j \le d$ and all $f \in \mathcal{F}$ vanishing outside of W_r. Then the measure $\nu \circ F^{-1}$ is absolutely continuous on U_r and has a density ϱ of class $BV(U_r)$. In particular, $\varrho \in L^{d/(d-1)}(U_r)$.*

If $\Delta_F(x) \neq 0$ μ-a.e. on W_r, then $\varrho \in W^{1,1}(U_r)$.

(ii) *If, in addition,*

$$u_i := \frac{I_{W_r}}{\Delta_F} \sum_{j\le d} [\partial_{v_j} M^{ij} + M^{ij}\beta_{v_j}] \in L^s(\nu) \quad \textit{for some } s > d, \tag{2.20}$$

then this density ϱ belongs to $W^{p,1}(U_r)$ with some $p > d$ and has a continuous version.

(iii) *If $s > 2d$, then for any $f \in \mathcal{F}$ the measure $(f \cdot \nu) \circ F^{-1}$ is absolutely continuous on U_r and has a bounded continuous density ϱ_f such that*

$$\sup_{y \in U_r} |\varrho_f(y)| \le C\Big(\|f\|_{L^{2d}(\nu)} + \sum_{j=1}^{d} \|\partial_{v_j} f\|_{L^{2d}(\nu)}\Big), \tag{2.21}$$

where C is a number that depends only on $d, s, r_0, \|u_i\|_{L^s(\nu)}$, and $\|I_{W_r} M^{ij}/\Delta_F\|_{L^s(\nu)}$), whenever $r \le r_0$ and $r_0 > 0$ is fixed.

Proof. (i) Let $\psi \in C_0^\infty(U_r)$. By (2.19) we have

$$\begin{aligned}
\int_{U_r} \partial_{y_i}\psi(y)\, \nu \circ F^{-1}(dy) &= \int_{W_r} \partial_{y_i}\psi(F(x))\Delta_F(x)\, \mu(dx)\\
&= \int_X \sum_{j,k \le d} M^{ij}(x)\sigma_{jk}(x)[\partial_{y_k}\psi(F(x))]\, \mu(dx) = \int_X \sum_{j \le d} \partial_{v_j}(\psi \circ F)(x) M^{ij}(x)\, \mu(dx)\\
&= -\sum_{j \le d} \int_X (\psi \circ F)(x) M^{ij}(x)\, d_{v_j}\mu(dx) - \sum_{j \le d} \int_X (\psi \circ F)(x)\partial_{v_j} M^{ij}(x)\, \mu(dx)\\
&= -\sum_{j \le d} \int_X (\psi \circ F)(x)[\partial_{v_j} M^{ij}(x) + M^{ij}(x)\beta_{v_j}(x)]\, \mu(dx).
\end{aligned}$$

The right-hand side can be written as the integral of ψ with respect to a bounded measure on U_r, hence the measure $\nu \circ F^{-1}$ on U_r has a density ϱ of class $BV(U_r)$. By the Sobolev embedding theorem $\varrho \in L^{d/(d-1)}(U_r)$, see, e.g., [13, Chapter 2], [33] or [79].

If the measure ν is equivalent to μ, which is the case where $\Delta_F > 0$ μ-a.e. and μ can be written as $\Delta_F^{-1} \cdot \nu$, the right-hand side can be written as the integral of $\psi g_i \varrho$, where g_i is the conditional expectation of the ν-integrable function $-u_i$ with respect to the measure ν and the σ-field generated by F (here we also take into account that $\psi(F) = \psi(F) I_{W_r}$, because ψ has support in U_r). Therefore, $\varrho \in W^{1,1}(U_r)$.

(ii) Note that for some function g_i we have

$$\partial_{y_i}\varrho = g_i \varrho.$$

By Jensen's inequality for conditional expectations the inclusion $|u_i|^s \in L^1(\nu)$ yields the inclusion $|g_i|^s \varrho \in L^1(U_r)$. We recall that this inequality states that

$$V(\mathrm{E}^{\mathcal{A}}\xi) \le \mathrm{E}^{\mathcal{A}}[V(\xi)]$$

for every convex function V. In particular, $|\mathrm{E}^{\mathcal{A}}\xi|^p \le \mathrm{E}^{\mathcal{A}}[|\xi|^p]$ for all $p \ge 1$.

We now show that $\partial_{y_i}\varrho$ is better integrable under the assumptions of the last assertion. Suppose that $\varrho \in L^p(U_r)$ for some $p \ge 1$. By Hölder's inequality we have $g_i\varrho \in L^{sp/(p+s)}(U_r)$. Therefore, $\varrho \in W^{p_1,1}(U_r)$ with $p_1 = sp/(p+s)$; if $p_1 < d$ by the Sobolev embedding this yields that $\varrho \in L^{p_2}(U_r)$ with

$$p_2 = \frac{dp_1}{d - p_1} = p\frac{ds}{ds - p(s-d)} \ge p\frac{ds}{ds - s + d} = \lambda p, \quad \lambda = \frac{ds}{ds - s + d} > 1.$$

If $p_1 = d$ then $\varrho \in L^q(U_r)$ for any $q < \infty$, hence $\partial_{y_i}\varrho \in W^{s-\varepsilon,1}(U_r)$ for any $\varepsilon > 0$. Therefore, in finitely many steps we arrive at the situation where $\partial_{y_i}\varrho \in W^{p,1}(U_r)$ with some $p > d$. Therefore, the Sobolev embedding ensures a continuous density.

(iii) It follows from the reasoning above that if we replace ν by the measure $f \cdot \nu$, the generalized partial derivative of the measure $(f \cdot \nu) \circ F^{-1}$ with respect to y_i will equal the function $\widehat{g}_i\varrho$, where $\widehat{g}_i$ is the conditional expectation of the function

$$\widehat{u}_i = fu_i + \frac{I_{W_r}}{\Delta_F} \sum_{j\le d} M^{ij}\partial_{v_j} f$$

with respect to the measure ν and the σ-field generated by F. Thus, we have

$$\partial_{y_i}\varrho_f = \widehat{g}_i\varrho.$$

However, now we already know that ϱ is bounded continuous on U_r and it follows from the previous step that its sup on U_r is estimated by a constant that depends on d, s, r and the $L^s(\nu)$-norm of u_i. Therefore, choosing $\varepsilon > 0$ such that $s = 2d(d+\varepsilon)/(d-\varepsilon)$ and letting $t = d + \varepsilon > d$, we conclude that the $L^t(U_r)$-norm of $\partial_{y_i}\varrho_f$ is estimated by a constant depending on the indicated quantities and the $L^t(\nu)$-norm of $\widehat{u}_i$. By Hölder's inequality

$$\|uw\|_t \le \|u\|_s\|w\|_{st/(s-t)}, \quad \frac{st}{s-t} = 2d.$$

We apply this inequality with $u = u_i$ and $w = f$ and also with $u = I_{W_r}M^{ij}/\Delta_F$ and $w = \partial_{v_j} f$. This gives a bound on the $W^{t,1}(U_r)$-norm of ϱ_f via

$$\|f\|_{L^{2d}(\nu)} + \sum_{j=1}^{d} \|\partial_{v_j} f\|_{L^{2d}(\nu)}$$

multiplied by a constant, which yields the announced estimate by the Sobolev embedding theorem. □

We now give a constructive sufficient condition for the continuity of densities of multidimensional distributions related to μ rather than ν. This requires, however, second derivatives of F. In the next proposition we assume that $\partial_{v_k}\partial_{v_j}F_i$ can be defined in the same sense as $\partial_{v_j}F_i$ above by using that $\psi(\partial_{v_j}F) \in \mathfrak{D}_{v_k}$ for smooth functions on $\mathbb{R}^d$ with compact support.

Theorem 2.7.2. (i) *Suppose that for every $r \in \mathbb{N}$ there is $\varepsilon_r > 0$ such that the functions*

$$\exp\Big(\varepsilon_r\Big|\sum_k (\gamma^{ik}\beta_{v_k} + \partial_{v_k}\gamma^{ik})\Big|\Big) \tag{2.22}$$

are μ-integrable on the set $\{|F| < r\}$. Then the measure $\mu \circ F^{-1}$ has a continuous density without zeros.

(ii) *Suppose that for every $r \in \mathbb{N}$ there is $p_r > d$ such that the functions*

$$\Big|\sum_k \gamma^{ik}\beta_{v_k}\Big|^{p_r}, \quad \Big|\sum_k \partial_{v_k}\gamma^{ik}\Big|^{p_r}$$

are μ-integrable on the set $\{|F| < r\}$. Then the measure $\mu \circ F^{-1}$ has a continuous density.

Proof. (i) We shall use the following result (see [19] or [13, Proposition 6.4.1]): if a nonnegative function ϱ on a ball $U \subset \mathbb{R}^d$ belongs to the Sobolev class $W^{1,1}(U)$ and there is $\varepsilon > 0$ such that $\varrho \exp(\varepsilon|\nabla\varrho|/\varrho) \in L^1(U)$, where we define $\nabla\varrho/\varrho = 0$ on the set $\{\varrho = 0\}$, then ϱ has a continuous version that is either identically zero or positive.

Let us fix $r \in \mathbb{N}$ and let U be the open ball of radius r in $\mathbb{R}^d$ centered at the origin. Let $\phi \in C_0^\infty(U)$. We have

$$\begin{aligned}
\int_U \partial_{y_i}\phi(y)\,\mu \circ F^{-1}(dy) &= \int_X \partial_{y_i}\phi(F(x))\,\mu(dx) = \int_X \sum_{k,j\le d} \gamma^{ik}\sigma_{kj}(\partial_{y_j}\phi)\circ F\,d\mu \\
&= \int_X \sum_{k\le d} \gamma^{ik}\partial_{v_k}(\phi\circ F)\,d\mu = -\int_{|F|<r} \sum_{k\le d} \phi\circ F[\partial_{v_k}\gamma^{ik} + \gamma^{ik}\beta_{v_k}]\,d\mu \\
&= -\int_U \phi(y)\eta_i(y)\,\mu\circ F^{-1}(dy),
\end{aligned}$$

where η_i is the conditional expectation of the function

$$\sum_k [\partial_{v_k}\gamma^{ik} + \gamma^{ik}\beta_{v_k}] I_{\{|F|<r\}}$$

with respect to the measure μ and the σ-field generated by F. It follows that the generalized derivative of the measure $\mu \circ F^{-1}$ on U in the variable y_i is the measure $\eta_i \cdot (\mu \circ F^{-1}) \ll \mu \circ F^{-1}$. Therefore, $\mu \circ F^{-1}$ on U has a density $\varrho \in W^{1,1}(U)$ and $\partial_{y_i}\varrho/\varrho = \eta_i$. By our assumption (2.22) and Jensen's inequality for conditional expectations (now applied to exp), we arrive at the condition mentioned above.

(ii) If we are given that $\mu \circ F^{-1}$ has a locally bounded density ϱ, then the previous relation can be written as

$$\int_U \partial_{y_i}\phi(y)\varrho(y)\,dy = -\int_U \phi(y)\eta_i(y)\varrho(y)\,dy,$$

which means that $\partial_{y_i}\varrho = \eta_i\varrho$ on U in the sense of distributions. We obtain again that $\varrho \in W^{1,1}(U)$, but now we conclude that $\partial_{y_i}\varrho \in L^{p_r}(U)$ by the same iteration of the Sobolev embedding theorem as in the previous proposition. Therefore, by the Sobolev embedding theorem ϱ has a continuous density (now it is not asserted that it is positive). □

In order to ensure (2.22) in terms of the original Malliavin matrix, we note that

$$\partial_{v_k}\gamma^{ik} = \partial_{v_k}(M^{ik}\Delta_F^{-1}) = (\partial_{v_k}M^{ik})\Delta_F^{-1} - (\partial_{v_k}\Delta_F)\Delta_F^{-2}.$$

The first term is a sum of functions of the form $\Delta_F^{-1}\partial_{v_k}\sigma^{ij}w$, where w is a product of $d-2$ matrix elements of the Malliavin matrix. The second term is a sum of functions of the form $\Delta_F^{-2}\partial_{v_k}\sigma^{ij}w$ with w as above. Therefore, it suffices to have the μ-integrability of the functions

$$\exp\Big(\frac{\varepsilon_r}{\Delta_F^2}|\partial_{v_k}\partial_{v_j}F_i|\,|\partial_{v_l}F_m|^{d-1}\Big),\ \exp\Big(\frac{\varepsilon_r}{\Delta_F}|\beta_{v_k}|\,|\partial_{v_l}F_m|^{d-1}\Big),$$

on the set $\{|F|<r\}$. For example, this holds if for some $\delta_r>0$ the exponents of

$$\delta_r\Delta_F^{-4},\quad \delta_r|\partial_{v_k}\partial_{v_j}F_i|^4,\quad \delta_r|\partial_{v_l}F_m|^{4d-4},\quad \delta_r\beta_{v_k}^2$$

are integrable.

Definition 2.7.3. *The surface measure σ^y is defined as a Radon measure such that*

$$\int_X f(x)\,\sigma^y(dx)=\varrho_f(y)\quad \forall f\in\mathcal{F}.$$

This definition means that

$$\int_X f(x)\,\sigma^y(dx)=\lim_{r\to 0}\frac{1}{|U_r|}\int_{\{|F-y|<r\}} f(x)\,\nu(dx),$$

where $|U_r|$ is the usual volume of the ball U_r. The existence of the limit in the right-hand side is the only condition required by the definition, and this condition is fulfilled in the situation of Theorem 2.7.1.

As in Section 4, we have to show that this relation defines a Radon measure. We need also an analog of Lemma 2.5.1.

Lemma 2.7.4. *Suppose that the hypotheses of case* (iii) *of Theorem* 2.7.1 *hold and*

$$\|f\|_{L^{2d}(\mu)}+\sum_{j=1}^d\|\partial_{v_j}f\|_{L^{2d}(\mu)}\le C_0\|f\|_{\mathcal{F}}\quad \forall f\in\mathcal{F}. \tag{2.23}$$

Then, for every open set $W\subset X$ and any $r>0$, we have

$$\nu(W\cap\{|F-y|<r\})\le C_0C_1r^dC_{\mathcal{F}}(W), \tag{2.24}$$

where C_1 depends on the same quantities as in assertion (iii) *of Theorem* 2.7.1.

Proof. Let $f\in\mathcal{F}$, $f\ge 0$ and $f\ge 1$ μ-a.e. on W. Then $f\ge 1$ ν-a.e. on W, hence as in the proof of Lemma 2.5.1 we obtain

$$\nu(W\cap\{|F-y|<r\})\le r^d\sup_{z:\ |z-y|<r}|\varrho_f(z)|,$$

where ϱ_f is the density of the measure $(f \cdot \nu) \circ F^{-1}$. According to assertion (iii) of Theorem 2.7.1 we can estimate the maximum of the continuous version of ϱ_f on W_r by

$$\|f\|_{L^{2d}(\nu)} + \sum_{j=1}^{d} \|\partial_{v_j} f\|_{L^{2d}(\nu)}$$

multiplied by some constant depending on the quantities indicated in that proposition. □

Theorem 2.7.5. *In the situation of Theorem* 2.7.1(iii), *the assertion of Theorem* 2.4.7 *is true. The assertion of Theorem* 2.5.2 *is true as well if the condition of the previous lemma holds and the respective quasi-continuous versions of* F_i *are considered.*

The proof is essentially the same, however, we should note that the assumptions are now much stronger.

As in case $d = 1$ we can equip $\mathcal{F}$ with the norm given by the left-hand side of (2.23). However, this is not always convenient since this norm depends on F and v. If the Gaussian case it may be preferable to use some Sobolev norm on $\mathcal{F}$. For example, if we take $v_i = D_H F_i$ as in Example 2.6.3, then $\partial_{v_i} f = (D_H f, D_H F_i)_H$, so that $\partial_{v_i} f \in L^{2d}(\nu)$ provided that $f \in W^{4d,1}(\mu)$, $F_i \in W^{8d,1}(\mu)$ and $\Delta_F^{-1} \in L^{8d-4}(\mu)$.

For Fomin differentiable measures, the above construction applies under much broader assumptions than in [78].

Remark 2.7.6. Since $\sigma^y = \varrho_1(y)\nu^y$, every ν-integrable $\mathcal{B}$-measurable function g is σ^y-integrable for $\nu \circ F^{-1}$-almost every y. This enables us to define surface measures for $g \cdot \nu$. Alternatively, we can use the trick described in Remark 2.5.3.

It should be noted the construction presented is chiefly oriented towards infinite-dimensional spaces, where typical measures are not doubling and differ also in other respects from measures usual in measure metric spaces. Nevertheless, it would be interesting to compare suitable Hausdorff measures on metric measure spaces and surface measures described above; measures differentiable along vector fields and all other objects considered above (differentiations, gradients, Sobolev classes, etc.) are meaningful on such spaces (see, e.g., [27], [40], [46], [64], and [65]). In particular, if we have a differentiation of the form $f \mapsto \Gamma(f,g)$ defined by a fixed function in the Dirichlet space (see [64]) built on a probability space (X, μ) admitting a "carré du champ" $\Gamma(f,g)$ such that $2\Gamma(f,g) = L(fg) - fLg - gLf$, where L is a Markov symmetric generator such that $\mathbb{E}\Gamma(f,g) = -\mathbb{E}fLg$, then we see that Lg is precisely the divergence of the considered field (a similar framework is considered in [70]). In relation to constructing vector fields possessing divergences, which is crucial for the presented approach, the recent paper [38] is of interest.

Finally, it would be interesting to continue investigation of surface measures and surface Sobolev and Besov classes connected, in particular, with restrictions of

Sobolev and Besov classes on the whole space. About fractional Sobolev classes over infinite-dimensional spaces, see [25], [52]–[54], and [55].

Bibliography

[1] H. Airault, P. Malliavin: Intégration géometrique sur l'espace de Wiener. Bull. Sci. Math. (2) 112(1), 3–52 (1988).

[2] L. Ambrosio, G. Da Prato, D.Pallara: *BV* functions in a Hilbert space with respect to a Gaussian measure. Rend. Acc. Lincei 21 (2010), 405–414.

[3] L. Ambrosio, S. Di Marino: Equivalent definitions of *BV* space and of total variation on metric measure spaces. J. Funct. Anal. 266, 4150–4188 (2014).

[4] L. Ambrosio, A. Figalli: Surface measures and convergence of the Ornstein–Uhlenbeck semigroup in Wiener spaces. Ann. Fac. Sci. Toulouse Math. (6) 20(2), 407–438 (2011).

[5] L. Ambrosio, A. Figalli, E. Runa: On sets of finite perimeter in Wiener spaces: reduced boundary and convergence to halfspaces. Atti Accad. Naz. Lincei Rend. Lincei Mat. Appl. 24(1), 111–122 (2013).

[6] L. Ambrosio, M. Miranda, S. Maniglia, D. Pallara: BV functions in abstract Wiener spaces. J. Funct. Anal. 258(3), 785–813 (2010).

[7] L. Ambrosio, M. Miranda, D. Pallara: Sets with nice perimeter in Wiener spaces, perimeter measure and boundary rectifiability. Discrete Contin. Dyn. Syst. 28 (2010), 591-606.

[8] L. Ambrosio, M. Miranda, D. Pallara: Some nice properties of *BV* functions on Wiener spaces. Anal. Geom. Metr. Spaces 3 (2015), 212–230.

[9] V.I. Bogachev: Smooth measures, the Malliavin calculus and approximation in infinite dimensional spaces. Acta Univ. Carolinae, Math. et Phys. 31(2), 9–23 (1990).

[10] V.I. Bogachev: Differentiable measures and the Malliavin calculus. J. Math. Sci. (New York) 87(4), 3577–3731 (1997).

[11] V.I. Bogachev: Gaussian measures. Amer. Math. Soc., Providence, Rhode Island, 1998.

[12] V.I. Bogachev: Measure theory. V. 1, 2. Springer, Berlin, 2007.

[13] V.I. Bogachev: Differentiable measures and the Malliavin calculus. Amer. Math. Soc., Providence, Rhode Island (2010).

[14] V.I. Bogachev: Gaussian measures on infinite-dimensional spaces. In: Real and Stochastic Analysis. Current Trends (M.M. Rao ed.), pp. 1–83. World Sci., Singapore (2014).

[15] V.I. Bogachev: Sobolev classes on infinite-dimensional spaces. In: Geometric Measure Theory and Real Analysis (L. Ambrosio ed.), pp. 1–56, Publications of the Scuola Normale Superiore, V. 17, Pisa, Edizioni della Normale, 2014.

[16] V.I. Bogachev: Distributions of polynomials on multidimensional and infinite-dimensional spaces with measures. Uspehi mat. Nauk 71(3) (2016) (in Russian); English transl.: Russian Math. Surveys, 71(3) (2016).

[17] V.I. Bogachev, M.N. Lukintsova: The Radon transform in infinite-dimensional spaces. Dokl. Akad. Nauk 443(3), 279–282 (2012), (in Russian); English transl.: Dokl. Math. 85(2), 208–211 (2012).

[18] V.I. Bogachev, I.I. Malofeev: Surface measures generated by differential measures. Potential Analysis 44(4), 767–792 (2016).

[19] V.I. Bogachev, E. Mayer-Wolf: Absolutely continuous flows generated by Sobolev class vector fields in finite and infinite dimensions. J. Funct. Anal. 167(1), 1–68 (1999).

[20] V.I. Bogachev, A.Yu. Pilipenko, E.A.Rebrova: Classes of functions of bounded variation on infinite-dimensional domains. Dokl. Akad. Nauk 451(2), 127–131 (2013) (in Russian); English transl.: Dokl. Math. 88(1), 391–395 (2013).
[21] V.I. Bogachev, A.Yu. Pilipenko, A.V.Shaposhnikov: Sobolev functions on infinite-dimensional domains. J. Math. Anal. Appl. 419, 1023–1044 (2014).
[22] V.I. Bogachev, E.A. Rebrova: Functions of bounded variation on infinite-dimensional spaces with measures. Dokl. Russian Acad. Sci. 449(2), 131–135 (2013) (in Russian); English transl.: Dokl. Math. 87(2), 144–147 (2013).
[23] N. Bouleau, F. Hirsch: Dirichlet forms and analysis on Wiener space. De Gruyter, Berlin – New York (1991).
[24] V. Caselles, A. Lunardi, M. Miranda, M. Novaga: Perimeter of sublevel sets in infinite dimensional spaces. Adv. Calc. Var. 5(1), 59–76 (2012).
[25] P. Celada, A. Lunardi: Traces of Sobolev functions on regular surfaces in infinite dimensions. J. Funct. Anal. 266, 1948–1987 (2014).
[26] S. Chaari, F. Cipriano, H.-H. Kuo, H. Ouerdiane: Surface measures on the dual space of the Schwartz space. Comm. Stoch. Anal. 4, 467–480 (2010).
[27] J. Cheeger: Differentiability of Lipschitz functions on metric measure spaces. Geom. Funct. Anal. 9, 428–517 (1999).
[28] G. Da Prato: Lunardi A. Maximal Sobolev regularity in Neumann problems for gradient systems in infinite dimensional domains. Ann. Inst. H. Poincaré Probab. Stat. 51(3), 1102–1123 (2015).
[29] G. Da Prato, A. Lunardi, L. Tubaro: Surface measures in infinite dimension. Rendic. Lincei 25(3), 309–330 (2014).
[30] Yu.L. Daletskiĭ: Stochastic differential geometry. Uspehi Matem. Nauk. 38(3), 87–111 (1983) (in Russian); English transl.: Russian Math. Surv. 38(3), 97–125 (1983).
[31] Yu.A. Davydov, M.A. Lifshits, N.V.Smorodina: Local properties of distributions of stochastic functionals. Translated from the Russian. Amer. Math. Soc., Providence, Rhode Island (1998) (Russian ed.: Moscow (1995)).
[32] E.I. Efimova, A.V. Uglanov: Green's formula on a Hilbert space. Mat. Sb. 119(2), 225–232 (1982) (in Russian); English transl.: Math. USSR Sb. 47, 215–222 (1984).
[33] C. Evans, R.F. Gariepy: Measure theory and fine properties of functions. CRC Press, Boca Raton – London, 1992.
[34] H. Federer: Geometric measure theory. Springer-Verlag, New York, 1969.
[35] D. Feyel: Hausdorff–Gauss measures. In: Stochastic Analysis and Related Topics, VII. Kusadasi 1998, pp. 59–76. Progr. in Probab. V. 98, Birkhäuser (2001).
[36] D. Feyel, A. de La Pradelle: Hausdorff measures on the Wiener space. Potential. Anal. 1(2), 177–189 (1992).
[37] M. Fukushima, M. Hino: On the space of BV functions and a related stochastic calculus in infinite dimensions. J. Funct. Anal. 183(1), 245–268 (2001).
[38] N. Gigli: Nonsmooth differential geometry – an approach tailored for spaces with Ricci curvature bounded from below. arXiv:1407.0809 (to appear in Memoirs of the AMS).
[39] V. Goodman: A divergence theorem for Hilbert space. Trans. Amer. Math. Soc. 164, 411–426 (1972).
[40] J. Heinonen: Nonsmooth calculus. Bull. Amer. Math. Soc. 44(2), 163–232 (2007).
[41] A. Hertle: Gaussian surface measures and the Radon transform on separable Banach spaces. Lecture Notes in Math. 794, 513–531 (1980).
[42] A. Hertle: Gaussian plane and spherical means in separable Hilbert spaces. Lecture Notes in Math. 945, 314–335 (1982).
[43] M. Hino: Sets of finite perimeter and the Hausdorff–Gauss measure on the Wiener space. J. Funct. Anal. 258(5), 1656–1681 (2010).

[44] M. Hino: Dirichlet spaces on H-convex sets in Wiener space. Bull. Sci. Math. 135, 667–683 (2011); Erratum: ibid. 137, 688–689 (2013).

[45] I. Holmes, A.N. Sengupta: A Gaussian Radon transform for Banach spaces. J. Funct. Anal. 263(11), 3689–3706 (2012).

[46] S. Keith: A differentiable structure for metric measure spaces. Adv. Math. 183(2), 271–315 (2004).

[47] H.-H. Kuo: Gaussian measures in Banach spaces. Springer-Verlag, Berlin – New York (1975).

[48] A. Lunardi, M. Miranda, D. Pallara: BV functions on convex domains in Wiener spaces. Potential Anal. 43 (2015), 23–48.

[49] P. Malliavin: Stochastic analysis. Springer-Verlag, Berlin (1997).

[50] V. Mihai, A.N. Sengupta: The Radon-Gauss transform. Soochow J. Math. 33(3), 415–433 (2007).

[51] M. Miranda, M. Novaga, D. Pallara: An introduction to BV functions in Wiener spaces. In: Variational methods for evolving objects (L. Ambrosio, Y. Giga, P. Rybka, Y. Tonegawa eds.), Advanced Studies in Pure Mathematics 67 (2015), 245–293.

[52] E.V. Nikitin: Fractional order Sobolev clsses on infinite-dimensional spaces. Dokl. Russian Acad. Sci. 452(2), 130–135 (2013) (in Russian); English tranls.: Dokl. Math. 88(2), 518–523 (2013).

[53] E.V. Nikitin: Besov classes on infinite-dimensional spaces. Matem. Zametki 93(6), 951–953 (2013) (in Russian); English tranls.: Math. Notes 93(6), 936–939 (2013).

[54] E.V. Nikitin: Comparison of two definitions of Besov classes on infinite–dimensional spaces. Matem. Zametki 95(1), 150–153 (2014) (in Russian); English tranls.: Math. Notes 95(1), 133–135 (2014).

[55] E. Pineda, W. Urbina: Some results on Gaussian Besov–Lipschitz spaces and Gaussian Triebel—Lizorkin spaces. J. Approx. Theory 161(2), 529–564 (2009).

[56] O.V. Pugachev: Surface measures in infinite-dimensional spaces. Matem. Zametki 63(1), 106–114 (1998) (in Russian); English transl.: Math. Notes 63(1), 94–101 (1998).

[57] O.V. Pugachev: The Gauss–Ostrogradskii formula in infinite-dimensional space. Matem. Sb. 189(5), 115–128 (1998) (in Russian); English transl.: Sb. Math. 189(5), 757–770 (1998).

[58] O.V. Pugachev: Construction of non-Gaussian surface measures by the Malliavin method. Matem. Zametki 65(3), 377–388 (1999) (in Russian); English transl.: Math. Notes 65(3-4), 315–325 (1999).

[59] O.V. Pugachev: Tightness of Sobolev capacities in infinite dimensional spaces. Infinite Dimensional Analysis, Quantum Probab. Relat. Topics 2(3), 427–440 (1999).

[60] O.V. Pugachev: Capacities and surface measures in locally convex spaces. Teor. Verojatn. i Primen. 53(1), 178–188 (2008) (in Russian); English transl.: Theory Probab. Appl. 53(1), 179–190 (2009).

[61] M. Röckner, B. Schmuland: Tightness of general $C_{1,p}$ capacities on Banach spaces. J. Funct. Anal. 108(1), 1–12 (1992).

[62] M. Röckner, R.-Ch. Zhu, X.-Ch. Zhu: The stochastic reflection problem on an infinite dimensional convex set and BV functions in a Gelfand triple. Ann. Probab. 40(4), 1759–1794 (2012).

[63] M. Röckner, R. Zhu, X. Zhu: BV functions in a Gelfand triple for differentiable measures and its applications. Forum Math. 27(3), 1657–1687 (2015).

[64] G. Savaré: Self-improvement of the Bakry–Émery condition and Wasserstein contraction of the heat flow in RCD(K, ∞) metric measure spaces. Discrete Contin. Dyn. Syst. 34(4), 1641–1661 (2014).

[65] A. Schioppa: On the relationship between derivations and measurable differentiable structures. Ann. Acad. Sci. Fenn. Math. 39(1), 275–304 (2014).

[66] A.V. Skorohod: Surface integrals and the Green formula in Hilbert space. Teor. Verojatnost. i Mat. Statist. N2, 172–175 (1970) (in Russian).

[67] A.V. Skorohod: Integration in Hilbert space. Transl. from the Russian. Springer-Verlag, Berlin – New York (1974).
[68] N.V. Smorodina: Differential calculus in configuration space and stable measures. I. Teor. Verojatn. i Primen. 33(3), 522–534 (1988) (in Russian); English transl.: Theory Probab. Appl. 33(3), 488–499 (1988).
[69] G. Stengle: A divergence theorem for Gaussian stochastic process expectations. J. Math. Anal. Appl. 21, 537–546 (1968).
[70] D. Stroock: The Malliavin calculus, a functional analytic approach. J. Funct. Anal. 44(2), 212–257 (1981).
[71] K.-T. Sturm: On the geometry of metric measure spaces. I, II. Acta Math. 196(1), 65–131, 133–177 (2006).
[72] J.F. Trèves: Linear partial differential equations with constant coefficients: existence, approximation and regularity of solutions. Gordon and Breach, New York, 1966.
[73] A.V. Uglanov: Surface integrals in a Banach space. Matem. Sb. 110(2), 189–217 (1979) (in Russian); English transl.: Math. USSR Sb. 38, 175–199 ().
[74] A.V. Uglanov: Surface integrals in Fréchet spaces. Matem. Sb. 189(11), 139–157 (1998) (in Russian); English transl.: Sb. Math. 189(11), 1719–1737 (1998).
[75] A.V. Uglanov: Integration on infinite-dimensional surfaces and its applications. Kluwer Acad. Publ., Dordrecht (2000).
[76] A.V. Uglanov: Calculus of variations on Banach spaces. Mat. Sb. 191(10), 105–118 (2000) (in Russian); English transl.: Sbornik Math. 191(10), 1527–1540 (2000).
[77] A.V. Uglanov: Surface integrals in locally convex spaces. Tr. Mosk. Mat. Ob-va. 62, 262–285 (2001) (in Russian); English transl.: Trans. Mosc. Math. Soc. 249–270 (2001).
[78] V.Yu. Yakhlakov: Surface measures on surfaces of finite codimension in a Banach space. Matem. Zametki 47(4), 147–156 (1990) (in Russian); English transl.: Math. Notes 47(3-4), 414–421 (1990).
[79] W. Ziemer: Weakly differentiable functions. Springer-Verlag, New York – Berlin (1989).

Fabio Cavalletti

An Overview of L^1 optimal transportation on metric measure spaces

3.1 Introduction

This note is a self-contained survey of the recent developments and achievements of the L^1-Optimal Transportation theory on metric measure spaces. We will focus on the general scheme adopted in the recent papers [20, 21] where the author, together with A. Mondino, proved a series of sharp (and in some cases even rigid and stable) geometric and functional inequalities in the setting of metric measure spaces enjoying a weak form of Ricci curvature lower bound. Roughly, the general scheme consists in reducing the initial problem to a family of easier one-dimensional problems; as it is probably the most relevant result obtained with this technique, we will review in detail how to proceed to obtain the Lévy-Gromov isoperimetric inequality for metric measure spaces verifying the Riemmanian Curvature Dimension condition (or, more generally, essentially non-branching metric measure spaces verifying the Curvature Dimension condition).

In [11, 18] a good analysis of the Monge problem in the metric setting treated, from a different perspective, similar questions whose answers were later also use in [20, 21]. We therefore believe the Monge problem and V.N. Sudakov's approach to it (see [53]) are a good starting point for our review, and to see how L^1-Optimal Transportation naturally yields a reduction of the problem to a family of one-dimensional problems.

It is worth stressing that the dimensional reduction proposed by V. N. Sudakov is only one possible strategy to solve the Monge problem. This problem has a long history and many authors contributed to obtain solutions in different frameworks with different approaches; here, we simply mention that the first existence result for the Monge problem was independently obtained in [15] and in [54]. We also mention the subsequent generalizations obtained in [1, 7, 28] and we refer to the monograph [55] for a more complete list of results.

3.1.1 Monge problem

The original problem posed by Monge in 1781 can be restated in modern language as follows: given two Borel probability measures μ_0 and μ_1 over $\mathbb{R}^d$, called marginal measures, find the optimal manner of transporting μ_0 to μ_1; the transportation of μ_0

Fabio Cavalletti: Università degli Studi di Pavia, Dipartimento di Matematica, E-mail: fabio.cavalletti@unipv.it

to μ_1 is understood as a map $T : \mathbb{R}^d \to \mathbb{R}^d$ assigning to each particle x a final position $T(x)$ fulfilling the following compatibility condition

$$T_\sharp \mu_0 = \mu_1, \qquad \text{i.e.} \quad \mu_0(T^{-1}(A)) = \mu_1(A), \quad \forall A \text{ Borel set}; \tag{3.1}$$

any map T verifying the previous condition will be called a transport map. The optimality requirement is stated as follows:

$$\int_{\mathbb{R}^d} |T(x) - x| \, \mu_0(dx) \le \int_{\mathbb{R}^d} |\hat{T}(x) - x| \, \mu_0(dx), \tag{3.2}$$

for any other $\hat{T}$ transport map. To prove the existence of a minimizer, the first difficulty arises while studying the minimization domain, that is the set of maps T verifying (3.1). Suppose $\mu_0 = f_0 \mathcal{L}^d$ and $\mu_1 = f_1 \mathcal{L}^d$ where $\mathcal{L}^d$ denotes the d-dimensional Lebesgue measure; a smooth injective map T is then a transport map if and only if

$$f_1(T(x)) |\det(DT)(x)| = f_0(x), \qquad \mu_0\text{-a.e. } x \in \mathbb{R}^d,$$

showing a strong non-linearity of the constrain. The first big leap in optimal transportation theory was achieved by Kantorovich considering a suitable relaxation of the problem: to each transport map associate the probability measure $(Id, T)_\sharp \mu_0$ over $\mathbb{R}^d \times \mathbb{R}^d$ and introduce the set of *transport plans*

$$\Pi(\mu_0, \mu_1) := \left\{ \pi \in \mathcal{P}(\mathbb{R}^d \times \mathbb{R}^d) : P_{1\sharp}\pi = \mu_0, \; P_{2\sharp}\pi = \mu_1 \right\};$$

where $P_i : \mathbb{R}^d \times \mathbb{R}^d \to \mathbb{R}^d$ is the projection on the i-th component, with $i = 1, 2$. By definition $(Id, T)_\sharp \mu_0 \in \Pi(\mu_0, \mu_1)$ and

$$\int_{\mathbb{R}^d} |T(x) - x| \, \mu_0(dx) = \int_{\mathbb{R}^d \times \mathbb{R}^d} |x - y| \, \left((Id, T)_\sharp \mu_0\right)(dxdy);$$

then it is natural to consider the minimization of the following functional (called Monge-Kantorovich minimization problem)

$$\Pi(\mu_0, \mu_1) \ni \pi \longmapsto \mathcal{J}(\pi) := \int_{\mathbb{R}^d \times \mathbb{R}^d} |x - y| \, \pi(dxdy). \tag{3.3}$$

$\Pi(\mu_0, \mu_1)$ is a convex subset of $\mathcal{P}(\mathbb{R}^d \times \mathbb{R}^d)$ and it is compact with respect to the weak topology - this is the big advantage of the new approach. Since the functional $\mathcal{J}$ is linear, the existence of a minimizer follows straightforwardly. Then a strategy to obtain a solution of the original Monge problem is to start from an optimal transport plan π and prove that it is indeed concentrated on the graph of a Borel map T; the latter is equivalent to $\pi = (Id, T)_\sharp \mu_0$.

To run this program one needs to deduce from optimality some condition on the geometry of the support of the transport plan. This was again obtained by Kantorovich: he introduced a dual formulation of (3.3) and found that for any probability measures μ_0 and μ_1 with finite first moment, there exists a 1-Lipschitz function $\phi : \mathbb{R}^d \to \mathbb{R}$ such that

$$\Pi(\mu_0, \mu_1) \ni \pi \text{ is optimal} \quad \Longleftrightarrow \quad \pi\big(\{(x, y) \in \mathbb{R}^{2d} : \phi(x) - \phi(y) = |x - y|\}\big) = 1.$$

At this point one needs to focus on the structure of the set

$$\Gamma := \big\{(x, y) \in \mathbb{R}^{2d} : \phi(x) - \phi(y) = |x - y|\big\}. \tag{3.4}$$

Definition 3.1.1. *A set $\Lambda \subset \mathbb{R}^{2d}$ is $|\cdot|$-cyclically monotone if and only if for any finite subset of Λ, $\{(x_1, y_1), \dots, (x_N, y_N)\} \subset \Lambda$ it holds*

$$\sum_{1 \le i \le N} |x_i - y_i| \le \sum_{1 \le i \le N} |x_i - y_{i+1}|,$$

where $y_{N+1} := y_1$.

Almost by definition, the set Γ is $|\cdot|$-cyclically monotone and whenever $(x, y) \in \Gamma$ considering $z_t := (1 - t)x + ty$ with $t \in [0, 1]$ it holds that $(z_s, z_t) \in \Gamma$, for any $s \le t$. In particular this suggests that Γ produces a family of disjoint lines of $\mathbb{R}^d$ along where the optimal transportation should move. This can be made rigorous considering the following "relation" between points: a point x is in relation with y if, using optimal geodesics selected by the above optimal transport problem, one can travel from x to y or viceversa. That is, consider $R := \Gamma \cup \Gamma^{-1}$ and define $x \sim y$ if and only if $(x, y) \in R$. Then $\mathbb{R}^d$ will be decomposed (up to a set of Lebesgue-measure zero) as $\mathcal{T} \cup Z$ where $\mathcal{T}$ will be called the *transport set* and Z the set of points not moved by the optimal transportation problem. The important property of $\mathcal{T}$ being that

$$\mathcal{T} = \bigcup_{q \in Q} X_q, \qquad X_q \text{ straight line}, \qquad X_q \cap X_{q'} = \emptyset, \quad \text{if } q \neq q'.$$

Here Q is a set of indices; a convenient way to index a straight line X_q is to select an element of X_q and call it, with an abuse of notation, q. With this choice the set Q can be understood as a subset of $\mathbb{R}^d$. Once a partition of the space is given, one obtains via the Disintegration Theorem a corresponding decomposition of marginal measures:

$$\mu_0 = \int_Q \mu_{0\,q}\, \mathfrak{q}(dq), \qquad \mu_1 = \int_Q \mu_{1\,q}\, \mathfrak{q}(dq);$$

where $\mathfrak{q}$ is a Borel probability measure over the set of indices $Q \subset \mathbb{R}^d$. If Q enjoys a measurability condition (see Theorem 3.2.8 for details), the conditional measures $\mu_{0\,q}$ and $\mu_{1\,q}$ are concentrated on the straight line with index q, i.e. $\mu_{0\,q}(X_q) = \mu_{1\,q}(X_q) = 1$, for $\mathfrak{q}$-a.e. $q \in Q$.

Then a classic way to construct an optimal transport maps is to

- consider T_q the monotone rearrangement along X_q of $\mu_{0\,q}$ to $\mu_{1\,q}$;
- define the transport map T as T_q on each X_q.

The map T will then be an optimal transport map moving μ_0 to μ_1; it is indeed easy to check that $(Id, T)_\sharp \mu_0 \in \Pi(\mu_0, \mu_1)$ and $(x, T(x)) \in \Gamma$ for μ_0-a.e. x.

The original Monge problem has been reduced to the following family of one-dimensional problems: for each $q \in Q$ find a minimizer of the following functional

$$\Pi(\mu_{0\,q}, \mu_{1\,q}) \ni \pi \longmapsto \mathfrak{I}(\pi) := \int_{X_q \times X_q} |x - y|\, \pi(dxdy),$$

that is concentrated on the graph of a Borel function. As X_q is isometric to the real line, whenever $\mu_{0\,q}$ does not contain any atom (i.e $\mu_{0\,q}(x) = 0$, for all $x \in X_q$) the monotone rearrangement T_q exists, and the existence of an optimal transport map T constructed as before follows. The existence of a solution has been reduced, therefore, to a regularity property of the disintegration of μ_0.

As already stressed before, this approach to the Monge problem, mainly due to V.N. Sudakov, was proposed in [53] and was later completed in the subsequent papers [15] and [54]. See also [23] for a complete Sudakov approach to the Monge problem when the Euclidean distance is replaced by any strictly convex norm, and [12] where any norm is considered. In all these papers, assuming μ_0 to be absolutely continuous with respect to $\mathcal{L}^d$ enables sufficient regularity to solve the problem.

The Monge problem can be actually stated, and solved, in a much more general framework. Given two Borel probability measures μ_0 and μ_1 over a complete and separable metric space (X, d), the notion of transportation map makes perfectly sense. Furthermore, the optimality condition (3.2) can be naturally formulated using the distance d as a cost function instead of the Euclidean norm:

$$\int_{\mathbb{R}^d} \mathsf{d}(T(x), x)\, \mu_0(dx) \leq \int_{\mathbb{R}^d} \mathsf{d}(\hat{T}(x), x)\, \mu_0(dx). \tag{3.5}$$

The problem can be relaxed to obtain a transport plan π solution of the corresponding Monge-Kantorovich minimization problem. Also, the Kantorovich duality applies yielding the existence of a 1-Lipschitz function $\phi : X \to \mathbb{R}$ such that

$$\Pi(\mu_0, \mu_1) \ni \pi \text{ is optimal} \quad \Longleftrightarrow \quad \pi(\Gamma) = 1,$$

where $\Gamma := \{(x, y) \in X \times X\colon \phi(x) - \phi(y) = \mathsf{d}(x, y)\}$ is d-cyclically monotone.
The strategy proposed for the Euclidean problem can be adopted: decompose X as $\mathfrak{T} \cup Z$; Z is the set of points not moved by the optimal transportation problem and $\mathfrak{T}$ is the transport set and it is partitioned, up to a set of measure zero, by a family of geodesics $\{X_q\}_{q \in Q}$. Using the Disintegration Theorem one obtains like before a reduction of the Monge problem to a family of one-dimensional problems

$$\Pi(\mu_{0\,q}, \mu_{1\,q}) \ni \pi \longmapsto \mathfrak{I}(\pi) := \int_{X_q \times X_q} \mathsf{d}(x, y)\, \pi(dxdy).$$

Therefore, since X_q with distance d is isometric to an interval of the real line with Euclidean distance, the problem is reduced to proving that for q-a.e. $q \in Q$ the conditional measure $\mu_{0\,q}$ does not have any atoms.

Clearly in showing such a result, besides the regularity of μ_0 itself, the regularity of the ambient space X plays a crucial role. In particular, together with the localization of the Monge problem to X_q there should come a localization of the regularity of the space. This is the case when the metric space (X, d) is endowed with a reference probability measure $\mathfrak{m}$ and the resulting metric measure space $(X, \mathsf{d}, \mathfrak{m})$ verifies a weak Ricci curvature lower bound.

In [11] we observed that if $(X, \mathsf{d}, \mathfrak{m})$ verifies the so-called measure contraction property MCP, then for q-a.e. $q \in Q$ the one-dimensional metric measure space $(X_q, \mathsf{d}, \mathfrak{m}_q)$ verifies MCP as well, where $\mathfrak{m}_q$ is the conditional measure of $\mathfrak{m}$ with respect to the family of geodesics $\{X_q\}_{q\in Q}$. Now the assumption $\mu_0 \ll \mathfrak{m}$ is sufficient to solve the Monge problem. It is worth mentioning that [11] was the first contribution where regularity of conditional measures were obtained in a purely non-smooth framework. The techniques introduced in [11] also allowed us to threat such regularity issues in the infinite dimensional setting of Wiener space; see [16].

This short introduction should suggest that L^1-Optimal Transportation allows for an efficient dimensional reduction together with a localization of the "smoothness" of the space for very general metric measure spaces. We now make a short introduction to the Lévy-Gromov isoperimetric inequality.

3.1.2 Lévy-Gromov isoperimetric inequality

The Lévy-Gromov isoperimetric inequality [35, Appendix C] can be stated as follows: if E is a (sufficiently regular) subset of a Riemannian manifold (M^N, g) with dimension N and Ricci bounded below by $K > 0$, then

$$\frac{|\partial E|}{|M|} \geq \frac{|\partial B|}{|S|}. \tag{3.6}$$

Here, B is a spherical cap in the model sphere S, i.e. the N-dimensional sphere with constant Ricci curvature equal to K, and $|M|$, $|S|$, $|\partial E|$, $|\partial B|$ denote the appropriate N or $N-1$ dimensional volume, and where B is chosen so that $|E|/|M| = |B|/|S|$. As $K > 0$ both M and S are compact and their volume is finite; hence the previous equality and (3.6) make sense. In other words, the Lévy-Gromov isoperimetric inequality states that isoperimetry in (M, g) is at least as strong as in the model space S.

A general introduction on the isoperimetric problem goes beyond the scope of this note; a complete description of isoperimetric inequality in spaces admitting singularities is quite a hard task covered mostly in [42, 44, 45]. See also [25, Appendix H] for more details. The approaches taken are also manifold: for a geometric measure theory approach see [43]; for the point of view of optimal transport see [29, 56]; for the connections with convex and integral geometry see [14]; for the recent quantitative forms see [24, 31] and finally for an overview of the more geometric aspects see [46, 48, 49].

The Lévy-Gromov isoperimetric inequality is also natural in the broader class of metric measure spaces (m.m.s.), i.e. triples $(X, \mathsf{d}, \mathfrak{m})$ where (X, d) is complete and separable and $\mathfrak{m}$ is a Radon measure over X. Indeed the volume of a Borel set is replaced by its $\mathfrak{m}$-measure, $\mathfrak{m}(E)$; the boundary area of the smooth framework can be replaced instead by the Minkowski content:

$$\mathfrak{m}^+(E) := \liminf_{\varepsilon \downarrow 0} \frac{\mathfrak{m}(E^\varepsilon) - \mathfrak{m}(E)}{\varepsilon}, \tag{3.7}$$

where $E^\varepsilon := \{x \in X : \exists y \in E \text{ such that } \mathsf{d}(x, y) < \varepsilon\}$ is the ε-neighborhood of E with respect to the metric d; the natural analogue of "dimension N and Ricci bounded below by $K > 0$" is encoded in the so-called Riemannian Curvature Dimension condition, $\mathsf{RCD}^*(K, N)$ for short. As normalization factors appear in (3.6), it is also more convenient to directly consider the case $\mathfrak{m}(X) = 1$.

The Lévy-Gromov isoperimetric problem for a m.m.s. $(X, \mathsf{d}, \mathfrak{m})$ with $\mathfrak{m}(X) = 1$ can be formulated as follows:

Find the largest function $\mathcal{I}_{K,N} : [0, 1] \to \mathbb{R}^+$ *such that for every Borel subset* $E \subset X$ *it holds*

$$\mathfrak{m}^+(E) \geq \mathcal{I}_{K,N}(\mathfrak{m}(E)),$$

with $\mathcal{I}_{K,N}$ *depending on* $N, K \in \mathbb{R}$ *with* $K > 0$ *and* $N > 1$.

Then in [20] (Theorem 1.2) the author and A. Mondino proved the non-smooth Lévy-Gromov isoperimetric inequality (3.6).

Theorem 3.1.2. (Lévy-Gromov in $\mathsf{RCD}^*(K, N)$-spaces, Theorem 1.2 of [20]) *Let* $(X, \mathsf{d}, \mathfrak{m})$ *be an* $\mathsf{RCD}^*(K, N)$ *space for some* $N \in \mathbb{N}$ *and* $K > 0$ *with* $\mathfrak{m}(X) = 1$. *Then for every Borel subset* $E \subset X$ *it holds*

$$\mathfrak{m}^+(E) \geq \frac{|\partial B|}{|S|},$$

where B *is a spherical cap in the model sphere* S *(the* N*-dimensional sphere with constant Ricci curvature equal to* K*) chosen so that* $|B|/|S| = \mathfrak{m}(E)$.

We refer to Theorem 1.2 of [20] (or Theorem 6.6) for the more general statement.

The link between Theorem 3.1.2 and the first part of the Introduction, where the Monge problem was discussed, stands in the techniques used to prove Theorem 3.1.2.

The main obstacle to Lévy-Gromov type inequalities in the non-smooth metric measure spaces setting is that the previously known proofs rely on regularity properties of isoperimetric regions and on powerful results of geometric measure theory (see for instance [35, 43]) that are not at our disposal in the framework of metric measure spaces. The recent paper of B. Klartag [38] allowed for a proof of the Lévy-Gromov isoperimetric inequality, still in the framework of smooth Riemannian manifolds, avoiding regularity of optimal shapes and using instead an optimal transportation argument involving L^1-Optimal Transportation and ideas of convex geometry. This approach goes back to Payne-Weinberger [47] and was later developed by Gromov-Milman [36], Lovász-Simonovits [40] and Kannan-Lovász-Simonovits [37]; it consists in reducing a multi-dimensional problem to easier one-dimensional problems. B. Klartag's observed that a suitable L^1-Optimal Transportation problem produces what he calls a *needle decomposition* (in our terminology, disintegration) that localize (or reduce) the proof of the isoperimetric inequality to the proof of a family of one-dimensional isoperimetric inequalities; also the regularity of the space is localized.

The approach of [38] does not rely on the regularity of the isoperimetric region, nevertheless it still heavily makes use of the smoothness of the ambient space to obtain the localization. In particular, it makes use of sharp properties of the geodesics in terms of Jacobi fields of estimates on the second fundamental forms of suitable level sets; all these are objects still not understood enough in general metric measure spaces in order to repeat the same arguments.

Hence to apply the localization technique to the Lévy-Gromov isoperimetric inequality in singular spaces, structural properties of geodesics and of L^1-optimal transportation have to be understood also in the general framework of metric measure spaces. Such a program started in the previous work of the author with S. Bianchini [11] and of the author [17, 18]. Finally with A. Mondino in [20] we obtained the general result underpinning the Lévy-Gromov isoperimetric inequality.

3.1.3 Outline

The chapter goes as follows: Section 3.2 contains all the basic material on Optimal Transportation and the theory of Lott-Sturm-Villani spaces, that is metric measure spaces verifying the Curvature Dimension condition, $\mathsf{CD}(K, N)$ for short. It also covers some basics on isoperimetric inequality, Disintegration Theorem and selection theorems we will use throughout the chapter. In Section 3.3 we prove all the structure results on the building block of L^1-Optimal Transportation, the d-cyclically monotone sets. Here no curvature assumption enters. In Section 3.4 we show that the aforementioned sets induce a partition of almost all transport, provided the space enjoys a stronger form of the essentially non-branching condition; we also show that each element of the partition is a geodesic (and therefore a one-dimensional set). Section 3.5 contains all the regularity results of conditional measures of the disintegration induced by the L^1-Optimal Transportation problem. In particular we will present three assumptions, each one implying the previous one, yielding three increasing levels of regularity of the conditional measures. Finally in Section 3.6 we collect the consequences of the regularity results of Section 3.5; in particular we first show the existence of a solution of the Monge problem under a very general regularity assumption

(Theorem 3.6.2) and finally we go back to the Lévy-Gromov isoperimetric inequality (Theorem 3.6.6).

3.2 Preliminaries

In what follows we say that a triple $(X, \mathsf{d}, \mathfrak{m})$ is a metric measure space, m.m.s. for short, if (X, d) is a complete and separable metric space and $\mathfrak{m}$ is a positive Radon measure over X. In this paper we will only be concerned with m.m.s. with $\mathfrak{m}$ a probability measure, that is $\mathfrak{m}(X) = 1$. The space of all Borel probability measures over X will be denoted by $\mathcal{P}(X)$.

A metric space is a geodesic space if and only if for each $x, y \in X$ there exists $\gamma \in \mathrm{Geo}(X)$ so that $\gamma_0 = x, \gamma_1 = y$, with

$$\mathrm{Geo}(X) := \{\gamma \in C([0,1], X) : \mathsf{d}(\gamma_s, \gamma_t) = |s-t|\mathsf{d}(\gamma_0, \gamma_1), \text{ for every } s, t \in [0,1]\}.$$

It follows from the metric version of the Hopf-Rinow Theorem (see Theorem 2.5.28 of [13]) that for complete geodesic spaces local completeness is equivalent to properness (a metric space is proper if every closed ball is compact).

We assume the ambient space (X, d) to be proper and geodesic, hence also complete and separable. Moreover we assume $\mathfrak{m}$ to be a proability measure, i.e. $\mathfrak{m}(X) = 1$.

We denote by $\mathcal{P}_2(X)$ the space of probability measures with finite second moment endowed with the L^2-Wasserstein distance W_2 defined as follows: for $\mu_0, \mu_1 \in \mathcal{P}_2(X)$ we set

$$W_2^2(\mu_0, \mu_1) = \inf_{\pi} \int_{X \times X} \mathsf{d}^2(x, y)\, \pi(dxdy), \tag{3.8}$$

where the infimum is taken over all $\pi \in \mathcal{P}(X \times X)$ with μ_0 and μ_1 as the first and the second marginal, called the set of transference plans. The set of transference plans realizing the minimum in (3.8) will be called the set of optimal transference plans. Assuming the space (X, d) to be geodesic, also the space $(\mathcal{P}_2(X), W_2)$ is geodesic.

Any geodesic $(\mu_t)_{t \in [0,1]}$ in $(\mathcal{P}_2(X), W_2)$ can be lifted to a measure $\nu \in \mathcal{P}(\mathrm{Geo}(X))$, so that $(\mathrm{e}_t)_\sharp\, \nu = \mu_t$ for all $t \in [0,1]$. Here for any $t \in [0,1]$, e_t denotes the evaluation map:

$$\mathrm{e}_t : \mathrm{Geo}(X) \to X, \qquad \mathrm{e}_t(\gamma) := \gamma_t.$$

Given $\mu_0, \mu_1 \in \mathcal{P}_2(X)$, we denote by $\mathrm{OptGeo}(\mu_0, \mu_1)$ the space of all $\nu \in \mathcal{P}(\mathrm{Geo}(X))$ for which $(\mathrm{e}_0, \mathrm{e}_1)_\sharp\, \nu$ minimizes in (3.8). If (X, d) is geodesic then the set $\mathrm{OptGeo}(\mu_0, \mu_1)$ is non-empty for any $\mu_0, \mu_1 \in \mathcal{P}_2(X)$. It is worth also introducing the subspace of $\mathcal{P}_2(X)$ formed by all those measures absolutely continuous with respect with $\mathfrak{m}$: it is denoted by $\mathcal{P}_2(X, \mathsf{d}, \mathfrak{m})$.

3.2.1 Geometry of metric measure spaces

Here we briefly recall the synthetic notions of lower Ricci curvature bounds; for more detail we refer to [9, 39, 51, 52, 56].

In order to formulate the curvature properties for $(X, \mathsf{d}, \mathfrak{m})$ we introduce the following distortion coefficients: given two numbers $K, N \in \mathbb{R}$ with $N \geq 0$, we set for $(t, \theta) \in [0, 1] \times \mathbb{R}_+$,

$$\sigma_{K,N}^{(t)}(\theta) := \begin{cases} \infty, & \text{if } K\theta^2 \geq N\pi^2, \\ \dfrac{\sin(t\theta\sqrt{K/N})}{\sin(\theta\sqrt{K/N})} & \text{if } 0 < K\theta^2 < N\pi^2, \\ t & \text{if } K\theta^2 < 0 \text{ and } N = 0, \text{ or if } K\theta^2 = 0, \\ \dfrac{\sinh(t\theta\sqrt{-K/N})}{\sinh(\theta\sqrt{-K/N})} & \text{if } K\theta^2 \leq 0 \text{ and } N > 0. \end{cases} \tag{3.9}$$

We also set, for $N \geq 1$, $K \in \mathbb{R}$ and $(t, \theta) \in [0, 1] \times \mathbb{R}_+$

$$\tau_{K,N}^{(t)}(\theta) := t^{1/N} \sigma_{K,N-1}^{(t)}(\theta)^{(N-1)/N}. \tag{3.10}$$

As we will consider only the case of essentially non-branching spaces, we recall the following definition.

Definition 3.2.1. *A metric measure space* $(X, \mathsf{d}, \mathfrak{m})$ is essentially non-branching *if and only if for any* $\mu_0, \mu_1 \in \mathcal{P}_2(X)$, μ_0, μ_1 *both absolutely continuous with respect to* $\mathfrak{m}$, *any element of* $\mathrm{OptGeo}(\mu_0, \mu_1)$ *is concentrated on a set of non-branching geodesics.*

A set $F \subset \mathrm{Geo}(X)$ is a set of non-branching geodesics if and only if for any $\gamma^1, \gamma^2 \in F$, it holds:

$$\exists\, \bar{t} \in (0, 1) \text{ such that } \forall t \in [0, \bar{t}] \quad \gamma_t^1 = \gamma_t^2 \quad \Rightarrow \quad \gamma_s^1 = \gamma_s^2, \quad \forall s \in [0, 1].$$

Definition 3.2.2 (CD condition). *An essentially non-branching m.m.s.* $(X, \mathsf{d}, \mathfrak{m})$ *verifies* $\mathsf{CD}(K, N)$ *if and only if for each pair* $\mu_0, \mu_1 \in \mathcal{P}_2(X, \mathsf{d}, \mathfrak{m})$ *there exists* $\nu \in \mathrm{OptGeo}(\mu_0, \mu_1)$ *such that*

$$\varrho_t^{-1/N}(\gamma_t) \geq \tau_{K,N}^{(1-t)}(\mathsf{d}(\gamma_0, \gamma_1))\varrho_0^{-1/N}(\gamma_0) + \tau_{K,N}^{(t)}(\mathsf{d}(\gamma_0, \gamma_1))\varrho_1^{-1/N}(\gamma_1), \qquad \nu\text{-a.e. } \gamma \in \mathrm{Geo}(X), \tag{3.11}$$

for all $t \in [0, 1]$, *where* $(\mathrm{e}_t)_\sharp\, \nu = \varrho_t \mathfrak{m}$.

For the general definition of $\mathsf{CD}(K, N)$ see [39, 51, 52].

Remark 3.2.3. *It is worth recalling that if* (M, g) *is a Riemannian manifold of dimension* n *and* $h \in C^2(M)$ *with* $h > 0$, *then the m.m.s.* $(M, g, h\, vol)$ *verifies* $\mathsf{CD}(K, N)$ *with* $N \geq n$ *if and only if (see Theorem 1.7 of [52])*

$$Ric_{g,h,N} \geq Kg, \qquad Ric_{g,h,N} := Ric_g - (N - n)\frac{\nabla_g^2 h^{\frac{1}{N-n}}}{h^{\frac{1}{N-n}}}.$$

In particular if $N = n$ *the generalized Ricci tensor* $Ric_{g,h,N} = Ric_g$ *makes sense only if* h *is constant.*

Another important case is when $I \subset \mathbb{R}$ *is any interval,* $h \in C^2(I)$ *and* $\mathcal{L}^1$ *is the one-dimensional Lebesgue measure; then the m.m.s.* $(I, |\cdot|, h\mathcal{L}^1)$ *verifies* $\mathsf{CD}(K, N)$ *if and only if*

$$\left(h^{\frac{1}{N-1}}\right)'' + \frac{K}{N-1} h^{\frac{1}{N-1}} \le 0, \tag{3.12}$$

and verifies $\mathsf{CD}(K, 1)$ *if and only if* h *is constant. Inequality* (3.12) *has also a non-smooth counterpart; if we drop the smoothness assumption on* h *it can be proven that the m.m.s.* $(I, |\cdot|, h\mathcal{L}^1)$ *verifies* $\mathsf{CD}(K, N)$ *if and only if*

$$h((1-s)t_0 + st_1)^{1/(N-1)} \ge \sigma^{(1-s)}_{K,N-1}(t_1 - t_0)h(t_0)^{1/(N-1)} + \sigma^{(s)}_{K,N-1}(t_1 - t_0)h(t_1)^{1/(N-1)}. \tag{3.13}$$

This is the formulation in the sense of distributions of the differential inequality (3.12). *Recall indeed that* $s \mapsto \sigma^{(s)}_{K,N-1}(\theta)$ *solves* $f'' + (t_1 - t_0)^2 \frac{K}{N-1} f = 0$ *in the classical sense.*

We also mention the more modern Riemannian curvature dimension condition $\mathrm{RCD}^*(K, N)$. In the infinite dimensional case, i.e. $N = \infty$, it was introduced in [5]. The class $\mathrm{RCD}^*(K, N)$ with $N < \infty$ has been proposed in [33] and deeply investigated in [3, 26] and [8]. We refer to these papers and references therein for a general account on the synthetic formulation of Ricci curvature lower bounds for metric measure spaces.

Here we only mention that the $\mathrm{RCD}^*(K, N)$ condition is an enforcement of the so-called reduced curvature dimension condition, denoted by $\mathsf{CD}^*(K, N)$, that has been introduced in [9]: in particular the additional condition is for the Sobolev space $W^{1,2}(X, \mathfrak{m})$ to be a Hilbert space, see [4, 5, 33].

The reduced $\mathsf{CD}^*(K, N)$ condition asks for the same inequality (3.11) of $\mathsf{CD}(K, N)$ but the coefficients $\tau^{(t)}_{K,N}(\mathsf{d}(\gamma_0, \gamma_1))$ and $\tau^{(1-t)}_{K,N}(\mathsf{d}(\gamma_0, \gamma_1))$ are replaced by $\sigma^{(t)}_{K,N}(\mathsf{d}(\gamma_0, \gamma_1))$ and $\sigma^{(1-t)}_{K,N}(\mathsf{d}(\gamma_0, \gamma_1))$, respectively.

Hence while the distortion coefficients of the $\mathsf{CD}(K, N)$ condition are formally obtained imposing one direction with linear distortion and $N - 1$ directions affected by curvature, the $\mathsf{CD}^*(K, N)$ condition imposes the same volume distortion in all the N directions.

For both definitions there is a local version that is of some relevance to our analysis. Here we state only the local formulation $\mathsf{CD}(K, N)$, being clear what would be the one for $\mathsf{CD}^*(K, N)$.

Definition 3.2.4 (CD_{loc} condition). *An essentially non-branching m.m.s.* $(X, \mathsf{d}, \mathfrak{m})$ *satisfies* $\mathsf{CD}_{loc}(K, N)$ *if for any point* $x \in X$ *there exists a neighborhood* $X(x)$ *of* x *such that for each pair* $\mu_0, \mu_1 \in \mathcal{P}_2(X, \mathsf{d}, \mathfrak{m})$ *supported in* $X(x)$ *there exists* $\nu \in \mathrm{OptGeo}(\mu_0, \mu_1)$ *such that* (3.11) *holds true for all* $t \in [0, 1]$. *The support of* $(\mathrm{e}_t)_\sharp\, \nu$ *is not necessarily contained in the neighborhood* $X(x)$.

One of the main properties of the reduced curvature dimension condition is the globalization one: under the essentially non-branching property, $\mathsf{CD}^*_{loc}(K, N)$ and $\mathsf{CD}^*(K, N)$

are equivalent (see [9, Corollary 5.4]), i.e. the CD^*-condition verifies the local-to-global property.

We also recall a few relations between CD and CD^*. It is known by [32, Theorem 2.7] that, if $(X, \mathsf{d}, \mathfrak{m})$ is a non-branching metric measure space verifying $\mathsf{CD}(K, N)$ and $\mu_0, \mu_1 \in \mathcal{P}(X)$ with μ_0 absolutely continuous with respect to $\mathfrak{m}$, then there exists a unique optimal map $T : X \to X$ such that $(id, T)_\sharp \mu_0$ realizes the minimum in (3.8) and the set $\mathrm{OptGeo}(\mu_0, \mu_1)$ contains only one element. The same proof holds if one replaces the non-branching assumption with the more general one of essentially non-branching, see for instance [34].

3.2.2 Isoperimetric profile function

Given a m.m.s. $(X, \mathsf{d}, \mathfrak{m})$ as above and a Borel subset $A \subset X$, let A^ε denote the ε-tubular neighborhood

$$A^\varepsilon := \{x \in X : \exists y \in A \text{ such that } \mathsf{d}(x, y) < \varepsilon\}.$$

The Minkowski (exterior) boundary measure $\mathfrak{m}^+(A)$ is defined by

$$\mathfrak{m}^+(A) := \liminf_{\varepsilon \downarrow 0} \frac{\mathfrak{m}(A^\varepsilon) - \mathfrak{m}(A)}{\varepsilon}. \tag{3.14}$$

The *isoperimetric profile*, denoted by $\mathcal{I}_{(X,\mathsf{d},\mathfrak{m})}$, is defined as the pointwise maximal function so that $\mathfrak{m}^+(A) \geq \mathcal{I}_{(X,\mathsf{d},\mathfrak{m})}(\mathfrak{m}(A))$ for every Borel set $A \subset X$; that is

$$\mathcal{I}_{(X,\mathsf{d},\mathfrak{m})}(v) := \inf\left\{\mathfrak{m}^+(A) : A \subset X \text{ Borel}, \ \mathfrak{m}(A) = v\right\}. \tag{3.15}$$

If $K > 0$ and $N \in \mathbb{N}$, by the Lévy-Gromov isoperimetric inequality (3.6) we know that, for N-dimensional smooth manifolds having Ricci $\geq K$, the isoperimetric profile function is bounded below by the one N-dimensional round sphere of suitable radius. In other words, the *model* isoperimetric profile function is that of $\mathbb{S}^N$. For $N \geq 1, K \in \mathbb{R}$ arbitrary real numbers the situation is more complicated, and just recently E. Milman [41] discovered what is the model isoperimetric profile. We refer to [41] for all the details. Here we just recall the relevance of isoperimetric profile functions for m.m.s. over $(\mathbb{R}, |\cdot|)$: given $K \in \mathbb{R}$, $N \in [1, +\infty)$ and $D \in (0, +\infty]$, consider the function

$$\mathcal{I}_{K,N,D}(v) := \inf\left\{\mu^+(A) : A \subset \mathbb{R}, \ \mu(A) = v, \ \mu \in \mathcal{F}_{K,N,D}\right\}, \tag{3.16}$$

where $\mathcal{F}_{K,N,D}$ denotes the set of $\mu \in \mathcal{P}(\mathbb{R})$ such that $\mathrm{supp}(\mu) \subset [0, D]$ and $\mu = h \cdot \mathcal{L}^1$ with $h \in C^2((0, D))$ satisfying

$$\left(h^{\frac{1}{N-1}}\right)'' + \frac{K}{N-1} h^{\frac{1}{N-1}} \leq 0 \quad \text{if } N \in (1, \infty), \qquad h \equiv \text{const} \quad \text{if } N = 1. \tag{3.17}$$

Then from [41, Theorem 1.2, Corollary 3.2] it follows that for N-dimensional smooth manifolds having Ricci $\geq K$, with $K \in \mathbb{R}$ arbitrary real number, and diameter D, the isoperimetric profile function is bounded below by $\mathcal{I}_{K,N,D}$ and the bound is sharp. This also justifies the notation.

Going back to non-smooth metric measure spaces (what follows is taken from [20]), it is necessary to consider the following broader family of measures:

$$\mathcal{F}^s_{K,N,D} := \{\mu \in \mathcal{P}(\mathbb{R}) : \operatorname{supp}(\mu) \subset [0, D],\ \mu = h_\mu \mathcal{L}^1,$$
$$h_\mu \text{ verifies (3.13) and is continuous if } N \in (1,\infty), \quad h_\mu \equiv \text{const if } N = 1\},$$

and the corresponding comparison *synthetic* isoperimetric profile:

$$\mathcal{I}^s_{K,N,D}(v) := \inf\left\{\mu^+(A) : A \subset \mathbb{R},\ \mu(A) = v,\ \mu \in \mathcal{F}^s_{K,N,D}\right\},$$

where $\mu^+(A)$ denotes the Minkowski content defined in (3.14). The term synthetic refers to $\mu \in \mathcal{F}^s_{K,N,D}$ meaning that the Ricci curvature bound is satisfied in its synthetic formulation: if $\mu = h \cdot \mathcal{L}^1$, then h verifies (3.13).

We have already seen that $\mathcal{F}_{K,N,D} \subset \mathcal{F}^s_{K,N,D}$; actually one can prove that $\mathcal{I}^s_{K,N,D}$ coincides with its smooth counterpart $\mathcal{I}_{K,N,D}$ for every volume $v \in [0,1]$ via a smoothing argument. We therefore need the following approximation result. In order to state it let us recall that a standard mollifier in $\mathbb{R}$ is a non negative $C^\infty(\mathbb{R})$ function ψ with compact support in $[0,1]$ such that $\int_{\mathbb{R}} \psi = 1$.

Lemma 3.2.5 (Lemma 6.2, [20]). *Let $D \in (0,\infty)$ and let $h : [0,D] \to [0,\infty)$ be a continuous function. Fix $N \in (1,\infty)$ and for $\varepsilon > 0$ define*

$$h_\varepsilon(t) := [h^{\frac{1}{N-1}} * \psi_\varepsilon(t)]^{N-1} := \left[\int_{\mathbb{R}} h(t-s)^{\frac{1}{N-1}} \psi_\varepsilon(s)\, ds\right]^{N-1} = \left[\int_{\mathbb{R}} h(s)^{\frac{1}{N-1}} \psi_\varepsilon(t-s)\, ds\right]^{N-1}, \tag{3.18}$$

where $\psi_\varepsilon(x) = \frac{1}{\varepsilon}\psi(x/\varepsilon)$ and ψ is a standard mollifier function. The following properties hold:

1. *h_ε is a non-negative C^∞ function with support in $[-\varepsilon, D+\varepsilon]$;*
2. *$h_\varepsilon \to h$ uniformly as $\varepsilon \downarrow 0$, in particular $h_\varepsilon \to h$ in L^1.*
3. *If h satisfies the convexity condition* (3.32) *corresponding to the above fixed $N > 1$ and some $K \in \mathbb{R}$ then also does h_ε. In particular h_ε satisfies the differential inequality* (3.17).

Using this approximation one can prove the following

Theorem 3.2.6 (Theorem 6.3, [20]). *For every $v \in [0,1]$, $K \in \mathbb{R}$, $N \in [1,\infty)$, $D \in (0,\infty]$ it holds $\mathcal{I}^s_{K,N,D}(v) = \mathcal{I}_{K,N,D}(v)$.*

3.2.3 Disintegration of measures

We include here a version of the Disintegration Theorem that we will use. We will follow Appendix A of [10] where a self-contained approach (and a proof) of the Disintegration Theorem in countably generated measure spaces can be found. An even more general version can be found in Section 452 of [30].

Recall that a σ-algebra is *countably generated* if there exists a countable family of sets so that the σ-algebra coincides with the smallest σ-algebra containing them.

Given a measurable space $(X, \mathcal{X})$, i.e. $\mathcal{X}$ is a σ-algebra of subsets of X, and a function $\mathfrak{Q} : X \to Q$, with Q general set, we can endow Q with the *push forward σ-algebra* $\mathcal{Q}$ of $\mathcal{X}$:

$$C \in \mathcal{Q} \quad \Longleftrightarrow \quad \mathfrak{Q}^{-1}(C) \in \mathcal{X},$$

which could be also defined as the biggest σ-algebra on Q such that $\mathfrak{Q}$ is measurable. Moreover given a probability measure $\mathfrak{m}$ on $(X, \mathcal{X})$, define a probability measure $\mathfrak{q}$ on $(Q, \mathcal{Q})$ by push forward via $\mathfrak{Q}$, i.e. $\mathfrak{q} := \mathfrak{Q}_\sharp\, \mathfrak{m}$.

This general scheme fits with the following situation: given a measure space $(X, \mathcal{X}, \mathfrak{m})$, suppose a partition of X is given in the form $\{X_q\}_{q \in Q}$, Q is the set of indices and $\mathfrak{Q} : X \to Q$ is the quotient map, i.e.

$$q = \mathfrak{Q}(x) \iff x \in X_q.$$

Following the previous scheme, we can also consider the quotient σ-algebra $\mathcal{Q}$ and the quotient measure $\mathfrak{q}$ obtaining the quotient measure space $(Q, \mathcal{Q}, \mathfrak{q})$.

Definition 3.2.7. *A* disintegration *of* $\mathfrak{m}$ consistent with $\mathfrak{Q}$ *is a map*

$$Q \ni q \longmapsto \mathfrak{m}_q \in \mathcal{P}(X, \mathcal{X})$$

such that the following hold:

1. *for all $B \in \mathcal{X}$, the map $\mathfrak{m}_\cdot(B)$ is $\mathfrak{q}$-measurable;*
2. *for all $B \in \mathcal{X}$, $C \in \mathcal{Q}$ satisfies the consistency condition*

$$\mathfrak{m}\left(B \cap \mathfrak{Q}^{-1}(C)\right) = \int_C \mathfrak{m}_q(B)\, \mathfrak{q}(dq).$$

A disintegration is strongly consistent with respect to $\mathfrak{Q}$ *if for $\mathfrak{q}$-a.e. $q \in Q$ we have $\mathfrak{m}_q(\mathfrak{Q}^{-1}(q)) = 1$. The measures $\mathfrak{m}_q$ are called* conditional probabilities.

When the map $\mathfrak{Q}$ is induced by a partition of X as before, we will directly say that the disintegration is consistent with the partition, meaning that the disintegration is consistent with the quotient map $\mathfrak{Q}$ associated to the partition $X = \cup_{q \in Q} X_q$.

We now describe the Disintegration Theorem.

Theorem 3.2.8 (Theorem A.7, Proposition A.9 of [10]). *Assume that $(X, \mathcal{X}, \rho)$ is a countably generated probability space and $X = \cup_{q \in Q} X_q$ is a partition of X.*

Then the quotient probability space $(Q, \mathcal{Q}, \mathfrak{q})$ is essentially countably generated and there exists a unique disintegration $q \mapsto \mathfrak{m}_q$ consistent with the partition $X = \cup_{q \in Q} X_q$.

If $\mathcal{X}$ contains all singletons, then the disintegration is strongly consistent if and only if there exists an $\mathfrak{m}$-section $S \in \mathcal{X}$ such that the σ-algebra $\mathcal{S}$ contains $\mathcal{B}(S)$.

We now expand on the statement of Theorem 3.2.8.
In the measure space $(Q, \mathcal{Q}, \mathfrak{q})$, the σ-algebra $\mathcal{Q}$ is *essentially countably generated* if, by definition, there exists a countable family of sets $Q_n \subset Q$ such that for any $C \in \mathcal{Q}$ there exists $\hat{C} \in \hat{\mathcal{Q}}$, where $\hat{\mathcal{Q}}$ is the σ-algebra generated by $\{Q_n\}_{n\in\mathbb{N}}$, such that $\mathfrak{q}(C \,\Delta\, \hat{C}) = 0$.

Uniqueness is understood in the following sense: if $q \mapsto \mathfrak{m}_q^1$ and $q \mapsto \mathfrak{m}_q^2$ are two consistent disintegrations then $\mathfrak{m}_q^1 = \mathfrak{m}_q^2$ for $\mathfrak{q}$-a.e. $q \in Q$.

Finally, a set S is a section for the partition $X = \cup_q X_q$ if for any $q \in Q$ there exists a unique $x_q \in S \cap X_q$. A set $S_\mathfrak{m}$ is an $\mathfrak{m}$-section if there exists $Y \subset X$ with $\mathfrak{m}(X \setminus Y) = 0$ such that the partition $Y = \cup_q (X_q \cap Y)$ has section $S_\mathfrak{m}$. Once a section (or an $\mathfrak{m}$-section) is given, one can obtain the measurable space $(S, \mathcal{S})$ by pushing forward the σ-algebra $\mathcal{X}$ on S via the map that associates to any $X_q \ni x \mapsto x_q = S \cap X_q$.

3.3 Transport set

The following setting is fixed once and for all:

$(X, \mathsf{d}, \mathfrak{m})$ is a fixed metric measure space with $\mathfrak{m}(X) = 1$ such that
the ambient metric space (X, d) is geodesic and proper (hence complete and separable).

Let $\phi : X \to \mathbb{R}$ be any 1-Lipschitz function. Here we present some useful results (all of them already presented in [11]) concerning the d-cyclically monotone set associated with ϕ:

$$\Gamma := \{(x, y) \in X \times X : \phi(x) - \phi(y) = \mathsf{d}(x, y)\}, \tag{3.19}$$

that can be seen as the set of pairs moved by ϕ with maximal slope. Recall that a set $\Lambda \subset X \times X$ is said to be d-cyclically monotone if for any finite set of points $(x_1, y_1), \dots, (x_N, y_N)$ it holds

$$\sum_{i=1}^{N} \mathsf{d}(x_i, y_i) \le \sum_{i=1}^{N} \mathsf{d}(x_i, y_{i+1}),$$

with the convention that $y_{N+1} = y_1$.

The following lemma is a consequence of the d-cyclically monotone structure of Γ.

Lemma 3.3.1. *Let $(x, y) \in X \times X$ be an element of Γ. Let $\gamma \in \mathrm{Geo}(X)$ be such that $\gamma_0 = x$ and $\gamma_1 = y$. Then*

$$(\gamma_s, \gamma_t) \in \Gamma,$$

for all $0 \le s \le t \le 1$.

Proof. Take $0 \le s \le t \le 1$ and note that

$$\begin{aligned}\phi(\gamma_s) - \phi(\gamma_t) &= \phi(\gamma_s) - \phi(\gamma_t) + \phi(\gamma_0) - \phi(\gamma_0) + \phi(\gamma_1) - \phi(\gamma_1) \\ &\ge \mathsf{d}(\gamma_0, \gamma_1) - \mathsf{d}(\gamma_0, \gamma_s) - \mathsf{d}(\gamma_t, \gamma_1) \\ &= \mathsf{d}(\gamma_s, \gamma_t).\end{aligned}$$

The claim follows. □

Then, it is natural to consider the set of geodesics $G \subset \mathrm{Geo}(X)$ such that

$$\gamma \in G \iff \{(\gamma_s, \gamma_t) : 0 \le s \le t \le 1\} \subset \Gamma,$$

that is $G := \{\gamma \in \mathrm{Geo}(X) : (\gamma_0, \gamma_1) \in \Gamma\}$. We now recall some basic definitions of the L^1-optimal transportation theory that will be needed to describe the structure of Γ.

Definition 3.3.2. *We define the set of* transport rays *by*

$$R := \Gamma \cup \Gamma^{-1},$$

where $\Gamma^{-1} := \{(x, y) \in X \times X : (y, x) \in \Gamma\}$. *The sets of* initial points *and* final points *are defined respectively by*

$$\begin{aligned}\mathfrak{a} :=& \{z \in X : \nexists x \in X, (x, z) \in \Gamma, \mathsf{d}(x, z) > 0\}, \\ \mathfrak{b} :=& \{z \in X : \nexists x \in X, (z, x) \in \Gamma, \mathsf{d}(x, z) > 0\}.\end{aligned}$$

The set of end points *is* $\mathfrak{a} \cup \mathfrak{b}$. *We define the subset of X,* transport set with end points*:*

$$\mathcal{T}_e = P_1(\Gamma \setminus \{x = y\}) \cup P_1(\Gamma^{-1} \setminus \{x = y\}).$$

where $\{x = y\}$ *stands for* $\{(x, y) \in X^2 : \mathsf{d}(x, y) = 0\}$.

A few comments are in order. Notice that R coincides with $\{(x, y) \in X \times X : |\phi(x) - \phi(y)| = \mathsf{d}(x, y)\}$; the name transport set with end points for $\mathcal{T}_e$ is motivated by how later on we will consider a more regular subset of $\mathcal{T}_e$ that will be called transport set; moreover if $x \in X$ is moved forward but not backward by ϕ, this is translated in $x \in \Gamma$ and $x \notin \Gamma^{-1}$; in any case it belongs to $\mathcal{T}_e$.

We also introduce the following notation to be used throughout the paper; we set $\Gamma(x) := P_2(\Gamma \cap (\{x\} \times X))$ and $\Gamma^{-1}(x) := P_2(\Gamma^{-1} \cap (\{x\} \times X))$. In general if $F \subset X \times X$, we set $F(x) = P_2(F \cap (\{x\} \times X))$.

Remark 3.3.3. *Here we discuss the measurability of the sets introduced in Definition 3.3.2. Since* ϕ *is 1-Lipschitz,* Γ *is closed and therefore* Γ^{-1} *and* R *are closed as well. Moreover by assumption the space is proper, since the sets* Γ, Γ^{-1}, R *are* σ*-compact (countable union of compact sets).*

Then we look at the sets of initial and final points:

$$\mathfrak{a} = P_2\left(\Gamma \cap \{(x, z) \in X \times X : \mathsf{d}(x, z) > 0\}\right)^c, \quad \mathfrak{b} = P_1\left(\Gamma \cap \{(x, z) \in X \times X : \mathsf{d}(x, z) > 0\}\right)^c.$$

Since $\{(x,z) \in X \times X : \mathsf{d}(x,z) > 0\} = \cup_n\{(x,z) \in X \times X : \mathsf{d}(x,z) \geq 1/n\}$, *it follows that both* $\mathfrak{a}$ *and* $\mathfrak{b}$ *are the complement of* σ*-compact sets. Hence* $\mathfrak{a}$ *and* $\mathfrak{b}$ *are Borel sets. Reasoning as before, it follows that* $\mathcal{T}_e$ *is a* σ*-compact set.*

Lemma 3.3.4. *Let* $\pi \in \Pi(\mu_0, \mu_1)$ *with* $\pi(\Gamma) = 1$*; then*

$$\pi(\mathcal{T}_e \times \mathcal{T}_e \cup \{x = y\}) = 1.$$

Proof. It is enough to observe that if $(z, w) \in \Gamma$ with $z \neq w$, then $w \in \Gamma(z)$ and $z \in \Gamma^{-1}(w)$; therefore

$$(z, w) \in \mathcal{T}_e \times \mathcal{T}_e.$$

Hence $\Gamma \setminus \{x = y\} \subset \mathcal{T}_e \times \mathcal{T}_e$. Since $\pi(\Gamma) = 1$, the claim follows. □

As a consequence, $\mu_0(\mathcal{T}_e) = \mu_1(\mathcal{T}_e)$ and any optimal map T such that $T_\sharp \mu_0 \llcorner_{\mathcal{T}_e} = \mu_1 \llcorner_{\mathcal{T}_e}$ can be extended to an optimal map T' with $T'_\sharp \mu_0 = \mu_1$ with the same cost by setting

$$T'(x) = \begin{cases} T(x), & \text{if } x \in \mathcal{T}_e \\ x, & \text{if } x \notin \mathcal{T}_e. \end{cases} \tag{3.20}$$

It can be proved that the set of transport rays R induces an equivalence relation on a subset of $\mathcal{T}_e$. It is sufficient to remove from $\mathcal{T}_e$ the branching points of geodesics. Then using curvature properties of the space, one can prove that such branching points all have $\mathfrak{m}$-measure zero.

3.3.1 Branching structures in the Transport set

What follows was first presented in [18]. Consider the sets of respectively forward and backward branching points

$$\begin{aligned} A_+ &:= \{x \in \mathcal{T}_e : \exists z, w \in \Gamma(x), (z, w) \notin R\}, \\ A_- &:= \{x \in \mathcal{T}_e : \exists z, w \in \Gamma(x)^{-1}, (z, w) \notin R\}. \end{aligned} \tag{3.21}$$

The sets $A_\pm$ are σ-compact sets. Indeed since (X, d) is proper, any open set is σ-compact. The main motivation for the definition of A_+ and A_- is contained below.

Theorem 3.3.5. *The set of transport rays* $R \subset X \times X$ *is an equivalence relation on the set*

$$\mathcal{T}_e \setminus (A_+ \cup A_-).$$

Proof. First, for all $x \in P_1(\Gamma)$, $(x, x) \in R$. If $x, y \in \mathcal{T}_e$ with $(x, y) \in R$, then by definition of R, it follows straightforwardly that $(y, x) \in R$.

Therefore, the only property needing a proof is transitivity. Let $x, z, w \in \mathcal{T}_e \setminus (A_+ \cup A_-)$ be such that $(x, z), (z, w) \in R$ with x, z and w distinct points. The claim is $(x, w) \in R$. So we have 4 different possibilities: the first one is

$$z \in \Gamma(x), \quad w \in \Gamma(z).$$

This immediately implies $w \in \Gamma(x)$ and therefore $(x, w) \in R$. The second possibility is

$$z \in \Gamma(x), \quad z \in \Gamma(w),$$

that can be rewritten as $(z, x), (z, w) \in \Gamma^{-1}$. Since $z \notin A_-$, necessarily $(x, w) \in R$. Third possibility:

$$x \in \Gamma(z), \quad w \in \Gamma(z),$$

and since $z \notin A_+$ it follows that $(x, w) \in R$. The last case is

$$x \in \Gamma(z), \quad z \in \Gamma(w),$$

and therefore $x \in \Gamma(w)$, hence $(x, w) \in R$ and the claim follows. □

Next, we show that each equivalence class of R is formed by a single geodesic.

Lemma 3.3.6. *For any $x \in \mathcal{T}$ and $z, w \in R(x)$ there exists $\gamma \in G \subset \mathrm{Geo}(X)$ such that*

$$\{x, z, w\} \subset \{\gamma_s : s \in [0, 1]\}.$$

If $\hat{\gamma} \in G$ enjoys the same property, then

$$\big(\{\hat{\gamma}_s : s \in [0, 1]\} \cup \{\gamma_s : s \in [0, 1]\}\big) \subset \{\tilde{\gamma}_s : s \in [0, 1]\}$$

for some $\tilde{\gamma} \in G$.

Since $G = \{\gamma \in \mathrm{Geo}(X) : (\gamma_0, \gamma_1) \in \Gamma\}$, Lemma 3.3.6 states that if we fix an element x in $\mathcal{T}_e \setminus (A_+ \cup A_-)$ and we pick two elements z, w in the same equivalence class of x, then these three points will align on a geodesic γ whose image is again all contained in the same equivalence class $R(x)$.

Proof. Assume that x, z and w are all distinct points (otherwise the claim follows trivially). We consider different cases.
First case: $z \in \Gamma(x)$ and $w \in \Gamma^{-1}(x)$.
By d-cyclical monotonicity

$$\mathsf{d}(z, w) \le \mathsf{d}(z, x) + \mathsf{d}(x, w) = \phi(w) - \phi(z) \le \mathsf{d}(z, w).$$

Hence z, x and w lie on a geodesic.
Second case: $z, w \in \Gamma(x)$.
Without loss of generality assume $\phi(x) \ge \phi(w) \ge \phi(z)$. Since in the proof of Lemma

3.4.2 we have already excluded the case $\phi(w) = \phi(z)$, we assume $\phi(x) > \phi(w) > \phi(z)$. Then if no geodesics $\gamma \in G$ with $\gamma_0 = x$ and $\gamma_1 = z$ and $\gamma_s = w$ exist, there will be $\gamma \in G$ with $(\gamma_0, \gamma_1) = (x, z)$ and $s \in (0, 1)$ such that

$$\phi(\gamma_s) = \phi(w), \qquad \gamma_s \in \Gamma(x), \qquad \gamma_s \neq w.$$

As observed in the proof of Lemma 3.4.2, this would imply that $(\gamma_s, w) \notin R$ and since $x \notin A_+$ this would be a contradiction. Hence the second case follows.

The remaining two cases follow under the same reasoning, exchanging the role of $\Gamma(x)$ with the one of $\Gamma^{-1}(x)$. The second part of the statement now easily follows. □

3.4 Cyclically monotone sets

Following Theorem 3.3.5 and Lemma 3.3.6, the next step is to prove that both A_+ and A_- have $\mathfrak{m}$-measure zero, (in other words, branching happens on rays with zero $\mathfrak{m}$-measure). From the statement of this property, it is clear that some regularity assumption on $(X, \mathsf{d}, \mathfrak{m})$ should play a role. We will indeed assume the space to enojoy a stronger form of essentially non-branching. Recall that the latter is formulated in terms of geodesics of $(\mathcal{P}_2(X), W_2)$, a d^2-cyclically monotone set, while we need regularity for the d-cyclically monotone set Γ. Hence it is necessary to include d^2-cyclically monotone sets as subsets of d-cyclically monotone sets.

We present here a strategy introduced by the author in [17, 18] from where all the material presented in this section is taken. Section 3.4.1 contains results from [11] while Section 3.4.2 is taken from [20].

Lemma 3.4.1 (Lemma 4.6 of [17]). *Let $\Delta \subset \Gamma$ be any set so that:*

$$(x_0, y_0), (x_1, y_1) \in \Delta \quad \Rightarrow \quad (\phi(y_1) - \phi(y_0)) \cdot (\phi(x_1) - \phi(x_0)) \geq 0.$$

Then Δ is d^2-cyclically monotone.

Proof. It follows directly from the hypothesis of the lemma that the set

$$\Lambda := \{(\phi(x), \phi(y)) : (x, y) \in \Delta\} \subset \mathbb{R}^2$$

is monotone in the Euclidean sense. Since $\Lambda \subset \mathbb{R}^2$, it is then a standard fact that Λ is also $|\cdot|^2$-cyclically monotone, where $|\cdot|$ denotes the modulus. We nevertheless include a short proof: there exists a maximal monotone multivalued function F such that $\Lambda \subset \mathrm{graph}(F)$ and its domain is an interval, say (a, b) with a and b possibly infinite; moreover, apart from countably many $x \in \mathbb{R}$, the set $F(x)$ is a singleton. Then the following function is well defined:

$$\Psi(x) := \int_c^x F(s)ds,$$

where c is any fixed element of (a, b). Then observe that

$$\Psi(z) - \Psi(x) \geq y(z - x), \qquad \forall\, z, x \in (a, b),$$

where y is any element of $F(x)$. In particular this implies that Ψ is convex and $F(x)$ is a subset of its sub-differential. In particular Λ is $|\cdot|^2$-cyclically monotone.
Then for $\{(x_i, y_i)\}_{i\leq N} \subset \Delta$, since $\Delta \subset \Gamma$, it holds

$$\begin{aligned}\sum_{i=1}^{N} \mathsf{d}^2(x_i, y_i) &= \sum_{i=1}^{N} |\phi(x_i) - \phi(y_i)|^2 \\ &\leq \sum_{i=1}^{N} |\phi(x_i) - \phi(y_{i+1})|^2 \\ &\leq \sum_{i=1}^{N} \mathsf{d}^2(x_i, y_{i+1}),\end{aligned}$$

where the last inequality is given by the 1-Lipschitz regularity of ϕ. The claim follows. □

To study the set of branching points is necessary to relate points of branching to geodesics. In the next Lemma, using Lemma 3.3.1, we observe that once a branching happens there exist two distinct geodesics, both contained in $\Gamma(x)$, that are not in relation in the sense of R.

Lemma 3.4.2. *Let $x \in A_+$. Then there exist two distinct geodesics $\gamma^1, \gamma^2 \in G$ such that*
- $(x, \gamma_s^1), (x, \gamma_s^2) \in \Gamma$ *for all* $s \in [0, 1]$;
- $(\gamma_s^1, \gamma_s^2) \notin R$ *for all* $s \in [0, 1]$;
- $\phi(\gamma_s^1) = \phi(\gamma_s^2)$ *for all* $s \in [0, 1]$.

Moreover both geodesics are non-constant.

Proof. From the definition of A_+ there exists $z, w \in \mathcal{T}_e$ such that $z, w \in \Gamma(x)$ and $(z, w) \notin R$. Since $z, w \in \Gamma(x)$, from Lemma 3.3.1 there exist two geodesics $\gamma^1, \gamma^2 \in G$ such that

$$\gamma_0^1 = \gamma_0^2 = x, \quad \gamma_1^1 = z, \quad \gamma_1^2 = w.$$

Since $(z, w) \notin R$, necessarily both z and w are different from x, and x is not a final point, that is $x \notin \mathfrak{b}$. So the previous geodesics are not constant. Since z and w can be exchanged, we can also assume that $\phi(z) \geq \phi(w)$. Since $z \in \Gamma(x)$, $\phi(x) \geq \phi(z)$ and by continuity there exists $s_2 \in (0, 1]$ such that

$$\phi(z) = \phi(\gamma_{s_2}^2).$$

Note that $z \neq \gamma_{s_2}^2$, otherwise $w \in \Gamma(z)$ and therefore $(z, w) \in R$. Moreover still $(z, \gamma_{s_2}^2) \notin R$. Indeed if the contrary was true, then

$$0 = |\phi(z) - \phi(\gamma_{s_2}^2)| = \mathsf{d}(z, \gamma_{s_2}^2),$$

that is a contradiction with $z \neq \gamma^2_{s_2}$.

Therefore, by continuity there exists $\delta > 0$ such that

$$\phi(\gamma^1_{1-s}) = \phi(\gamma^2_{s_2(1-s)}), \qquad d(\gamma^1_{1-s}, \gamma^2_{s_2-s}) > 0,$$

for all $0 \leq s \leq \delta$.

Hence reapplying the previous argument $(\gamma^1_{1-s}, \gamma^2_{s_2(1-s)}) \notin R$. The curve γ^1 and γ^2 of the claim are then obtained by properly restricting and rescaling the geodesics γ^1 and γ^2 considered so far. □

The previous correspondence between branching points and pairs of branching geodesics can be proved to be measurable. We will make use of the following selection result, Theorem 5.5.2 of [50]. We again refer to [50] for some preliminaries on analytic sets.

Theorem 3.4.3. *Let X and Y be Polish spaces, $F \subset X \times Y$ analytic, and $\mathcal{A}$ be the σ-algebra generated by the analytic subsets of X. Then there is an $\mathcal{A}$-measurable section $u : P_1(F) \to Y$ of F.*

Recall that given $F \subset X \times Y$, a *section u of F* is a function from $P_1(F)$ to Y such that $\mathrm{graph}(u) \subset F$.

Lemma 3.4.4. *There exists an $\mathcal{A}$-measurable map $u : A_+ \mapsto G \times G$ such that if $u(x) = (\gamma^1, \gamma^2)$ then*
- *$(x, \gamma^1_s), (x, \gamma^2_s) \in \Gamma$ for all $s \in [0, 1]$;*
- *$(\gamma^1_s, \gamma^2_s) \notin R$ for all $s \in [0, 1]$;*
- *$\phi(\gamma^1_s) = \phi(\gamma^2_s)$ for all $s \in [0, 1]$.*

Moreover both geodesics are non-constant.

Proof. Since $G = \{\gamma \in \mathrm{Geo}(X) : (\gamma_0, \gamma_1) \in \Gamma\}$ and $\Gamma \subset X \times X$ is closed, the set G is a complete and separable metric space. Consider now the set

$$\begin{aligned} F := \{(x, \gamma^1, \gamma^2) \in \mathcal{T}_e \times G \times G : (x, \gamma^1_0), (x, \gamma^2_0) \in \Gamma\} \\ \cap \Big(X \times \{(\gamma^1, \gamma^2) \in G \times G : d(\gamma^1_1, \gamma^2_1) > 0\}\Big) \\ \cap \Big(X \times \{(\gamma^1, \gamma^2) \in G \times G : d(\gamma^1_0, \gamma^2_0) > 0\}\Big) \\ \cap \Big(X \times \{(\gamma^1, \gamma^2) \in G \times G : d(\gamma^1_0, \gamma^1_1) > 0\}\Big) \\ \cap \Big(X \times \{(\gamma^1, \gamma^2) \in G \times G : \phi(\gamma^1_i) = \phi(\gamma^2_i),\ i = 0, 1\}\Big). \end{aligned}$$

It follows from Remark 3.3.3 that F is σ-compact. To avoid possible intersections in interior points of γ^1 with γ^2 we consider the following map:

$$h : G \times G \to [0, \infty)$$

$$(\gamma^1, \gamma^2) \mapsto h(\gamma^1, \gamma^2) := \min_{s \in [0,1]} \mathsf{d}(\gamma_s^1, \gamma_s^2).$$

From the compactness of $[0, 1]$, we deduce the continuity of h. Therefore

$$\hat{F} := F \cap \{(x, \gamma^1, \gamma^2) \in X \times G \times G : h(\gamma^1, \gamma^2) > 0\}$$

is a Borel set and from Lemma 3.4.2,

$$\hat{F} \cap (\{x\} \times G \times G) \neq \emptyset$$

for all $x \in A_+$. By Theorem 3.4.3 we infer the existence of an $\mathcal{A}$-measurable selection u of $\hat{F}$. Since $A_+ = P_1(\hat{F})$ and if $u(x) = (\gamma^1, \gamma^2)$, then

$$\mathsf{d}(\gamma_s^1, \gamma_s^2) > 0, \qquad \phi(\gamma_s^1) = \phi(\gamma_s^2),$$

for all $s \in [0, 1]$, and therefore $(\gamma_s^1, \gamma_s^2) \notin R$ for all $s \in [0, 1]$. The claim follows. □

We are ready to prove the following

Proposition 3.4.5. *Let $(X, \mathsf{d}, \mathfrak{m})$ be a m.m.s. such that for any $\mu_0, \mu_1 \in \mathcal{P}(X)$ with $\mu_0 \ll \mathfrak{m}$ any optimal transference plan for W_2 is concentrated on the graph of a function. Then*

$$\mathfrak{m}(A_+) = \mathfrak{m}(A_-) = 0.$$

Proof. **Step 1.**
Suppose by contradiction that $\mathfrak{m}(A_+) > 0$. By definition of A_+, thanks to Lemma 3.4.2 and Lemma 3.4.4, for every $x \in A_+$ there exist two non-constant geodesics $\gamma^1, \gamma^2 \in G$ such that

- $(x, \gamma_s^1), (x, \gamma_s^2) \in \Gamma$ for all $s \in [0, 1]$;
- $(\gamma_s^1, \gamma_s^2) \notin R$ for all $s \in [0, 1]$;
- $\phi(\gamma_s^1) = \phi(\gamma_s^2)$ for all $s \in [0, 1]$.

Moreover the map $A_+ \ni x \mapsto u(x) := (\gamma^1, \gamma^2) \in G^2$ is $\mathcal{A}$-measurable.

By inner regularity of compact sets (or by Lusin's Theorem), possibly selecting a subset of A_+ still with strictly positive $\mathfrak{m}$-measure, we can assume that the previous map is continuous and in particular the functions

$$A_+ \ni x \mapsto \phi(\gamma_j^i) \in \mathbb{R}, \qquad i = 1, 2,\ j = 0, 1$$

are all continuous. Set

$$\alpha_x := \phi(\gamma_0^1) = \phi(\gamma_0^2), \qquad \beta_x := \phi(\gamma_1^1) = \phi(\gamma_1^2)$$

and note that $\alpha_x > \beta_x$. Now we want to show the existence of a subset $B \subset A_+$, still with $\mathfrak{m}(B) > 0$, such that

$$\sup_{x \in B} \beta_x < \inf_{x \in B} \alpha_x.$$

By continuity of α and β, a set B verifying the previous inequality can be obtained by considering the set $A_+ \cap B_r(x)$, for $x \in A_+$ with r sufficiently small. Since $\mathfrak{m}(A_+) > 0$, for $\mathfrak{m}$-a.e. $x \in A_+$ the set $A_+ \cap B_r(x)$ has positive $\mathfrak{m}$-measure. So the existence of $B \subset A_+$ enjoying the aforementioned properties follows.

Step 2.
Let $I = [c, d]$ be a non trivial interval such that

$$\sup_{x \in B} \beta_X < c < d < \inf_{x \in B} \alpha_X.$$

Then, by construction, for all $x \in B$ the image of the composition of the geodesics γ^1 and γ^2 with ϕ contains the interval I:

$$I \subset \{\phi(\gamma_s^i) : s \in [0, 1]\}, \qquad i = 1, 2.$$

Fix any point inside I, say c, and consider for any $x \in B$ the value $s(x)$ such that $\phi(\gamma^1_{s(x)}) = \phi(\gamma^2_{s(x)}) = c$. We can now define on B two transport maps T^1 and T^2 by

$$B \ni x \mapsto T^i(x) := \gamma^i_{s(x)}, \qquad i = 1, 2.$$

Accordingly we define the transport plan

$$\eta := \frac{1}{2}\left((Id, T^1)_\sharp \mathfrak{m}_B + (Id, T^2)_\sharp \mathfrak{m}_B\right),$$

where $\mathfrak{m}_B := \mathfrak{m}(B)^{-1}\mathfrak{m}\llcorner_B$.

Step 3.
The support of η is d^2-cyclically monotone. To prove it we will use Lemma 3.4.1. The measure η is concentrated on the set

$$\Delta := \{(x, \gamma^1_{s(x)}) : x \in B\} \cup \{(x, \gamma^2_{s(x)}) : x \in B\} \subset \Gamma.$$

Take any two pairs $(x_0, y_0), (x_1, y_1) \in \Delta$ and notice that by definition:

$$\phi(y_1) - \phi(y_0) = 0.$$

Therefore, trivially $(\phi(y_1) - \phi(y_0))\,(\phi(x_1) - \phi(x_0)) = 0$, and Lemma 3.4.1 can be applied to Δ. Hence η is optimal with $(P_1)_\sharp\eta \ll \mathfrak{m}$ and is not induced by a map; this is a contradiction with the assumption. It follows that $\mathfrak{m}(A_+) = 0$. The claim for A_- follows in the same manner. □

Remark 3.4.6. *If the space is itself non-branching, then Proposition 3.4.5 can be proved more directly under the assumption* (**A.1**)*, that will be introduced at the beginning of Section 3.5. Recall that* $(X, \mathsf{d}, \mathfrak{m})$ *is non-branching if for any* $\gamma^1, \gamma^2 \in \mathrm{Geo}$ *such that*

$$\gamma_0^1 = \gamma_0^2, \qquad \gamma_t^1 = \gamma_t^2,$$

for some $t \in (0, 1)$*, implies that* $\gamma_1^1 = \gamma_1^2$*. In particular the following statement holds*

Let $(X, \mathsf{d}, \mathfrak{m})$ be non-branching and assume moreover (**A.1**) to hold. Then

$$\mathfrak{m}(A_+) = \mathfrak{m}(A_-) = 0.$$

For the proof of this statement (that goes beyond the scope of this note) we refer to [11], Lemma 5.3. The same comment will also apply to the next Theorem 3.4.7.

The set

$$\mathcal{T} := \mathcal{T}_e \setminus (A_+ \cup A_-) \tag{3.22}$$

will be called the *transport set*. Since $\mathcal{T}_e$, A_+ and A_- are σ-compact sets, notice that $\mathcal{T}$ is a countable intersection of σ-compact sets and in particular Borel.

Theorem 3.4.7 (Theorem 5.5, [18]). *Let $(X, \mathsf{d}, \mathfrak{m})$ be such that for any $\mu_0, \mu_1 \in \mathcal{P}(X)$ with $\mu_0 \ll \mathfrak{m}$ any optimal transference plan for W_2 is concentrated on the graph of a function. Then the set of transport rays $R \subset X \times X$ is an equivalence relation on the transport set $\mathcal{T}$ and*

$$\mathfrak{m}(\mathcal{T}_e \setminus \mathcal{T}) = 0.$$

To summarize, we have shown that given a d-monotone set Γ, the set of all those points moved by Γ, denoted $\mathcal{T}_e$, can be written, neglecting a set of $\mathfrak{m}$-measure zero, as the union of a family of disjoint geodesics. The next step is to decompose the reference measure $\mathfrak{m}$ restricted to $\mathcal{T}$ with respect to the partition given by R, where each equivalence class is given by

$$[x] = \{y \in \mathcal{T} : (x, y) \in R\}.$$

Denoting the set of equivalence classes by Q, we can apply the Disintegration Theorem (see Theorem 3.2.8) to the measure space $(\mathcal{T}, \mathcal{B}(\mathcal{T}), \mathfrak{m})$ and obtain the disintegration of $\mathfrak{m}$ consistent with the partition of $\mathcal{T}$ in rays:

$$\mathfrak{m}_{\llcorner \mathcal{T}} = \int_Q \mathfrak{m}_q \, \mathfrak{q}(dq),$$

where $\mathfrak{q}$ is the quotient measure.

3.4.1 Structure of the quotient set

In order to use the strength of the Disintegration Theorem to localize the measure, one needs to obtain a *strongly consistent* disintegration. Following the last part of Theorem 3.2.8, it is necessary to build a section S of $\mathcal{T}$ together with a measurable quotient map with image S.

Proposition 3.4.8 (Q is locally contained in level sets of ϕ). *It is possible to construct a Borel quotient map $\mathfrak{Q} : \mathcal{T} \to Q$ such that the quotient set $Q \subset X$ can be written locally* *as a level set of ϕ in the following sense:*

$$Q = \bigcup_{i \in \mathbb{N}} Q_i, \qquad Q_i \subset \phi^{-1}(\alpha_i),$$

where $\alpha_i \in \mathbb{Q}$, Q_i is analytic and $Q_i \cap Q_j = \emptyset$, for $i \neq j$.

Proof. **Step 1.**
For each $n \in \mathbb{N}$, consider the set $\mathcal{T}_n$ of those points x having ray $R(x)$ longer than $1/n$, i.e.

$$\mathcal{T}_n := P_1\{(x, z, w) \in \mathcal{T}_e \times \mathcal{T}_e \times \mathcal{T}_e : z, w \in R(x),\ \mathsf{d}(z, w) \geq 1/n\} \cap \mathcal{T}.$$

It is easily seen that $\mathcal{T} = \bigcup_{n \in \mathbb{N}} \mathcal{T}_n$ and that $\mathcal{T}_n$ is Borel: the set $\mathcal{T}_e$ is σ-compact and therefore its projection is again σ-compact.

Moreover if $x \in \mathcal{T}_n$, $y \in \mathcal{T}$ and $(x, y) \in R$ then also $y \in \mathcal{T}_n$: for $x \in \mathcal{T}_n$ there exists $z, w \in \mathcal{T}_e$ with $z, w \in R(x)$ and $\mathsf{d}(z, w) \geq 1/n$. Since $x \in \mathcal{T}$ necessarily $z, w \in \mathcal{T}$. Since R is an equivalence relation on $\mathcal{T}$ and $y \in \mathcal{T}$, it follows that $z, w \in R(y)$. Hence $y \in \mathcal{T}_n$. In particular, $\mathcal{T}_n$ is the union of all those maximal rays of $\mathcal{T}$ with length at least $1/n$.

Using the same notation, we have $\mathcal{T} = \cup_{n \in \mathbb{N}} \mathcal{T}_n$ with $\mathcal{T}_n$ Borel, saturated with respect to R; each ray of $\mathcal{T}_n$ is longer than $1/n$ and $\mathcal{T}_n \cap \mathcal{T}_{n'} = \emptyset$ if $n \neq n'$.

Now we consider the following saturated subsets of $\mathcal{T}_n$: for $\alpha \in \mathbb{Q}$

$$\begin{aligned} \mathcal{T}_{n,\alpha} := {} & P_1\Big(R \cap \Big\{(x, y) \in \mathcal{T}_n \times \mathcal{T}_n : \phi(y) = \alpha - \frac{1}{3n}\Big\}\Big) \\ & \cap P_1\Big(R \cap \Big\{(x, y) \in \mathcal{T}_n \times \mathcal{T}_n : \phi(y) = \alpha + \frac{1}{3n}\Big\}\Big), \end{aligned} \tag{3.23}$$

and we claim that

$$\mathcal{T}_n = \bigcup_{\alpha \in \mathbb{Q}} \mathcal{T}_{n,\alpha}. \tag{3.24}$$

We show the above identity by double inclusion. First note that $(\supset)$ holds trivially. For the converse inclusion $(\subset)$ observe that for each $\alpha \in \mathbb{Q}$, the set $\mathcal{T}_{n,\alpha}$ coincides with the family of those rays $R(x) \cap \mathcal{T}_n$ such that there exist $y^+, y^- \in R(x)$ satisfying

$$\phi(y^+) = \alpha - \frac{1}{3n}, \qquad \phi(y^-) = \alpha + \frac{1}{3n}. \tag{3.25}$$

Then we need to show that any $x \in \mathcal{T}_n$ also verifies $x \in \mathcal{T}_{n,\alpha}$ for a suitable $\alpha \in \mathbb{Q}$. Fix $x \in \mathcal{T}_n$; since $R(x)$ is longer than $1/n$, there exist $z, y^+, y^- \in R(x) \cap \mathcal{T}_n$ such that

$$\phi(y^-) - \phi(z) = \frac{1}{2n}, \qquad \phi(z) - \phi(y^+) = \frac{1}{2n}.$$

Consider now the geodesic $\gamma \in G$ such that $\gamma_0 = y^-$ and $\gamma_1 = y^+$. By continuity of $[0, 1] \ni t \mapsto \phi(\gamma_t)$ it follows the existence of $0 < s_1 < s_2 < s_3 < 1$ such that

$$\phi(\gamma_{s_3}) = \phi(\gamma_{s_2}) - \frac{1}{3n}, \qquad \phi(\gamma_{s_1}) = \phi(\gamma_{s_2}) + \frac{1}{3n}, \qquad \phi \in \mathbb{Q}.$$

This concludes the proof of identity (3.24).

Step 2.

By the above construction, one can check that for each $\alpha \in \mathbb{Q}$ the level set $\phi^{-1}(\alpha)$ is a quotient set for $\mathcal{T}_{n,\alpha}$, i.e. $\mathcal{T}_{n,\alpha}$ is formed by disjoint geodesics each one intersecting $\phi^{-1}(\alpha)$ in exactly one point. Equivalently, $\phi^{-1}(\alpha)$ is a section for the partition of $\mathcal{T}_n$ induced by R.

Moreover $\mathcal{T}_{n,\alpha}$ is obtained as the projection of a Borel set and it is therefore analytic.

Since $\mathcal{T}_{n,\alpha}$ is saturated with respect to R either $\mathcal{T}_{n,\alpha} \cap \mathcal{T}_{n,\alpha'} = \emptyset$ or $\mathcal{T}_{n,\alpha} = \mathcal{T}_{n,\alpha'}$. Hence, removing the unnecessary α, we can assume that $\mathcal{T} = \bigcup_{n\in\mathbb{N},\alpha\in\mathbb{Q}} \mathcal{T}_{n,\alpha}$, is a partition. Then we characterize $\mathfrak{Q} : \mathcal{T} \to \mathcal{T}$ defining its graph as follows:

$$\mathrm{graph}(\mathfrak{Q}) := \bigcup_{n\in\mathbb{N},\alpha\in\mathbb{Q}} \mathcal{T}_{n,\alpha} \times \left(\phi^{-1}(\alpha) \cap \mathcal{T}_{n,\alpha}\right).$$

Notice that graph($\mathfrak{Q}$) is analytic and therefore $\mathfrak{Q} : \mathcal{T} \to Q$ is Borel (see Theorem 4.5.2 of [50]). The claim follows. □

Corollary 3.4.9. *The following strongly consistent disintegration formula holds true:*

$$\mathfrak{m}_{\llcorner\mathcal{T}} = \int_Q \mathfrak{m}_q\, \mathfrak{q}(dq), \qquad \mathfrak{m}_q(\mathfrak{Q}^{-1}(q)) = 1, \ \mathfrak{q}\text{-a.e. } q \in Q. \tag{3.26}$$

Proof. From Proposition 3.4.8 there exists an analytic quotient set Q with Borel quotient map $\mathfrak{Q} : \mathcal{T} \to Q$. In particular Q is a section and the push-forward σ-algebra of $\mathcal{B}(\mathcal{T})$ on Q contains $\mathcal{B}(Q)$. From Theorem 3.2.8 (3.26) follows. □

Remark 3.4.10. *One can improve the regularity of the disintegration formula* (3.26) *as follows. From inner regularity of Borel measures there exists $S \subset Q$ σ-compact such that $\mathfrak{q}(Q \setminus S) = 0$. The subset $R^{-1}(S) \subset \mathcal{T}$ is again σ-compact, indeed*

$$\begin{aligned} R^{-1}(S) &= \{x \in \mathcal{T} : (x, q) \in R,\ q \in S\} = P_1(\{(x, q) \in \mathcal{T} \times S : (x, q) \in R\}) \\ &= P_1(\mathcal{T} \times S \cap R) = P_1(\mathcal{T}_e \times S \cap R) \end{aligned}$$

and the regularity follows. Notice that $R^{-1}(S)$ is formed by non-branching rays and $\mathfrak{m}(\mathcal{T} \setminus R^{-1})(S)) = \mathfrak{q}(Q \setminus S) = 0$. Hence we have proved that the transport set with end points $\mathcal{T}_e$ admits a saturated, partitioned by disjoint rays, σ-compact subset of full measure with σ-compact quotient set. Since in what follows we will not use the definition (3.22), *we will denote this set by $\mathcal{T}$ and its quotient set by Q.*

For ease of notation $X_q := \mathfrak{Q}^{-1}(q)$. The next goal will be to deduce regularity properties for the conditional measures $\mathfrak{m}_q$. The next function will be of some help.

Definition 3.4.11 (Definition 4.5, [11]). *[Ray map] Define the* ray map

$$g : Dom(g) \subset Q \times \mathbb{R} \to \mathcal{T}$$

via the formula:

$$\begin{aligned} graph(g) &:= \Big\{(q,t,x) \in Q \times [0,+\infty) \times \mathcal{T} : (q,x) \in \Gamma,\ \mathsf{d}(q,x) = t\Big\} \\ &\cup \Big\{(q,t,x) \in Q \times (-\infty,0] \times \mathcal{T} : (x,q) \in \Gamma,\ \mathsf{d}(x,q) = t\Big\} \\ &= graph(g^{+}) \cup graph(g^{-}). \end{aligned}$$

Hence the ray map associates to each $q \in Q$ and $t \in \mathrm{Dom}\,(g(q,\cdot)) \subset \mathbb{R}$ the unique element $x \in \mathcal{T}$ such that $(q,x) \in \Gamma$ at distance t from q if t is positive or the unique element $x \in \mathcal{T}$ such that $(x,q) \in \Gamma$ at distance $-t$ from q if t is negative. By definition $\mathrm{Dom}(g) := g^{-1}(\mathcal{T})$. Notice that from Remark 3.4.10 it is not restrictive to assume graph(g) to be σ-compact. In particular the map g is Borel.

Next we list a few (trivial) regularity properties enjoyed by g.

Proposition 3.4.12. *The following holds.*

- *g is a Borel map.*
- *$t \mapsto g(q,t)$ is an isometry and if $s,t \in \mathrm{Dom}\,(g(q,\cdot))$ with $s \le t$ then $(g(q,s), g(q,t)) \in \Gamma$;*
- *$\mathrm{Dom}(g) \ni (q,t) \mapsto g(q,t)$ is bijective on $\mathfrak{Q}^{-1}(Q) = \mathcal{T}$, and its inverse is*

$$x \mapsto g^{-1}(x) = \big(\mathfrak{Q}(x), \pm\mathsf{d}(x,\mathfrak{Q}(x))\big)$$

where $\mathfrak{Q}$ is the quotient map previously introduced and the positive or negative sign depends on $(x,\mathfrak{Q}(x)) \in \Gamma$ or $(\mathfrak{Q}(x), x) \in \Gamma$.

Observe that from Lemma 3.3.1, $\mathrm{Dom}\,(g(q,\cdot))$ is a convex subset of $\mathbb{R}$ (i.e. an interval) for any $q \in Q$. Using the ray map g, we will review in Section 3.5 how to prove that the $\mathfrak{q}$-a.e. conditional measure $\mathfrak{m}_q$ is absolutely continuous with respect to the 1-dimensional Hausdorff measure on X_q provided $(X, \mathsf{d}, \mathfrak{m})$ enjoys weak curvature properties. The other main use of the ray map g was presented in Section 7 of [11] where it was used to build the 1-dimensional metric currents in the sense of Ambrosio-Kirchheim (see [6]) associated to $\mathcal{T}$.

It is worth noticing that so far, besides the assumption of Proposition 3.4.5, no extra assumptions on the geometry of the space were used. In particular, given two probability measures μ_0 and μ_1 with finite first moment, the associated transport set allows for the decomposition of the reference measure $\mathfrak{m}$ in one-dimensional conditional measures $\mathfrak{m}_q$, i.e. formula (3.26) holds.

3.4.2 Balanced transportation

Here we want to underline that the disintegration (or one-dimensional localization) of $\mathfrak{m}$ induced by the L^1-Optimal Transportation problem between μ_0 and μ_1 is actually a localization of the Monge problem. We will present this fact by considering a function $f : X \to \mathbb{R}$ such that

$$\int_X f(x)\,\mathfrak{m}(dx) = 0, \qquad \int_X |f(x)|\mathsf{d}(x, x_0)\,\mathfrak{m}(dx) < \infty,$$

and considering $\mu_0 := f_+\,\mathfrak{m}$ and $\mu_1 := f_-\,\mathfrak{m}$, where $f_\pm$ denote the positive and the negative parts of f. We can also assume $\mu_0, \mu_1 \in \mathcal{P}(X)$ and study the Monge minimization problem between μ_0 and μ_1. This setting is equivalent to study the general Monge problem assuming both $\mu_0, \mu_1 \ll \mathfrak{m}$; indeed, note that μ_0 and μ_1 can always be assumed to be concentrated on disjoint sets (see [11] for details).

If ϕ is an associated Kantorovich potential producing as before the transport set $\mathcal{T}$, we have a disintegration of $\mathfrak{m}$ as follows:

$$\mathfrak{m}_{\llcorner\mathcal{T}} = \int_Q \mathfrak{m}_q\,\mathfrak{q}(dq), \qquad \mathfrak{m}_q(X_q) = 1, \ \mathfrak{q}\text{-a.e. } q \in Q.$$

Then the natural localization of the Monge problem would be to consider for every $q \in Q$ the Monge minimization problem between

$$\mu_{0\,q} := f_+\,\mathfrak{m}_q, \quad \mu_{1\,q} := f_-\,\mathfrak{m}_q,$$

in the metric space (X_q, d) (that is isometric via the ray map g to an interval of $\mathbb{R}$ with the Euclidean distance). To check that this family of problems makes sense we need to prove the following

Lemma 3.4.13. *It holds that for* $\mathfrak{q}$*-a.e.* $q \in Q$ *one has* $\int_X f\,\mathfrak{m}_q = 0$.

Proof. Since for both μ_0 and μ_1 the set $\mathcal{T}_e \setminus \mathcal{T}$ is negligible ($\mu_0, \mu_1 \ll \mathfrak{m}$), for any Borel set $C \subset Q$

$$\begin{aligned} \mu_0(\mathfrak{Q}^{-1}(C)) &= \pi\Big((\mathfrak{Q}^{-1}(C) \times X) \cap \Gamma \setminus \{x = y\}\Big) \\ &= \pi\Big((X \times \mathfrak{Q}^{-1}(C)) \cap \Gamma \setminus \{x = y\}\Big) \\ &= \mu_1(\mathfrak{Q}^{-1}(C)), \end{aligned} \tag{3.27}$$

where the second equality follows from the fact that $\mathcal{T}$ does not branch: indeed since $\mu_0(\mathcal{T}) = \mu_1(\mathcal{T}) = 1$, then $\pi\big((\Gamma \setminus \{x = y\}) \cap \mathcal{T} \times \mathcal{T}\big) = 1$ and therefore if $x, y \in \mathcal{T}$ and $(x, y) \in \Gamma$, then necessarily $\mathfrak{Q}(x) = \mathfrak{Q}(y)$; that is, they belong to the same ray. It follows that

$$(\mathfrak{Q}^{-1}(C) \times X) \cap (\Gamma \setminus \{x = y\}) \cap (\mathcal{T} \times \mathcal{T}) = (X \times \mathfrak{Q}^{-1}(C)) \cap (\Gamma \setminus \{x = y\}) \cap (\mathcal{T} \times \mathcal{T}),$$

and (3.27) follows.

Since f has null mean value it holds $\int_X f_+(x)\mathfrak{m}(dx) = -\int_X f_-(x)\mathfrak{m}(dx)$, which combined with (3.27) implies that for each Borel $C \subset Q$

$$\begin{aligned}\int_C \int_{X_q} f(x)\mathfrak{m}_q(dx)\mathfrak{q}(dq) &= \int_C \int_{X_q} f_+(x)\mathfrak{m}_q(dx)\mathfrak{q}(dq) - \int_C \int_{X_q} f_-(x)\mathfrak{m}_q(dx)\mathfrak{q}(dq) \\ &= \left(\int_X f_+(x)\mathfrak{m}(dx)\right)^{-1} \left(\mu_0(\mathfrak{Q}^{-1}(C)) - \mu_1(\mathfrak{Q}^{-1}(C))\right) \\ &= 0.\end{aligned}$$

Therefore for $\mathfrak{q}$-a.e. $q \in Q$ the integral $\int f\, \mathfrak{m}_q$ vanishes and the claim follows. □

It can be proven in greater generality and without assuming $\mu_1 \ll \mathfrak{m}$ that the Monge problem is localized once a strongly consistent disintegration of $\mathfrak{m}$ restricted to the transport ray is obtained. See [11] for details.

3.5 Regularity of conditional measures

We now review regularity and curvature properties of $\mathfrak{m}_q$. This section contains a collection of results spread across [11, 17, 18] and [20]. Here, we try to provide a unified presentation. We will inspect three increasing levels of regularity: for $\mathfrak{q}$-a.e. $q \in Q$

- **(R.1)** $\mathfrak{m}_q$ has no atomic part, i.e. $\mathfrak{m}_q(\{x\}) = 0$, for any $x \in X_q$;
- **(R.2)** $\mathfrak{m}_q$ is absolutely continuous with respect to $\mathcal{H}^1 \llcorner_{X_q} = g(q,\cdot)_\sharp \mathcal{L}^1$;
- **(R.3)** $\mathfrak{m}_q = g(q,\cdot)_\sharp(h_q \mathcal{L}^1)$ verifies $\mathsf{CD}(K,N)$, i.e. the m.m.s. $(\mathbb{R}, |\cdot|, h_q \mathcal{L}^1)$ verifies $\mathsf{CD}(K,N)$.

We will review how to obtain **(R.1)**, **(R.2)**, **(R.3)** starting from the following three *increasing* regularity assumptions on the space:

- **(A.1)** if $C \subset \mathcal{T}$ is compact with $\mathfrak{m}(C) > 0$, then $\mathfrak{m}(C_t) > 0$ for uncountably many $t \in \mathbb{R}$;
- **(A.2)** if $C \subset \mathcal{T}$ is compact with $\mathfrak{m}(C) > 0$, then $\mathfrak{m}(C_t) > 0$ for a set of $t \in \mathbb{R}$ with $\mathcal{L}^1$-positive measure;
- **(A.3)** the m.m.s. $(X, \mathsf{d}, \mathfrak{m})$ verifies $\mathsf{CD}(K,N)$.

Given a compact set $C \subset X$, we indicate with C_t its translation along the transport set at distance with sign t, see the following Definition 3.5.1.

We will see that: **(A.1)** implies **(R.1)**, **(A.2)** implies **(R.2)** and **(A.3)** implies **(R.3)**. Actually we will also show a variant of **(A.3)** (assuming MCP instead of CD) implies a variant of **(R.3)** (MCP instead of CD).

Even if we do not always state it, assumptions **(A.1)** and **(A.2)** are not hypothesis on the smoothness of the space but on the regularity of the set Γ and therefore on the Monge problem itself; they should both be read as: *for μ_0 and μ_1 probability measures over X, assume the existence of a 1-Lipschitz Kantorovich potential ϕ such that the associated transport set $\mathcal{T}$ verifies* (**A.1**) *(or* (**A.2**)*).*

3.5.1 Atomless conditional probabilities

The results presented here are taken from [11].

Definition 3.5.1. *Let $C \subset \mathcal{T}$ be a compact set. For $t \in \mathbb{R}$ define the* t-translation C_t of C *by*

$$C_t := g\big(\{(q, s+t) \colon (q,s) \in g^{-1}(C)\}\big).$$

Since $C \subset \mathcal{T}$ is compact, $g^{-1}(C) \subset Q \times \mathbb{R}$ is σ-compact (graph(g) is σ-compact) and the same holds true for

$$\{(q, s+t) \colon (q,s) \in g^{-1}(C)\}.$$

Since

$$C_t = P_3(\text{graph}(g) \cap \{(q, s+t) \colon (q,s) \in g^{-1}(C)\} \times \mathcal{T}),$$

it follows that C_t is σ-compact (projection of σ-compact sets is again σ-compact). Moreover the set $B := \{(t,x) \in \mathbb{R} \times \mathcal{T} \colon x \in C_t\}$ is Borel and therefore by Fubini's Theorem the map $t \mapsto \mathfrak{m}(C_t)$ is Borel. It follows that (**A.1**) makes sense.

Proposition 3.5.2 (Proposition 5.4, [11]). *Assume* (**A.1**) *to hold and the space to be non-branching. Then* (**R.1**) *holds true, that is for* $\mathfrak{q}$*-a.e.* $q \in Q$ *the conditional measure* $\mathfrak{m}_q$ *has no atoms.*

Proof. The partition in trasport rays and the associated disintegration are well defined, see Remark 3.4.6. From the regularity of the disintegration and the fact that $\mathfrak{q}(Q) = 1$, we can assume that the map $q \mapsto \mathfrak{m}_q$ is weakly continuous on a compact set $K \subset Q$ with $\mathfrak{q}(Q \setminus K) < \varepsilon$ such that the length of the ray X_q, denoted by $L(X_q)$, is strictly larger than ε for all $q \in K$. It is enough to prove the proposition on K.

Step 1.

From the continuity of $K \ni q \mapsto \mathfrak{m}_q \in \mathcal{P}(X)$ w.r.t. the weak topology, it follows that the map

$$q \mapsto C(q) := \big\{x \in X_q : \mathfrak{m}_q(\{x\}) > 0\big\} = \cup_n \big\{x \in X_q : \mathfrak{m}_q(\{x\}) \geq 2^{-n}\big\}$$

is σ-closed, i.e. its graph is a countable union of closed sets: in fact, if $(q_m, x_m) \to (y, x)$ and $\mathfrak{m}_{q_m}(\{x_m\}) \geq 2^{-n}$, then $\mathfrak{m}_q(\{x\}) \geq 2^{-n}$ by upper semi-continuity on compact sets.

Hence K is Borel, and by the Lusin Theorem (Theorem 5.8.11 of [50]) it is the countable union of Borel graphs: setting in this case $c_i(q) = 0$, we can consider them as Borel functions on K and order them w.r.t. Γ in the following sense:

$$\mathfrak{m}_{q,\text{atomic}} = \sum_{i \in \mathbb{Z}} c_i(q) \delta_{x_i(q)}, \quad (x_i(q), x_{i+1}(q)) \in \Gamma,\ i \in \mathbb{Z},$$

with $K \ni q \mapsto x_i(q)$ Borel.

Step 2.

Define the sets

$$S_{ij}(t) := \Big\{ q \in K : x_i(q) = g\big(g^{-1}(x_j(q)) + t\big) \Big\},$$

Since $K \subset Q$, to define $S_{ij}(t)$ we are using the graph$(g) \cap Q \times \mathbb{R} \times \mathcal{T}$, which is σ-compact: hence graph(S_{ij}) is analytic. For $A_j := \{x_j(q), q \in K\}$ and $t \in \mathbb{R}^+$ we have that

$$\begin{aligned} \mathfrak{m}((A_j)_t) &= \int_K \mathfrak{m}_q((A_j)_t)\, \mathfrak{q}(dq) = \int_K \mathfrak{m}_{q,\text{atomic}}((A_j)_t)\, \mathfrak{q}(dq) \\ &= \sum_{i \in \mathbb{Z}} \int_K c_i(q) \delta_{x_i(q)}\big(g(g^{-1}(x_j(q)) + t)\big)\, \mathfrak{q}(dq) = \sum_{i \in \mathbb{Z}} \int_{S_{ij}(t)} c_i(q)\, \mathfrak{q}(dq), \end{aligned}$$

and we have used that $A_j \cap X_q$ is a singleton. Then for fixed $i, j \in \mathbb{N}$, again from the fact that $A_j \cap X_q$ is a singleton

$$S_{ij}(t) \cap S_{ij}(t') = \begin{cases} S_{ij}(t) & t = t', \\ \emptyset & t \neq t', \end{cases}$$

and therefore the cardinality of the set $\big\{t : \mathfrak{q}(S_{ij}(t)) > 0\big\}$ has to be countable. On the other hand,

$$\mathfrak{m}((A_j)_t) > 0 \quad \Rightarrow \quad t \in \bigcup_i \big\{t : \mathfrak{q}(S_{ij}(t)) > 0\big\},$$

contradicting (**A.1**). □

3.5.2 Absolute continuity

The results presented here are taken from [11]. The condition (**A.2**) can be stated also in the following way: for every compact set $C \subset \mathcal{T}$

$$\mathfrak{m}(C) > 0 \quad \Rightarrow \quad \int_{\mathbb{R}} \mathfrak{m}(C_t) dt > 0.$$

Lemma 3.5.3. *Let* $\mathfrak{m}$ *be a Radon measure and*

$$\mathfrak{m}_q = r_q\, g(q, \cdot)_\sharp \mathcal{L}^1 + \omega_q, \quad \omega_q \perp g(q, \cdot)_\sharp \mathcal{L}^1$$

be the Radon-Nikodym decomposition of $\mathfrak{m}_q$ *w.r.t.* $g(q, \cdot)_\sharp \mathcal{L}^1$. *Then there exists a Borel set* $C \subset X$ *such that*

$$\mathcal{L}^1\Big(P_2\big(g^{-1}(C) \cap (\{q\} \times \mathbb{R})\big)\Big) = 0,$$

and $\omega_q = \mathfrak{m}_{q \llcorner C}$ *for* $\mathfrak{q}$*-a.e.* $q \in Q$.

Proof. Consider the measure $\lambda = g_\sharp(\mathfrak{q} \otimes \mathcal{L}^1)$, and compute the Radon-Nikodym decomposition

$$\mathfrak{m} = \frac{D\mathfrak{m}}{D\lambda}\lambda + \omega.$$

Then there exists a Borel set C such that $\omega = \mathfrak{m}\llcorner_C$ and $\lambda(C) = 0$. The set C proves the Lemma. Indeed $C = \cup_{q\in Q} C_q$ where $C_q = C \cap R(q)$ is such that $\mathfrak{m}_q\llcorner_{C_q} = \omega_q$ and $g(q,\cdot)_\sharp\mathcal{L}^1(C_q) = 0$ for $\mathfrak{q}$-a.e. $q \in Q$. □

Theorem 3.5.4 (Theorem 5.7, [11]). *Assume* (**A.2**) *to hold and the space to be non-branching. Then* (**R.2**) *holds true, that is for* $\mathfrak{q}$*-a.e.* $q \in Q$ *the conditional measure* $\mathfrak{m}_q$ *is absolute continuous with respect to* $g(q,\cdot)_\sharp\mathcal{L}^1$.

The proof is based on the following simple observation.
Let η be a Radon measure on $\mathbb{R}$. Suppose that for all $A \subset \mathbb{R}$ Borel with $\eta(A) > 0$ it holds

$$\int_{\mathbb{R}^+} \eta(A+t)dt = \eta \otimes \mathcal{L}^1\big(\{(x,t) : t \geq 0, x - t \in A\}\big) > 0.$$

Then $\eta \ll \mathcal{L}^1$.

Proof. The proof will use Lemma 3.5.3: take C the set constructed in Lemma 3.5.3 and suppose by contradiction that

$$\mathfrak{m}(C) > 0 \quad \text{and} \quad \mathfrak{q} \otimes \mathcal{L}^1(g^{-1}(C)) = 0.$$

In particular, for all $t \in \mathbb{R}$ it follows that

$$\mathfrak{q} \otimes \mathcal{L}^1(g^{-1}(C_t)) = 0.$$

By the Fubini-Tonelli Theorem

$$\begin{aligned}
0 < \int_{\mathbb{R}^+} \mathfrak{m}(C_t)\,dt &= \int_{\mathbb{R}^+}\Big(\int_{g^{-1}(C_t)} (g^{-1})_\sharp\mathfrak{m}(dq\,d\tau)\Big)dt \\
&= \big((g^{-1})_\sharp\mathfrak{m} \otimes \mathcal{L}^1\big)\Big(\Big\{(q,\tau,t) : (q,\tau) \in g^{-1}(\mathcal{T}), (q,\tau-t) \in g^{-1}(C)\Big\}\Big) \\
&\leq \int_{Q\times\mathbb{R}} \mathcal{L}^1\big(\{\tau - g^{-1}(C \cap \mathfrak{Q}^{-1}(q))\}\big)\,(g^{-1})_\sharp\mathfrak{m}(dq\,d\tau) \\
&= \int_{Q\times\mathbb{R}} \mathcal{L}^1\big(g^{-1}(C \cap \mathfrak{Q}^{-1}(q))\big)\,(g^{-1})_\sharp\mathfrak{m}(dq\,d\tau) \\
&= \int_{Q} \mathcal{L}^1\big(g^{-1}(C \cap \mathfrak{Q}^{-1}(y))\big)\,\mathfrak{q}(dy) = 0.
\end{aligned}$$

That gives a contradiction. □

The proof of Theorem 3.5.4 inspired the definition of *inversion points* and of *inversion plan* as presented in [19], in particular see **Step 2.** of the proof of Theorem 5.3 of [19].

3.5.3 Weak Ricci curvature bounds: MCP(K, N)

The presentation of the following results is taken from [18]. The same results were proved in [11] using more involved arguments and different notation.

In this section we assume in addition the metric measure space to satisfy the measure contraction property MCP(K, N). Recall that the space is also assumed to be non-branching.

Lemma 3.5.5. *For each Borel $C \subset \mathcal{T}$ and $\delta \in \mathbb{R}$ the set*

$$(C \times \{\phi = \delta\}) \cap \Gamma,$$

is d^2*-cyclically monotone.*

Proof. The proof follows straightforwardly from Lemma 3.4.1. The set $(C \times \{\phi = c\}) \cap \Gamma$ is trivially a subset of Γ and whenever

$$(x_0, y_0), (x_1, y_1) \in (C \times \{\phi = \delta\}) \cap \Gamma,$$

then $(\phi(y_1) - \phi(y_0)) \cdot (\phi(x_1) - \phi(x_0)) = 0$. □

We can deduce the following

Corollary 3.5.6. *For each Borel $C \subset \mathcal{T}$ and $\delta \in \mathbb{R}$ define*

$$C_\delta := P_1((C \times \{\phi = \delta\}) \cap \Gamma).$$

If $\mathfrak{m}(C_\delta) > 0$, there exists a unique $\nu \in \mathrm{OptGeo}$ such that

$$(e_0)_\sharp \nu = \mathfrak{m}(C_\delta)^{-1} \mathfrak{m}\llcorner_{C_\delta}, \qquad (e_0, e_1)_\sharp(\nu)\Big((C \times \{\phi = \delta\}) \cap \Gamma \Big) = 1. \tag{3.28}$$

From Corollary 3.5.6, we infer the existence of a map $T_{C,\delta}$ depending on C and δ such that

$$\left(Id, T_{C,\delta}\right)_\sharp \left(\mathfrak{m}(C_\delta)^{-1} \mathfrak{m}\llcorner_{C_\delta}\right) = (e_0, e_1)_\sharp \nu.$$

Taking advantage of the ray map g, we define a convex combination between the identity map and $T_{C,\delta}$ as follows:

$$C_\delta \ni x \mapsto \left(T_{C,\delta}\right)_t (x) \in \{z \in \Gamma(x) : \mathsf{d}(x, z) = t \cdot \mathsf{d}(x, T_{C,\delta}(x))\}.$$

Since $C \subset \mathcal{T}$, the map $\left(T_{C,\delta}\right)_t$ is well defined for all $t \in [0, 1]$. We then define the evolution of any subset A of C_δ in the following way:

$$[0, 1] \ni t \mapsto \left(T_{C,\delta}\right)_t (A).$$

In particular from now on we will adopt the following notation:

$$A_t := \left(T_{C,\delta}\right)_t (A), \qquad \forall A \subset C_\delta,\ A \text{ compact}.$$

So for any Borel $C \subset \mathcal{T}$ compact and $\delta \in \mathbb{R}$ we have defined an evolution for compact subsets of C_δ. The definition of the evolution depends both on C and δ.

Remark 3.5.7. *Here we spend a few lines on the measurability of the maps involved in the definition of evolution of sets (assuming for simplicity C to be compact). First note that since Γ is closed and C is compact, we can prove that also C_δ is compact. Indeed from the compactness of C we obtain that ϕ is bounded on C and then, since C is bounded, it follows that also $C \times \{\phi = c\} \cap \Gamma$ is bounded. Since X is proper, compactness follows. Moreover*

$$graph(T_{C,\delta}) = (C \times \{\phi = \delta\}) \cap \Gamma,$$

hence $T_{C,\delta}$ is continuous. Moreover

$$\left(T_{C,\delta}\right)_t(A) = P_2\left(\{(x,z) \in \Gamma \cap (A \times X) : \mathsf{d}(x,z) = t \cdot \mathsf{d}(x, T_{C,\delta}(x))\}\right),$$

hence if A is compact, the same hold for $\left(T_{C,\delta}\right)_t(A)$ and

$$[0,1] \ni t \mapsto \mathfrak{m}(\left(T_{C,\delta}\right)_t(A))$$

is $\mathfrak{m}$-measurable.

The next result gives quantitative information on the behavior of the map $t \mapsto \mathfrak{m}(A_t)$. The statement will be given assuming the lower bound on the generalized Ricci curvature K to be positive. Analogous estimates holds for any $K \in \mathbb{R}$.

Proposition 3.5.8. *For each compact $C \subset \mathcal{T}$ and $\delta \in \mathbb{R}$ such that $\mathfrak{m}(C_\delta) > 0$, it holds*

$$\mathfrak{m}(A_t) \geq (1-t) \cdot \inf_{x \in A} \left(\frac{\sin\left((1-t)\mathsf{d}(x, T_{C,\delta}(x))\sqrt{K/(N-1)}\right)}{\sin\left(\mathsf{d}(x, T_{C,\delta}(x))\sqrt{K/(N-1)}\right)} \right)^{N-1} \mathfrak{m}(A), \tag{3.29}$$

for all $t \in [0,1]$ and $A \subset C_\delta$ compact set.

Proof. The proof of (3.29) is obtained by the standard method of approximation with Dirac deltas of the second marginal. Even though similar arguments already appeared many times in literature, in order to be self-contained, we include all the details. For ease of notation $T = T_{C,\delta}$ and $C = C_\delta$.

Step 1.
Consider a sequence $\{y_i\}_{i \in \mathbb{N}} \subset \{\phi = \delta\}$ dense in $T(C)$. For each $I \in \mathbb{N}$, define the family of sets

$$E_{i,I} := \{x \in C : \mathsf{d}(x, y_i) \leq \mathsf{d}(x, y_j), j = 1, \ldots, I\},$$

for $i = 1, \ldots, I$. Then for all $I \in \mathbb{N}$, by the same argument of Lemma 3.5.5, the set

$$\Lambda_I := \bigcup_{i=1}^{I} E_{i,I} \times \{y_i\} \subset X \times X,$$

is d^2-cyclically monotone. Consider then $A_{i,I} := A \cap E_{i,I}$ and the approximate evolution

$$A_{i,I,t} := \{z \in X : \mathsf{d}(z, y_i) = (1-t)\mathsf{d}(x, y_i),\ x \in A_{i,I}\};$$

notice that $A_{i,I,0} = A_{i,I}$. Then by MCP(K, N) it holds

$$\mathfrak{m}(A_{i,I,t}) \geq (1-t) \cdot \inf_{x \in A_{i,I}} \left(\frac{\sin\left((1-t)\mathsf{d}(x,x_i)\sqrt{K/(N-1)}\right)}{\sin\left(\mathsf{d}(x,x_i)\sqrt{K/(N-1)}\right)} \right)^{N-1} \mathfrak{m}(A_{i,I}).$$

Taking the sum over $i \leq I$ in the previous inequality implies

$$\sum_{i \leq I} \mathfrak{m}(A_{i,I,t}) \geq (1-t) \cdot \inf_{x \in A} \left(\frac{\sin\left((1-t)\mathsf{d}(x,T_I(x))\sqrt{K/(N-1)}\right)}{\sin\left(\mathsf{d}(x,T_I(x))\sqrt{K/(N-1)}\right)} \right)^{N-1} \mathfrak{m}(A),$$

where $T_I(x) := y_i$ for $x \in E_{i,I}$. From d^2-cyclically monotonicity and the non-branching of the space, up to a set of measure zero, the map T_I is well defined, i.e. $\mathfrak{m}(E_{i,I} \cap E_{j,I}) = 0$ for $i \neq j$. It follows that for each $I \in \mathbb{N}$ we can remove a set of measure zero from A and obtain

$$A_{i,I,t} \cap A_{j,I,t} = \emptyset, \quad i \neq j.$$

As before consider also the interpolated map $T_{I,t}$ and observe that $A_{I,t} = T_{I,t}(A)$. Since also A is compact we obtain

$$\mathfrak{m}(A_{I,t}) \geq (1-t) \cdot \min_{x \in A} \left(\frac{\sin\left((1-t)\mathsf{d}(x,T_I(x))\sqrt{K/(N-1)}\right)}{\sin\left(\mathsf{d}(x,T_I(x))\sqrt{K/(N-1)}\right)} \right)^{N-1} \mathfrak{m}(A).$$

Step 2.
Since C is a compact set, for every $I \in \mathbb{N}$ the set Λ_I is compact as well and it is a subset of $C \times \{\phi = \delta\}$ that can be assumed to be compact as well. By compactness, there exists a subsequence I_n and a compact set $\Theta \subset C \times \{\phi = \delta\}$ compact such that

$$\lim_{n \to \infty} \mathsf{d}_{\mathcal{H}}(\Lambda_{I_n}, \Theta) = 0,$$

where $\mathsf{d}_{\mathcal{H}}$ is the Hausdorff distance. Since the sequence $\{y_i\}_{i \in \mathbb{N}}$ is dense in $\{\phi = \delta\}$ and $C \subset \mathcal{T}$ is compact, by definition of $E_{i,I}$, necessarily for every $(x, y) \in \Theta$ it holds

$$\phi(x) + \phi(y) = \mathsf{d}(x,y), \quad \phi(y) = \delta.$$

Hence $\Theta \subset \Gamma \cap C \times \{\phi = \delta\}$ and this in particular implies, by upper semicontinuity of $\mathfrak{m}$ along converging sequences of closed sets, that

$$\mathfrak{m}(A_t) \geq \limsup_{n \to \infty} \mathfrak{m}(A_{I_n,t}).$$

The claim follows. □

As the goal is to localize curvature conditions, we first need to prove that almost every conditional probability is absolutely continuous with respect to the one dimensional

Hausdorff measure restricted to the correct geodesic. One way is to prove that Proposition 3.5.8 implies (**A.2**) and then apply Theorem 3.5.4 to obtain (**R.2**) (approach used in [11]). Another option is to repeat verbatim the proof of Theorem 3.5.4 substituting the translation with the evolution considered in Proposition 3.5.8 and to observe that the claim follows (approach used in [18]). So we take for granted the following.

Proposition 3.5.9. *Assume the non-branching m.m.s.* $(X, \mathsf{d}, \mathfrak{m})$ *to satisfy* $\mathsf{MCP}(K, N)$. *Then* (**R.2**) *holds true, that is for* $\mathfrak{q}$*-a.e.* $q \in Q$ *the conditional measure* $\mathfrak{m}_q$ *is absolute continuous with respect to* $g(q, \cdot)_\sharp \mathcal{L}^1$.

To fix the notation, we now have proved the existence of a Borel function $h : \mathrm{Dom}\,(g) \to \mathbb{R}_+$ such that

$$\mathfrak{m} \llcorner \mathcal{T} = g_\sharp \left(h \, \mathfrak{q} \otimes \mathcal{L}^1 \right) \tag{3.30}$$

Using standard arguments, estimate (3.29) can be localized at the level of the density h: for each compact set $A \subset \mathcal{T}$

$$\int\limits_{P_2(g^{-1}(A_t))} h(q,s)\mathcal{L}^1(ds)$$

$$\geq (1-t)\left(\inf_{\tau \in P_2(g^{-1}(A))} \frac{\sin((1-t)|\tau - \sigma|\sqrt{K/(N-1)})}{\sin(|\tau-\sigma|\sqrt{K/(N-1)})} \right)^{N-1} \int\limits_{P_2(g^{-1}(A))} h(q,s)\mathcal{L}^1(ds),$$

for $\mathfrak{q}$-a.e. $q \in Q$ such that $g(q, \sigma) \in \mathcal{T}$. Then using a change of variables, one obtains that for $\mathfrak{q}$-a.e. $q \in Q$:

$$h(q, s + |s-\sigma|t) \geq \left(\frac{\sin((1-t)|s-\sigma|\sqrt{K/(N-1)})}{\sin(|s-\sigma|\sqrt{K/(N-1)})} \right)^{N-1} h(y,s),$$

for $\mathcal{L}^1$-a.e. $s \in P_2(g^{-1}(R(q)))$ and $\sigma \in \mathbb{R}$ such that $s + |\sigma - s| \in P_2(g^{-1}(R(q)))$. We can rewrite the estimate in the following way:

$$h(q,\tau) \geq \left(\frac{\sin((\sigma-\tau)\sqrt{K/(N-1)})}{\sin((\sigma-s)\sqrt{K/(N-1)})} \right)^{N-1} h(q,s),$$

for $\mathcal{L}^1$-a.e. $s \leq \tau \leq \sigma$ such that $g(q,s), g(q,\tau), g(q,\sigma) \in \mathcal{T}$. Since an evolution can be also considered backwardly, we have proved the result below.

Theorem 3.5.10 (Localization of MCP, Theorem 9.5 of [11]). *Assume the non-branching m.m.s.* $(X, \mathsf{d}, \mathfrak{m})$ *to satisfy* $\mathsf{MCP}(K, N)$. *For* $\mathfrak{q}$*-a.e.* $q \in Q$ *it holds:*

$$\left(\frac{\sin((\sigma_+ - \tau)\sqrt{K/(N-1)})}{\sin((\sigma_+ - s)\sqrt{K/(N-1)})} \right)^{N-1} \leq \frac{h(q,\tau)}{h(q,s)} \leq \left(\frac{\sin((\tau - \sigma_-)\sqrt{K/(N-1)})}{\sin((s - \sigma_-)\sqrt{K/(N-1)})} \right)^{N-1},$$

for $\sigma_- < s \leq \tau < \sigma_+$ *such that their image via* $g(q,\cdot)$ *is contained in* $R(q)$.

In particular from Theorem 3.5.10 we deduce that

$$\{t \in \mathrm{Dom}\,(g(q,\cdot)) : h(q,t) > 0\} = \mathrm{Dom}\,(g(q,\cdot)), \tag{3.31}$$

such a set is convex and $t \mapsto h(q,t)$ is locally Lipschitz continuous.

3.5.4 Weak Ricci curvature bounds: CD(K, N)

The results presented here are taken from [20].

We now turn to proving that the conditional probabilities inherit the synthetic Ricci curvature lower bounds, that is, (**A.3**) implies (**R.3**). Actually, it is enough to assume the space locally satisfies such a lower bound to obtain a global synthetic Ricci curvature lower bound on almost every 1-dimensional metric measure space.

Since under the essentially non-branching condition $\mathsf{CD}_{loc}(K,N)$ implies $\mathsf{MCP}(K,N)$ and existence and uniqueness of optimal transport maps, see [22], we can already assume (3.30) and (3.31) to hold. In particular $t \mapsto h_q(t)$ is locally Lipschitz continuous, where for ease of notation $h_q = h(q,\cdot)$.

Theorem 3.5.11 (Theorem 4.2 of [20]). *Let $(X, \mathsf{d}, \mathfrak{m})$ be an essentially non-branching m.m.s. verifying the $\mathsf{CD}_{loc}(K,N)$ condition for some $K \in \mathbb{R}$ and $N \in [1,\infty)$.*

Then for any 1-Lipschitz function $\phi : X \to \mathbb{R}$, the associated transport set Γ induces a disintegration of $\mathfrak{m}$ restricted to the transport set verifying the following inequality: if $N > 1$

for $\mathfrak{q}$-a.e. $q \in Q$ the following curvature inequality holds:

$$h_q((1-s)t_0 + st_1)^{1/(N-1)} \geq \sigma^{(1-s)}_{K,N-1}(t_1 - t_0)h_q(t_0)^{1/(N-1)} + \sigma^{(s)}_{K,N-1}(t_1 - t_0)h_q(t_1)^{1/(N-1)}, \tag{3.32}$$

for all $s \in [0,1]$ and for all $t_0, t_1 \in \mathrm{Dom}\,(g(q,\cdot))$ with $t_0 < t_1$. If $N = 1$, for $\mathfrak{q}$-a.e. $q \in Q$ the density h_q is constant.

Proof. We first consider the case $N > 1$.

Step 1.

Thanks to Proposition 3.4.8, without any loss of generality we can assume that the quotient set Q (identified with the set $\{g(q,0) : q \in Q\}$) is locally a subset of a level set of the map ϕ inducing the transport set, i.e. there exists a countable partition $\{Q_i\}_{i\in\mathbb{N}}$ with $Q_i \subset Q$ Borel set such that

$$\{g(q,0) : q \in Q_i\} \subset \{x \in X : \phi(x) = \alpha_i\}.$$

It is clearly sufficient to prove (3.32) on each Q_i; so fix $\bar{i} \in \mathbb{N}$ and for ease of notation assume $\alpha_{\bar{i}} = 0$ and $Q = Q_{\bar{i}}$. As $\mathrm{Dom}\,(g(q,\cdot))$ is a convex subset of $\mathbb{R}$, we can also restrict to a uniform subinterval

$$(a_0, a_1) \subset \mathrm{Dom}\,(g(q,\cdot)), \qquad \forall\, q \in Q_i,$$

for some $a_0, a_1 \in \mathbb{R}$. Again without any loss of generality we also assume $a_0 < 0 < a_1$.

Consider any $a_0 < A_0 < A_1 < a_1$ and $L_0, L_1 > 0$ such that $A_0 + L_0 < A_1$ and $A_1 + L_1 < a_1$. Then define the following two probability measures

$$\mu_0 := \int_Q g(q,\cdot)_\sharp \left(\frac{1}{L_0}\mathcal{L}^1 \llcorner_{[A_0,A_0+L_0]}\right) \mathfrak{q}(dq), \ \mu_1 := \int_Q g(q,\cdot)_\sharp \left(\frac{1}{L_1}\mathcal{L}^1 \llcorner_{[A_1,A_1+L_1]}\right) \mathfrak{q}(dq).$$

Since $g(q,\cdot)$ is an isometry one can also represent μ_0 and μ_1 in the following way:

$$\mu_i := \int_Q \frac{1}{L_i}\mathcal{H}^1 \llcorner_{\{g(q,t):\ t\in[A_i,A_i+L_i]\}} \mathfrak{q}(dq)$$

for $i = 0, 1$. Both μ_i are absolutely continuous with respect to $\mathfrak{m}$ and $\mu_i = \varrho_i \mathfrak{m}$ with

$$\varrho_i(g(q,t)) = \frac{1}{L_i} h_q(t)^{-1}, \qquad \forall t \in [A_i, A_i + L_i].$$

Moreover from Lemma 3.4.1 it follows that the curve $[0,1] \ni s \mapsto \mu_s \in \mathcal{P}(X)$ defined by

$$\mu_s := \int_Q \frac{1}{L_s}\mathcal{H}^1 \llcorner_{\{g(q,t):\ t\in[A_s,A_s+L_s]\}} \mathfrak{q}(dq)$$

where

$$L_s := (1-s)L_0 + sL_1, \qquad A_s := (1-s)A_0 + sA_1$$

is the unique L^2-Wasserstein geodesic connecting μ_0 to μ_1. Again one has $\mu_s = \varrho_s \mathfrak{m}$ and can also write its density in the following way:

$$\varrho_s(g(q,t)) = \frac{1}{L_s} h_q(t)^{-1}, \qquad \forall t \in [A_s, A_s + L_s].$$

Step 2.
By $\mathsf{CD}_{loc}(K,N)$ and the essentially non-branching property one has: for $\mathfrak{q}$-a.e. $q \in Q_i$

$$(L_s)^{\frac{1}{N}} h_q((1-s)t_0 + st_1)^{\frac{1}{N}} \geq \tau^{(1-s)}_{K,N}(t_1 - t_0)(L_0)^{\frac{1}{N}} h_q(t_0)^{\frac{1}{N}} + \tau^{(s)}_{K,N}(t_1 - t_0)(L_1)^{\frac{1}{N}} h_q(t_1)^{\frac{1}{N}},$$

for $\mathcal{L}^1$-a.e. $t_0 \in [A_0, A_0 + L_0]$, t_1 obtained as the image of t_0 through the monotone rearrangement of $[A_0, A_0 + L_0]$ to $[A_1, A_1 + L_1]$, and every $s \in [0,1]$. If $t_0 = A_0 + \tau L_0$, then $t_1 = A_1 + \tau L_1$. Also A_0 and $A_1 + L_1$ should be taken close enough to verify the local curvature condition.

Then we can consider the previous inequality for $s = 1/2$, include the explicit formula for t_1, and obtain:

$$\begin{aligned}
&(L_0 + L_1)^{\frac{1}{N}} h_q(A_{1/2} + \tau L_{1/2})^{\frac{1}{N}} \\
&\geq \sigma^{(1/2)}_{K,N-1}(A_1 - A_0 + \tau|L_1 - L_0|)^{\frac{N-1}{N}} \left\{(L_0)^{\frac{1}{N}} h_q(A_0 + \tau L_0)^{\frac{1}{N}} + (L_1)^{\frac{1}{N}} h_q(A_1 + \tau L_1)^{\frac{1}{N}}\right\},
\end{aligned}$$

for $\mathcal{L}^1$-a.e. $\tau \in [0,1]$, where we used the notation $A_{1/2} := \frac{A_0+A_1}{2}, L_{1/2} := \frac{L_0+L_1}{2}$. Now observing that the map $s \mapsto h_q(s)$ is continuous, the previous inequality also holds for $\tau = 0$:

$$(L_0 + L_1)^{\frac{1}{N}} h_q(A_{1/2})^{\frac{1}{N}} \geq \sigma^{(1/2)}_{K,N-1}(A_1 - A_0)^{\frac{N-1}{N}} \left\{ (L_0)^{\frac{1}{N}} h_q(A_0)^{\frac{1}{N}} + (L_1)^{\frac{1}{N}} h_q(A_1)^{\frac{1}{N}} \right\}, \quad (3.33)$$

for all $A_0 < A_1$ with $A_0, A_1 \in (a_0, a_1)$, all sufficiently small L_0, L_1 and $\mathfrak{q}$-a.e. $q \in Q$, with the exceptional set depending on A_0, A_1, L_0 and L_1.

Notice that (3.33) depends in a continuous way on A_0, A_1, L_0 and L_1. It follows that there exists a common exceptional set $N \subset Q$ such that $\mathfrak{q}(N) = 0$ and for each $q \in Q \setminus N$, for all A_0, A_1, L_0 and L_1 the inequality (3.33) holds true. Then one can make the following (optimal) choice

$$L_0 := L \frac{h_q(A_0)^{\frac{1}{N-1}}}{h_q(A_0)^{\frac{1}{N-1}} + h_q(A_1)^{\frac{1}{N-1}}}, \qquad L_1 := L \frac{h_q(A_1)^{\frac{1}{N-1}}}{h_q(A_0)^{\frac{1}{N-1}} + h_q(A_1)^{\frac{1}{N-1}}},$$

for any $L > 0$ sufficiently small, and obtain that

$$h_q(A_{1/2})^{\frac{1}{N-1}} \geq \sigma^{(1/2)}_{K,N-1}(A_1 - A_0) \left\{ h_q(A_0)^{\frac{1}{N-1}} + h_q(A_1)^{\frac{1}{N-1}} \right\}. \quad (3.34)$$

Now one can observe that (3.34) is precisely the inequality requested for $\mathsf{CD}^*_{loc}(K, N-1)$ to hold. As stated in Section 3.2.1, the reduced curvature-dimension condition verifies the local-to-global property. In particular, see [22, Lemma 5.1, Theorem 5.2], if a function verifies (3.34) locally, then it also satisfies it globally. Hence h_q also verifies the inequality requested for $\mathsf{CD}^*(K, N-1)$ to hold, i.e. for $\mathfrak{q}$-a.e. $q \in Q$, the density h_q verifies (3.32).

Step 3.
For the case $N = 1$, repeat the same construction of **Step 1.** and obtain for $\mathfrak{q}$-a.e. $q \in Q$

$$(L_s) h_q((1-s)t_0 + s t_1) \geq (1-s) L_0 h_q(t_0) + s L_1 h_q(t_1),$$

for any $s \in [0,1]$ and L_0 and L_1 sufficiently small. As before, we deduce for $s = 1/2$ that

$$\frac{L_0 + L_1}{2} h_q(A_{1/2}) \geq \frac{1}{2} \left(L_0 h_q(A_0) + L_1 h_q(A_1) \right).$$

Now taking $L_0 = 0$ or $L_1 = 0$, it follows that necessarily h_q has to be constant. □

According to Remark 3.2.3, Theorem 3.5.11 can be alternatively stated as follows.

If $(X, \mathsf{d}, \mathfrak{m})$ is an essentially non-branching m.m.s. verifying $\mathsf{CD}_{loc}(K, N)$ and $\phi : X \to \mathbb{R}$ is a 1-Lipschitz function, then the corresponding decomposition of the space in maximal rays $\{X_q\}_{q \in Q}$ produces a disintegration $\{\mathfrak{m}_q\}_{q \in Q}$ of $\mathfrak{m}$ so that for $\mathfrak{q}$-a.e. $q \in Q$,

$$\text{the m.m.s. } (\mathrm{Dom}\,(g(q,\cdot)), |\cdot|, h_q \mathcal{L}^1) \quad \text{verifies} \quad \mathsf{CD}(K, N).$$

Accordingly, one says that the disintegration $q \mapsto \mathfrak{m}_q$ is a $\mathsf{CD}(K, N)$ disintegration.

The disintegration obtained with L^1-Optimal Transportation is also balanced in the sense of Section 3.4.2. This additional information together with what proved so far is summarized below.

Theorem 3.5.12 (Theorem 5.1 of [20]). *Let $(X, \mathsf{d}, \mathfrak{m})$ be an essentially non-branching metric measure space verifying the $\mathsf{CD}_{loc}(K, N)$ condition for some $K \in \mathbb{R}$ and $N \in [1, \infty)$. Let $f : X \to \mathbb{R}$ be $\mathfrak{m}$-integrable such that $\int_X f\,\mathfrak{m} = 0$ and assume the existence of $x_0 \in X$ such that $\int_X |f(x)|\,\mathsf{d}(x, x_0)\,\mathfrak{m}(dx) < \infty$.*

Then the space X can be written as the disjoint union of two sets Z and $\mathcal{T}$ with $\mathcal{T}$ admitting a partition $\{X_q\}_{q\in Q}$ and a corresponding disintegration of $\mathfrak{m}\llcorner_{\mathcal{T}}$, $\{\mathfrak{m}_q\}_{q\in Q}$ such that:

- *For any $\mathfrak{m}$-measurable set $B \subset \mathcal{T}$ it holds*

$$\mathfrak{m}(B) = \int_Q \mathfrak{m}_q(B)\,\mathfrak{q}(dq),$$

  where $\mathfrak{q}$ is a probability measure over Q defined on the quotient σ-algebra $\mathcal{Q}$.
- *For $\mathfrak{q}$-almost every $q \in Q$, the set X_q is a geodesic and $\mathfrak{m}_q$ is supported on it. Moreover $q \mapsto \mathfrak{m}_q$ is a $\mathsf{CD}(K, N)$ disintegration.*
- *For $\mathfrak{q}$-almost every $q \in Q$, it holds $\int_{X_q} f\,\mathfrak{m}_q = 0$ and $f = 0$ $\mathfrak{m}$-a.e. in Z.*

The proof is just a collection of already proven statements. We include it for the reader's convenience.

Proof. Consider

$$\mu_0 := f_+\mathfrak{m}\frac{1}{\int f_+\mathfrak{m}}, \qquad \mu_1 := f_-\mathfrak{m}\frac{1}{\int f_-\mathfrak{m}},$$

where $f_\pm$ stands for the positive and negative part of f, respectively. From the summability assumption on f it follows the existence of $\phi : X \to \mathbb{R}$, 1-Lipschitz Kantorovich potential for the pair of marginal probabilities μ_0, μ_1. Since the m.m.s. $(X, \mathsf{d}, \mathfrak{m})$ is essentially non-branching, the transport set $\mathcal{T}$ is partitioned by the rays:

$$\mathfrak{m}_{\mathcal{T}} = \int_Q \mathfrak{m}_q\,\mathfrak{q}(dq), \qquad \mathfrak{m}_q(X_q) = 1, \quad \mathfrak{q} - \text{a.e. } q \in Q;$$

moreover $(X, \mathsf{d}, \mathfrak{m})$ verifies CD_{loc} and therefore Theorem 3.5.11 implies that $q \mapsto \mathfrak{m}_q$ is a $\mathsf{CD}(K, N)$ disintegration. Lemma 3.4.13 implies that

$$\int_{X_q} f(x)\,\mathfrak{m}_q(dx) = 0.$$

Moreover, note that f has necessarily to be zero in $X \setminus \mathcal{T}$. Take indeed any $B \subset X \setminus \mathcal{T}$ compact with $\mathfrak{m}(B) > 0$ and assume $f \neq 0$ over B. Then possibly taking a subset, we can assume $f > 0$ over B and therefore $\mu_0(B) > 0$. Since

$$\mu_0 = \int_Q \mu_{0\,q}\mathfrak{q}(dq), \qquad \mu_{0\,q}(X_q) = 1,$$

necessarily B cannot be a subset of $X \setminus \mathcal{T}$ yielding a contradiction. All the claims are proved. □

3.6 Applications

Here we will summarize some applications of the results proved so far, in particular of Proposition 3.5.2 and Theorem 3.5.11.

3.6.1 Solution of the Monge problem

We review how the regularity of conditional probabilities of the one-dimensional disintegration allows for the construction of a solution to the Monge problem. In particular we will see how Proposition 3.5.2 leads to an optimal map T. First we recall the one dimensional result for the Monge problem [56].

Theorem 3.6.1. *Let μ_0, μ_1 be probability measures on $\mathbb{R}$, μ_0 with no atoms, and let*

$$H(s) := \mu_0((-\infty, s)), \quad F(t) := \mu_1((-\infty, t)),$$

be the left-continuous distribution functions of μ_0 and μ_1 respectively. Then the following holds.

1. *The non decreasing function $T : \mathbb{R} \to \mathbb{R} \cup [-\infty, +\infty)$ defined by*

$$T(s) := \sup \{ t \in \mathbb{R} : F(t) \le H(s) \}$$

   maps μ_0 to μ_1. Moreover any other non decreasing map T' such that $T'_\sharp \mu_0 = \mu_1$ coincides with T on the support of μ_0 up to a countable set.
2. *If $\varphi : [0, +\infty] \to \mathbb{R}$ is non decreasing and convex, then T is an optimal transport relative to the cost $c(s,t) = \varphi(|s-t|)$. Moreover T is the unique optimal transference map if φ is strictly convex.*

Theorem 3.6.2 (Theorem 6.2 of [11]). *Let $(X, \mathsf{d}, \mathfrak{m})$ be a non-branching metric measure space and consider $\mu_0, \mu_1 \in \mathcal{P}(X)$ with finite first moment. Assume the existence of a Kantorovich potential ϕ such that the associated transport set $\mathcal{T}$ verifies* (**A.1**). *Assume $\mu_0 \ll \mathfrak{m}$.*

Then there exists a Borel map $T : X \to X$ such that

$$\int_X \mathsf{d}(x, T(X))\, \mu_0(dx) = \min_{\pi \in \Pi(\mu_0, \mu_1)} \int_{X \times X} \mathsf{d}(x, y)\, \pi(dxdy).$$

Theorem 3.6.2 was presented in [11] assuming the space to be non-branching, while here we assume essentially non-branching.

Proof. **Step 1.** One dimensional reduction of μ_0.
Let $\phi : X \to \mathbb{R}$ be the Kantorovich potential from the assumptions and $\mathcal{T}$ the corresponding transport set. Accordingly

$$\mathfrak{m}\llcorner_{\mathcal{T}} = \int_Q \mathfrak{m}_q\, \mathfrak{q}(dq),$$

with $\mathfrak{m}_q(X_q) = 1$ for $\mathfrak{q}$-a.e. $q \in Q$. Moreover from **(A.1)** for $\mathfrak{q}$-a.e. $q \in Q$ the conditional $\mathfrak{m}_q$ has no atoms, i.e. $\mathfrak{m}_q(\{z\}) = 0$ for all $z \in X$. From Lemma 3.3.4, we can assume that $\mu_0(\mathcal{T}_e) = \mu_1(\mathcal{T}_e) = 1$. Since $\mu_0 = \varrho_0 \mathfrak{m}$, with $\varrho_0 : X \to [0, \infty)$, from Theorem 3.3.5 we have $\mu_0(\mathcal{T}) = 1$. Hence

$$\mu_0 = \int_Q \varrho_0 \mathfrak{m}_q\, \mathfrak{q}(dq) = \int_Q \mu_{0\,q}\, \mathfrak{q}_0(dq), \qquad \mu_{0\,q} := \varrho_0 \mathfrak{m}_q \left(\int_X \varrho_0(x) \mathfrak{m}_q(dx) \right)^{-1},$$

and $\mathfrak{q}_0 = \mathfrak{Q}_\sharp \mu_0$. In particular $\mu_{0,\,q}$ has no atoms and $\mu_{0\,q}(X_q) = 1$.

Step 2. One dimensional reduction of μ_1.
As we are not making any assumption on μ_1 we cannot exclude that $\mu_1(\mathcal{T}_e \setminus \mathcal{T}) > 0$ and therefore to localize μ_1 one cannot proceed as for μ_0. Consider therefore an optimal transport plan π with $\pi(\Gamma) = 1$. Since $\pi(\mathcal{T} \times \mathcal{T}_e) = 1$ and a partition of $\mathcal{T}$ is given, we can consider the following family of sets $\{X_q \times \mathcal{T}_e\}_{q \in Q}$ as a partition of $\mathcal{T} \times \mathcal{T}_e$; note indeed that $X_q \times \mathcal{T}_e \cap X_{q'} \cap \mathcal{T}_e = \emptyset$ if $q \neq q'$. The domain of the quotient map $\mathfrak{Q} : \mathcal{T} \to Q$ can be trivially extended to $\mathcal{T} \times \mathcal{T}_e$ by saying that $\mathfrak{Q}(x, z) = \mathfrak{Q}(x)$ and observing that

$$\mathfrak{Q}_\sharp\, \pi(I) = \pi\left(\mathfrak{Q}^{-1}(I)\right) = \pi\left(\mathfrak{Q}^{-1}(I) \times \mathcal{T}_e\right) = \mu_0(\mathfrak{Q}^{-1}(I)) = \mathfrak{q}_0(I).$$

In particular this implies that

$$\pi = \int_Q \pi_q\, \mathfrak{q}_0(dq), \qquad \pi_q(X_q \times \mathcal{T}_e) = 1, \quad \text{for } \mathfrak{q}_0\text{-a.e. } q \in Q.$$

Then applying the projection

$$\mu_0 = P_{1\,\sharp}\pi = \int_Q P_{1\,\sharp}(\pi_q)\, \mathfrak{q}_0(dq),$$

and by uniqueness of disintegration $P_{1\,\sharp}(\pi_q) = \mu_{0\,q}$ for $\mathfrak{q}_0$-a.e. $q \in Q$. Then we can find a localization of μ_1 as follows:

$$\mu_1 = P_{2\,\sharp}\pi = \int_Q P_{2\,\sharp}(\pi_q)\, \mathfrak{q}_0(dq) = \int_Q \mu_{1\,q}\, \mathfrak{q}_0(dq),$$

where by definition we posed $\mu_{1\,q} := P_{2\,\sharp}(\pi_q)$ and by construction $\mu_{1\,q}(X_q) = \mu_{0\,q}(X_q) = 1$.

Step 3. Solution to the Monge problem.
For each $q \in Q$ consider the distribution functions

$$H(q,t) := \mu_{0\,q}((-\infty,t)), \quad F(q,t) := \mu_{1\,q}((-\infty,t)),$$

where for ease of notation $\mu_{i\,q} = g(q,\cdot)^{-1}_{\sharp}\mu_{i\,q}$ for $i = 0, 1$. Then define $\hat{T}$, as Theorem 3.6.1 suggests, by

$$\hat{T}(q,s) := \Big(q, \sup\{t : F(q,t) \le H(q,s)\}\Big).$$

Note that since H is continuous ($\mu_{0\,q}$ has no atoms), the map $s \mapsto \hat{T}(q,s)$ is well-defined. Then define the transport map $T : \mathcal{T} \to X$ as $g \circ \hat{T} \circ g^{-1}$. It is fairly easy to observe that

$$T_\sharp\,\mu_0 = \int_Q \left(g \circ \hat{T} \circ g^{-1}\right)_\sharp \mu_{0\,q}\,\mathfrak{q}_0(dq) = \int_Q \mu_{1\,q}\,\mathfrak{q}_0(dq) = \mu_1;$$

moreover $(x, T(x)) \in \Gamma$ and therefore the graph of T is d-cyclically monotone; therefore the map T is optimal. Extend T to X as the identity.

It remains to show that it is Borel. First observe that, possibly taking a compact subset of Q, the map $q \mapsto (\mu_{0\,q}, \mu_{1\,q})$ can be assumed to be weakly continuous; it follows that the maps

$$\mathrm{Dom}\,(g) \ni (q,t) \mapsto H(q,t) := \mu_{0\,q}((-\infty,t)), \quad (q,t) \mapsto F(q,t) := \mu_{1\,q}((-\infty,t))$$

are lower semicontinuous. Then for A Borel,

$$\hat{T}^{-1}(A \times [t,+\infty)) = \big\{(q,s) : q \in A, H(q,s) \ge F(q,t)\big\} \in \mathcal{B}(Q \times \mathbb{R}),$$

and therefore the same applies for T. □

If $(X, \mathsf{d}, \mathfrak{m})$ verifies MCP then it also verifies (**A.1**), see Proposition 3.5.9. So we have the following

Corollary 3.6.3 (Corollary 9.6 of [11]). *Let $(X, \mathsf{d}, \mathfrak{m})$ be a non-branching metric measure space verifying $\mathsf{MCP}(K,N)$. Let μ_0 and μ_1 be probability measures with finite first moment and $\mu_0 \ll \mathfrak{m}$. Then there exists a Borel optimal transport map $T : X \to X$ solution to the Monge problem.*

Corollary 3.6.3 in particular implies the existence of solutions to the Monge problem in the Heisenberg group when μ_0 is assumed to be absolutely continuous with respect to the left-invariant Haar measure.

Theorem 3.6.4 (Monge problem in the Heisenberg group). *Consider $(\mathbb{H}^n, \mathsf{d}_c, \mathcal{L}^{2n+1})$, the n-dimensional Heisenberg group endowed with the Carnot-Carathéodory distance*

d_c *and the* $(2n+1)$*-Lebesgue measure that coincide with the Haar measure on* $(\mathbb{H}^n, \mathsf{d}_c)$ *under the identification* $\mathbb{H}^n \simeq \mathbb{R}^{2n+1}$. *Let* μ_0 *and* μ_1 *be two probability measures with finite first moment and* $\mu_0 \ll \mathcal{L}^{2n+1}$. *Then there exists a Borel optimal transport map* $T : X \to X$ *solution to the Monge problem.*

Remark 3.6.5. *The techniques used so far were successfully used also to threat the more general case of infinite dimensional spaces with curvature bound, see [16] where the existence of solutions for the Monge minimization problem in the Wiener space is proved. Note that the material presented in the previous sections can be obtained also without assuming the existence of a* 1*-Lipschitz Kantorovich potential (e.g. the Wiener space); the decomposition of the space in geodesics and the associated disintegration of the reference measures can be obtained starting from a generic* d*-cyclically monotone set. For all the details see [11].*

3.6.2 Isoperimetric inequality

We now turn to the second main application of techniques reviewed so far, the Lévy-Gromov isoperimetric inequality in singular spaces. The results of this section are taken from [20, 21].

Theorem 3.6.6 (Theorem 1.2 of [20]). *Let* $(X, \mathsf{d}, \mathfrak{m})$ *be a metric measure space with* $\mathfrak{m}(X) = 1$, *verifying the essentially non-branching property and* $\mathsf{CD}_{loc}(K, N)$ *for some* $K \in \mathbb{R}$, $N \in [1, \infty)$. *Let* D *be the diameter of* X, *possibly assuming the value* ∞.

Then for every $v \in [0, 1]$,

$$\mathcal{I}_{(X,\mathsf{d},\mathfrak{m})}(v) \geq \mathcal{I}_{K,N,D}(v),$$

where $\mathcal{I}_{K,N,D}$ *is the model isoperimetric profile defined in* (3.16).

Proof. First of all we can assume $D < \infty$ and therefore $\mathfrak{m} \in \mathcal{P}_2(X)$: indeed from the Bonnet-Myers Theorem if $K > 0$ then $D < \infty$, and if $K \leq 0$ and $D = \infty$ then the model isoperimetric profile (3.16) trivializes, i.e. $\mathcal{I}_{K,N,\infty} \equiv 0$ for $K \leq 0$.

For $v = 0, 1$ one can take as competitor the empty set and the whole space respectively, so it trivially holds

$$\mathcal{I}_{(X,\mathsf{d},\mathfrak{m})}(0) = \mathcal{I}_{(X,\mathsf{d},\mathfrak{m})}(1) = \mathcal{I}_{K,N,D}(0) = \mathcal{I}_{K,N,D}(1) = 0.$$

Fix then $v \in (0, 1)$ and let $A \subset X$ be an arbitrary Borel subset of X such that $\mathfrak{m}(A) = v$. Consider the $\mathfrak{m}$-measurable function $f(x) := \chi_A(x) - v$ and notice that $\int_X f\, \mathfrak{m} = 0$. Thus f verifies the hypothesis of Theorem 3.5.12 and noticing that f is never null, we can decompose $X = Y \cup \mathcal{T}$ with

$$\mathfrak{m}(Y) = 0, \qquad \mathfrak{m}\llcorner_{\mathcal{T}} = \int_Q \mathfrak{m}_q\, \mathfrak{q}(dq),$$

with $\mathfrak{m}_q = g(q,\cdot)_\sharp\,(h_q \cdot \mathcal{L}^1)$; moreover, for $\mathfrak{q}$-a.e. $q \in Q$, the density h_q verifies (3.32) and

$$\int_X f(z)\,\mathfrak{m}_q(dz) = \int_{\mathrm{Dom}\,(g(q,\cdot))} f(g(q,t)) \cdot h_q(t)\,\mathcal{L}^1(dt) = 0.$$

Therefore

$$v = \mathfrak{m}_q(A \cap \{g(q,t) : t \in \mathbb{R}\}) = (h_q\mathcal{L}^1)(g(q,\cdot)^{-1}(A)), \qquad \text{for } \mathfrak{q}\text{-a.e. } q \in Q. \tag{3.35}$$

For every $\varepsilon > 0$ we then have

$$\begin{aligned}
\frac{\mathfrak{m}(A^\varepsilon) - \mathfrak{m}(A)}{\varepsilon} &= \frac{1}{\varepsilon}\int_{\mathcal{T}} \chi_{A^\varepsilon\setminus A}\,\mathfrak{m}(dx) = \frac{1}{\varepsilon}\int_Q \left(\int_X \chi_{A^\varepsilon\setminus A}\,\mathfrak{m}_q(dx)\right)\mathfrak{q}(dq) \\
&= \int_Q \frac{1}{\varepsilon}\left(\int_{\mathrm{Dom}\,(g(q,\cdot))} \chi_{A^\varepsilon\setminus A}\,h_q(t)\,\mathcal{L}^1(dt)\right)\mathfrak{q}(dq) \\
&= \int_Q \left(\frac{(h_q\mathcal{L}^1)(g(q,\cdot)^{-1}(A^\varepsilon)) - (h_q\mathcal{L}^1)(g(q,\cdot)^{-1}(A))}{\varepsilon}\right)\mathfrak{q}(dq) \\
&\geq \int_Q \left(\frac{(h_q\mathcal{L}^1)((g(q,\cdot)^{-1}(A))^\varepsilon) - (h_q\mathcal{L}^1)(g(q,\cdot)^{-1}(A))}{\varepsilon}\right)\mathfrak{q}(dq),
\end{aligned}$$

where the last inequality is given by the inclusion $(g(q,\cdot)^{-1}(A))^\varepsilon \cap \mathrm{supp}(h_q) \subset g(q,\cdot)^{-1}(A^\varepsilon)$.
Recalling (3.35) together with $h_q\mathcal{L}^1 \in \mathcal{F}^s_{K,N,D}$, by Fatou's Lemma we get

$$\begin{aligned}
\mathfrak{m}^+(A) &= \liminf_{\varepsilon\downarrow 0} \frac{\mathfrak{m}(A^\varepsilon) - \mathfrak{m}(A)}{\varepsilon} \\
&\geq \int_Q \left(\liminf_{\varepsilon\downarrow 0} \frac{(h_q\mathcal{L}^1)((g(q,\cdot)^{-1}(A))^\varepsilon) - (h_q\mathcal{L}^1)(g(q,\cdot)^{-1}(A))}{\varepsilon}\right)\mathfrak{q}(dq) \\
&= \int_Q \left((h_q\mathcal{L}^1)^+(g(q,\cdot)^{-1}(A))\right)\mathfrak{q}(dq) \\
&\geq \int_Q \mathcal{I}^s_{K,N,D}(v)\,\mathfrak{q}(dq) \\
&= \mathcal{I}_{K,N,D}(v),
\end{aligned}$$

where in the last equality we used Theorem 3.2.6. □

From the definition of $\mathcal{I}_{K,N,D}$, see (3.16), and the smooth results of E. Milman in [41], the estimates proved in Theorem 3.6.6 are sharp.

Furthermore, the 1-dimensional localization technique allows for rigidity in the following sense: if for some $v \in (0,1)$ it holds $\mathcal{I}_{(X,\mathsf{d},\mathfrak{m})}(v) = \mathcal{I}_{K,N,\pi}(v)$, then $(X,\mathsf{d},\mathfrak{m})$ is a spherical suspension. It is worth underlining that to obtain such a result $(X,\mathsf{d},\mathfrak{m})$ is assumed to be in the more regular class of RCD-spaces.

Furthermore, one can prove an almost rigidity statement: if $(X,\mathsf{d},\mathfrak{m})$ is an $\mathsf{RCD}^*(K,N)$ space such that $\mathcal{I}_{(X,\mathsf{d},\mathfrak{m})}(v)$ is close to $\mathcal{I}_{K,N,\pi}(v)$ for some $v \in (0,1)$, this forces X to be close, in the measure-Gromov-Hausdorff distance, to a spherical suspension. What follows is Corollary 1.6 of [20].

Theorem 3.6.7. (Almost equality in Lévy-Gromov implies mGH-closeness to a spherical suspension) *For every $N \in [2,\infty)$, $v \in (0,1)$, $\varepsilon > 0$ there exists $\bar\delta = \bar\delta(N,v,\varepsilon) > 0$ such that the following hold. For every $\delta \in [0,\bar\delta]$, if $(X,\mathsf{d},\mathfrak{m})$ is an* $\mathsf{RCD}^*(N-1-\delta, N+\delta)$ *space satisfying*

$$\mathcal{I}_{(X,\mathsf{d},\mathfrak{m})}(v) \le \mathcal{I}_{N-1,N,\pi}(v) + \delta,$$

then there exists an $\mathsf{RCD}^*(N-2,N-1)$ *space* $(Y,\mathsf{d}_Y,\mathfrak{m}_Y)$ *with* $\mathfrak{m}_Y(Y)=1$ *such that*

$$\mathsf{d}_{mGH}(X,[0,\pi]\times_{\sin}^{N-1} Y) \le \varepsilon.$$

We refer to [20] for the precise rigidity statement (Theorem 1.4, [20]) and for the proof of Theorem 1.4 and Corollary 1.6 of [20]. See also [20] for the precise definition of spherical suspension. We conclude by recalling that 1-dimensional localization was used also in [21] to obtain sharp versions of several functional inequalities (e.g. Brunn-Minkowski, spectral gap, Log-Sobolev etc.) in the class of $\mathsf{CD}(K,N)$-spaces. See [21] for details.

Bibliography

[1] L. Ambrosio: Lecture notes on the Optimal Transport problems. Mathematical aspects of evolving interfaces - CIME Springer Lecture Notes, 1–52, (2003).

[2] L. Ambrosio, N. Fusco and D. Pallara: Functions of Bounded Variation and Free Discontinuity Problems, Oxford Mathematical Monographs, (2000).

[3] L. Ambrosio, N Gigli and G. Savaré: Bakry-Émery curvature-dimension condition and Riemannian Ricci curvature bounds. Ann. Probab., 43, 1, (2015), 339–404.

[4] ———: Calculus and heat flow in metric measure spaces and applications to spaces with Ricci bounds from below, Invent. Math., 195, 2, (2014), 289–391.

[5] ———: Metric measure spaces with Riemannian Ricci curvature bounded from below, Duke Math. J., 163, (2014), 1405–1490.

[6] L. Ambrosio and B Kirchheim: Currents in metric spaces, Acta Math., 185, 1, (2000), 1–80.

[7] L. Ambrosio, B Kirchheim and A. Pratelli: Existence of optimal transports with crystalline norms. Duke Math. J., 125, 2, (2004), 207–241.

[8] L. Ambrosio, A. Mondino and G. Savaré: Nonlinear diffusion equations and curvature conditions in metric measure spaces, Preprint arXiv:1509.07273.

[9] K. Bacher and K.-T. Sturm: Localization and tensorization properties of the curvature-dimension condition for metric measure spaces, J. Funct. Anal., 259, (2010), 28–56.

[10] S. Bianchini and L. Caravenna: On the extremality, uniqueness and optimality of transference plans, Bull. Inst. Math. Acad. Sin. (N.S.), 4(4), (2009), 353–455.
[11] S. Bianchini and F. Cavalletti: The Monge problem for distance cost in geodesic spaces, Commun. Math. Phys., 318, (2013), 615 – 673.
[12] S. Bianchini and S. Daneri: On Sudakov's type decomposition of transference plans with norm costs, Mem. Amer. Math. Soc., in press.
[13] D. Burago, Y. Burago and S. Ivanov: A course in Metric geometry, Graduate Studies in Mathematics, AMS, 2001.
[14] Y.D. Burago and V.A Zalgaller: Geometric inequalities, Grundlehren der Mathematischen Wissenschaften [Fundamental Principles of Mathematical Sciences], 285. Springer, Berlin (1988).
[15] L. Caffarelli, M. Feldman and R.J. McCann: Constructing optimal maps for Monge's transport problem as a limit of strictly convex costs. J. Amer. Math. Soc., 15, (2002), 1–26.
[16] F. Cavalletti: The Monge problem in Wiener space Calc. Var. PDE, 45, (2012), 101–124.
[17] ———: Decomposition of geodesics in the Wasserstein space and the globalization property. Geom. Funct. Anal., 24, (2014), 493 – 551.
[18] ———: Monge problem in metric measure spaces with Riemannian curvature-dimension condition, Nonlinear Anal., 99, (2014), 136–151.
[19] F. Cavalletti and A. Mondino: Measure rigidity of Ricci curvature lower bounds, Adv. Math. 286 (2016) 430–480.
[20] ———: Sharp and rigid isoperimetric inequalities in metric-measure spaces with lower Ricci curvature bounds, to appear, Invent. math. DOI 10.1007/s00222-016-0700-6
[21] ———: Sharp geometric and functional inequalities in metric measure spaces with lower Ricci curvature bounds, to appear in Geom. Top. arXiv: 1502.06465, (2015).
[22] F. Cavalletti and K.-T. Sturm: Local curvature-dimension condition implies measure-contraction property. J. Funct. Anal., 262, 5110 – 5127, 2012.
[23] L. Caravenna: A proof of Sudakov theorem with strictly convex norms, Math. Z., 268, 1, (2011), 371–407.
[24] M. Cicalese, G.P. Leonardi: A selection principle for the sharp quantitative isoperimetric inequality, Arch. Rat. Mech. Anal., 206, 2, (2012), 617–643.
[25] M. Eichmair and J. Metzger: Unique isoperimetric foliations of asymptotically flat manifolds in all dimensions, Invent. Math., 194, (2013), 591–630.
[26] M Erbar, Kuwada and K.T. Sturm: On the Equivalence of the Entropic Curvature-Dimension Condition and Bochner's Inequality on Metric Measure Space, Invent. Math., 201, 3, (2015), 993–1071.
[27] L. C. Evans: Partial Differential Equations, Graduate Studies in Mathematics, vol. 19, AMS, 1998.
[28] M. Feldman and R.J. McCann: Monge's transport problem on a Riemannian manifold, Trans. Amer. Math. Soc., 354, (2002), 1667–1697.
[29] A. Figalli, F. Maggi and A. Pratelli: A mass transportation approach to quantitative isoperimetric inequalities, Invent. Math., 182, 1, (2010), 167–211.
[30] D. H. Fremlin: Measure Theory, volume 4. Torres Fremlin, 2002.
[31] N. Fusco, F. Maggi and A. Pratelli: The sharp quantitative isoperimetric inequality, Ann. of Math., 168, (2008), 941–980.
[32] N. Gigli: Optimal maps in non branching spaces with Ricci curvature bounded from below, Geom. Funct. Anal., 22 (2012) no. 4, 990–999.
[33] N. Gigli: On the differential structure of metric measure spaces and applications, Mem. Am. Math. Soc., 236 (2015) no. 1113.

[34] N. Gigli, T. Rajala and K.T. Sturm: Optimal maps and exponentiation on finite dimensional spaces with Ricci curvature bounded from below to appear in J. Geom. Analysis.

[35] M. Gromov: Metric structures for Riemannian and non Riemannian spaces, Modern Birkhäuser Classics, (2007).

[36] M. Gromov and V. Milman: Generalization of the spherical isoperimetric inequality to uniformly convex Banach spaces. Compositio Math., 62, 3, (1987), 263–282.

[37] R. Kannan, L. Lovász and M. Simonovits: Isoperimetric problems for convex bodies and a localization lemma, Discrete Comput. Geom., 13, 3-4, (1995), 541–559.

[38] B. Klartag: Needle decomposition in Riemannian geometry, to appear in Mem. AMS.

[39] J. Lott and C. Villani: Ricci curvature for metric-measure spaces via optimal transport, Ann. of Math. (2) 169 (2009), 903–991.

[40] L. Lovász and M. Simonovits: Random walks in a convex body and an improved volume algorithm, Random Structures Algorithms, 4, 4, (1993), 359–412.

[41] E. Milman: Sharp Isoperimetric Inequalities and Model Spaces for Curvature-Dimension-Diameter Condition, J. Europ. Math. Soc. to appear.

[42] E. Milman and L. Rotem: Complemented Brunn-Minkowski Inequalities and Isoperimetry for Homogeneous and Non-Homogeneous Measures, Advances in Math., 262, 867–908, (2014).

[43] F. Morgan: Geometric Measure Theory (A Beginner's Guide), Elsevier/Academic Press, Amsterdam, Fourth edition, (2009).

[44] ———: In polytopes, small balls about some vertex minimize perimeter. J. Geom. Anal. 17, 97–106, (2007).

[45] F. Morgan and M. Ritoré: Isoperimetric regions in cones, Trans. Amer. Math. Soc., 354, (2002), 2327–2339.

[46] R. Osserman: The isoperimetric inequality, Bull. Am. Math. Soc., 84 (6), 1182–1238, (1978)

[47] L.E. Payne and H.F. Weinberger" An optimal Poincaré inequality for convex domains Arch. Rational Mech. Anal., 5, (1960), 286–292.

[48] M. Ritoré: Geometric flows, isoperimetric inequalities and hyperbolic geometry, Mean curvature flow and isoperimetric inequalities, Adv. Courses Math. CRM Barcelona, 45-113, Birkhäuser, Basel, (2010).

[49] A. Ros: The isoperimetric problem, Lecture series at the Clay Mathematics Institute, Summer School on the Global Theory of Minimal Surfaces, MSRI, Berkeley, California, (2001).

[50] A. M. Srivastava: A course on Borel sets. Springer, 1998.

[51] K.T. Sturm: On the geometry of metric measure spaces. I, Acta Math. 196 (2006), 65–131.

[52] K.T. Sturm: On the geometry of metric measure spaces. II, Acta Math. 196 (2006), 133–177.

[53] V.N. Sudakov: Geometric problems in the theory of dimensional distributions. Proc. Steklov Inst. Math. 146 (1979), 1–178.

[54] N.S. Trudinger and X.J. Wang: On the Monge mass transfer problem, Calc. Var. PDE, 13, (2001), 19–31.

[55] C. Villani: Topics in Optimal Transportation, Graduate Studies in Mathematics, AMS, 2003.

[56] C. Villani: Optimal transport. Old and new, Grundlehren der Mathematischen Wissenschaften, 338, Springer-Verlag, Berlin, (2009).

Guido De Philippis, Andrea Marchese, and Filip Rindler

On a conjecture of Cheeger

4.1 Introduction

In [7] Cheeger, proved that in every doubling metric measure space (X, ρ, μ) satisfying a Poincaré inequality Lipschitz functions are differentiable μ-almost everywhere. More precisely, he showed the existence of a family $\{(U_i, \varphi_i)\}_{i\in\mathbb{N}}$ of Borel charts (that is, $U_i \subset X$ is a Borel set, $X = \bigcup_i U_i$ up to a μ-negligible set, and $\varphi_i : X \to \mathbb{R}^{d(i)}$ is Lipschitz) such that for every Lipschitz map $f : X \to \mathbb{R}$ at μ-almost every $x_0 \in U_i$ there exists a unique (co-)vector $df(x_0) \in \mathbb{R}^{d(i)}$ with

$$\limsup_{x\to x_0} \frac{\left|f(x) - f(x_0) - df(x_0)\cdot(\varphi(x) - \varphi(x_0))\right|}{\rho(x, x_0)} = 0.$$

This fact was later axiomatized by Keith [15], leading to the notion of *Lipschitz differentiability space*, see Section 4.2 below.

Cheeger also conjectured that the push-forward of the reference measure μ under every chart ϕ_i has to be absolutely continuous with respect to the Lebesgue measure, that is,

$$(\varphi_i)_\#(\mu \llcorner U_i) \ll \mathcal{L}^{d(i)},$$

see [7, Conjecture 4.63]. Some consequences of this fact concerning existence of bi-Lipschitz embeddings of X into some $\mathbb{R}^N$ are detailed in [7, Section 14], also see [8, 9].

Let us assume that $(X, \rho, \mu) = (\mathbb{R}^d, \rho_{\mathcal{E}}, \nu)$ with $\rho_{\mathcal{E}}$ the Euclidean distance and ν a positive Radon measure, is a Lipschitz differentiability space when equipped with the (single) identity chart (note that it follows a-posteriori from the validity of Cheeger's conjecture that no mapping into a higher-dimensional space can be a chart in a Lipschitz differentiability structure of $\mathbb{R}^d$). In this case the validity of Cheeger's conjecture reduces to the validity of the (weak) converse of Rademacher's theorem, which states that a positive Radon measure ν on $\mathbb{R}^d$ with the property that all Lipschitz functions are differentiable ν-almost everywhere must be absolutely continuous with respect to $\mathcal{L}^d$. Actually, it is well known to experts that this converse of Rademacher's theorem implies Cheeger's conjecture in any metric space, see for instance [15, Section 2.4], [6, Remark 6.11], and [12].

Guido De Philippis: SISSA, Via Bonomea 265, 34136 Trieste, Italy. E-mail: guido.dephilippis@sissa.it
Andrea Marchese: Universität Zürich, Winterthurerstrasse 190, CH-8057 Zürich, Switzerland. E-mail: andrea.marchese@math.uzh.ch
Filip Rindler: Mathematics Institute, University of Warwick, Coventry CV4 7AL, UK. E-mail: F.Rindler@warwick.ac.uk

The (strong) converse of Rademacher's theorem has been known to be true in $\mathbb{R}$ since the work of Zahorski [21], where he characterized the sets $E \subset \mathbb{R}$ that are sets of non-differentiability points of some Lipschitz function. In particular, he proved that for every Lebesgue negligible set $E \subset \mathbb{R}$ there exists a Lipschitz function which is nowhere differentiable on E.

The same result for maps $f: \mathbb{R}^d \to \mathbb{R}^d$ has been proved by Alberti, Csörnyei & Preiss for $d = 2$ as a consequence of a deep structural result for negligible sets in the plane [1, 2]. In 2011, Csörnyei & Jones [14] announced the extension of the above result to every Euclidean space. For Lipschitz maps $f: \mathbb{R}^d \to \mathbb{R}^m$ with $m < d$ the situation is fundamentally different and there exists a null set such that every Lipschitz function is differentiable at at least one point from that set, see [17, 18]. We finally remark that the weak converse of Rademacher's theorem in $\mathbb{R}^2$ can also be obtained by combining the results of [4] and [5], see [5, Remark 6.2 (iv)].

Recently, a result concerning the singular structure of measures satisfying a differential constraint was proved in [10]. When combined with the main result of [5] this proves the weak converse of Rademacher's theorem in any dimension, see [10, Theorem 1.14].

In this note we detail how the results in [5, 10] in conjunction with Bate's result on the existence of a sufficient number of independent Alberti representations in a Lipschitz differentiability space [6] imply Cheeger's conjecture; see Section 4.2 for the relevant definitions.

Theorem 4.1.1. *Let (X, ρ, μ) be a Lipschitz differentiability space and let (U, ϕ) be a d-dimensional chart. Then, $\phi_\#(\mu \llcorner U) \ll \mathcal{L}^d$.*

Note that by the same arguments of this paper Cheeger's conjecture would also follow from the results announced in [1] and [14].

After we finished writing this note we learned that similar results have been proved by Kell and Mondino [16] and by Gigli and Pasqualetto [13].

4.2 Setup

4.2.1 Lipschitz differentiability spaces

Throughout this chapter, the triple (X, ρ, μ) will always denote a *metric measure space*, that is, (X, ρ) is a separable, complete metric space and $\mu \in \mathcal{M}_+(X)$ is a positive Radon measure on X.

We call a pair (U, φ) such that $U \subset X$ is a Borel set and $\varphi: X \to \mathbb{R}^d$ is Lipschitz, a *d-dimensional chart* or simply a *d-chart*. A function $f: X \to \mathbb{R}$ is said to be *differentiable with respect to a d-chart* (U, φ) at $x_0 \in U$ if there exists a unique (co-)vector $df(x_0) \in$

$\mathbb{R}^d$ such that

$$\limsup_{x \to x_0} \frac{|f(x) - f(x_0) - df(x_0) \cdot (\varphi(x) - \varphi(x_0))|}{\rho(x, x_0)} = 0.$$

We call a metric measure space (X, ρ, μ) a *Lipschitz differentiability space* (also called a metric measure space that admits a *measurable differentiable structure*) if there exists a countable family of $d(i)$-charts (U_i, φ_i) $(i \in \mathbb{N})$ such that $X = \bigcup_i U_i$ and any Lipschitz map $f: X \to \mathbb{R}$ is differentiable with respect to every (U_i, φ_i) at μ-almost every point $x_0 \in U_i$.

4.2.2 Alberti representations

We denote by $\Gamma(X)$ the set of *curves* in X, that is, the set of all Lipschitz maps $\gamma\colon \operatorname{Dom}\gamma \to X$, for which the domain $\operatorname{Dom}\gamma \subset \mathbb{R}$ is non-empty and compact. Note that we are not requiring $\operatorname{Dom}\gamma$ to be an interval and thus the set $\Gamma(X)$ is sometimes also called the set of *curve fragments* on X. We equip $\Gamma(X)$ with the Hausdorff metric $\operatorname{dist}_{\mathcal{H}}$ on graphs and we consider it as a subspace of the Polish space

$$\mathcal{K} = \{ K \subset \mathbb{R} \times X : K \text{ compact} \}, \tag{4.1}$$

endowed with the Hausdorff metric. Moreover, by arguing as in [19, Lemma 2.20], it is easy to see that $\Gamma(X)$ is an F_σ-subset of $\mathcal{K}$, i.e. a countable union of closed sets.

The decomposition of a measure into a family of 1-dimensional Hausdorff measures supported on curves leads to the notion of Alberti representation. First introduced in [4] for the study of the rank-one property of BV-derivatives, this decomposition has turned out to be a key tool in the study of differentiability properties of Lipschitz functions, see for instance [1, 2, 5, 6].

Definition 4.2.1. *Let (X, ρ, μ) be a metric measure space. An* Alberti representation *of μ on a μ-measurable set $A \subset X$ is a parameterized family $(\mu_\gamma)_{\gamma \in \Gamma(X)}$ of positive Borel measures $\mu_\gamma \in \mathcal{M}_+(X)$ with*

$$\mu_\gamma \ll \mathcal{H}^1 \llcorner \operatorname{Im}\gamma,$$

together with a Borel probability measure $\pi \in \mathcal{P}(\Gamma(X))$ such that

$$\mu(B) = \int \mu_\gamma(B)\, \mathrm{d}\pi(\gamma) \qquad \textit{for all Borel sets } B \subset A. \tag{4.2}$$

Here, the measurability of the integrand is part of the requirement of being an Alberti representation.

Remark 4.2.2. *Note that this definition is slightly different from the one in [6, Definition 2.2] since there the set $\Gamma(X)$ consists of* bi-Lipschitz *curves. Clearly, the existence of a representation in the sense of [6] implies the existence of a representation in our* *sense and this will suffice for our purposes. Let us, however, point out that the converse holds true as well. Indeed, the part of γ that contributes to the integral in (4.2) can be decomposed into countably many bi-Lipschitz pieces, see [19, Remark 2.17].*

We will further need the notion of *independent* Alberti representations of a measure. Let $C \subset \mathbb{R}^d$ be a closed, convex, one-sided cone, i.e. a set of the form

$$C := \left\{ v \in \mathbb{R}^d \,:\, v \cdot w \geq (1 - \theta)\|v\| \right\}$$

for some $w \in \mathbb{S}^{d-1}$ and $\theta \in (0, 1)$. With a Lipschitz map $\phi \colon X \to \mathbb{R}^d$, we say that an Alberti representation $\int \nu_\gamma \, d\pi(\gamma)$ has *φ-directions in C* if

$$(\varphi \circ \gamma)'(t) \in C \setminus \{0\} \qquad \text{for } \pi\text{-a.e. curve } \gamma \text{ and } \mathcal{H}^1\text{-a.e. } t \in \operatorname{Dom} \gamma.$$

A number of m Alberti representations of μ are *φ-independent* if there are linearly independent cones $C_1, \ldots, C_m$ such that the i'th Alberti representation has φ-directions in C_i. Here, linear independence of the cones $C_1, \ldots, C_m$ means that any collection of vectors $v_i \in C_i \setminus \{0\}$ is linearly independent. In the case $X = \mathbb{R}^d$ we will always consider $\phi = \mathrm{Id}$.

One of the main results of [6] asserts that a Lipschitz differentiability space necessarily admits many independent Alberti representations, also cf. [5, Theorem 1.1]. Recall that according to Remark 4.2.2 any representation in the sense of [6] is also a representation in the sense of Definition 4.2.1.

Theorem 4.2.3. *Let (X, ρ, μ) be a Lipschitz differentiability space with a d-chart (U, φ). Then, there exists a countable decomposition*

$$U = \bigcup_{k \in \mathbb{N}} U_k, \qquad U_k \subset U \text{ Borel sets,}$$

such that every $\mu \llcorner U_k$ has d φ-independent Alberti representations.

A proof of this theorem can be found in [6, Theorem 6.6].

4.2.3 One-dimensional currents

To use the results of [10] we need a link between Alberti representations and 1-dimensional currents. Recall that a 1-*dimensional current* T in $\mathbb{R}^d$ is a continuous linear functional on the space of smooth and compactly supported differential 1-forms on $\mathbb{R}^d$. The *boundary* of T, ∂T is the distribution (0-current) defined via $\langle \partial T, f \rangle := \langle T, df \rangle$ for every smooth and compactly supported function $f \colon \mathbb{R}^d \to \mathbb{R}$. The *mass* of T, denoted by $\mathbf{M}(T)$, is the supremum of $\langle T, \omega \rangle$ over all 1-forms ω such that $|\omega| \leq 1$ everywhere. A current T is called *normal* if both T and ∂T have finite mass; we denote the set of normal 1-currents by $\mathbf{N}_1(\mathbb{R}^d)$.

By the Radon–Nikodým theorem, a 1-dimensional current T with finite mass can be written in the form $T = \vec{T}\|T\|$ where $\|T\|$ is a finite positive measure and $\vec{T}$ is a vector field in $\mathrm{L}^1(\mathbb{R}^d, \|T\|)$ with $|\vec{T}(x)| = 1$ for $\|T\|$-almost every $x \in \mathbb{R}^d$. In particular, the action of T on a smooth and compactly supported 1-form ω is given by

$$\langle T, \omega \rangle = \int_{\mathbb{R}^d} \langle \omega(x), \vec{T}(x) \rangle \, \mathrm{d}\|T\|(x) \, .$$

An *integer-multiplicity rectifiable* 1-*current* (in the following called simply rectifiable 1-current) $T = [\![E, \tau, m]\!]$ is a 1-current which acts on 1-forms ω as

$$\langle T, \omega \rangle = \int_E \langle \omega(x), \tau(x) \rangle \, m(x) \, \mathrm{d}\mathcal{H}^1(x) \, ,$$

where E is a 1-rectifiable set, $\tau(x)$ is a unit vector spanning the approximate tangent space $\mathrm{Tan}(E, x)$ and m is an integer-valued function such that $\int_E m \; \mathrm{d}\mathcal{H}^1 < \infty$. More information on currents can be found in [11].

The relation between Alberti representations and normal 1-currents is partially encoded in the following decomposition theorem, due to Smirnov [20].

Theorem 4.2.4. *Let $T = \vec{T}\|T\| \in \mathbf{N}_1(\mathbb{R}^d)$ be a normal* 1*-current with $|\vec{T}(x)| = 1$ for $\|T\|$-almost every x. Then, there exists a family of rectifiable* 1*-currents*

$$T_\gamma = [\![E_\gamma, \tau_\gamma, 1]\!], \qquad \gamma \in \Gamma,$$

where Γ is a measure space endowed with a finite positive Borel measure $\pi \in \mathcal{M}_+(\Gamma)$, such that the following assertions hold:

(i) T can be decomposed as

$$T = \int_\Gamma T_\gamma \, \mathrm{d}\pi(\gamma)$$

and

$$\mathbf{M}(T) = \int_\Gamma \mathbf{M}(T_\gamma) \, \mathrm{d}\pi(\gamma) = \int_\Gamma \mathcal{H}^1(E_\gamma) \, \mathrm{d}\pi(\gamma) \, ;$$

(ii) $\tau_\gamma(x) = \vec{T}(x)$ for $\mathcal{H}^1$-almost every $x \in E_\gamma$ and for π-almost every $\gamma \in \Gamma$;

(iii) $\|T\|$ can be decomposed as

$$\|T\| = \int_\Gamma \mu_\gamma \, \mathrm{d}\pi(\gamma) \, ,$$

where each μ_γ is the restriction of $\mathcal{H}^1$ to the 1*-rectifiable set E_γ.*

An Alberti representation of an Euclidean measure splits it into measures concentrated on “fragments” of curves. In general, these fragments cannot be glued together to obtain a 1-dimensional normal current since the boundary may have infinite mass. Nevertheless, the “holes” of every curve appearing in an Alberti representation of a measure $\nu \in \mathcal{M}_+(\mathbb{R}^d)$ can be “filled” in such a way as to produce a normal 1-current T with $\nu \ll \|T\|$. Moreover, if the representation has directions in a cone C then the constructed normal current T has orienting vector $\vec{T}$ in $C \setminus \{0\}$ almost everywhere (with respect to $\|T\|$). Indeed, we have the following lemma, which is essentially [5, Corollary 6.5]; it can be interpreted as a partial converse to Theorem 4.2.4:

Lemma 4.2.5. *Let $\nu \in \mathcal{M}_+(\mathbb{R}^d)$ be a finite Radon measure. If there is an Alberti representation $\nu = \int \nu_\gamma \, d\pi(\gamma)$ with directions in a cone C, then there exists a normal 1-current $T \in \mathbf{N}_1(\mathbb{R}^d)$ such that $\vec{T}(x) \in C \setminus \{0\}$ for $\|T\|$-almost every $x \in \mathbb{R}^d$ and $\nu \ll \|T\|$.*

Proof. For the purpose of illustration we sketch the proof.

Step 1. Given ν as in the statement, we claim that there exists a normal 1-current $T = \vec{T}\|T\|$ with $\mathbf{M}(T) \le 1$ and $\mathbf{M}(\partial T) \le 2$ such that $\vec{T}(x) \in C$, for $\|T\|$-almost every x and that ν is not singular with respect to $\|T\|$.

The claim follows from the proof of [5, Lemma 6.12]. For the sake of completeness let us present the main line of reasoning. By arguing as in Step 1 of the proof of [5, Lemma 6.12], to every $\gamma \in \Gamma(\mathbb{R}^d)$ with $\gamma'(t) \in C$ and a Borel measure $\nu_\gamma \ll \mathcal{H}^1 \llcorner \operatorname{Im} \gamma$ we can associate a 1-Lipschitz map $\psi_{\nu_\gamma} : [0,1] \to \mathbb{R}^d$ satisfying

$$\nu_\gamma(\operatorname{Im}(\psi_{\nu_\gamma})) > 0 \qquad \text{and} \qquad \psi'_{\nu_\gamma}(t) \in C \setminus \{0\} \quad \text{for } \mathcal{H}^1\text{-a.e. } t \in [0,1].$$

This map can moreover be chosen such that $\gamma \mapsto \psi_{\nu_\gamma}$ coincides with a Borel measurable map π-almost everywhere once we endow the set of curves with the topology of uniform convergence, see Step 3 in the proof of [5, Lemma 6.12].

Let $T_{\nu_\gamma} := [\![\operatorname{Im} \psi_{\nu_\gamma}, \tau_{\psi_{\nu_\gamma}}, 1]\!]$ be the rectifiable 1-current associated to ψ_{ν_γ} and set

$$T := \int T_{\nu_\gamma} \, d\pi(\gamma) .$$

Since ψ_{ν_γ} is 1-Lipschitz, $\mathcal{H}^1(\operatorname{Im} \psi_{\nu_\gamma}) \le 1$ and thus $\mathbf{M}(T) \le 1$. Moreover, for all smooth compactly supported functions $f : \mathbb{R}^d \to \mathbb{R}$ we have

$$\langle \partial T, f \rangle = \langle T, df \rangle = \int f(\psi_{\nu_\gamma}(1)) - f(\psi_{\nu_\gamma}(0)) \, d\pi(\gamma) ,$$

so that $\mathbf{M}(\partial T) \le 2$.

By assumption, $\vec{T}(x) \in C \setminus \{0\}$ for $\|T\|$-almost every $x \in \mathbb{R}^d$. To show that $\|T\|$ and ν are not mutually singular, for π-almost every γ set

$$\nu'_\gamma := \nu_\gamma \llcorner \operatorname{Im} \psi_{\nu_\gamma} \qquad \text{and} \qquad \nu' := \int \nu'_\gamma \, d\pi(\gamma) ,$$

so that $\nu' \neq 0$ and $\nu' \le \nu$. We will now establish that $\nu' \ll \|T\|$, for which we will prove that ν and $\|T\|$ are not mutually singular. Let $E \subset \mathbb{R}^d$ be such that $\|T\|(E) = 0$. Using

$$T = \int [\![\operatorname{Im} \psi_{\nu_\gamma}, \tau_{\psi_{\nu_\gamma}}, 1]\!] \, d\pi(\gamma) \qquad \text{with} \qquad \tau_{\psi_{\nu_\gamma}} = \frac{\psi'_{\nu_\gamma}}{|\psi'_{\nu_\gamma}|} \in C ,$$

we get

$$\mathcal{H}^1(\operatorname{Im}\psi_{v_\gamma} \cap E) = 0 \qquad \text{for } \pi\text{-a.e. } \gamma.$$

Since by definition $\nu_\gamma \ll \mathcal{H}^1 \llcorner \operatorname{Im}\gamma$, we have that $\nu'_\gamma \ll \mathcal{H}^1 \llcorner \operatorname{Im}\psi_{\nu_\gamma}$. Thus, $\nu'(E) = 0$.

Step 2. Let us define

$$\mathcal{T} := \left\{ T \in \mathbf{N}_1(\mathbb{R}^d) \,:\, \mathbf{M}(T) \leq 1, \mathbf{M}(\partial T) \leq 2 \text{ and } \vec{T} \in C \ \|T\|\text{-a.e.} \right\}$$

and

$$\mathcal{T}_\nu := \left\{ T \in \mathcal{T} \,:\, \nu \text{ and } T \text{ are not singular} \right\}.$$

Note that if $C = \{ v \in \mathbb{R}^d \,:\, v \cdot w \geq (1-\theta)\|v\| \}$ for some $w \in \mathbb{S}^{d-1}$, $\theta \in (0,1)$, then $\vec{T} \in C$ almost everywhere implies that

$$\|T\| \geq T \cdot w \geq (1-\theta)\|T\| \tag{4.3}$$

as measures (here we are identifying T with an $\mathbb{R}^d$-valued Radon measure and use the pointwise scalar product). Moreover, as a consequence of the Radon–Nikodým theorem, for every $T \in \mathcal{T}_\nu$ we may write

$$\nu = g_{\|T\|}\|T\| + \nu^s_{\|T\|} \qquad \text{with} \qquad \nu^s_{\|T\|} \perp \|T\|\,, \ \int g_{\|T\|} \, \mathrm{d}\|T\| > 0\,.$$

Let us set $M := \sup_{T \in \mathcal{T}_\nu} \int g_{\|T\|} \, \mathrm{d}\|T\| > 0$ and let $T_k \in \mathcal{T}_\nu$ be a sequence with

$$\int g_{\|T_k\|} \, \mathrm{d}\|T_k\| \to M.$$

Define

$$T := \sum_k 2^{-k} T_k$$

and note that $T \in \mathcal{T}$. Moreover, by (4.3), $\|T_k\| \ll \|T\|$ for all $k \in \mathbb{N}$, so that there exist $h_k \colon \mathbb{R}^d \to \mathbb{R}$ with

$$\int_E h_k \, \mathrm{d}\|T\| = \int_E g_{\|T_k\|} \, \mathrm{d}\|T_k\| \leq \nu(E) \qquad \text{for all Borel sets } E \subset \mathbb{R}^d.$$

In particular, $T \in \mathcal{T}_\nu$ and $h_k \leq g_{\|T\|}$. Set $m_k = \max_{1 \leq j \leq k} h_j$. By the monotone convergence theorem, $m_k \to m_\infty \leq g_{\|T\|}$ in $\mathrm{L}^1(\mathbb{R}^d, \|T\|)$ and

$$M \leq \lim_{k\to\infty} \int m_k \, \mathrm{d}\|T\| = \int m_\infty \, \mathrm{d}\|T\| \leq \int g_{\|T\|} \, \mathrm{d}\|T\| \leq M.$$

Hence, M is actually a maximum and it is attained by T.

We now claim that $\nu \ll \|T\|$. Indeed, assume by contradiction that $\nu = g_{\|T\|} \, \mathrm{d}\|T\| + \nu^s_{\|T\|}$ with $\nu^s_{\|T\|} \neq 0$. Since the Alberti representation of ν induces an Alberti representation of $\nu^s_{\|T\|}$, we can apply Step 1 to find a normal 1-current

$$S \in \mathcal{T}_{\nu^s_{\|T\|}} \subset \mathcal{T}_\nu$$

such that $\nu^s_{\|T\|}$ and $\|S\|$ are not mutually singular. In particular, if $\nu = g_{\|S\|}\,\mathrm{d}\|S\| + \nu^s_{\|S\|}$, then there exists a Borel set $F \subset \mathbb{R}^d$ such that

$$\|T\|(F) = 0 \qquad \text{and} \qquad \int_F g_{\|S\|}\,\mathrm{d}\|S\| > 0. \tag{4.4}$$

Let us define $W := (T + S)/2$ and note that by (4.3) it holds that $\|T\|, \|S\| \ll \|W\|$ so that $W \in \mathcal{T}_\nu$. Moreover, there are functions $h_T, h_S \le g_{\|W\|}$ such that

$$\int_E h_T\,\mathrm{d}\|W\| = \int_E g_{\|T\|}\,\mathrm{d}\|T\|\,, \qquad \int_E h_S\,\mathrm{d}\|W\| = \int_E g_{\|S\|}\,\mathrm{d}\|S\|$$

for all Borel sets E. However, for F as in (4.4) we obtain

$$M \ge \int_{\mathbb{R}^d} g_{\|W\|}\,\mathrm{d}\|W\| \ge \int_{\mathbb{R}^d} g_{\|T\|}\,\mathrm{d}\|T\| + \int_F g_{\|S\|}\,\mathrm{d}\|S\| > M,$$

a contradiction. □

4.3 Proof of Cheeger's conjecture

The key tool to prove Cheeger's conjecture is the following result from [10, Corollary 1.12]:

Theorem 4.3.1. *Let $T_1 = \vec{T}_1\|T_1\|, \dots, T_d = \vec{T}_d\|T_d\| \in \mathbf{N}_1(\mathbb{R}^d)$ be 1-dimensional normal currents. Let $\nu \in \mathcal{M}_+(\mathbb{R}^d)$ be a positive Radon measure such that*
(i) $\nu \ll \|T_i\|$ for $i = 1, \dots, d$, and
(ii) $\operatorname{span}\{\vec{T}_1(x), \dots, \vec{T}_d(x)\} = \mathbb{R}^d$ for ν-almost every x.
Then, $\nu \ll \mathcal{L}^d$.

Combining the above result with Lemma 4.2.5 we immediately get the following:

Lemma 4.3.2. *Let $\nu \in \mathcal{M}_+(\mathbb{R}^d)$ have d independent Alberti representations. Then, $\nu \ll \mathcal{L}^d$.*

Proof. Denote by $C_1, \dots, C_d$ independent cones such that there are d Alberti representations having directions in these cones. By Lemma 4.2.5 there are d normal 1-dimensional currents $T_1 = \vec{T}_1\|T_1\|, \dots, T_d = \vec{T}_d\|T_d\| \in \mathbf{N}_1(\mathbb{R}^d)$ such that

$$\nu \ll \|T_i\| \qquad \text{for } i = 1, \dots, d,$$

and $\vec{T}_i(x) \in C_i$ for ν-almost every $x \in \mathbb{R}^d$. By the independence of the cones,

$$\operatorname{span}\{\vec{T}_1(x), \dots, \vec{T}_d(x)\} = \mathbb{R}^d \qquad \text{for } \nu\text{-a.e. } x \in \mathbb{R}^d.$$

This implies $\nu \ll \mathcal{L}^d$ via Theorem 4.3.1. □

In order to use the above result to prove Theorem 8.1.1 one also needs the following "push-forward lemma".

Lemma 4.3.3. *Let (X, ρ, μ) be a Lipschitz differentiability space with a d-chart (U, φ). If $\mu \llcorner U$ has d φ-independent Alberti representations, then also the push-forward $\varphi_{\#}(\mu \llcorner U) \in \mathcal{M}_+(\mathbb{R}^d)$ has d independent Alberti representations.*

Proof. It is enough to show that if there exists a representation of the form $\mu \llcorner U = \int \mu_\gamma \, d\pi(\gamma)$ with φ-directions in a cone C (i.e. such that $(\phi \circ \gamma)'(t) \in C \setminus \{0\}$ for almost all $t \in \operatorname{Dom} \gamma$ and for π-almost every γ), then we can build an Alberti representation

$$\varphi_{\#}(\mu \llcorner U) = \int \nu_{\bar\gamma} \, d\bar\pi(\bar\gamma) \qquad \text{with} \qquad \bar\pi \in \mathcal{P}(\Gamma(\mathbb{R}^d)),$$

with $\bar\gamma'(t) \in C \setminus \{0\}$ for $\bar\pi$-almost every $\bar\gamma$ and almost every $t \in \operatorname{Dom} \bar\gamma$. To this end consider the map $\Phi \colon \Gamma(X) \to \Gamma(\mathbb{R}^d)$ given by $\Phi(\gamma) := \varphi \circ \gamma$ and let $\bar\pi := \Phi_{\#}\pi \in \mathcal{M}_+(\Gamma(\mathbb{R}^d))$. Note that, by the very definition of the push-forward measure, for $\bar\pi$-almost every $\bar\gamma$ it holds that $\bar\gamma = \phi \circ \gamma$ for some $\gamma \in \Gamma(X)$.

By considering π as a probability measure defined on the Polish space $\mathcal{K}$ defined in (4.1), and noting that π is concentrated on $\Gamma(X)$, we can apply the disintegration theorem for measures [3, Theorem 5.3.1] to show that for $\bar\pi$-almost every $\bar\gamma$ there exists a Borel probability measure $\eta_{\bar\gamma}$ concentrated on $\Phi^{-1}(\bar\gamma)$ and such that

$$\pi(A) = \int \eta_{\bar\gamma}(A) \, d\bar\pi(\bar\gamma) \qquad \text{for all Borel sets } A \subset \Gamma(X).$$

Note also that, by the disintegration theorem, the map $\bar\gamma \mapsto \eta_{\bar\gamma}$ is Borel measurable. Let us now set

$$\nu_{\bar\gamma} := \int_{\Phi^{-1}(\bar\gamma)} \varphi_{\#}(\mu_\gamma) \, d\eta_{\bar\gamma}(\gamma).$$

Clearly, we have the representation

$$\varphi_{\#}(\mu \llcorner U) = \int \nu_{\bar\gamma} \, d\bar\pi(\bar\gamma)$$

and $\bar\gamma'(t) = (\phi \circ \gamma)'(t) \in C \setminus \{0\}$ for $\bar\pi$-almost every $\bar\gamma$ and almost every $t \in \operatorname{Dom} \bar\gamma$. Hence, to conclude the proof we only have to show that

$$\nu_{\bar\gamma} \ll \mathcal{H}^1 \llcorner \operatorname{Im} \bar\gamma \qquad \text{for } \bar\pi\text{-a.e. } \bar\gamma.$$

Let E be a set with $\mathcal{H}^1(E \cap \operatorname{Im} \bar\gamma) = 0$. Since $\bar\gamma'(t) \neq 0$ for almost every $t \in \operatorname{Dom} \gamma$, the area formula implies that $\mathcal{L}^1(\bar\gamma^{-1}(E)) = 0$. If $\gamma \in \Phi^{-1}(\bar\gamma)$, say $\bar\gamma = \phi \circ \gamma$, then

$$\mathcal{H}^1(\phi^{-1}(E) \cap \operatorname{Im} \gamma) \leq \mathcal{H}^1(\gamma(\bar\gamma^{-1}(E))) = 0 \qquad \text{for all } \gamma \in \Phi^{-1}(\bar\gamma).$$

Hence, $\mu_\gamma(\phi^{-1}(E)) = 0$ for all $\gamma \in \Phi^{-1}(\bar\gamma)$ which immediately gives

$$\nu_{\bar\gamma}(E) = \int_{\Phi^{-1}(\bar\gamma)} \mu_\gamma(\phi^{-1}(E)) \, d\eta_{\bar\gamma}(\gamma) = 0 .$$

This concludes the proof. □

Proof of Theorem 8.1.1. Let (U, φ) be a d-chart. By Theorem 4.2.3 there are d φ-independent Alberti representations of $\mu \llcorner U_k$, where $U = \bigcup_{k \in \mathbb{N}} U_k$ is the decomposition from Bate's theorem. Then, via Lemma 4.3.3, the push-forward $\varphi_\#(\mu \llcorner U_k)$ also has d independent Alberti representations. Finally, Lemma 4.3.2 yields $\varphi_\#(\mu \llcorner U_k) \ll \mathcal{L}^d$ and this concludes the proof. □

Acknowledgment: The authors would like to thank Nicola Gigli for his suggestion to write this note and Andrea Schioppa for several useful discussions. G. D. P. is supported by the MIUR SIR-grant "Geometric Variational Problems" (RBSI14RVEZ). A. M. is supported by the ERC-grant "Regularity of area-minimizing currents" (306247). F. R. gratefully acknowledges the support from an EPSRC Research Fellowship on "Singularities in Nonlinear PDEs" (EP/L018934/1).

Bibliography

[1] G. Alberti, M. Csörnyei, and D. Preiss: Structure of null sets in the plane and applications. In Proceedings of the Fourth European Congress of Mathematics (Stockholm, 2004), pages 3–22. European Mathematical Society, 2005.

[2] G. Alberti, M. Csörnyei, and D. Preiss: Differentiability of lipschitz functions, structure of null sets, and other problems. In Proceedings of the International Congress of Mathematicians 2010 (Hyderabad 2010), pages 1379–1394. European Mathematical Society, 2010.

[3] L. Ambrosio, N. Gigli, and G. Savaré: Gradient flows in metric spaces and in the space of probability measures. Lectures in Mathematics ETH Zürich. Birkhäuser, 2005.

[4] G. Alberti: Rank one property for derivatives of functions with bounded variation. Proc. Roy. Soc. Edinburgh Sect. A, 123:239–274, 1993.

[5] G. Alberti and A. Marchese: On the differentiability of lipschitz functions with respect to measures in the Euclidean space. Geom. Funct. Anal., 26:1–66, 2016.

[6] D. Bate: Structure of measures in Lipschitz differentiability spaces. J. Amer. Math. Soc., 28:421–482, 2015.

[7] J. Cheeger: Differentiability of Lipschitz functions on metric measure spaces. Geom. Funct. Anal., 9:428–517, 1999.

[8] J. Cheeger and B. Kleiner: On the differentiability of Lipschitz maps from metric measure spaces to Banach spaces. In Inspired by S. S. Chern, volume 11 of Nankai Tracts Math., pages 129–152. World Scientific, 2006.

[9] J. Cheeger and B. Kleiner: Differentiability of Lipschitz maps from metric measure spaces to Banach spaces with the Radon-Nikodým property. Geom. Funct. Anal., 19:1017–1028, 2009.

[10] G. De Philippis and F. Rindler: On the structure of $\mathcal{A}$-free measures and applications. Ann. of Math., 2016. to appear, arXiv:1601.06543.

[11] H. Federer: Geometric measure theory. Die Grundlehren der mathematischen Wissenschaften, Band 153. Springer-Verlag New York Inc., New York, 1969.

[12] J. Gong: Rigidity of derivations in the plane and in metric measure spaces. Illinois J. Math., 56:1109–1147, 2012.

[13] N. Gigli and E. Pasqualetto: Behaviour of the reference measure on RCD spaces under charts. arXiv:1607.05188, 2016.
[14] P. Jones: Product formulas for measures and applications to analysis and geometry, 2011. Talk given at the conference "Geometric and algebraic structures in mathematics", Stony Brook University, May 2011, http://www.math.sunysb.edu/Videos/dennisfest/.
[15] S. Keith: A differentiable structure for metric measure spaces. Adv. Math., 183:271–315, 2004.
[16] M. Kell and A. Mondino: On the volume measure of non-smooth spaces with Ricci curvature bounded below. arXiv:1607.02036, 2016.
[17] D. Preiss: Differentiability of lipschitz functions on banach spaces. J. Funct. Anal., 91:312–345, 1990.
[18] D. Preiss and G. Speight: Differentiability of Lipschitz functions in Lebesgue null sets. Invent. Math., 199:517–559, 2015.
[19] A. Schioppa: Derivations and Alberti representations. Adv. Math., 293:436–528, 2016.
[20] S. K. Smirnov: Decomposition of solenoidal vector charges into elementary solenoids, and the structure of normal one-dimensional flows. Algebra i Analiz, 5:206–238, 1993. translation in St. Petersburg Math. J. 5 (1994), 841–867.
[21] Z. Zahorski: Sur l'ensemble des points de non-dérivabilité d'une fonction continue. Bull. Soc. Math. France, 74:147–178, 1946.

Tom Leinster and Mark W. Meckes

The magnitude of a metric space: from category theory to geometric measure theory

5.1 Introduction

Magnitude is a numerical isometric invariant of metric spaces. Its definition arises by viewing a metric space as a kind of *enriched category* – an abstract structure that appears more algebraic than geometric in nature – and adapting a construction from the intersection of category theory and homotopy theory. One would hardly expect, from such a provenance, that magnitude would have any strong relationship to geometry as usually conceived. Surprisingly, however, magnitude turns out to encode many invariants from integral geometry and geometric measure theory, including volume, capacity, dimension, and intrinsic volumes. This paper will give an overview of the theory of magnitude, from its category-theoretic genesis to its connections with these geometric quantities.

We begin with a brief overview of the history of magnitude so far. The grandparent of magnitude is the Euler characteristic of a topological space, which is a natural analogue of the cardinality of a finite set. To each category there is associated a topological space called its *classifying space*. In [16], a formula was found for the Euler characteristic of the classifying space of a suitably nice finite category; applying this formula to less nice categories (for which the Euler characteristic of the classifying space need not exist) yielded a new cardinality-like invariant of categories, again called the *Euler characteristic* of a finite category.

Categories are a special case of a more general family of structures, *enriched categories*, which encompass both categories with additional structure (like linear categories) and, surprisingly, metric spaces. In [19, 23], the definition of Euler characteristic of a category was generalized to enriched categories, renamed *magnitude*, then re-specialized to finite metric spaces. The first paper to be written on magnitude [23] focused on the asymptotic behavior of the magnitudes of finite approximations to specific compact subsets of Euclidean space. The results there hinted strongly that magnitude is closely related to geometric quantities including volume and fractal dimension; numerical computations in [40] gave further evidence of these relationships.

Tom Leinster: School of Mathematics, University of Edinburgh, James Clerk Maxwell Building, Peter Guthrie Tait Road, Edinburgh EH9 3FD, United Kingdom, E-mail: Tom.Leinster@ed.ac.uk
Mark W. Meckes: Department of Mathematics, Applied Mathematics, and Statistics, Case Western Reserve University, 10900 Euclid Ave., Cleveland, Ohio 44106, U.S.A., email: mark.meckes@case.edu

In [41], a definition was proposed for the magnitude of certain compact metric spaces, and connections were found between magnitude and some intrinsic volumes of Riemannian manifolds. Shortly thereafter, the paper [19] appeared which laid out for the first time the general theory of the magnitude of finite metric spaces; and [27] which put the asymptotic approach of [23] for studying magnitude of compact spaces on firm footing, and showed that it also coincides with the definition used in [41].

The paper [28] introduced yet another equivalent approach to magnitude for compact spaces, which makes magnitude more accessible to a wide variety of analytic techniques. Using a result from potential theory, [28] showed in particular that magnitude can be used to recover the Minkowski dimension of a compact set in Euclidean space. Following the approach of [28], the paper [3] applied Fourier analysis to show that magnitude also recovers volume in Euclidean space, and applied PDE techniques to compute precisely magnitudes of Euclidean balls.

This paper aims to serve as a guide to the path from the definition of the Euler characteristic of a finite category, to the geometric results of [28] and [3] on magnitude in Euclidean space. It also includes a number of new results, in particular a significant partial result toward a conjecture from [19] relating magnitude in ℓ_1^n to a family of intrinsic volumes adapted to the ℓ_1 metric, as well as generalizations of several regularity results for magnitude from Euclidean space to more general normed spaces. In order to reach the results of geometric interest as quickly as possible, we omit many results from the papers named above, and depart significantly at some points from the historical development of ideas. We give complete proofs only for the new results, and for a few known results for which we take a more direct approach than in previous papers.

Section 5.2 begins with the definition of the Euler characteristic of a finite category, and leads up to the magnitude of a finite metric space and its basic properties. Section 5.3 covers the definition of the magnitude of a compact space, its basic properties, and the results on magnitude of manifolds. Section 5.4 covers magnitude in (quasi)normed spaces, particularly ℓ_1^n and Euclidean space, and contains the new results of this paper. Finally, in section 5.5, we discuss a number of open problems about magnitude.

Before moving on, we need to mention two threads in the story of magnitude which have been ignored above and will make only brief appearances in this paper. The first is the magnitude of a graph, viewed as a metric space with the shortest-path distance between vertices. This subject has been developed in [20], which in particular investigated its relationship to classical, combinatorial graph invariants, and [8], which found that the magnitude of graphs is the Euler characteristic associated to a graded homology theory for graphs. The second thread is the connection of magnitude to quantifying biodiversity and maximum entropy problems. This is actually related with the historically first appearance of the magnitude of a metric space in the literature, in [35], and was developed in [17, 22]; section 5.3.3 will take half a step in the direction of these connections.

5.2 Finite metric spaces

Here we explain the origins of the notion of magnitude. There is a simple combinatorial definition of the magnitude or Euler characteristic of a finite category (section 5.2.2), which extends in a natural way to a more general class of structures, the enriched categories (section 5.2.3). As we show, this general invariant is closely related to several existing invariants of size. Specializing it in a different direction gives the definition of the magnitude of a finite metric space (sections 5.2.4 and 5.2.5).

In order to do any of this, we first need to define the magnitude of a matrix.

5.2.1 The magnitude of a matrix

Recall that a **semiring** is a "ring without negatives", that is, an abelian group (written additively) with an associative operation of multiplication that distributes over addition. Let k be a commutative semiring (always assumed to have a multiplicative identity 1) and A a finite set, and let $Z \in k^{A\times A}$ be a square matrix over k indexed by the elements of A. A **weighting** on Z is a column vector $w \in k^A$ satisfying $Zw = e$, where e is the column vector of 1s, and a **coweighting** on Z is a row vector $v \in k^A$ satisfying $vZ = e^{\mathrm{T}}$. That is,

$$\sum_{b\in A} Z(a,b)w_b = 1 \text{ for every } a \in A$$

and

$$\sum_{a\in A} v_a Z(a,b) = 1 \text{ for every } b \in A.$$

If w is a weighting and v a coweighting on Z then

$$\sum_{a\in A} w_a = e^{\mathrm{T}}w = vZw = ve = \sum_{a\in A} v_a.$$

When Z admits both a weighting and a coweighting, we may therefore define the **magnitude** $|Z|$ of Z to be the common quantity $\sum_a w_a = \sum_a v_a$, for any weighting w and coweighting v.

An important special case is when Z is invertible. Then Z has a unique weighting and a unique coweighting, and its magnitude is the sum of the entries of Z^{-1}:

$$|Z| = \sum_{a,b\in A} Z^{-1}(a,b). \tag{5.1}$$

An even more special case is that of positive definite matrices:

Proposition 5.2.1. *Let $Z \in \mathbb{R}^A$ be a positive definite matrix. Then*

$$|Z| = \sup_{0\neq x\in\mathbb{R}^A} \frac{(\sum_a x_a)^2}{x^{\mathrm{T}}Zx},$$

and the supremum is attained exactly when x is a scalar multiple of the unique weighting on Z.

This follows swiftly from the Cauchy–Schwarz inequality [19, Proposition 2.4.3].

5.2.2 The Euler characteristic of a finite category

A category can be viewed as a directed graph (allowing multiple parallel edges) together with an associative, unital operation of composition. The vertices of the graph are the objects of the category, and for each pair (a, b) of vertices, the edges from a to b in the graph are the maps from a to b in the category, which form a set $\mathrm{Hom}(a, b)$. Thus, composition defines a function $\mathrm{Hom}(a, b) \times \mathrm{Hom}(b, c) \to \mathrm{Hom}(a, c)$ for each a, b, c, and there is a loop $1_a \in \mathrm{Hom}(a, a)$ on each vertex a. Although in many categories of interest, the collections of objects and maps form infinite sets or even proper classes, we will be considering **finite categories**: those with only finitely many objects and maps.

Let **A** be a finite category, with set of objects ob **A**. The **Euler characteristic** of **A** is the magnitude of the matrix $Z_{\mathbf{A}} \in \mathbb{Q}^{\mathrm{ob}\,\mathbf{A}\times\mathrm{ob}\,\mathbf{A}}$ given by $Z_{\mathbf{A}}(a, b) = \#\mathrm{Hom}(a, b)$ (where # denotes cardinality), whenever this magnitude is defined.

For example, if **A** has no maps other than identities then $Z_{\mathbf{A}}$ is the identity and the Euler characteristic of **A** is simply the number of objects. More generally, any partially ordered set $(P, \leq)$ gives rise to a category **A** whose objects are the elements of P, and with one map $a \to b$ when $a \leq b$ and none otherwise. In a theory made famous by Rota [30], every finite partially ordered set P has associated with it a Möbius function μ, which is defined on pairs (a, b) of elements of P such that $a \leq b$, and takes values in $\mathbb{Z}$. It generalizes the classical Möbius function, and the construction above for categories generalizes it further still: $\mu(a, b) = Z_{\mathbf{A}}^{-1}(a, b)$ whenever $a \leq b$, and the definition of Euler characteristic of a category extends the existing definition for ordered sets [16, Proposition 4.5].

To any small category **A** there is assigned a topological space, called its **classifying space**. The name "Euler characteristic" is largely justified by the following result.

Theorem 5.2.2 ([16, Proposition 2.11]). *Let* **A** *be a finite category. Under appropriate conditions (which imply, in particular, that the Euler characteristic of the classifying space of* **A** *is defined), the Euler characteristic of the category* **A** *is equal to the Euler characteristic of its classifying space.*

Euler characteristic for finite categories enjoys many properties analogous to those enjoyed by topological Euler characteristic [16, Section 2]. For instance, categorical Euler characteristic is invariant under equivalence (mirroring homotopy invariance in the topological setting), and is additive with respect to disjoint union of categories and multiplicative with respect to products. There is even an analogue of the topological formula for the Euler characteristic of the total space of a fibration.

Schanuel [33] argued that Euler characteristic for topological spaces is closely analogous to cardinality for sets. For instance, it has analogous additivity and multiplicativity properties, it satisfies the inclusion-exclusion principle (under hypotheses), and, indeed, it reduces to cardinality for finite discrete spaces. Similarly, the results described above suggest that Euler characteristic for finite categories is the categorical analogue of cardinality.

5.2.3 Enriched categories

A **monoidal category** is a category $\mathcal{V}$ equipped with an associative binary operation $\otimes$ (which is formally a functor $\mathcal{V}\times\mathcal{V} \to \mathcal{V}$) and a unit object $\mathbb{1} \in \mathcal{V}$. The associativity and unit axioms are only required to hold up to suitably coherent isomorphism; see [26] for details.

Typical examples of monoidal categories $(\mathcal{V}, \otimes, \mathbb{1})$ are the categories $(\text{Set}, \times, \{*\})$ of sets with cartesian product and $(\mathbf{FDVect}_K, \otimes, K)$ of finite-dimensional vector spaces over a field K. A less obvious example is the ordered set $([0, \infty], \geq)$. As a category, its objects are the nonnegative reals together with ∞, there is one map $x \to y$ when $x \geq y$, and there are none otherwise. It is monoidal with $\otimes = +$ and $\mathbb{1} = 0$.

Let $\mathcal{V} = (\mathcal{V}, \otimes, \mathbb{1})$ be a monoidal category. The definition of category enriched in $\mathcal{V}$, or $\mathcal{V}$-category, is obtained from the definition of ordinary category by requiring that the hom-sets are no longer sets but objects of $\mathcal{V}$. Thus, a (small) **$\mathcal{V}$-category A** consists of a set ob **A** of objects, an object $\text{Hom}(a, b)$ of $\mathcal{V}$ for each $a, b \in \text{ob}\,\mathbf{A}$, and operations of composition and identity satisfying appropriate axioms [10]. The composition consists of a map

$$\text{Hom}(a, b) \otimes \text{Hom}(b, c) \to \text{Hom}(a, c)$$

in $\mathcal{V}$ for each $a, b, c \in \text{ob}\,\mathbf{A}$, while the identities are provided by a map $\mathbb{1} \to \text{Hom}(a, a)$ for each $a \in \text{ob}\,\mathbf{A}$.

Examples 5.2.3. 1. *When* $\mathcal{V} = \text{Set}$ *(with monoidal structure as above), a* $\mathcal{V}$*-category is an ordinary (small) category.*

2. *When* $\mathcal{V} = \mathbf{Vect}_K$*, a* $\mathcal{V}$*-category is a* ***linear category****, that is, a category in which each hom-set carries the structure of a vector space, and composition is bilinear.*
3. *When* $\mathcal{V} = [0, \infty]$*, a* $\mathcal{V}$*-category is a* ***generalized metric space*** *[14, 15]. That is, a* $\mathcal{V}$*-category consists of a set A of objects or points together with, for each* $a, b \in A$*, a real number* $\text{Hom}(a, b) = d(a, b) \in [0, \infty]$*, satisfying the axioms*

$$d(a, b) + d(b, c) \geq d(a, c), \qquad d(a, a) = 0$$

$(a, b, c \in A)$. Such spaces are more general than classical metric spaces in three ways: ∞ is permitted as a distance, the separation axiom $d(a, b) = 0 \implies a = b$ is dropped, and, most significantly, d is not required to be symmetric.

4. *The category $\mathcal{V} = ([0, \infty], \geq)$ can alternatively be given the monoidal structure $(\max, 0)$. A $\mathcal{V}$-category is then a generalized ultrametric space, that is, a generalized metric space satisfying the stronger triangle inequality $\max\{d(a, b), d(b, c)\} \geq d(a, c)$.*

To define the magnitude of an enriched category, we start with a monoidal category $(\mathcal{V}, \otimes, \mathbb{1})$ together with a commutative semiring k and a map $|\cdot| : \text{ob}\,\mathcal{V} \to k$, with the property that $|X| = |Y|$ whenever $X \cong Y$, and satisfying the multiplicativity axioms $|X \otimes Y| = |X| \cdot |Y|$ and $|\mathbb{1}| = 1$.

Definition 5.2.4. *Let* **A** *be a $\mathcal{V}$-category with only finitely many objects.*

1. *The **similarity matrix** of* **A** *is the* ob **A** × ob **A** *matrix $Z_{\mathbf{A}}$ over k defined by $Z_{\mathbf{A}}(a, b) = \big|\text{Hom}(a, b)\big|$.*
2. *A **(co)weighting** on* **A** *is a (co)weighting on $Z_{\mathbf{A}}$, and* **A** ***has magnitude*** *if $Z_{\mathbf{A}}$ does. Its **magnitude** is then $|\mathbf{A}| = |Z_{\mathbf{A}}|$.*

Examples 5.2.5.

1. *Let $\mathcal{V}$ be the monoidal category $(\mathbf{FinSet}, \times, \{*\})$ of finite sets. Let $k = \mathbb{Q}$, and for $X \in \mathbf{FinSet}$, let $|X| \in \mathbb{Q}$ be the cardinality of X. Then we obtain a notion of magnitude for finite categories; it is exactly the Euler characteristic of section 5.2.2.*
2. *Let $\mathcal{V}$ be the monoidal category $\mathbf{FDVect}_K$ of finite-dimensional vector spaces over a field K. Let $k = \mathbb{Q}$, and for $X \in \mathbf{FDVect}_K$, put $|X| = \dim X \in \mathbb{Q}$. Then we obtain a notion of magnitude for linear categories with finitely many objects and finite-dimensional hom-spaces. As shown in [5], this invariant is closely related to the Euler form of an associative algebra, defined homologically.*
3. *Let $\mathcal{V} = [0, \infty]$, with monoidal structure $(+, 0)$. Let $k = \mathbb{R}$, and for $x \in [0, \infty]$, put $|x| = e^{-x}$. (We have little choice about this: the multiplicativity axioms force $|x| = C^x$ for some constant C, at least assuming that $|\cdot|$ is to be measurable. We will address the one degree of freedom here through the introduction of magnitude functions in the next section.) Then we obtain a notion of the magnitude $|A| \in \mathbb{R}$ of a finite metric space $|A|$, examined in detail later.*
4. *Let $\mathcal{V} = [0, \infty]$, now with monoidal structure $(\max, 0)$. Let $k = \mathbb{R}$, and define $|\cdot| : [0, \infty] \to \mathbb{R}$ to be either the indicator function of $[0, 1]$ or that of $[0, 1)$. It is shown in Section 8 of [28] that these are essentially the only possibilities for $|\cdot|$, and that the resulting magnitude of a finite ultrametric space is simply the number of balls of radius 1 (closed or open, respectively) needed to cover it. It is also shown that this leads naturally to the notion of ε-entropy or ε-capacity.*

The multiplicativity condition $|X \otimes Y| = |X| \cdot |Y|$ on objects of $\mathcal{V}$ has so far not been used. However, it implies a similar multiplicativity condition on categories enriched in $\mathcal{V}$. In the case of metric spaces, this reduces to Proposition 5.2.8 below; for the general statement, see [19, Proposition 1.4.3].

5.2.4 The magnitude of a finite metric space

Concretely, the magnitude $|A|$ of a finite metric space (A, d) is the magnitude of the matrix $Z = Z_A \in \mathbb{R}^{A \times A}$ given by $Z_A(a, b) = e^{-d(a,b)}$, if that is defined. Taking advantage of the symmetry of Z_A to simplify slightly, this means the following. A vector $w \in \mathbb{R}^A$ is a **weighting** for A if $Z_A w = e$, where $e \in \mathbb{R}^A$ is the column vector of 1s, and if a weighting for A exists, then the **magnitude** of A is

$$|A| = \sum_{a \in A} w_a.$$

This is not a classical invariant, or one that appears to have previously been explored mathematically prior to the work cited in the introduction. Neither is it wholly new. In a probabilistic analysis of the benefits of highly diverse ecosystems, Solow and Polasky [35] derived a lower bound on the benefit and identified one term, which they called the "effective number of species", as especially interesting. Although it was not thoroughly investigated in [35], this term is exactly our magnitude. The reader is referred to [17, 19, 21, 22] for more information about this connection.

Not every finite metric space possesses a weighting or, therefore, has well-defined magnitude. One large and important class of spaces which always does is the subject of section 5.2.5. The next two results give additional examples.

From now on, to simplify the statements of results, all metric spaces and all compact sets in a metric space are assumed to be nonempty.

Proposition 5.2.6 (([23, Theorem 2] and [19, Proposition 2.1.3])). *Let (A, d) be a finite metric space, and suppose that whenever $a, b \in A$ with $a \neq b$, we have $d(a, b) > \log(\#A - 1)$. Then A possesses a positive weighting, and $|A|$ is therefore defined.*

A metric space (A, d) is called **homogeneous** if its isometry group acts transitively on the points of A.

Proposition 5.2.7 ([36]; see also [19, Proposition 2.1.5]). *If (A, d) is a finite homogeneous metric space and $a_0 \in A$ is any fixed point, then A possesses a positive weighting and*

$$|A| = \frac{(\#A)^2}{\sum_{a,b \in A} e^{-d(a,b)}} = \frac{\#A}{\sum_{a \in A} e^{-d(a,a_0)}}.$$

For metric spaces (A, d_A) and (B, d_B), we denote by $A \times_1 B$ the set $A \times B$ equipped with the metric

$$d\big((a,b),(a',b')\big) = d_A(a,a') + d_B(b,b').$$

Proposition 5.2.8 ([19, Proposition 2.3.6]). *Suppose that (A, d_A) and (B, d_B) are finite metric spaces with weightings $w \in \mathbb{R}^A$ and $v \in \mathbb{R}^B$ respectively. Then $x \in \mathbb{R}^{A\times B}$ given by $x_{(a,b)} = w_a v_b$ is a weighting for $A \times_1 B$, and $|A \times_1 B| = |A|\,|B|$.*

Proposition 5.2.8 has a generalization, Theorem 2.3.11 of [19], which is an analogue for magnitude of the formula for the Euler characteristic of the total space of a fibration.

As noted earlier, there is an arbitrary choice of scale implicit in the definition of magnitude: we could choose any other base for the exponent in place of e^{-1}. To deal with this, we will often work with the whole family of metric spaces $\{tA\}_{t>0}$, where tA denotes the metric space (A, td). We will sometimes also let $0A$ denote a one-point space. The (partially defined) function $t \mapsto |tA|$ is called the **magnitude function** of A.

Proposition 5.2.9 ([19, Proposition 2.2.6]). *Let (A, d) be a finite metric space.*
1. *$|tA|$ is defined for all but finitely many $t > 0$.*
2. *For sufficiently large t, $|tA|$ is an increasing function of t.*
3. *$\lim_{t\to\infty} |tA| = \#A$.*

Proposition 5.2.9 supports the interpretation of the magnitude $|tA|$ as the "effective number of points" in A, when viewed at a scale determined by t. (We recall Solow and Polasky's interpretation of $|A|$ as the "effective number of species".) However, the hypotheses of the propositions above also highlight the counterintuitive behaviors that magnitude may exhibit. In particular, there exists a metric space A such that each of the following holds:

1. $|tA|$ is undefined for some $t > 0$.
2. $|tA|$ is decreasing for some $t > 0$.
3. $|tA| < 0$ for some $t > 0$.
4. There exists a $B \subseteq A$ such that $|tB| > |tA|$ for some $t > 0$.

We need not look that hard to find such an ill-behaved space: the complete bipartite graph $K_{3,2}$, equipped with the shortest path metric, has all these unpleasant properties; see Example 2.2.7 of [19]. In the next section we will consider a class of spaces which avoids most of these pathologies.

We end this section by noting that the issue of scale can be dealt with in a more elegant way if A is the vertex set of a graph and d is the shortest path metric, or more generally, whenever d is integer-valued. By (5.1), in this situation $|tA|$ is a rational function of $q = e^{-t}$. More directly, if one restricts attention to such spaces, the semiring k in the previous section can be taken to be the ring $\mathbb{Q}(q)$ of rational functions in a formal variable q. Then the matrix $Z_A \in (\mathbb{Q}(q))^{A\times A}$ is always invertible, so the magnitude $|A|$ is always defined as an element of $\mathbb{Q}(q)$; see section 2 of [20].

5.2.5 Positive definite metric spaces

As noted in section 5.2.1, a positive definite matrix Z always has magnitude, given by Proposition 5.2.1. We will now explore the consequences of this observation for magnitude of metric spaces.

A finite metric space (A, d) is said to be **positive definite** if the associated matrix Z_A is positive definite, and is said to be of **negative type** if Z_{tA} is positive semidefinite for every $t > 0$. It can be shown [27, Theorem 3.3] that if (A, d) is of negative type, then in fact Z_{tA} is positive definite, and hence tA is a positive definite space. A general metric space is said to be positive definite or of negative type, respectively, if every finite subspace is.

The strange turn of terminology here is due to the negative sign in e^{-d}. Negative type has several other equivalent formulations, and is an important property in the theory of metric embeddings (see, e.g., [4, 6, 39]). The fact that negative type appears naturally when considering magnitude is a hint that magnitude does in fact connect with more classical topics in geometry.

The following result is an immediate consequence of Proposition 5.2.1 and the definition of magnitude.

Proposition 5.2.10 ([19, Proposition 2.4.3]). *If A is a finite positive definite metric space, then the magnitude $|A|$ is defined, and*

$$|A| = \max_{0 \neq x \in \mathbb{R}^A} \frac{\left(\sum_{a\in A} x_a\right)^2}{x^{\mathrm{T}} Z_A x},$$

and the supremum is attained exactly when x is a scalar multiple of the unique weighting on A.

A first application of Proposition 5.2.10 is Proposition 5.2.6, which is proved by showing that for large enough t, Z_{tA} is positive definite.

Corollary 5.2.11 (Corollaries 2.4.4 and 2.4.5 of [19]). *If A is a finite positive definite metric space and $\emptyset \neq B \subseteq A$, then $1 \leq |B| \leq |A|$.*

Proposition 5.2.10 will also be one of our main tools in the extension of magnitude to compact spaces in section 5.3.

Proposition 5.2.10 and its consequences would be of little interest without a large supply of interesting examples of positive definite spaces. Many are collected in the following result; we refer to [27, Theorem 3.6] for references and further examples.

Theorem 5.2.12. *The following metric spaces are of negative type, and thus magnitude is defined for all their finite subsets.*

1. *ℓ_p^n, the set $\mathbb{R}^n$ equipped with the metric derived from the ℓ_p-norm, for $n \geq 1$ and $1 \leq p \leq 2$;*
2. *Lebesgue space $L_p[0,1]$, for $1 \leq p \leq 2$;*
3. *round spheres (with the geodesic distance);*
4. *real and complex hyperbolic space;*
5. *ultrametric spaces;*
6. *weighted trees.*

Furthermore, some natural operations on positive definite spaces yield new positive definite spaces.

Proposition 5.2.13 ([19, Lemma 2.4.2])**.**

1. *Every subspace of a positive definite metric space is positive definite.*
2. *If A and B are positive definite metric spaces, then $A \times_1 B$ is positive definite.*

On the other hand, many spaces of geometric interest are *not* of negative type, and many natural operations fail to preserve positive definiteness; see [27, Section 3.2] for examples and references.

5.3 Compact metric spaces

Despite strong and growing interest in the geometry of finite metric spaces (see e.g. [25]), it is natural to try to define an invariant of metric spaces, like magnitude, more generally. The most obvious context is that of compact spaces. The general definition of the magnitude of an enriched category does not help us here; nonetheless, several strategies present themselves, including approximating a compact space by finite subspaces and generalizing the notion of a weighting to compact spaces. In section 5.3.1 we will see that there is a canonical (hence "correct") extension of magnitude from finite metric spaces to compact positive definite spaces, which can be formulated in several ways. In section 5.3.2 we will investigate a generalization of weightings to compact spaces, and see that this approach to defining magnitude agrees with the former one. This approach is of more limited scope, but often gives the easiest approach to computing magnitude; using it, we will see that magnitude knows about at least some intrinsic volumes of certain Riemannian manifolds. Finally, section 5.3.3 will introduce another invariant, maximum diversity, which is closely related to magnitude, and will be a crucial tool in proving the connection between magnitude and Minkowski dimension.

5.3.1 Compact positive definite spaces

To justify the "correctness" of our definition of magnitude for compact positive definite spaces, we need a topology on the family of (isometry classes of) compact metric spaces. Recall that the **Hausdorff metric** d_H on the family of compact subsets of a metric space X is given by

$$d_H(A, B) = \max\{\sup_{a\in A} d(a, B), \sup_{b\in B} d(b, A)\}.$$

The **Gromov–Hausdorff distance** between two compact metric spaces A and B is

$$d_{GH}(A, B) = \inf d_H(\phi(A), \psi(B)),$$

where the infimum is over all metric spaces X and isometric embeddings $\phi : A \to X$ and $\psi : B \to X$. This defines a metric on the family of isometry classes of compact metric spaces; see [7, Chapter 3].

The following result follows from the proof of [27, Theorem 2.6], although our definitions are organized rather differently in that paper. We give a more streamlined version of the argument from [27].

Proposition 5.3.1. *The quantity*

$$M(A) = \sup\left\{\left|A'\right| \;\middle|\; A' \subseteq A,\ A' \text{ finite}\right\} \tag{5.2}$$

is lower semicontinuous as a function of A (taking values in $[0, \infty]$), on the class of compact positive definite metric spaces equipped with the Gromov–Hausdorff topology.

Proof. Suppose first that $d_{GH}(A, B) < \delta$ for *finite* positive definite spaces A and B, and let $w \in \mathbb{R}^A$ be a weighting for A. There is a function $f : A \to B$ such that $\left|d(f(a), f(a')) - d(a, a')\right| < 2\delta$ for all $a, a' \in A$. Define $v \in \mathbb{R}^B$ by $v_b = \sum_{a\in f^{-1}(b)} w_a$, and $Z_f \in \mathbb{R}^{A\times A}$ by $Z_f(a, a') = e^{-d(f(a), f(a'))}$. Then $v^{\mathrm{T}} Z_B v = w^{\mathrm{T}} Z_f w$, and so

$$\left|w^{\mathrm{T}} Z_A w - v^{\mathrm{T}} Z_B v\right| = \left|w^{\mathrm{T}}(Z_A - Z_f)w\right| \le \|w\|_1^2 \left\|Z_A - Z_f\right\|_\infty < 2\|w\|_1^2 \,\delta.$$

Thus by Proposition 5.2.10,

$$|B| \ge \frac{(\sum_b v_b)^2}{v^{\mathrm{T}} Z_B v} \ge \frac{(\sum_a w_a)^2}{w^{\mathrm{T}} Z_A w + 2\|w\|_1^2 \,\delta} = \frac{|A|^2}{|A| + 2\|w\|_1^2 \,\delta} \ge |A| - 2\|w\|_1^2 \,\delta. \tag{5.3}$$

Now for general A, assume for simplicity that $M(A) < \infty$ (the case $M(A) = \infty$ is handled similarly). Given $\varepsilon > 0$, pick a finite subset $A' \subset A$ such that $\left|A'\right| \ge M(A) - \varepsilon$, and let $w \in \mathbb{R}^{A'}$ be a weighting for A'. If $d_{GH}(A, B) < \delta$, then there is a finite subset $B' \subseteq B$ such that $d_{GH}(A', B') < \delta$, and so by (5.3),

$$M(B) \ge \left|B'\right| \ge \left|A'\right| - 2\|w\|_1^2 \,\delta \ge M(A) - \varepsilon - 2\|w\|_1^2 \,\delta.$$

Therefore $M(B) \ge M(A) - 2\varepsilon$ when $d_{GH}(A, B)$ is sufficiently small. □

Corollary 5.2.11 implies that $M(A) = |A|$ when A itself is finite and positive definite. Proposition 5.3.1 thus implies first of all that magnitude is lower semicontinuous (l.s.c.) on the class of finite positive definite metric spaces. It follows that there is a canonical extension of magnitude to the class of compact positive definite metric spaces, namely, the *maximal* l.s.c. extension. Proposition 5.3.1 furthermore implies that this extension is precisely the function M in (5.2). For a compact positive definite metric space (A, d), we therefore define the **magnitude** $|A|$ to be the value of the supremum $M(A)$ in (5.2).

Thus magnitude is lower semicontinuous on the class of compact positive spaces. This cannot be improved to continuity in general, even for the class of finite spaces of negative type. Examples 2.2.8 and 2.4.9 in [19] discuss a space A of negative type with six points, such that $|tA| = 6/(1 + 4e^{-t})$; thus $\lim_{t\to 0^+} |tA| = 6/5$, whereas the space tA itself converges to a one-point space. On the other hand, magnitude is continuous when restricted to certain classes of spaces, as we will see in Corollary 5.3.13 and Theorem 5.4.15 below.

Proposition 5.3.1, Proposition 5.2.8, and Corollary 5.2.11 yield the following results.

Proposition 5.3.2 ([19, Lemma 3.1.3]). *If A is a compact positive definite metric space and $\emptyset \neq B \subseteq A$, then $1 \le |B| \le |A|$.*

Proposition 5.3.3 ([27, Corollary 2.7]). *Let A be a compact positive definite metric space, and let $\{A_k\}$ be any sequence of compact subsets of A such that $A_k \xrightarrow{k\to\infty} A$ in the Hausdorff topology. Then $|A| = \lim_{k\to\infty} |A_k|$.*

Proposition 5.3.4 ([19, Proposition 3.1.4]). *If A and B are compact positive definite metric spaces, then $|A \times_1 B| = |A|\,|B|$.*

Proposition 5.3.1 justifies the above definition of magnitude as the "correct" one for a compact positive definite space A. Nevertheless, for both aesthetic and practical reasons, it is desirable to be able to work directly with A itself, as opposed to approximations of A by finite subspaces. Two different, more direct approaches to defining magnitude for compact positive definite spaces were developed in [27, 28]. In essence, these papers introduced two different topologies on the space $\left\{ w \in \mathbb{R}^A \;\middle|\; \operatorname{supp} w \text{ is finite} \right\}$. The topology used in [27] has the advantage of being more familiar, whereas the topology in [28] has the advantage of being better suited to the analysis of magnitude. In particular, the topology used in [28] can be dualized in a way that presents a new set of tools to study magnitude. In the pursuit of our goal of proceeding as quickly as possible to geometric results, here we will go straight to the dual version.

Recall that a **positive definite kernel** on a space X is a function $K : X \times X \to \mathbb{C}$ such that, for every finite set $A \subseteq X$, the matrix $[K(a, b)]_{a,b\in A} \in \mathbb{C}^{A\times A}$ is positive definite. Given a positive definite kernel on X, the **reproducing kernel Hilbert space**

(RKHS) $\mathcal{H}$ on X with kernel K is the completion of the linear span of the functions $k_x(y) = K(x, y)$ with respect to the inner product given by

$$\langle k_x, k_y \rangle_{\mathcal{H}} = K(x, y)$$

(see [2]). If $f \in \mathcal{H}$, then $f(x) = \langle f, k_x \rangle_{\mathcal{H}}$ for every $x \in X$, and consequently

$$|f(x)| \le \|f\|_{\mathcal{H}} \|k_x\|_{\mathcal{H}} = \|f\|_{\mathcal{H}} \sqrt{K(x, x)} \tag{5.4}$$

by the Cauchy–Schwarz inequality.

Now if (X, d) is a positive definite metric space, then $K(x, y) = e^{-d(x,y)}$ is a positive definite kernel on X. We will refer to the corresponding RKHS as the RKHS $\mathcal{H}$ for X.

Theorem 5.3.5 ([28, Theorem 4.1 and Proposition 4.2]). *Let X be a positive definite metric space, and let $A \subseteq X$ be compact. Then $|A| < \infty$ if and only if there exists a function $h \in \mathcal{H}$ such that $h \equiv 1$ on A. In that case,*

$$|A| = \inf\left\{ \|h\|_{\mathcal{H}}^2 \;\middle|\; h \in \mathcal{H},\ h \equiv 1 \text{ on } A \right\}.$$

The infimum is achieved for a unique function h. If $f \in \mathcal{H}$ also satisfies $f \equiv 1$ on A, then $|A| = \langle f, h \rangle_{\mathcal{H}}$.

Proof. First observe that if $w \in \mathbb{R}^B$ for a finite subset $B \subseteq X$, and $f_w = \sum_{b \in B} w_b e^{-d(\cdot, b)}$, then

$$w^{\mathrm{T}} Z_B w = \sum_{a,b \in B} w_a e^{-d(a,b)} w_b = \|f_w\|_{\mathcal{H}}^2 . \tag{5.5}$$

Now suppose that $|A| < \infty$. If $B \subseteq A$ is finite and $w \in \mathbb{R}^B$, then by Proposition 5.2.10, (5.5), and the definition of $|A|$,

$$\left(\sum_{b \in B} w_b \right)^2 \le |B| \, \|f_w\|_{\mathcal{H}}^2 \le |A| \, \|f_w\|_{\mathcal{H}}^2 .$$

Thus the linear functional $f_w \mapsto \sum_{b \in B} w_b$ on the subspace $\left\{ f_w \;\middle|\; w \in \mathbb{R}^B,\ B \subseteq A \text{ finite} \right\} \subseteq \mathcal{H}$ has norm at most $\sqrt{|A|}$. Therefore there is a function $h \in \mathcal{H}$ with $\|h\|_{\mathcal{H}}^2 = |A|$ such that

$$\sum_{b \in B} w_b = \langle f_w, h \rangle_{\mathcal{H}} = \sum_{b \in B} w_b h(b)$$

for every f_w; taking $f_w = e^{-d(\cdot, a)}$ for $a \in A$ yields $h(a) = 1$.

Next suppose that there exists an $h \in \mathcal{H}$ such that $h \equiv 1$ on A. Then for any finite subset $B \subseteq A$ and $w \in \mathbb{R}^B$, by the Cauchy–Schwarz inequality,

$$\left| \sum_{b \in B} w_b \right| = |\langle h, f_w \rangle| \le \|h\|_{\mathcal{H}} \|f_w\|_{\mathcal{H}} .$$

Equation (5.5) and Proposition 5.2.10 then imply that $|B| \le \|h\|_{\mathcal{H}}^2$, and so by definition $|A| \le \|h\|_{\mathcal{H}}^2$.

The above arguments prove both the "if and only if" statement and the infimum expression for $|A|$. The last two statements follow from elementary Hilbert space geometry. □

We will call the unique function h which achieves the infimum in Theorem 5.3.5 the **potential function** of A. Theorem 5.3.5 will prove its worth in sections 5.4.3 and 5.4.4 below.

For now, we consider what has happened to weightings, which were central to the original category-inspired definition of magnitude, but have vanished from the scene in Theorem 5.3.5. Weightings of finite subspaces of X are naturally identified with elements of the dual space $\mathcal{H}^*$, if we restrain ourselves from the usual impulse to identify $\mathcal{H}^*$ with $\mathcal{H}$ itself. We can then identify a weighting of a compact subspace A with finite magnitude as an element of $\mathcal{H}^*$, specifically the element of $\mathcal{H}^*$ represented by the potential function h. See [28] for details.

5.3.2 Weight measures

Proposition 5.3.1 may justify the definition of magnitude adopted in the previous section as the canonical correct definition, but it has two deficiencies. First, it applies only to positive definite spaces, and second, it lies quite far from the original category-inspired definition, being fundamentally based instead on the reformulation in Proposition 5.2.10. The second drawback is to some extent addressed in the last paragraph of the previous section, though still only for positive definite spaces.

In this section we discuss another approach to defining magnitude for compact metric spaces, first used in [41], which more closely parallels the original definition for finite spaces.

A **weight measure** on a compact metric space (A, d) is a finite signed Borel measure μ on A such that

$$\int_A e^{-d(a,b)}\, d\mu(b) = 1$$

for every $a \in A$.

A finite metric space A possesses a weight measure μ if and only if it possesses a weighting $w \in \mathbb{R}^A$, with the correspondence given by $w_a = \mu(\{a\})$. The magnitude of A is in that case

$$|A| = \sum_{a\in A} w_a = \mu(A).$$

This suggests defining the magnitude of a compact metric space to be $|A| = \mu(A)$ whenever A possesses a weight measure μ. The following result shows that doing so agrees with the definition adopted in the previous section, whenever both definitions apply.

Proposition 5.3.6 ([27, Theorem 2.3]). *Suppose that A is a compact positive definite metric space with weight measure μ. Then $|A| = \mu(A)$.*

Proof. For any finite signed measure μ on A and $f \in \mathcal{H}$,

$$\left| \int f \, d\mu \right| \leq \|f\|_\infty \, \|\mu\|_{TV} \leq \|f\|_{\mathcal{H}} \, \|\mu\|_{TV}$$

by (5.4) (since $K(x, x) = 1$ here), where $\|\mu\|_{TV}$ denotes the total variation norm of μ. Therefore $f \mapsto \int f \, d\mu$ is a bounded linear functional on $\mathcal{H}$, represented by some $g \in \mathcal{H}$. So for each $a \in A$,

$$1 = \int e^{-d(a,b)} \, d\mu(b) = \left\langle e^{-d(\cdot,b)}, g \right\rangle_{\mathcal{H}} = g(a).$$

Then by the last statement of Theorem 5.3.5, if h is the potential function of A, then

$$|A| = \langle g, h \rangle_{\mathcal{H}} = \int h \, d\mu = \mu(A). \qquad \square$$

In fact it can be shown that $g = h$ in the proof above.

We therefore define the **magnitude** of a compact metric space A with a weight measure μ to be $|A| := \mu(A)$, with Proposition 5.3.6's assurance that when A is positive definite, this definition is consistent with the previous one.

A first nontrivial example is a compact interval $[a, b] \subseteq \mathbb{R}$. A straightforward computation (see [41, Theorem 2]) shows that

$$\mu_{[a,b]} = \frac{1}{2}(\delta_a + \lambda_{[a,b]} + \delta_b) \tag{5.6}$$

is a weight measure for $[a, b]$, where δ_x denotes the point mass at x and $\lambda_{[a,b]}$ denotes Lebesgue measure restricted to $[a, b]$. It follows that

$$\big|[a, b]\big| = 1 + \frac{b - a}{2}. \tag{5.7}$$

See [32] for a contention that (up to the $\frac{1}{2}$ scaling factor) this is the "correct" size of an interval. In any case, the appearance of the length $(b - a)$ gives the first compelling evidence that magnitude knows about genuinely "geometric" information for infinite spaces.

The following easy consequence of Fubini's theorem further extends the reach of Propositions 5.2.8 and 5.3.4.

Proposition 5.3.7. *If μ_A and μ_B are weight measures on compact metric spaces A and B, then $\mu_A \otimes \mu_B$ is a weight measure on $A \times_1 B$, and so $|A \times_1 B| = |A| \, |B|$.*

The chief drawback to the definition of magnitude in terms of weight measures is that many interesting spaces do not possess weight measures. For example, the results of

[3] imply that balls in ℓ_2^3 do not possess weight measures (rather, their weightings turn out to be higher-order distributions), and numerical computations in [40] suggest that squares and discs in ℓ_2^2 also do not possess weight measures.

On the other hand, the following result can be interpreted as saying that compact positive definite spaces "almost" possess weight measures.

Proposition 5.3.8 ([27, Theorems 2.3 and 2.4]). *If A is a compact positive definite metric space, then*

$$|A| = \sup\left\{ \frac{\mu(A)^2}{\int_A \int_A e^{-d(a,b)}\, d\mu(a)\, d\mu(b)} \;\middle|\; \mu \in M(A),\ \int_A \int_A e^{-d(a,b)}\, d\mu(a)\, d\mu(b) \neq 0 \right\},$$

where M(A) denotes the space of finite signed Borel measures on A. The supremum is attained if and only if A possesses a weight measure; in that case it is attained precisely by scalar multiples of weight measures.

One positive result about the existence of weight measures is the following.

Proposition 5.3.9 ([27, Lemma 2.8 and Corollary 2.10]). *Suppose (A, d) is a compact positive definite space, and that each finite $A' \subseteq A$ possesses a weighting with positive components. Then A possesses a positive weight measure.*

The hypothesis of Proposition 5.3.9 is satisfied, for example, by all compact subsets of $\mathbb{R}$ and by all compact ultrametric spaces (see Theorem 5.4.1 below and [19, Proposition 2.4.18]). Since Proposition 5.3.9 applies only to positive definite spaces, it does not extend the scope of magnitude beyond that of the previous section. Nevertheless, the existence of a positive weight measure makes it much easier to compute magnitude, and has other theoretical consequences which will come up in the next section.

The following generalization of Proposition 5.2.7 gives another large class of spaces which possess weight measures.

Lemma 5.3.10 ([41, Theorem 1]). *Let A be a compact homogeneous metric space. Then A possesses a weight measure, which is a scalar multiple of the unique isometry-invariant probability measure μ on A. Furthermore,*

$$|A| = \left(\int_A \int_A e^{-d(a,b)}\, d\mu(a)\, d\mu(b) \right)^{-1}.$$

Using Lemma 5.3.10, Willerton explicitly computed the magnitudes of round spheres with the geodesic metric: for n even, the magnitude of the n-sphere with radius R is

$$\frac{2}{1+e^{-\pi R}} \left[1 + \left(\frac{R}{1}\right)^2\right] \left[1 + \left(\frac{R}{3}\right)^2\right] \cdots \left[1 + \left(\frac{R}{n-1}\right)^2\right],$$

and there is a similar formula for odd n; see [41, Theorem 7].

Lemma 5.3.10 is particularly useful in analyzing the magnitude function of a homogeneous space A, since it implies that tA possesses a weight measure for every $t > 0$, which is moreover independent of t (up to normalization). In the particular case of a homogeneous Riemannian manifold, Willerton proved the following asymptotic results. (We note that most homogeneous manifolds are *not* of negative type, so that tA need not be positive definite; see [12].)

Theorem 5.3.11 ([41, Theorem 11]). *Suppose that (M, d) is an n-dimensional homogeneous Riemannian manifold equipped with its geodesic distance d. Then*

$$|tM| = \frac{1}{n!\omega_n}\left(\mathrm{vol}(M)t^n + \frac{n+1}{6}\,\mathrm{tsc}(M)t^{n-2} + O(t^{n-4})\right) \quad \text{as } t \to \infty,$$

where vol *denotes Riemannian volume,* tsc *denotes total scalar curvature, and ω_n is the volume of the n-dimensional unit ball in ℓ_2^n.*

In particular, if M is a homogeneous Riemannian surface, then

$$|tM| = \frac{\mathrm{area}(M)}{2\pi}t^2 + \chi(M) + O(t^{-2}) \quad \text{as } t \to \infty,$$

where $\chi(M)$ denotes the Euler characteristic of M.

Theorem 5.3.11 shows in particular that the magnitude function of a homogeneous Riemannian manifold determines both its volume and its total scalar curvature.

We note that most Riemannian manifolds are neither homogeneous nor positive definite, and it is so far not clear how to define their magnitude.

5.3.3 Maximum diversity

Proposition 5.3.8 suggests considering, for a compact metric space (A, d), the quantity

$$\begin{aligned} |A|_+ &:= \sup\left\{ \frac{\mu(A)^2}{\int_A \int_A e^{-d(a,b)}\, d\mu(a)\, d\mu(b)} \;\middle|\; \mu \in M_+(A), \mu \neq 0 \right\} \\ &= \sup_{\mu \in P(A)} \left(\int_A \int_A e^{-d(a,b)}\, d\mu(a)\, d\mu(b) \right)^{-1}, \end{aligned} \tag{5.8}$$

where $M_+(A)$ is the space of finite positive Borel measures on A, and $P(A)$ is the space of Borel probability measures on A. We refer to $|A|_+$ as the **maximum diversity** of A, for reasons that will be described shortly. Maximum diversity lacks the category-theoretic motivation of magnitude, but it turns out to have its own interesting interpretations, and to be both intimately related to magnitude and easier to analyze in certain respects.

Regarding interpretation, suppose that A is finite, the points of A represent species in some ecosystem, and that $e^{-d(a,b)} \in (0,1]$ represents the "similarity" of two species $a, b \in A$. If $\mu \in P(A)$ gives the relative abundances of species, then

$$\left(\int_A \int_A e^{-d(a,b)}\, d\mu(a)\, d\mu(b) \right)^{-1}$$

gives a way of quantifying the "diversity" of the ecosystem which is sensitive to both the abundances of the species and the similarities between them; see [21] for extensive discussion of a much larger family of diversities that this fits into. It is this interpretation that motivates the name "maximum diversity".

There are multiple connections between magnitude and maximum diversity. The most obvious is that, by Proposition 5.3.8, $|A|_+ \le |A|$ for any compact positive definite space A. Moreover, Proposition 5.3.8 implies that $|A|_+ = |A|$ if A is positive definite and possesses a positive weight measure; Proposition 5.3.9 and Lemma 5.3.10 indicate some families of such spaces. Finally, as we will see in Corollary 5.4.23 below, if $A \subseteq \ell_2^n$, then the inequality $|A|_+ \le |A|$ can be reversed, up to a (dimension-dependent) multiplicative constant. We will see applications of all these connections below.

A more subtle connection between maximum diversity and magnitude, which we will not discuss here, is proved in the main result of [17, 22].

We now move on to ways in which maximum diversity is better behaved than magnitude. One is that the supremum in (5.8) is always achieved, unlike the one in Proposition 5.3.8. This is a consequence of the compactness of $P(A)$ in the weak-* topology; see [27, Proposition 2.9] (this fact is used in the proof of Proposition 5.3.9 above). Another is the following improvement, for maximum diversity, of Proposition 5.3.1.

Proposition 5.3.12 ([27, Proposition 2.11]). *The maximum diversity $|A|_+$ is continuous as a function of A, on the class of compact metric spaces equipped with the Gromov–Hausdorff topology.*

Corollary 5.3.13 ([27, Corollary 2.12]). *The magnitude $|A|$ is continuous as a function of A, on the class of compact positive definite metric spaces which possess positive weight measures, equipped with the Gromov–Hausdorff topology.*

In particular, magnitude is continuous on the class of compact subsets of $\mathbb{R}$, and on the class of compact ultrametric spaces.

The next result shows how the asymptotic behavior of $|tA|_+$ is relatively easy to analyze. Recall that the **covering number** $N(A, \varepsilon)$ is the minimum number of ε-balls required to cover A, and that the **Minkowski dimension** of A may be defined as

$$\dim_{\mathrm{Mink}} A := \lim_{\varepsilon \to 0^+} \frac{N(A,\varepsilon)}{\log(1/\varepsilon)} \tag{5.9}$$

whenever this limit exists. The idea of the proof of Proposition 5.3.14 below is simply that when t is large and ε is small, the supremum over $P(A)$ defining $|tA|_+$ is approximately attained by a measure uniformly supported on the centers of a maximal family of disjoint ε-balls in A.

Proposition 5.3.14 ([28, Theorem 7.1]). *If A is a compact metric space, then*

$$\lim_{t\to\infty} \frac{\log |tA|_+}{\log t} = \dim_{\mathrm{Mink}} A. \tag{5.10}$$

Proposition 5.3.14 should be interpreted as saying that the limit on the left hand side of (5.10) exists if and only if $\dim_{\mathrm{Mink}} A$ exists. Moreover, if the limit is replaced with a lim inf or lim sup, the left hand side of (5.10) is equal to the so-called lower or upper Minkowski dimension of A, respectively, defined by modifying (5.9) in the same way.

Since $|A|_+ \leq |A|$ for any compact positive definite space, Proposition 5.3.14 gives a lower bound for the growth rate of the magnitude function for a compact space of negative type. Moreover, in Euclidean space ℓ_2^n, Proposition 3.12 and the rough equivalence of magnitude and maximum diversity mentioned above will be used to show that Minkowski dimension can be recovered from magnitude; see Theorem 5.4.24 below. (Proposition 7.5 of [28] proves the same fact for compact homogeneous metric spaces, using Lemma 5.3.10 above.)

5.4 Magnitude in normed spaces

In this section we will specialize magnitude to compact subsets of finite-dimensional vector spaces with translation-invariant metrics. It is in these settings that we find the strongest connections between magnitude and geometry. In section 5.4.1, we find a quite complete description of the magnitude of an arbitrary compact set $A \subseteq \mathbb{R}$; in particular, $|A|$ depends only on the Lebesgue measure of A and the sizes of the "gaps" in A (Corollary 5.4.3). In section 5.4.2, we show that in ℓ_1^n, magnitude can be used to recover ℓ_1 analogues of the classical intrinsic volumes of a convex body (Theorem 5.4.6). In section 5.4.3, we apply Fourier analysis to the study of magnitude, when $\mathbb{R}^n$ is equipped with a norm (or more generally, a p-norm) which makes it a positive definite metric space. In particular, we find that magnitude is continuous on convex bodies in such spaces (Theorem 5.4.15). Finally, in section 5.4.4, we specialize these tools to the most familiar normed space, the Euclidean space ℓ_2^n. In that setting the Fourier-analytic perspective of section 5.4.3 uncovers connections with partial differential equations and potential theory. Among other results, we will see that in Euclidean space, magnitude knows about volume (Theorem 5.4.14) and Minkowski dimension (Theorem 5.4.24), although there are frustratingly few compact sets in ℓ_2^n whose exact magnitudes are known (see Theorem 5.4.21).

Corollary 5.4.3 and the material of section 5.4.2 are new. Most of the results of section 5.4.3 were previously proved for Euclidean space, but are new in the generality discussed here.

5.4.1 Magnitude in $\mathbb{R}$

In the real line $\mathbb{R}$, magnitude can be analyzed in great detail thanks to the order structure underlying the metric structure. Namely, if $a < b < c$, then $Z(a, c) = Z(a, b)Z(b, c)$, where we recall that $Z(a, b) = e^{-d(a,b)}$. This simple fact lies behind the proof of the next result.

Theorem 5.4.1 ([23, Theorem 4] and [19, Proposition 2.4.13]). *Given real numbers $a_1 < a_2 < \cdots < a_N$, the weighting w of $A = \{a_1, \ldots, a_N\}$ is given by*

$$w_{a_i} = \frac{1}{2}\left(\tanh\frac{a_i - a_{i-1}}{2} + \tanh\frac{a_{i+1} - a_i}{2}\right)$$

for $2 \le i \le N - 1$, and

$$w_{a_1} = \frac{1}{2}\left(1 + \tanh\frac{a_2 - a_1}{2}\right), \qquad w_{a_N} = \frac{1}{2}\left(1 + \tanh\frac{a_N - a_{N-1}}{2}\right).$$

Consequently,

$$|A| = 1 + \sum_{i=2}^{N} \tanh\frac{a_i - a_{i-1}}{2}.$$

Theorem 5.4.1, together with Proposition 5.3.3, was used to give the first derivation of the magnitude of an interval; see [23, Theorem 7] and [19, Theorem 3.2.2].

As mentioned above, by Proposition 5.3.9, Theorem 5.4.1 implies that every compact subset of $\mathbb{R}$ possesses a weight measure. Furthermore, as noted in Corollary 5.3.13, this implies that magnitude on $\mathbb{R}$ is continuous with respect to the Gromov–Hausdorff topology.

The last part of the following corollary appears, with additional technical assumptions, as [41, Lemma 3].

Corollary 5.4.2. *Suppose that $A, B \subseteq \mathbb{R}$ are compact with $a = \sup A \le \inf B = b$. Then*

$$|A \cup B| = |A| + |B| - 1 + \tanh\frac{b - a}{2}.$$

Consequently, if $C \subseteq \mathbb{R}$ is compact and $[a, b] \subseteq C$, then

$$\left|C \setminus (a, b)\right| = |C| - \frac{b - a}{2} + \tanh\frac{b - a}{2}.$$

Proof. The first claim follows immediately from Theorem 5.4.1 in the case that A and B are finite, and then follows for general compact sets by continuity. The second equality follows by writing $C = A \cup [a, b] \cup B$, where $A = C \cap (-\infty, a]$ and $B = C \cap [b, \infty)$, then applying the first equality twice and (5.7). □

Corollary 5.4.2, together with continuity and the knowledge of the magnitude of a compact interval, can be used to compute the magnitude of any compact set $A \subseteq \mathbb{R}$, since A can be written as

$$A = [a, b] \setminus \bigcup_i (a_i, b_i), \tag{5.11}$$

where $\{(a_i, b_i)\}$ is a finite or countable collection of disjoint subintervals of $[a, b]$.

Corollary 5.4.3. *If $A \subseteq \mathbb{R}$ is compact, then*

$$|A| = 1 + \frac{\mathrm{vol}_1 A}{2} + \sum_i \tanh \frac{b_i - a_i}{2},$$

where a_i and b_i are as in (5.11).

Another proof of Corollary 5.4.3 can be given using [19, Proposition 3.2.3]. As an application of Corollary 5.4.3, we obtain the magnitude of the length ℓ ternary Cantor set C_ℓ (see [23, Theorem 10], [41, Theorem 4]):

$$|C_\ell| = 1 + \frac{1}{2} \sum_{i=1}^{\infty} \tanh \frac{\ell}{2 \cdot 3^i}.$$

5.4.2 Magnitude in the ℓ_1-norm

The magnitude of subsets of $\mathbb{R}^n$ is generally most tractable when we equip $\mathbb{R}^n$ with the ℓ_1-norm. Although that may not be the norm of primary geometric interest, it provides a testing ground for questions that are more difficult to settle in Euclidean space.

We have already seen that ℓ_1^n, like ℓ_2^n, is of negative type (Theorem 5.2.12). The key difference is Proposition 5.3.4, the multiplicativity of magnitude with respect to the ℓ_1 product. Since we already know the magnitude of intervals, this immediately allows us to calculate the magnitude of boxes in ℓ_1^n. Unions of boxes can then be used to approximate more complex subsets, as we shall see.

Explicitly, a box $\prod_{i=1}^n [a_i, a_i + L_i]$ in ℓ_1^n has magnitude $\prod_{i=1}^n (1 + L_i/2)$. It follows that $|tA| \to 1$ as $t \to 0^+$ for boxes A. But then monotonicity of magnitude (Proposition 5.3.2) implies a more general result:

Proposition 5.4.4. *If $A \subseteq \ell_1^n$ is compact, then* $\lim_{t\to 0^+} |tA| = 1$.

(In ℓ_2^n, this is much harder to prove; see Theorem 5.4.18.) Proposition 5.4.4 and Theorem 5.4.17 together imply that the magnitude function $t \mapsto |tA|$ is continuous on $[0, \infty)$.

Our formula for the magnitude of a box in ℓ_1^n can be rewritten in terms of the intrinsic volumes $V_0, V_1, \dots$ (defined in, for instance, Chapter 7 of [11] or Chapter 4 of [34]). Recall that $V_i(A)$ is the canonical i-dimensional measure of a convex set $A \subseteq \mathbb{R}^n$, and that the intrinsic volumes are characterized by Steiner's polynomial formula

$$\mathrm{vol}(A + r\mathcal{B}^n) = \sum_{i=0}^{n} \omega_{n-i} V_i(A) r^{n-i}$$

(Proposition 9.2.2 of [11] or Equation 4.1 of [34]), where $\mathcal{B}^n$ is the unit Euclidean n-ball and $\omega_j = \mathrm{vol}(\mathcal{B}^j)$. For boxes $A \subseteq \ell_1^n$, the formula above can be rewritten as

$$|A| = \sum_{i=0}^{n} \frac{V_i(A)}{2^i}, \tag{5.12}$$

either by direct calculation or by noting that $\big|[0, L]\big| = 1 + V_1([0, L])/2$ and using the multiplicative property of the intrinsic volumes (Theorem 9.7.1 of [11]). Hence the magnitude function of a box A is a polynomial

$$|tA| = \sum_{i=0}^{n} \frac{V_i(A)}{2^i} t^i$$

whose coefficients are (up to known factors) the intrinsic volumes of A, and whose degree is its dimension. In particular, the magnitude function of a box determines all of its intrinsic volumes and its dimension.

In fact, such a result is true for a much larger class of subsets of ℓ_1^n than just boxes. To show this, we must adapt the classical notion of intrinsic volume to ℓ_1^n, following [18].

First recall that a metric space A is **geodesic** if for any $a, b \in A$ there exists a distance-preserving map $\gamma\colon [0, d(a, b)] \to A$ such that $\gamma(0) = a$ and $\gamma(d(a, b)) = b$. The geodesic subsets of ℓ_2^n are the convex sets. The geodesic subsets of ℓ_1^n, called the **ℓ_1-convex sets** [18], include the convex sets and much else besides (such as L shapes). In this setting, there is a Steiner-type theorem in which balls are replaced by cubes (Theorem 6.2 of [18]): for any ℓ_1-convex compact set $A \subseteq \ell_1^n$, writing $\mathcal{C}^n = [-1/2, 1/2]^n$,

$$\mathrm{vol}(A + r\mathcal{C}^n) = \sum_{i=0}^{n} V_i'(A) r^{n-i} \tag{5.13}$$

where $V_0'(A), \dots, V_n'(A)$ depend only on A.

The functions $V_0', V_1', \dots$ on the class of ℓ_1-convex compact sets are called the **ℓ_1-intrinsic volumes** [18]. They are valuations (that is, finitely additive), continuous with respect to the Hausdorff metric, and invariant under isometries of ℓ_1^n. There is a well-developed integral geometry of ℓ_1-convex sets [18], closely parallel to the classical integral geometry of convex sets; for instance, there is a Hadwiger-type theorem for ℓ_1-intrinsic volumes.

Although the intrinsic and ℓ_1-intrinsic volumes are not in general equal, they coincide for boxes A, giving

$$|A| = \sum_{i=0}^{n} \frac{V_i'(A)}{2^i}, \qquad |tA| = \sum_{i=0}^{n} \frac{V_i'(A)}{2^i} t^i \tag{5.14}$$

(the latter because V_i' is homogeneous of degree i). It is this relationship, not (5.12), that generalizes from boxes to a much larger class of sets.

Conjecture 5.4.5 ([19, Conjecture 3.4.10]). *For all compact ℓ_1-convex sets $A \subseteq \ell_1^n$,*

$$|A| = \sum_{i=0}^{n} \frac{V_i'(A)}{2^i}.$$

We will prove the following parts of this conjecture:

Theorem 5.4.6.
1. $|A| \leq \sum_{i=0}^{n} 2^{-i} V_i'(A)$ *for all compact ℓ_1-convex sets $A \subseteq \ell_1^n$.*
2. $|A| = \sum_{i=0}^{n} 2^{-i} V_i'(A)$ *for all convex bodies $A \subseteq \ell_1^n$.*
3. $|A| = \sum_{i=0}^{2} 2^{-i} V_i'(A)$ *for all compact convex sets $A \subseteq \ell_1^2$.*

(A **convex body** is a compact convex set with nonempty interior.)

For the proof, we will use some special classes of box. A **pixel** in $\mathbb{R}^n$ is a unit cube $\prod_{i=1}^{n} [a_i, a_i + 1]$ with integer coordinates a_i. More generally, a **subpixel** is a box $\prod_{i=1}^{n} [a_i, b_i]$ with $a_i \in \mathbb{Z}$ and $b_i \in \{a_i, a_i + 1\}$. Note that the intersection of two subpixels is either a subpixel or empty.

Equation (5.6) and Proposition 5.3.7 imply that for any box $B = \prod_i [a_i, b_i]$ in ℓ_1^n, the product measure $\mu_B = \prod_{i=1}^{n} \mu_{[a_i, b_i]}$ is a weight measure on B.

Lemma 5.4.7. *There is a unique function*

$$\begin{array}{ccc} \{\textit{finite unions of subpixels in } \mathbb{R}^n\} & \to & \{\textit{signed Borel measures on } \mathbb{R}^n\} \\ A & \mapsto & \mu_A \end{array}$$

extending the definition above for subpixels and satisfying $\operatorname{supp} \mu_A \subseteq A$, $\mu_\emptyset = 0$, *and* $\mu_{A \cup B} = \mu_A + \mu_B - \mu_{A \cap B}$ *whenever A and B are finite unions of subpixels.*

Proof. By the extension theorem of Groemer (Theorem 6.2.1 of [34]), it suffices to show that for any subpixels $B_1, \ldots, B_m$ such that $B_1 \cup \cdots \cup B_m$ is a subpixel,

$$\mu_{B_1 \cup \cdots \cup B_m} = \sum_{k \geq 0} (-1)^k \sum_{1 \leq j_0 < \cdots < j_k \leq m} \mu_{B_{j_0} \cap \cdots \cap B_{j_k}}.$$

But $B_1 \cup \cdots \cup B_m$ is only a subpixel if some B_j contains all the others, and in that case the sum telescopes and the proof is trivial. □

A subset A of ℓ_1^n is **1-pixelated** if it is a finite union of pixels; then λA is said to be **λ-pixelated**. A set is **pixelated** if it is λ-pixelated for some $\lambda > 0$.

Proposition 5.4.8. *Let $A \subseteq \ell_1^n$ be an ℓ_1-convex pixelated set. Then μ_A as given in Lemma 5.4.7 is a weight measure on A.*

Proof. We may harmlessly assume that A is 1-pixelated. The result holds when either $n = 0$ or A is a single pixel. So, we may assume inductively that $n \geq 1$, that A contains at least two pixels, and that the result holds for ℓ_1-convex 1-pixelated sets of smaller dimension or fewer pixels than A.

Fix $a \in A$. We may assume without loss of generality that at least two of the pixels in A differ in their last coordinates, that $\sup_{b\in A} b_n = 1$, and that a belongs to some pixel of A whose center has negative last coordinate. Write A_- for the union of the pixels in A whose centers have negative last coordinates, and similarly A_+. Thus, $a \in A_-$ and the center of every pixel in A_+ has last coordinate $1/2$. Both A_- and A_+ are ℓ_1-convex 1-pixelated sets (by Lemma 3.3 of [18]), and $A_- \cap A_+$ is a finite union of subpixels (though need not be pixelated).

We have to show that

$$\int_{\mathbb{R}^n} Z(a, b)\, d\mu_A(b) = 1.$$

Since $\mu_A = \mu_{A_+} + \mu_{A_-} - \mu_{A_+ \cap A_-}$ and μ_{A_-} is a weight measure on A_- (by inductive hypothesis), an equivalent statement is that

$$\int_{\mathbb{R}^n} Z(a, b)\, d\mu_{A_+}(b) = \int_{\mathbb{R}^n} Z(a, b)\, d\mu_{A_- \cap A_+}(b). \tag{5.15}$$

Write $\pi\colon \mathbb{R}^n \to \mathbb{R}^{n-1}$ for orthogonal projection onto the first $(n-1)$ coordinates, and write $a' = (\pi(a), 0) = (a_1, \dots, a_{n-1}, 0)$. Then $Z(a, b) = Z(a, a')Z(a', b)$ for $b \in A_+$, so (5.15) is equivalent to

$$\int_{\mathbb{R}^n} Z(a', b)\, d\mu_{A_+}(b) = \int_{\mathbb{R}^n} Z(a', b)\, d\mu_{A_- \cap A_+}(b).$$

We analyze each side in turn. First, $A_+ = (\pi A_+) \times [0, 1]$, so it follows from Proposition 5.3.7 that $\mu_{A_+} = \mu_{\pi A_+} \otimes \mu_{[0,1]}$. Using this and the fact that $\mu_{[0,1]}$ is a weight measure on $[0, 1]$, we find that the left-hand side is equal to

$$\int_{\mathbb{R}^{n-1}} Z(\pi(a), c)\, d\mu_{\pi A_+}(c). \tag{5.16}$$

Next, $\mu_{A_- \cap A_+}$ is supported on $\mathbb{R}^{n-1} \times \{0\}$, and $\pi(A_- \cap A_+) = \pi A_- \cap \pi A_+$ (by Corollary 2.5 of [18]), which together imply that the right-hand side is equal to

$$\int_{\mathbb{R}^{n-1}} Z(\pi(a), c)\, d\mu_{\pi A_- \cap \pi A_+}(c). \tag{5.17}$$

Hence it suffices to show that the integrals (5.16) and (5.17) are equal. Since $\mu_{\pi A} = \mu_{\pi A_-} + \mu_{\pi A_+} - \mu_{\pi A_- \cap \pi A_+}$, an equivalent statement is that

$$\int\limits_{\mathbb{R}^{n-1}} Z(\pi(a), c)\, d\mu_{\pi A}(c) = \int\limits_{\mathbb{R}^{n-1}} Z(\pi(a), c)\, d\mu_{\pi A_-}(c). \tag{5.18}$$

But πA and πA_- are 1-pixelated sets of dimension $n-1$, and are ℓ_1-convex (by Corollary 1.12 of [18]), so our inductive hypothesis implies that $\mu_{\pi A}$ and $\mu_{\pi A_-}$ are weight measures on them. Since $\pi(a) \in \pi A_- \subseteq \pi A$, both sides of (5.18) are equal to 1, completing the proof. □

Our proof of Theorem 5.4.6 rests on the following result:

Proposition 5.4.9. *$|A| = \sum_{i=0}^{n} 2^{-i} V_i'(A)$ for all pixelated ℓ_1-convex sets $A \subseteq \ell_1^n$.*

Proof. Assume that A is 1-pixelated, and write A as a union $\bigcup_{j=1}^{m} B_j$ of pixels. Also write $W = \sum_{i=0}^{n} 2^{-i} V_i'$; then $|B| = W(B)$ whenever B is a box or the empty set. Propositions 5.3.6 and 5.4.8 together with the valuation property of W give

$$\begin{aligned} |A| = \mu_A(\mathbb{R}^n) &= \sum_{k\geq 0} (-1)^k \sum_{1\leq j_0<\cdots<j_k\leq m} \mu_{B_{j_0}\cap\cdots\cap B_{j_k}}(\mathbb{R}^n) \\ &= \sum_{k\geq 0} (-1)^k \sum_{1\leq j_0<\cdots<j_k\leq m} |B_{j_0}\cap\cdots\cap B_{j_k}| \\ &= \sum_{k\geq 0} (-1)^k \sum_{1\leq j_0<\cdots<j_k\leq m} W(B_{j_0}\cap\cdots\cap B_{j_k}) = W(A), \end{aligned}$$

as required. □

Proof of Theorem 5.4.6. For part (1), let $A \subseteq \ell_1^n$ be a compact ℓ_1-convex set. For each $\lambda > 0$, let A_λ be the smallest λ-pixelated set containing A. Then A_λ is ℓ_1-convex (by Proposition 3.1 of [18]), and $A_\lambda \to A$ in the Hausdorff metric as $\lambda \to 0$. The result now follows from Proposition 5.4.9, continuity of the ℓ_1-intrinsic volumes, and the monotonicity of magnitude (Proposition 5.3.2).

For (2), let $A \subseteq \ell_1^n$ be a compact convex set with 0 in its interior. Given $\varepsilon > 0$, we can choose $\alpha < 1$ such that $d_H(\alpha A, A) < \varepsilon$. But by convexity, αA is a subset of the interior of A, so we can choose $\lambda > 0$ such that $\alpha A_\lambda \subseteq A$. Thus, we have a pixelated ℓ_1-convex subset $B = \alpha A_\lambda$ of A satisfying $d_H(B, A) < \varepsilon$. Arguing as in part (1) but approximating from the inside rather than the outside, we obtain the opposite inequality $|A| \geq \sum \frac{V_i'(A)}{2^i}$. (Alternatively, use Theorem 5.4.15 below.)

For (3), the only nontrivial case remaining is that of a line segment, which is straightforward. □

5.4.3 The Fourier-analytic perspective

In the real line, the study of magnitude is facilitated by the order structure of $\mathbb{R}$; in ℓ_1^n we can exploit the algebraic structure of ℓ_1 products. In general normed spaces the most obvious special feature is translation-invariance. It will therefore come as no surprise that Fourier analysis is our key tool in that setting. This approach was developed in [28] for ℓ_2^n, but with some additional effort we can work not only with more general norms but with the broader class of p-(quasi)norms for $0 < p \le 1$.

Let $0 < p \le 1$. A **p-norm** on a real vector space V is a function $\|\cdot\| : V \to \mathbb{R}$ such that

- $\|v\| \ge 0$ for every $v \in V$, with equality only if $v = 0$;
- $\|tv\| = |t| \, \|v\|$ for every $t \in \mathbb{R}$ and $v \in V$;
- $\|v + w\|^p \le \|v\|^p + \|w\|^p$ for every $v, w \in V$.

Thus a 1-normed space is simply a normed space. A principal example of a p-normed space for $p < 1$ is $L_p[0, 1]$ with $\|f\| = \left(\int_0^1 |f(x)|^p \, dx\right)^{1/p}$.

If $(V, \|\cdot\|)$ is a p-normed space, then $d_p(v, w) = \|v - w\|^p$ is a metric on V. Conversely, if d is any translation-invariant, symmetric, positively homogeneous metric on a real vector space V, then $\|v\| = d(v, 0)$ defines a p-norm on V, where $p \in (0, 1]$ is the degree of homogeneity of d.

The following classical result, which goes back to Lévy [24] (see also [13, Theorem 6.6]), identifies which finite-dimensional p-normed spaces are positive definite metric spaces (and hence, by homogeneity, of negative type).

Theorem 5.4.10. *Let $0 < p \le 1$, let $\|\cdot\|$ be a p-norm on $\mathbb{R}^n$, and equip $\mathbb{R}^n$ with the metric $d_p(x, y) = \|x - y\|^p$. Then $(\mathbb{R}^n, d_p)$ is a positive definite metric space if and only if there is a linear map $T : \mathbb{R}^n \to L_p[0, 1]$ such that $\|Tx\|_p = \|x\|$ for every $x \in \mathbb{R}^n$.*

Theorem 5.4.10 implies in particular that $L_p[0, 1]$ and ℓ_p^n are positive definite with the metric d_p for $0 < p \le 1$. We recall from Theorem 5.2.12 that $L_q[0, 1]$ and ℓ_q^n are also positive definite, with the usual metric, for $1 \le q \le 2$.

To simplify the statements of results:

> For the rest of this section, $\|\cdot\|$ will always denote a p-norm on $\mathbb{R}^n$ such that $(\mathbb{R}^n, d_p)$ is a positive definite metric space.

We will make use of the function $F_p : \mathbb{R}^n \to \mathbb{R}$ defined by $F_p(x) = e^{-\|x\|^p}$, and denote by $\mathcal{B} = \{x \in \mathbb{R}^n \mid \|x\| = 1\}$ the unit ball of $\|\cdot\|$. For $f \in L_1(\mathbb{R}^n)$, we adopt the convention that the Fourier transform of f is given by $\widehat{f}(x) = \int_{\mathbb{R}^n} f(y) e^{-2\pi i \langle x, y \rangle} \, dy$.

A key observation is that F_p is the Fourier transform of a p-stable probability distribution. Proposition 5.4.11 collects some crucial facts which follow from results from the literature on stable random processes.

Proposition 5.4.11.

1. *There is a constant $c > 0$ (depending on the p-norm $\|\cdot\|$) such that $\widehat{F_p}(x) \geq c(1 + \|x\|_2)^{-(1+p)n}$ for every $x \in \mathbb{R}^n$.*
2. *For each $x \in \mathbb{R}^n$, $\widehat{F_p}(tx)$ is nonincreasing as a function of $t \geq 0$. In particular, $\left\|\widehat{F_p}\right\|_\infty = \widehat{F_p}(0) = \Gamma\left(\frac{n}{p} + 1\right) \operatorname{vol} \mathcal{B}$.*

Proof. It follows from Theorem 5.4.10 and Bochner's theorem that $\widehat{F_p}$ is the density of a p-stable distribution μ on $\mathbb{R}^n$.

By a theorem of Lévy (see [13, Lemma 6.4]), there is a symmetric measure σ on S^{n-1} such that

$$\|x\|^p = \int_{S^{n-1}} |\langle x, \theta\rangle|^p \, d\sigma(\theta);$$

since $\|x\| \neq 0$ for $x \neq 0$, the support of σ is not contained in any proper subspace of $\mathbb{R}^n$. Then σ is a positive scalar multiple of the spherical part of the Lévy measure of μ (cf. [31, Section 14]). Since σ is symmetric and not supported in a proper subspace of $\mathbb{R}^n$, the linear span of its support is all of $\mathbb{R}^n$, and [38, Theorem 1.1(iii)] then implies the first claim.

Corollary 4.2 of [9] implies that every symmetric stable distribution on $\mathbb{R}^n$ is *unimodal* in the sense defined in [9] and hence *n-unimodal* in the sense defined in [29] (see discussion on p. 80 and p. 84 of [9]). The second claim then follows from [29, Theorem 6]. □

As in section 5.3.1, for a finite set $B \subseteq \mathbb{R}^n$ and $w \in \mathbb{R}^B$, we write $f_w(x) = \sum_{b\in B} w_b F_p(x-b)$. Recall that the RKHS $\mathcal{H}$ of the metric space $(\mathbb{R}^n, d_p)$ is the completion of the span of such functions f_w with respect to the norm given by

$$\|f_w\|_{\mathcal{H}}^2 = \sum_{a,b\in B} w_a w_b F_p(a-b) = \int_{\mathbb{R}^n} \widehat{F_p}(x) \left| \sum_{b\in B} w_b e^{2\pi i\langle x,b\rangle} \right|^2 dx = \int_{\mathbb{R}^n} \frac{1}{\widehat{F_p}(x)} \left|\widehat{f_w}(x)\right|^2 dx.$$

Observe that the Fourier inversion theorem may be used here since $\widehat{F_p}$ is the density of a random variable, hence integrable.

From here, standard arguments imply the following.

Proposition 5.4.12. *The RKHS of $(\mathbb{R}^n, d_p)$ is*

$$\mathcal{H} = \left\{ f \in L_2(\mathbb{R}^n) \,\middle|\, \int_{\mathbb{R}^n} \frac{1}{\widehat{F_p}(x)} \left|\widehat{f}(x)\right|^2 dx < \infty \right\},$$

with norm given by

$$\|f\|_{\mathcal{H}}^2 = \int_{\mathbb{R}^n} \frac{1}{\widehat{F_p}(x)} \left|\widehat{f}(x)\right|^2 dx.$$

The Schwartz space $\mathcal{S}(\mathbb{R}^n)$ is contained in $\mathcal{H}$.

The dual space of $\mathcal{H}$ is naturally identified with the space of tempered distributions

$$\left\{ \phi \in \mathcal{S}'(\mathbb{R}^n) \;\middle|\; \widehat{\phi} \in L_2\big(\widehat{F_p}(x)\, dx\big) \right\}.$$

Thus weightings for compact subsets of $(\mathbb{R}^n, d_p)$ can be identified as tempered distributions satisfying a weak smoothness condition, although we will not make use of this fact here. Note that, since $\widehat{F_p}$ is integrable, this space of distributions includes all finite signed measures on $\mathbb{R}^n$, so that weight measures fit gracefully into this perspective.

This concrete identification of the RKHS of $(\mathbb{R}^n, d_p)$, together with Proposition 5.4.11, make it possible to use Fourier analysis to prove a number of nice properties of magnitude in these spaces, including the following fundamental fact.

Proposition 5.4.13. *Let $A \subseteq (\mathbb{R}^n, d_p)$ be compact. Then*

$$\frac{\operatorname{vol} A}{\Gamma\big(\frac{n}{p}+1\big)\operatorname{vol}\mathcal{B}} \le |A| < \infty.$$

Proof. By Proposition 5.4.12, $\mathcal{H}$ contains functions which are uniformly equal to 1 on A, so the finiteness follows from Theorem 5.3.5.

For the lower bound, let h be the potential function of A. By Theorem 5.3.5, Proposition 5.4.12, Proposition 5.4.11(2), and Plancherel's theorem,

$$|A| = \|h\|_{\mathcal{H}}^2 \ge \frac{\|\widehat{h}\|_2^2}{\Gamma(\frac{n}{p}+1)\operatorname{vol}\mathcal{B}} = \frac{\|h\|_2^2}{\Gamma(\frac{n}{p}+1)\operatorname{vol}\mathcal{B}} \ge \frac{\operatorname{vol} A}{\Gamma(\frac{n}{p}+1)\operatorname{vol}\mathcal{B}}. \qquad \square$$

The finiteness statement in Proposition 5.4.13 was proved in Theorem 3.4.8 and Proposition 3.5.3 of [19] for ℓ_1^n and ℓ_2^n, and in somewhat greater generality in [27, Theorem 4.3]. The lower bound was proved in [19, Theorem 3.5.6] for $p = 1$ and [27, Theorem 4.5] for the general case.[5.1] The proof here follows the approach used in [28] for ℓ_2^n (see Proposition 5.6 and the remarks following Corollary 5.3 there).

We now consider the behavior of magnitude functions in $(\mathbb{R}^n, d_p)$. We must be careful about a subtle notational issue when $p < 1$. Recall that for a metric space (A, d) and $t > 0$, we denote by tA the metric space (A, td), which in the present context is different from the usual interpretation of tA. Therefore we will introduce the notation $t \cdot A = \{ta \mid a \in A\}$ for $A \subseteq \mathbb{R}^n$. Not that when $A \subseteq \mathbb{R}^n$ is equipped with the metric $d_p(x, y) = \|x - y\|^p$ associated to a p-norm, the metric space tA is isometric to the set $t^{1/p} \cdot A \subseteq \mathbb{R}^n$ equipped with d_p.

The next result shows that magnitude knows about volume in all finite-dimensional positive definite p-normed spaces. This generalizes [3, Theorem 1] for Euclidean space ℓ_2^n.

5.1 Theorem 4.5 in the published version of [27] is misstated in the case $p < 1$; see the current arXiv version for a correct statement.

Theorem 5.4.14. *If $A \subseteq (\mathbb{R}^n, d_p)$ is compact, then*

$$\lim_{t\to\infty} \frac{|tA|}{t^{n/p}} = \lim_{t\to\infty} \frac{|t\cdot A|}{t^n} = \frac{\operatorname{vol} A}{\Gamma\left(\frac{n}{p}+1\right)\operatorname{vol}\mathcal{B}}.$$

Proof. Proposition 5.4.13 implies that

$$|t\cdot A| \geq \frac{\operatorname{vol}(t\cdot A)}{\Gamma\left(\frac{n}{p}+1\right)\operatorname{vol}\mathcal{B}} = \frac{t^n \operatorname{vol} A}{\Gamma\left(\frac{n}{p}+1\right)\operatorname{vol}\mathcal{B}}$$

for every $t > 0$. Now suppose that $h \in \mathcal{H}$ satisfies $h \equiv 1$ on A, and let $h_t(x) = h(x/t)$. Then by Theorem 5.3.5 and Proposition 5.4.12,

$$\frac{|t\cdot A|}{t^n} \leq \int_{\mathbb{R}^n} \frac{1}{\widehat{F_p}(x)} \left|\widehat{h_t}(x)\right|^2 dx = \int_{\mathbb{R}^n} \frac{1}{\widehat{F_p}(x/t)} \left|\widehat{h}(x)\right|^2 dx. \tag{5.19}$$

Proposition 5.4.11(2), the monotone convergence theorem, and Plancherel's theorem imply that

$$\lim_{t\to\infty} \int_{\mathbb{R}^n} \frac{1}{\widehat{F_p}(x/t)} \left|\widehat{h}(x)\right|^2 dx = \frac{\|h\|_2^2}{\Gamma\left(\frac{n}{p}+1\right)\operatorname{vol}\mathcal{B}}.$$

By Theorem 5.4.12, there exist functions $h \in \mathcal{H}$ with $h \equiv 1$ on A such that $\|h\|_2^2$ is arbitrarily close to $\operatorname{vol} A$ (cf. the proof of [3, Theorem 1]), which completes the proof. □

The next theorem is the major known continuity result (as opposed to mere *semi*continuity) for magnitude.

Theorem 5.4.15. *Denote by $\mathcal{K}_n$ the class of nonempty compact subsets of $\mathbb{R}^n$, equipped with the Hausdorff metric d_H induced by d_p, and suppose that $A \in \mathcal{K}_n$ is star-shaped with respect to some point in its interior. Then magnitude, as a function $\mathcal{K}_n \to \mathbb{R}$, is continuous at A.*

Proof. By Proposition 5.3.1, we only need to show that magnitude is upper semicontinuous at A. Letting h be the potential function of A, (5.19) and Proposition 5.4.12 imply that $|t\cdot A| \leq t^n |A|$ for $t \geq 1$. By translation-invariance, we may assume that A is star-shaped about 0 and $r^{1/p}\cdot\mathcal{B} \subseteq A$ for some $r > 0$. Now if $B \in \mathcal{K}_n$ and $d_H(A,B) < \varepsilon$, then

$$B \subseteq A + \varepsilon^{1/p}\cdot\mathcal{B} \subseteq \left(1 + \left(\frac{\varepsilon}{r}\right)^{1/p}\right)\cdot A,$$

and so $|B| \leq \left(1 + \left(\frac{\varepsilon}{r}\right)^{1/p}\right)^n |A|$. Thus magnitude is upper semicontinuous at A. □

The family of sets A in Theorem 5.4.15 is slightly larger than what are sometimes called "star bodies", and of course includes all convex bodies. It is unknown, however, whether magnitude is continuous when restricted to compact convex sets which are not required to have nonempty interior.

The final result in this section shows that, in positive definite p-normed spaces, magnitude can be computed from potential functions simply by integrating, as opposed to computing the (more complicated) $\mathcal{H}$-norm.

Theorem 5.4.16. *Let $A \subseteq (\mathbb{R}^n, d_p)$ be compact, and suppose that the potential function $h \in \mathcal{H}$ of A is integrable. Then*

$$|A| = \frac{1}{\Gamma\left(\frac{n}{p}+1\right)\operatorname{vol}\mathcal{B}} \int_{\mathbb{R}^n} h(x)\,dx.$$

Proof. Fix an even function $f \in \mathcal{S}(\mathbb{R}^n)$ with $f \equiv 1$ on some open neighborhood of the origin. Set $f_k(x) = f(x/k)$ and $\phi_k = \widehat{f_k}/\widehat{F_p}$ for $k \in \mathbb{N}$. Then $\phi_k \in L_1(\mathbb{R}^n)$ and

$$\left\|\widehat{\phi_k}\right\|_\infty \le \|\phi_k\|_1 = \int_{\mathbb{R}^n} \frac{1}{\widehat{F_p}(x/k)}\left|\widehat{f}(x)\right|\,dx \le \int_{\mathbb{R}^n} \frac{1}{\widehat{F_p}(x)}\left|\widehat{f}(x)\right|\,dx < \infty$$

by Proposition 5.4.11(2). Furthermore, for every $x \in \mathbb{R}^n$,

$$\widehat{\phi_k}(x) = \int_{\mathbb{R}^n} e^{-2\pi i\langle x, y/k\rangle}\frac{\widehat{f}(y)}{\widehat{F_p}(y/k)}\,dy \xrightarrow{k\to\infty} \int_{\mathbb{R}^n} \frac{\widehat{f}(y)}{\widehat{F_p}(0)}\,dy = \frac{f(0)}{\Gamma\left(\frac{n}{p}+1\right)\operatorname{vol}\mathcal{B}} = \frac{1}{\Gamma\left(\frac{n}{p}+1\right)\operatorname{vol}\mathcal{B}}$$

by the dominated convergence theorem.

By the last part of Theorem 5.3.5, for sufficiently large k,

$$|A| = \langle h, f_k\rangle_{\mathcal{H}} = \int_{\mathbb{R}^n} \widehat{h}(x)\phi_k(x)\,dx = \int_{\mathbb{R}^n} h(x)\widehat{\phi_k}(x)\,dx$$

by Parseval's identity, and the claim now follows by the dominated convergence theorem. □

5.4.4 Magnitude in Euclidean space

Finally, we specialize the tools of section 5.4.3 to the setting of Euclidean space ℓ_2^n, where they become even more powerful, allowing one to prove much more refined results about continuity, asymptotics, and exact values of magnitude than in more general normed spaces.

We will write simply $F(x) = e^{-\|x\|_2}$, and let $\mathcal{B}_R^n = \left\{x \in \mathbb{R}^n \;\middle|\; \|x\|_2 \le R\right\}$. In this setting we have the explicit formula

$$\widehat{F}(x) = \frac{n!\,\omega_n}{(1+4\pi^2\|x\|_2^2)^{(n+1)/2}}, \tag{5.20}$$

where $\omega_n = \operatorname{vol}_n(\mathcal{B}_1^n)$ (see [37, Theorem 1.14]). This implies that the RKHS $\mathcal{H}$ for ℓ_2^n is the classical Sobolev space

$$H^{(n+1)/2}(\mathbb{R}^n) = \left\{ f \in L_2(\mathbb{R}^n) \;\middle|\; \int_{\mathbb{R}^n} (1+4\pi^2\|x\|_2^2)^{(n+1)/2}\left|\widehat{f}(x)\right|^2 dx < \infty \right\},$$

and that $\|f\|_{\mathcal{H}}^2 = \frac{1}{n!\omega_n}\|f\|_{H^{(n+1)/2}}^2$.

A first application of this observation is the following, proved for ℓ_2^n in [28, Corollary 5.5].

Theorem 5.4.17. *If A is a compact subset of ℓ_1^n or ℓ_2^n, then the magnitude function $t \mapsto |tA|$ is continuous on $(0, \infty)$.*

Sketch of proof. For ℓ_2^n, using (5.20) one can show that $|tA| \geq \frac{1}{t}|A|$ for $t \geq 1$, along the lines of (5.19). For ℓ_1^n, if we let $G(x) = e^{-\|x\|_1} = \prod_{i=1}^n e^{-|x_i|}$, then the $n = 1$ case of (5.20) implies that

$$\widehat{G}(x) = \prod_{i=1}^{n} \frac{2}{1 + 4\pi^2 x_i^2},$$

and a similar argument yields that $|tA| \geq t^{-n}|A|$ for $t \geq 1$.

In either case, (5.19) shows that $|tA| \leq t^n |A|$. Together, these estimates imply that the magnitude function of A is continuous on $(0, \infty)$; see Theorem 5.4 and Corollary 5.5 of [28]. □

The most significant consequence of (5.20) is that when n is odd, $1/\widehat{F}$ is the symbol of a differential operator on $\mathbb{R}^n$. In particular, when $f : \mathbb{R}^n \to \mathbb{R}$ is smooth,

$$\|f\|_{H^{(n+1)/2}}^2 = \int_{\mathbb{R}^n} f(x)\left[(I - \Delta)^{(n+1)/2} f\right](x)\, dx, \tag{5.21}$$

where I is the identity operator and Δ is the Laplacian on $\mathbb{R}^n$. This opens the door to using differential equations techniques to study magnitude. A first application is the proof of the following result.

Theorem 5.4.18 ([3, Theorem 1]). *If $A \subseteq \ell_2^n$ is compact, then* $\lim_{t\to 0^+} |tA| = 1$.

Sketch of proof. By Proposition 5.3.2, it suffices to show that $\limsup_{R\to 0^+} \left|\mathcal{B}_R^n\right| \leq 1$; it further suffices, by embedding ℓ_2^n in ℓ_2^{n+1} if necessary, to assume that n is odd. For $0 < R < 1$ we can choose smooth functions f_R such that

$$f_R(x) = \begin{cases} 1 & \text{if } \|x\|_2 \leq R, \\ e^R e^{-\|x\|_2} & \text{if } \|x\|_2 \geq \sqrt{R} \end{cases}$$

and the derivatives of f_R are sufficiently small for $R \leq \|x\|_2 \leq \sqrt{R}$ that, using (5.21),

$$\|f_R\|_{H^{(n+1)/2}}^2 = \|f_0\|_{H^{(n+1)/2}}^2 + o(1) = n!\omega_n + o(1)$$

when $R \to 0$; see the proof of [3, Theorem 1]. By Theorem 5.3.5, this completes the proof. □

Together with Theorem 5.4.17, this shows that the magnitude function of a compact $A \subseteq \ell_2^n$ is continuous on $[0, \infty)$. Recall that this result is false for general metric spaces

A of negative type [19, Example 2.2.8], but it does also hold for $A \subseteq \ell_1^n$ (Proposition 5.4.4). A monotone convergence argument would prove the same result in a p-normed space if $\sup_x \widehat{F_p}(x)/\widehat{F_p}(2x) < \infty$.

More significantly, we obtain the following conditions on the potential function of a compact set $A \subseteq \ell_2^n$, which provide the starting point for the only known approach for explicit computation of magnitude for a convex body in ℓ_2^n when $n > 1$. This result follows by considering the Euler–Lagrange equation of the minimization problem in Theorem 5.3.5, and applying elliptic regularity.

Theorem 5.4.19 (Proposition 5.7 and Corollary 5.8 of [28]). *Suppose that n is odd and $A \subseteq \ell_2^n$ is compact. Then the potential function h of A is C^∞ on $\mathbb{R}^n \setminus A$, and satisfies*

$$(I - \Delta)^{(n+1)/2} h(x) = 0 \tag{5.22}$$

on $\mathbb{R}^n \setminus A$.

To indicate the usefulness of this observation, we show how Theorem 5.4.19 can be used to quickly compute the magnitude of an interval in $[a, b] \subseteq \mathbb{R}$. By Theorem 5.4.19, the potential function h satisfies $h - h'' = 0$ outside $[a, b]$. The boundary conditions $h(x) = 1$ for $a \le x \le b$ and $h(x) \to 0$ when $|x| \to \infty$ (since $h \in H^1(\mathbb{R}^n)$) imply that

$$h(x) = \begin{cases} e^{x-a} & \text{if } x < a, \\ 1 & \text{if } a \le x \le b, \\ e^{b-x} & \text{if } x > b. \end{cases}$$

Then by Theorem 5.4.16,

$$\big|[a, b]\big| = \frac{1}{2} \int_{\mathbb{R}} h(x)\, dx = 1 + \frac{b-a}{2},$$

in agreement with (5.7). A more involved, but still elementary computation yields another proof of Corollary 5.4.3.

For higher dimensions, Barceló and Carbery [3] analyzed the minimization problem in more depth, and proved the following result using standard techniques of the theory of partial differential equations.

Proposition 5.4.20 (See Proposition 2 and Lemma 4 of [3]). *Suppose that n and m are positive integers, and $A \subseteq \mathbb{R}^n$ is a convex body.*

1. *There is a unique function $f \in H^m(\mathbb{R}^n)$ such that*

$$(I - \Delta)^m f(x) = 0 \text{ on } \mathbb{R}^n \setminus A$$

   weakly and $f \equiv 1$ on A.
2. *If ∂A is piecewise C^1 and $f \in H^m(\mathbb{R}^n)$, then all derivatives of f up to order $m - 1$ vanish on ∂A (in the sense of traces of Sobolev functions).*

Together with Theorems 5.4.16 and 5.4.19, Proposition 5.4.20 reduces the computation of magnitudes (in many cases) to the solution of a PDE boundary value problem. In general, of course, solving a PDE boundary value problem is no simple matter. But in the case that $A = \mathcal{B}_R^n$ is a Euclidean ball, rotational symmetry reduces the partial differential equation to an ordinary differential equation on $[R, \infty)$, albeit of high degree. Barceló and Carbery gave an algorithm for solving the resulting ODE boundary value problem, and hence determining the potential function h of $\mathcal{B}_R^n$, for every odd dimension n and radius $R > 0$. From there, Theorem 5.4.16 can be used to compute the magnitude $\mathcal{B}_R^n$. (In [3] the magnitude was found by computing $\|h\|_{H^{(n+1)/2}}^2$ using (5.21), since Theorem 5.4.16 had not yet been proved; Theorem 5.4.16 makes the computation much simpler.) This approach yields the following.

Theorem 5.4.21 (Theorems 2, 3, and 4 of [3]). *For every $R > 0$,*

$$\left|\mathcal{B}_R^3\right| = 1 + 2R + R^2 + \frac{1}{6}R^3$$

and

$$\left|\mathcal{B}_R^5\right| = \frac{24 + 72R^2 + 35R^3 + 9R^4 + R^5}{8(R+3)} + \frac{1}{120}R^5.$$

In general, when n is odd, the magnitude $\left|\mathcal{B}_R^n\right|$ is a rational function of $R > 0$ with rational coefficients.

Barceló and Carbery also give an explicit formula for $\left|\mathcal{B}_R^7\right|$. We recall that they also determined the asymptotics of $\left|\mathcal{B}_R^n\right|$ when $R \to 0$ and $R \to \infty$ in [3, Theorem 1], stated above in Theorems 5.4.18 and 5.4.14. To date, odd-dimensional balls are the only convex bodies in Euclidean space whose exact magnitudes are known.

It was previously conjectured in [23] that for a compact convex set $A \subseteq \ell_2^n$,

$$|A| = \sum_{i=0}^{n} \frac{V_i(A)}{i!\omega_i}, \tag{5.23}$$

where $V_0, \dots, V_n$ denotes the classical intrinsic volumes. Theorem 5.4.21 implies that (5.23) holds for balls in ℓ_2^3, but is false in dimensions $n \geq 5$.

To put this conjecture in context, observe that (5.23) is a Euclidean version of Conjecture 5.4.5 in ℓ_1^n. Note that $2^i = i!\omega_i'$, where ω_i' denotes the volume of the unit ball in ℓ_1^i, tightening the analogy between (5.23) and Conjecture 5.4.5. At the time that (5.23) was proposed, it was known to hold for $n = 1$, and was supported by numerical computations in $n = 2$ [40]. Furthermore, some cases of Conjecture 5.4.5 in ℓ_1^n (contained in Theorem 5.4.6) were known to be true.

Several interesting questions remain open, most obviously whether (5.23) holds for $n \leq 4$. Noting that (5.23) is equivalent to

$$|tA| = \sum_{i=0}^{n} \frac{V_i(A)}{i!\omega_i} t^i, \tag{5.24}$$

Proposition 5.4.13 and Theorem 5.4.14 say that (5.24) is true to top order for sets of positive volume when $t \to \infty$, and Theorem 5.4.18 shows that (5.23) predicts the correct behavior when $t \to 0$. One could ask whether (5.24) is approximately true in some sharper asymptotic senses. Note that Theorem 5.3.11 is a Riemannian analogue of a weak asymptotic version of (5.24). Barceló and Carbery also raise the question of whether (5.23) holds if magnitude is replaced by a suitable modification which coincides with magnitude in ℓ_2^n for $n \le 3$. We mention another related question in section 5.5.

The final major consequence of the concrete identification of $\mathcal{H}$ for Euclidean space is the realization that magnitude and maximum diversity, in the setting of ℓ_2^n, are actually classical notions of capacity, well-known in potential theory. The formal similarity between magnitude and maximum diversity on the one hand, and capacity on the other, is clear from the definitions (cf. section 1.1 of [3]). But in ℓ_2^n, magnitude and maximum diversity almost precisely reproduce classically studied forms of capacity.

Specifically, (5.20) and [1, Theorem 2.2.7] imply that for a compact set $A \subseteq \ell_2^n$, $|A|_+ = \frac{1}{n!\omega_n} C_{(n+1)/2}(A)$, where

$$C_\alpha(A) := \inf\left\{ \|f\|_{H^\alpha}^2 \;\middle|\; f \in \mathcal{S}(\mathbb{R}^n),\ f \ge 1 \text{ on } A \right\},$$

is the **Bessel capacity** of A of order α. An alternative notion of capacity, which naturally arises in the study of removability of singularities (see [1, Section 2.7]), is

$$N_\alpha(A) := \inf\left\{ \|f\|_{H^\alpha}^2 \;\middle|\; f \in \mathcal{S}(\mathbb{R}^n),\ f \ge 1 \text{ on a neighborhood of } A \right\}.$$

By Theorem 5.3.5, $|A| \le \frac{1}{n!\omega_n} N_{(n+1)/2}(A)$. In fact one would expect $|A| = \frac{1}{n!\omega_n} N_{(n+1)/2}(A)$; this appears not to be the case for arbitrary compact A, but happily the comparison we have will be enough for our purpose.

Before moving on, we pause to observe that, although we have just seen that magnitude and its cousin maximum diversity fit into classical families of capacities, they both just fail to fit into the parameter range which is of relevance for classical applications. As alluded to above, capacities are frequently used to quantify “exceptional” sets; sets of capacity 0 are a frequent substitute for sets of measure 0 when studying singularities. However, $C_\alpha(A)$ and $N_\alpha(A)$ are bounded below by positive constants whenever $\alpha > n/2$. So from the point of view of classical potential theory, magnitude and maximum diversity are rather pathological. Nevertheless, the following result from potential theory, whose main classical application is to show that $C_\alpha(A) = 0$ if and only if $N_\alpha(A) = 0$, also applies in our setting.

Proposition 5.4.22 ([1, Theorem 3.3.4]). *For each n and each $\alpha > 0$ there is a constant $\kappa_{n,\alpha} \ge 1$ such that, for every compact set $A \subseteq \ell_2^n$,*

$$C_\alpha(A) \le N_\alpha(A) \le \kappa_{n,\alpha} C_\alpha(A).$$

Corollary 5.4.23 ([28, Corollary 6.2]). *For each n there is a constant $\kappa_n \geq 1$ such that, for every compact set $A \subseteq \ell_2^n$,*

$$|A|_+ \leq |A| \leq \kappa_n |A|_+ .$$

The significance of Corollary 5.4.23 is that, although maximum diversity is no easier to compute explicitly than magnitude, in some ways its rough behavior is easier to analyze. For example, it is natural to conjecture that the magnitude function $t \mapsto |tA|$ is increasing for a compact space A of negative type. It is unknown whether this is true. On the other hand, it is obvious that $t \mapsto |tA|_+$ is increasing, and Corollary 5.4.23 therefore implies that the magnitude function of a compact set $A \subseteq \ell_2^n$ is at least bounded above and below by constant multiples of an increasing function.

A more substantial consequence of Corollary 5.4.23 is the following result, which, like Theorem 5.4.14, shows that the category-theoretically inspired notion of magnitude turns out to encode quantities of fundamental importance in geometry.

Theorem 5.4.24 ([28, Corollary 7.4]). *If $A \subseteq \ell_2^n$ is compact, then*

$$\lim_{t\to\infty} \frac{\log |tA|}{\log t} = \dim_{\mathrm{Mink}} A.$$

Theorem 5.4.24, which should be interpreted in the same sense as Proposition 5.3.14, follows immediately from Proposition 5.3.14 and Corollary 5.4.23. Another interesting aspect of this result is that, as noted above, classically Proposition 5.4.22 is of interest primarily for sets of capacity 0, or more generally for small sets; here it is instead applied to large sets.

5.5 Open problems

There are many interesting open problems about magnitude. These include extending partial results discussed above, as well as some quite basic questions about the behavior of magnitude. We mention several of them below.

1. Does every compact positive definite space (or space of negative type) have finite magnitude?
   Proposition 5.4.13 implies that every compact subset of a finite dimensional positive definite normed (or p-normed) space has finite magnitude, so that the obvious place to look for a counterexample is in infinite dimensions. Essentially the only infinite-dimensional spaces whose magnitudes are known are boxes in ℓ_1, which just miss being a counterexample:

$$\left|\prod_{i=1}^{\infty}[0, r_i]\right| = \prod_{i=1}^{\infty}\left(1 + \frac{r_i}{2}\right).$$

The condition $\|r\|_1 < \infty$, which both guarantees that this infinite-dimensional box lies in ℓ_1 and is compact, is also equivalent to the finiteness of the product on the right-hand side.

2. Is magnitude continuous on the class of compact sets in a positive definite normed (or p-normed) space? What if we assume the space is finite-dimensional, or we restrict to geodesic sets, or convex sets?
   Recall that magnitude is continuous on convex bodies in a finite-dimensional positive definite p-normed space (Theorem 5.4.15), but is not continuous on the class of compact spaces of negative type (Examples 2.2.8 and 2.4.9 of [19]).
3. Is Conjecture 5.4.5 true? Is it at least true for compact convex sets $A \subseteq \ell_1^n$?
   In light of Theorem 5.4.6, Conjecture 5.4.5 is equivalent to the continuity of magnitude on compact, geodesic (i.e., ℓ_1-convex) sets in ℓ_1^n. Similarly, if magnitude is continuous on compact convex sets in ℓ_1^n, then Theorem 5.4.6 would imply that Conjecture 5.4.5 holds for compact convex sets.
4. Does the magnitude function of a convex body $A \subseteq \ell_2^n$ determine its intrinsic volumes? What about a homogeneous compact Riemannian manifold?
5. Does it hold that

$$\left|t(A \cup B)\right| + \left|t(A \cap B)\right| - |tA| - |tB| \to 0$$

   as $t \to \infty$ for compact, convex sets $A, B \subseteq \ell_2^n$ (or in more general normed spaces) such that $A \cup B$ is convex?
   For convex bodies in ℓ_1^n, the left-hand side of the above is 0 for every t, as a consequence of Theorem 5.4.6; the same would be true in ℓ_2^n if (5.23) were true.
6. Does Theorem 5.4.24 hold for arbitrary compact spaces of negative type?
   Theorem 5.4.24 applies to compact subsets of ℓ_2^n. As mentioned earlier, Proposition 7.5 of [28] shows that the conclusion of Theorem 5.4.24 also holds for compact homogeneous metric spaces. In addition, Theorem 5.4.14 implies that the conclusion of Theorem 5.4.24 holds for compact subsets of positive n-dimensional volume in an n-dimensional positive definite p-normed space; hence it holds, for example, for compact convex sets in any positive definite p-normed space.

Bibliography

[1] D. R. Adams and L. I. Hedberg: *Function Spaces and Potential Theory*, volume 314 of *Grundlehren der Mathematischen Wissenschaften*. Springer-Verlag, Berlin, 1996.

[2] N. Aronszajn: Theory of reproducing kernels. *Trans. Amer. Math. Soc.*, 68:337–404, 1950.

[3] J. A. Barceló and A. Carbery: On the magnitudes of compact sets in Euclidean spaces. arXiv:1507.02502, 2015. To appear in *Amer. J. Math.*

[4] Y. Benyamini and J. Lindenstrauss: *Geometric Nonlinear Functional Analysis. Vol. 1*, volume 48 of *American Mathematical Society Colloquium Publications*. American Mathematical Society, Providence, RI, 2000.

[5] J. Chuang, A. King, and T. Leinster: On the magnitude of a finite dimensional algebra. Theory and Applications of Categories, 31:63–72, 2016.

[6] M. M. Deza and M. Laurent: *Geometry of Cuts and Metrics*, volume 15 of *Algorithms and Combinatorics*. Springer-Verlag, Berlin, 1997.

[7] M. Gromov: *Metric Structures for Riemannian and Non-Riemannian Spaces*. Birkhäuser, Boston, 2001.

[8] R. Hepworth and S. Willerton: Categorifying the magnitude of a graph. To appear in Homology, Homotopy, and Applications, 2015.

[9] M. Kanter: Unimodality and dominance for symmetric random vectors. *Trans. Amer. Math. Soc.*, 229:65–85, 1977.

[10] G. M. Kelly: *Basic Concepts of Enriched Category Theory*, volume 64 of *London Mathematical Society Lecture Note Series*. Cambridge University Press, Cambridge, 1982. Also *Reprints in Theory and Applications of Categories* 10 (2005), 1–136.

[11] D. A. Klain and G.-C. Rota: *Introduction to Geometric Probability*. Lezioni Lincee. Cambridge University Press, Cambridge, 1997.

[12] S. L. Kokkendorff: Does negative type characterize the round sphere? *Proc. Amer. Math. Soc.*, 135(11):3695–3702 (electronic), 2007.

[13] A. Koldobsky: *Fourier Analysis in Convex Geometry*, volume 116 of *Mathematical Surveys and Monographs*. American Mathematical Society, Providence, RI, 2005.

[14] F. W. Lawvere: Metric spaces, generalized logic and closed categories. *Rendiconti del Seminario Matematico e Fisico di Milano*, XLIII:135–166, 1973. Also *Reprints in Theory and Applications of Categories* 1:1–37, 2002.

[15] F. W. Lawvere: Taking categories seriously. *Revista Colombiana de Matemáticas*, XX:147–178, 1986. Also *Reprints in Theory and Applications of Categories* 8:1–24, 2005.

[16] T. Leinster: The Euler characteristic of a category. *Documenta Mathematica*, 13:21–49, 2008.

[17] T. Leinster: A maximum entropy theorem with applications to the measurement of biodiversity. arXiv:0910.0906, 2009.

[18] T. Leinster: Integral geometry for the 1-norm. *Advances in Applied Mathematics*, 49:81–96, 2012.

[19] T. Leinster: The magnitude of metric spaces. *Documenta Mathematica*, 18:857–905, 2013.

[20] T. Leinster: The magnitude of a graph. To appear in Math. Proc. Cam. Phil. Soc., 2014.

[21] T. Leinster and C. Cobbold: Measuring diversity: the importance of species similarity. *Ecology*, 93:477–489, 2012.

[22] T. Leinster and M. W. Meckes: Maximizing diversity in biology and beyond. *Entropy*, 18(3):88, 2016.

[23] T. Leinster and S. Willerton: On the asymptotic magnitude of subsets of Euclidean space. *Geom. Dedicata*, 164(1):287–310, 2013.

[24] P. Lévy: *Théorie de l'addition de variables aléatoires*. Gauthier-Villars, Paris, 1937.

[25] N. Linial: Finite metric-spaces—combinatorics, geometry and algorithms. In *Proceedings of the International Congress of Mathematicians, Vol. III (Beijing, 2002)*, pages 573–586. Higher Ed. Press, Beijing, 2002.

[26] S. Mac Lane: *Categories for the Working Mathematician*. Graduate Texts in Mathematics 5. Springer, New York, 1971.

[27] M. W. Meckes: Positive definite metric spaces. *Positivity*, 17:733–757, 2013. Corrected version available at arxiv.org/abs/1012.5863v5.

[28] M. W. Meckes: Magnitude, diversity, capacities, and dimensions of metric spaces. *Potential Analysis*, 42:549–572, 2015.

[29] R. A. Olshen and L. J. Savage: A generalized unimodality. *J. Appl. Probability*, 7:21–34, 1970.

[30] G.-C. Rota: On the foundations of combinatorial theory I: theory of Möbius functions. *Zeitschrift für Wahrscheinlichkeitstheorie und Verwandte Gebiete*, 2:340–368, 1964.

[31] K.-I. Sato: *Lévy Processes and Infinitely Divisible Distributions*, volume 68 of *Cambridge Studies in Advanced Mathematics*. Cambridge University Press, Cambridge, 2013. Translated from the 1990 Japanese original, Revised edition of the 1999 English translation.

[32] S. H. Schanuel: What is the length of a potato? An introduction to geometric measure theory. In *Categories in Continuum Physics*, Lecture Notes in Mathematics 1174. Springer, Berlin, 1986.

[33] S. H. Schanuel: Negative sets have Euler characteristic and dimension. In *Category Theory (Como, 1990)*, Lecture Notes in Mathematics 1488, pages 379–385. Springer, Berlin, 1991.

[34] R. Schneider: *Convex Bodies: the Brunn–Minkowski Theory*, volume 151 of *Encyclopedia of Mathematics and its Applications*. Cambridge University Press, Cambridge, expanded edition, 2014.

[35] A. R. Solow and S. Polasky: Measuring biological diversity. *Environmental and Ecological Statistics*, 1:95–107, 1994.

[36] D. Speyer: Re: Metric spaces. Comment at golem.ph.utexas.edu/category/2008/02/metric_spaces.html, 2008.

[37] E. M. Stein and G. Weiss: *Introduction to Fourier Analysis on Euclidean Spaces*, volume 32 of *Princeton Mathematical Series*. Princeton University Press, Princeton, N.J., 1971.

[38] T. Watanabe: Asymptotic estimates of multi-dimensional stable densities and their applications. *Trans. Amer. Math. Soc.*, 359(6):2851–2879 (electronic), 2007.

[39] J. H. Wells and L. R. Williams: *Embeddings and Extensions in Analysis*. Springer-Verlag, New York-Heidelberg, 1975. Ergebnisse der Mathematik und ihrer Grenzgebiete, Band 84.

[40] S. Willerton: Heuristic and computer calculations for the magnitude of metric spaces. arxiv.org/abs/0910.5500, 2009.

[41] S. Willerton: On the magnitude of spheres, surfaces and other homogeneous spaces. *Geom. Dedicata*, 168:291–310, 2014.

Christian Léonard

On the convexity of the entropy along entropic interpolations

Introduction

Displacement convexity of relative entropy plays a crucial role in the Lott-Sturm-Villani (LSV) theory of curvature lower bounds of metric measure spaces [31, 41, 42, 44] and the related Ambrosio-Gigli-Savaré gradient flow approach [1, 2].

Let us explain shortly what is meant by displacement convexity of the relative entropy. Let $(\boldsymbol{X}, d, m)$ be a metric measure space and $\mathrm{P}(\boldsymbol{X})$ denote the space of all Borel probability measures on $\boldsymbol{X}$. The distance d is the basic ingredient to construct the displacement interpolations and the positive measure m is used as the relative entropy reference measure, defined by $H(\mu|m) := \int_{\boldsymbol{X}} \log(d\mu/dm)\, d\mu \in (\infty, \infty]$, $\mu \in \mathrm{P}(\boldsymbol{X})$. Displacement convexity means that along any displacement interpolation $[\mu_0, \mu_1]^{\mathrm{disp}} = (\mu_t)_{0\le t\le 1}$ between two probability measures μ_0 and μ_1, the entropy

$$H(t) := H(\mu_t|m), \quad 0 \le t \le 1$$

of the interpolation as a function of time admits some convexity lower bound.
Let us recall briefly what displacement interpolations are. The Wasserstein pseudo-distance W_2 is the square root of the optimal quadratic transport cost between μ_0 and $\mu_1 \in \mathrm{P}(\boldsymbol{X})$ given by $W_2^2(\mu_0, \mu_1) := \inf_\pi \int_{\boldsymbol{X}^2} d^2\, d\pi$ where the infimum is taken over all the couplings $\pi \in \mathrm{P}(\boldsymbol{X}^2)$ of μ_0 and μ_1. It becomes a distance on the subset $\mathrm{P}_2(\boldsymbol{X})$ of all $\mu \in \mathrm{P}(\boldsymbol{X})$ such that $\int_{\boldsymbol{X}} d^2(x_o, x)\, \mu(dx) < \infty$, for some $x_o \in \boldsymbol{X}$. Displacement interpolations are the geodesics of the metric space $(\mathrm{P}_2(\boldsymbol{X}), W_2)$. They were introduced in McCann's PhD thesis [33] together with the notion of displacement convexity of the entropy. Related convex inequalities turn out to be functional and geometric inequalities (Brunn-Minkowski, Prekopa-Leindler, Borell-Brascamp-Lieb, logarithmic Sobolev, Talagrand inequalities), see [12, 34].
As a typical result, it is known [12, 38, 43] that a Riemannian manifold has a nonnegative Ricci curvature if and only if $t \mapsto H(\mu_t|\,\mathrm{vol})$ is convex along any displacement interpolation $(\mu_t)_{0\le t\le 1}$.

An important hypothesis of the LSV theory is that the metric space is geodesic. This rules discrete spaces out. Something new must be found to develop an analogue of this theory in the discrete setting. Several attempts in this direction were proposed recently. Maas and Mielke [18, 19, 32, 35, 36] have discovered a Riemannian distance

Christian Léonard: Modal-X. Université Paris Ouest. Bât.G, 200 av. de la République. 92001 Nanterre, France

on the set of probability measures such that the evolution of continuous-time Markov chains on a graph are gradient flows of some entropy with respect to this distance. Bonciocat and Sturm in [7] and Gozlan, Roberto, Samson and Tetali in [23] proposed two different types of interpolations on the set of probability measures on a discrete metric measure graph which play the role of displacement interpolations in the LSV theory. The author recently proposed in [29] a general procedure for constructing displacement interpolations on graphs by slowing down another type of interpolation, called *entropic interpolation*.

Entropic interpolations are the main actors of the present paper. Analogously to the displacement interpolations which solve dynamical transport problems, they solve dynamical entropy minimization problems. Although related to the purpose of this article, [29] is mainly concerned with the slowing down asymptotic to build displacement – rather than entropic – interpolations. Displacement interpolations are connected to optimal transport, while entropic interpolations are connected to minimal entropy.

In this article, we consider the entropic interpolations in their own right, without slowing down. The goal remains the same: studying convexity of the entropy along these interpolations to derive curvature lower bounds of some Markov generators and state spaces. Our main result states that along any entropic interpolation $[\mu_0, \mu_1] = (\mu_t)_{0\le t\le 1}$ between two probability measures μ_0 and μ_1, the first and second derivatives of the relative entropy $H(t) := H(\mu_t|m)$ of μ_t with respect to some reference measure m can be written as

$$H'(t) = \int_X (\overrightarrow{\Theta}\psi_t - \overleftarrow{\Theta}\phi_t)\, d\mu_t, \qquad H''(t) = \int_X (\overrightarrow{\Theta}_2\psi_t + \overleftarrow{\Theta}_2\phi_t)\, d\mu_t. \tag{6.1}$$

The arrows on Θ and Θ_2 are related to the forward and backward directions of time. The functions ϕ and ψ depend on the endpoints μ_0 and μ_1, but the nonlinear operators $\overrightarrow{\Theta}$, $\overleftarrow{\Theta}$, $\overrightarrow{\Theta}_2$ and $\overleftarrow{\Theta}_2$ only depend on some reference m-stationary Markov process. For instance, when this process is the Brownian diffusion with generator $L = (\Delta - \nabla V \cdot \nabla)/2$ on a Riemannian manifold with $m = e^{-V}\,\mathrm{vol}$ its reversing measure, we show that

$$\overrightarrow{\Theta} = \overleftarrow{\Theta} = \Gamma/2, \quad \overrightarrow{\Theta}_2 = \overleftarrow{\Theta}_2 = \Gamma_2/2 \tag{6.2}$$

are respectively half the carré du champ $\Gamma u := L(u^2) - 2uLu$ and the iterated carré du champ $\Gamma_2 u := L\Gamma u - 2\Gamma(u, Lu)$ which were introduced by Bakry and Émery in their article [5] on hypercontractive diffusion processes. In the discrete case, these operators are not linked to Γ and Γ_2 anymore but they depend on L in a different manner, see (6.16) and (6.17). In the Riemannian case, Γ_2 is related to the Ricci curvature via Bochner's formula (6.34). This is the main connection between geometry and displacement convexity. Because of the tight relation (6.2) between (Θ, Θ_2) and (Γ, Γ_2) in the Riemannian case, it is natural to expect that (Θ, Θ_2) has also some geometric content in the discrete case. Does it allow for an efficient definition of curvature of graphs? This question is left open in the present paper. However, based on Θ and Θ_2, we give a unified proof of the logarithmic Sobolev inequality on manifolds and a modified version on graphs. This is a clue in favor of the relevance of the entropic interpolations.

An advantage of the entropic calculus, compared to its displacement analogue, is that entropic interpolations are regular in general. This is in contrast with displacement interpolations. Otto's informal calculus provides heuristics that need to be rigorously proved by means of alternate methods on the other hand, entropic interpolations allow for a direct rigorous calculus. As a matter of fact, it is proved in [26, 29] that displacement interpolations are semiclassical limits of entropic interpolations: entropic interpolations are regular approximations of displacement interpolations.

The entropic approach is not only available for reversible reference Markov dynamics, but also for stationary dynamics (in the reversible case, the time arrows on Θ and Θ_2 disappear).

The article's point of view is probabilistic. Its basic ingredients are measures on path spaces: typically, Brownian diffusions on manifolds and random walks on graphs. The Markov property plays a crucial role and we take advantage of its time symmetry. Recall that it states that conditionally on the knowledge of the present state, past and future events are independent. In particular, a time-reversed Markov process is still Markov. Time-reversal stands in the core of our approach (even when the reference stochastic process is assumed to be reversible); it explains the forward and backward arrows on Θ and Θ_2. In contrast with analytic approaches, very few probabilistic attempts have been implemented to explore geometric and functional inequalities. In this respect, let us cite the contributions of Cattiaux [10] and Fontbona and Jourdain [22] where stochastic calculus is central. In the present article, stochastic calculus is secondary. The symmetry of the Markov property allows us to proceed with basic measure theoretical technics.

The drawback of the entropic interpolations in the discrete setting is that the second derivative of the entropy along the interpolations is difficult to handle in practice, see Proposition 6.4.4 for instance. As a consequence, no specific examples based on random walks on graphs are treated in the present article. There is still some (hard) work to do to proceed in this direction.

Outline of the chapter

The article is organized as follows. Entropic interpolations are discussed in Section 6.1. In particular, we describe their dynamics. This allows us to write their equation of motion and derive our main abstract result (6.1) in Section 6.2. This abstract result is exemplified with Brownian diffusion processes in Section 6.3 and random walks on a countable graph in Section 6.4. A unified proof of an entropy-entropy production inequality (logarithmic Sobolev inequality), in both the Riemannian and discrete graph settings, is given in Section 6.5 in connection with convergence to equilibrium. The stationary non-reversible case is investigated. In Section 6.6, we propose heuristics based on thought experiments in order to grasp the link between entropic and displacement interpolations. Finally, we address several open questions related to curvature and entropic interpolations, keeping in mind the LSV theory of geodesic measure spaces as a guideline. This article is a preliminary step and these open questions should be seen as part of a program toward an analogue of the LSV theory that is available in a discrete setting.

Notation

For any measurable space Y, we denote by $\mathrm{P}(Y)$ and $\mathrm{M}_+(Y)$ the sets of all probability measures and positive measures on Y. Let $\Omega = D([0,1], \boldsymbol{X})$ be the space of all right-continuous and left-limited paths from the time interval $[0,1]$ taking their values in a Polish state space $\boldsymbol{X}$ endowed with its Borel σ-field. The canonical process $X = (X_t)_{0\le t\le 1}$ is defined by $X_t(\omega) = \omega_t \in \boldsymbol{X}$ for all $0 \le t \le 1$ and $\omega = (\omega_t)_{0\le t\le 1} \in \Omega$. As usual, Ω is endowed with the canonical σ-field $\sigma(X_t, 0 \le t \le 1)$ generated by the canonical process[6.1]. For any $\mathcal{T} \subset [0,1]$ and any positive measure Q on Ω, we denote $X_{\mathcal{T}} = (X_t)_{t\in\mathcal{T}}$, $Q_{\mathcal{T}} = (X_{\mathcal{T}})_\# Q$, $\sigma(X_{\mathcal{T}})$ is the σ-field generated by $X_{\mathcal{T}}$. In particular, for each $t \in [0,1]$, $Q_t = (X_t)_\# Q \in \mathrm{M}_+(\boldsymbol{X})$ and $Q_{01} = (X_0, X_1)_\# Q \in \mathrm{M}_+(\boldsymbol{X}^2)$.

6.1 Entropic interpolation

Entropic interpolations are defined and their equations of motion are derived. The continuous and discrete space cases are both imbedded in the same abstract setting. This prepares the next section where the first and second derivatives of $H(t) := H(\mu_t|m)$ are calculated.

(f,g)-transform of a Markov process

The h-transform of a Markov process was introduced by Doob [16] in 1957. It is revisited and slightly extended in two directions:

1. We consider a two-sided version of the h-transform, which we call (f,g)-transform, taking advantage of the invariance of the Markov property with respect to time reversal;

6.1 The canonical process is the standard naming by probabilists of the evaluation process of the analysts. The usual Skorokhod topology of the path space is useless in this article; only its measurable structure is used.

2. We extend the notion of Markov property to unbounded positive measures on Ω, having in mind two typical examples:
   - The reversible Wiener process on $\mathbb{R}^n$, i.e. the law of the Brownian motion with Lebesgue measure as its initial law;
   - The reversible simple random walk on a countable locally finite graph.

Assumptions on *R*

We specify some $R \in \mathrm{M}_+(\Omega)$ which we call the *reference path measure* and we assume that it is Markov and m-stationary[6.2] where $m \in \mathrm{M}_+(\boldsymbol{X})$ is a σ-finite positive measure on $\boldsymbol{X}$. See Definitions .0.4 and .0.7 at the appendix 6.6.

Definition 6.1.1 ((f, g)-transform). *Let $f_0, g_1 : \boldsymbol{X} \to [0, \infty)$ be two nonnegative functions such that $f_0 \in L^p(m)$ and $g_1 \in L^{p^*}(m)$ with $p, p^* \in [1, \infty]$ and $1/p + 1/p^* = 1$. The (f, g)-transform $P \in \mathrm{P}(\Omega)$ of R associated with the couple of functions (f_0, g_1) is defined by*

$$P := f_0(X_0)g_1(X_1)\, R \in \mathrm{P}(\Omega) \tag{6.3}$$

with $\int_{\boldsymbol{X}^2} f_0(x)g_1(y)\, R_{01}(dxdy) = 1$ for P to be a probability measure.

Note that $f_0(X_0)g_1(X_1) \in L^1(R)$. Indeed, denoting $F = f_0(X_0)$ and $G = g_1(X_1)$, we see with $R_0 = R_1 = m$ that $\|F\|_{L^p(R)} = \|f_0\|_{L^p(m)}$, $\|G\|_{L^{p^*}(R)} = \|g_1\|_{L^{p^*}(m)}$ *and* $E_R(FG) \le \|F\|_p \|G\|_{p^*} = \|f_0\|_p \|g_1\|_{p^*} < \infty$.

Remarks 6.1.2. *Let us write some easy facts about (f, g)-transforms and h-transforms.*
(a) Taking $f_0 = \rho_0$ and $g_1 = 1$ gives us $P = \rho_0(X_0)R$ which is the path measure with initial marginal $\mu_0 = \rho_0\, m \in \mathrm{P}(\boldsymbol{X})$ and the same forward Markov dynamics as R.
(b) Symmetrically, choosing $f_0 = 1$ and $g_1 = \rho_1$ corresponds to $P = \rho_1(X_1)R$ which is the path measure with final marginal $\mu_1 = \rho_1\, m \in \mathrm{P}(\boldsymbol{X})$ and the same backward Markov dynamics as R.
(c) Doob's h-transform is an extension of item (b). It is defined by $P = h(X_\tau)\, R_{[0,\tau]}$ where τ is a stopping time and h is such that P is a probability measure.

Let us denote the backward and forward Markov transition kernels of R by

$$\begin{cases} \overleftarrow{r}(s, \cdot; t, z) & := & R(X_s \in \cdot \mid X_t = z), & 0 \le s \le t \le 1, \\ \overrightarrow{r}(t, z; u, \cdot) & := & R(X_u \in \cdot \mid X_t = z), & 0 \le t \le u \le 1, \end{cases}$$

6.2 This stationarity assumption is not necessary and one could replace m by R_t everywhere. It simply makes things more readable and shortens some computations.

and for m-almost all $z \in \boldsymbol{X}$,

$$\begin{cases} f_t(z) & := & E_R(f_0(X_0) \mid X_t = z) & = & \int_{\boldsymbol{X}} f_0(x) \overleftarrow{r}(0, dx; t, z) \\ g_t(z) & := & E_R(g_1(X_1) \mid X_t = z) & = & \int_{\boldsymbol{X}} \overrightarrow{r}(t, z; 1, dy) g_1(y). \end{cases} \tag{6.4}$$

The relation between (f_0, g_1) and the endpoint marginals $\mu_0, \mu_1 \in \mathrm{P}(\boldsymbol{X})$ is

$$\begin{cases} \rho_0 & = & f_0 g_0, & m\text{-a.e.} \\ \\ \rho_1 & = & f_1 g_1, & m\text{-a.e.} \end{cases} \tag{6.5}$$

where the density functions ρ_0 and ρ_1 are defined by $\begin{cases} \mu_0 & = & \rho_0\, m \\ \mu_1 & = & \rho_1\, m \end{cases}$. The system of equations (6.5) was exhibited by Schrödinger in 1931 [39] in connection with the entropy minimization problem ($\mathrm{S}_{\mathrm{dyn}}$) below.

Entropic interpolation

Let us introduce the main notion of this article.

Definition 6.1.3 (Entropic interpolation). *Let $P \in \mathrm{P}(\Omega)$ be the (f, g)-transform of R given by* (6.3). *Its flow of marginal measures*

$$\mu_t := (X_t)_\# P \in \mathrm{P}(\boldsymbol{X}), \quad 0 \le t \le 1,$$

*is called the R-*entropic interpolation *between μ_0 and μ_1 in $\mathrm{P}(\boldsymbol{X})$ and is denoted by*

$$[\mu_0, \mu_1]^R := (\mu_t)_{t \in [0,1]}.$$

When the context is clear, we simply write $[\mu_0, \mu_1]$, dropping the superscript R.

Next theorem tells us that μ_t is absolutely continuous with respect to the reference measure m on $\boldsymbol{X}$. We denote by

$$\rho_t := d\mu_t / dm, \quad 0 \le t \le 1,$$

its density with respect to m.

Identities (6.5) extend to all $0 \le t \le 1$ as next result shows.

Theorem 6.1.4. *Let $P \in \mathrm{P}(\Omega)$ be the (f, g)-transform of R given by* (6.3). *Then, P is Markov and for all $0 \le t \le 1$, $\mu_t = \rho_t\, m$ with*

$$\rho_t = f_t g_t, \quad m\text{-a.e.} \tag{6.6}$$

Proof. Without getting into detail, the proof works as follows. As $m = R_t$,

$$\rho_t(z) := \frac{dP_t}{dR_t}(z) = E_R\left[\frac{dP}{dR}|X_t = z\right] = E_R\left[f_0(X_0)g_1(X_1)|X_t = z\right]$$
$$= E_R\left[f_0(X_0)|X_t = z\right] E_R\left[g_1(X_1)|X_t = z\right] =: f_t g_t(z).$$

The Markov property of R at the last but one equality is essential in this proof. For more detail, see [28, Thm. 3.4]. □

With (6.6), we see that an equivalent analytical definition of $[\mu_0, \mu_1]^R$ is as follows.

Theorem 6.1.5. *The entropic interpolation* $[\mu_0, \mu_1]^R$ *satisfies* $\mu_t = \rho_t\, m$ *with*

$$\rho_t(z) = \int_{\mathbf{X}} f_0(x)\, \overleftarrow{r}(0, dx; t, z) \int_{\mathbf{X}} \overrightarrow{r}(t, z; 1, dy)\, g_1(y), \quad \textit{for } m\textit{-a.e. } z \in \mathbf{X}, \tag{6.7}$$

for each $0 \le t \le 1$.

Marginal flows of bridges are (possibly degenerate) entropic interpolations

Let us have a look at the R-entropic interpolation between the Dirac measures δ_x and δ_y.

(a) When $R_{01}(x, y) > 0$, $[\delta_x, \delta_y]^R$ is the time marginal flow $(R_t^{xy})_{0\le t\le 1}$ of the bridge

$$R^{xy}(\cdot) := R(\cdot \mid X_0 = x, X_1 = y), \quad x, y \in \mathbf{X}.$$

Indeed, taking $\mu_0 = \delta_x$ and $\mu_1 = \delta_y$, a solution (f_0, g_1) of (6.5) is $f_0 = \mathbf{1}_x / R_{01}(x, y)$ and $g_1 = \mathbf{1}_y$. It follows that the corresponding (f, g)-transform is $f_0(X_0)g_1(X_1)\, R = R_{01}(x, y)^{-1}\mathbf{1}_{\{X_0 = x, X_1 = y\}}\, R = R^{xy}$.

(b) When $R_{01}(x, y) = 0$, $[\delta_x, \delta_y]^R$ is undefined. Indeed, Definition 6.1.3 implies that $\mu_0, \mu_1 \ll m$. Let us take the sequences of functions $f_0^n = c_n^{-1}\mathbf{1}_{B(x,1/n)}$ and $g_1^n = \mathbf{1}_{B(y,1/n)}$ with $B(x, r)$ the open ball centered at x with radius r and $c_n = R_{01}(B(x, 1/n) \times B(y, 1/n)) > 0$ the normalizing constant which is assumed to be positive for all $n \ge 1$. The corresponding (f, g)-transform of R is the conditioned probability measure $P^n = R(\cdot \mid X_0 \in B(x, 1/n), X_1 \in B(y, 1/n))$ which converges as n tends to infinity to the bridge R^{xy} under some assumptions, see [11] for instance. As a natural extension, one can see $[\delta_x, \delta_y]^R$ as the time marginal flow $t \mapsto R_t^{xy}$ of the bridge R^{xy}.

(c) When the transition kernels are absolutely continuous with respect to m, i.e. $\overrightarrow{r}(t, z; 1, dy) = \overrightarrow{r}(t, z; 1, y)\, m(dy)$, for all $(t, z) \in [0, 1) \times \mathbf{X}$ and $\overleftarrow{r}(0, dx; t, z) = \overleftarrow{r}(0, x; t, z)\, m(dx)$, for all $(t, z) \in (0, 1] \times \mathbf{X}$, then the density functions are equal: $\overrightarrow{r} = \overleftarrow{r} := r$. This comes from the m-stationarity of R since $R((X_t, X_{t'}) \in dz dz') = R(X_t \in dz) R(X_{t'} \in dz' \mid X_t = z) = m(dz)\overrightarrow{r}(t, z; t', z')m(dz')$

and similarly $R((X_t, X_{t'}) \in dzdz') = R(X_{t'} \in dz')R(X_t \in dz \mid X_{t'} = z') = m(dz')\overleftarrow{r}(t,z;t',z')m(dz)$. Now, (6.7) extends as

$$\frac{dR_t^{xy}}{dm}(z) = \frac{r(0,x;t,z)r(t,z;1,y)}{r(0,x;1,y)}, \qquad 0 < t < 1.$$

(d) Note that in general, for any intermediate time $0 < t < 1$, R_t^{xy} is not a Dirac mass. This is in contrast with the standard displacement interpolation $[\delta_x, \delta_y]^{\text{disp}}$ whose typical form is $\delta_{\gamma_t^{xy}}$ where γ^{xy} is the constant speed geodesic between x and y when this geodesic is unique.

Any entropic interpolation is a mixture of marginal flows of bridges. This is expressed in (6.63). For a proof, see [28].

Schrödinger problem

The notion of entropic interpolation is related to the entropy minimization problem (S_{dyn}) below. In order to state this problem properly, let us first recall an informal definition of the relative entropy

$$H(p|r) := \int \log(dp/dr)\, dp \in (-\infty, \infty]$$

of the probability measure p with respect to the reference σ-finite measure r; see appendix 6.6 for further detail. The dynamical Schrödinger problem associated with the reference path measure $R \in \mathrm{M}_+(\Omega)$ is

$$H(P|R) \to \min; \qquad P \in \mathrm{P}(\Omega) : P_0 = \mu_0,\ P_1 = \mu_1. \qquad (\text{S}_{\text{dyn}})$$

It consists of minimizing the relative entropy $H(P|R)$ of the path probability measure P with respect to the reference path measure R subject to the constraints $P_0 = \mu_0$ and $P_1 = \mu_1$ where $\mu_0, \mu_1 \in \mathrm{P}(\boldsymbol{X})$ are prescribed initial and final marginals.

We shall need a little bit more than $f_0, g_1 \in L^2(m)$ what follows. The following set of assumptions implies that some relative entropies are finite, and will be invoked in Theorem 6.1.7 below.

Assumptions 6.1.6. *In addition to $f_0, g_1 \in L^2(m)$ and the normalization condition $\int_{\boldsymbol{X}^2} f_0(x)g_1(y)\, R_{01}(dxdy) = 1$, the functions f_0 and g_1 entering the definition of the (f,g)-transform P of R given in (6.3) satisfy*

$$\int_{\boldsymbol{X}^2} [\log_+ f_0(x) + \log_+ g_1(y)] f_0(x) g_1(y)\, R_{01}(dxdy) < \infty, \qquad (6.8)$$

where $\log_+ h := \mathbf{1}_{\{h>1\}} \log h$, and as a convention $0 \log 0 = 0$.

Theorem 6.1.7. *Under the Assumptions 6.1.6, the (f, g)-transform P of R which is defined in* (6.3) *is the unique solution of* $(\mathrm{S_{dyn}})$ *where the prescribed constraints $\mu_0 = \rho_0\, m$ and $\mu_1 = \rho_1\, m$ are chosen to satisfy* (6.5).

Proof. See [28, Thm. 3.3]. □

Note that the solution of $(\mathrm{S_{dyn}})$ may not be a (f, g)-transform of R. More detail about the Schrödinger problem can be found in the survey paper [28].

$[\mu, \mu]^R$ is not constant in time

One must be careful when employing the term interpolation with regard to entropic interpolations because in general when $\mu_0 = \mu_1 =: \mu$ the interpolation $[\mu, \mu]^R$ is not constant in time. Let us give two examples.

(a) Suppose that $R_{01}(x, x) > 0$. Then the solution of $(\mathrm{S_{dyn}})$ with $\mu_0 = \mu_1 = \delta_x$ is the bridge R^{xx} and $[\delta_x, \delta_x]^R$ is the marginal flow $(R_t^{xx})_{0 \le t \le 1}$ which is not constant in general.

(b) Consider R to be the reversible Brownian motion on a compact connected Riemannian manifold $\boldsymbol{X}$ without boundary. Starting from μ at time $t = 0$, the entropic minimizer P is such that $\mu_t := P_t$ gets closer to the invariant volume measure $m = \mathrm{vol}$ on the time interval [0,1/2] and then goes back to μ on the remaining time period [1/2,1]. Let us show this. On the one hand, under our assumptions on the manifold, assuming that the Ricci curvature is strictly positive and with Theorem 6.3.5 below, it is easy to show that the function $t \in [0, 1] \mapsto H(t) := H(\mu_t | \mathrm{vol})$ is strictly convex whenever $\mu \neq \mathrm{vol}$. And on the other hand, a time reversal argument tells us that $H(t) = H(1 - t)$, $0 \le t \le 1$. Therefore, the only constant entropic interpolation is $[\mathrm{vol}, \mathrm{vol}]^R$.

$[\mu, \mu]^{\mathrm{disp}}$ is constant in time

Unlike the entropic interpolation $[\mu, \mu]^R$, McCann's displacement interpolation $[\mu, \mu]^{\mathrm{disp}}$ has the pleasant property of being constant in time. See (6.61) below for the definition of the displacement interpolation and compare with the representation (6.63) of the entropic interpolation.

Forward and backward stochastic derivatives of a Markov measure

Since P is Markov, its dynamics is characterized by either its forward stochastic derivative and its initial marginal μ_0, or its backward stochastic derivative and its final marginal μ_1. Before computing these derivatives, let us recall some basic facts about these notions.

Let $Q \in \mathrm{M}_+(\Omega)$ be a Markov measure. Its forward stochastic derivative $\partial + \overrightarrow{L}^Q$ is defined by

$$[\partial_t + \overrightarrow{L}^Q_t](u)(t,z) := \lim_{h\downarrow 0} h^{-1} E_Q\big(u(t+h, X_{t+h}) - u(t, X_t) \mid X_t = z\big)$$

for any measurable function $u : [0,1] \times \boldsymbol{X} \to \mathbb{R}$ in the set $\mathrm{Dom}\, \overrightarrow{L}^Q$ for which this limit exists Q_t-a.e. for all $0 \le t < 1$. In fact this definition is only approximate, we give it here as a support for understanding the relation between the forward and backward derivatives. For a precise statement see [25, §. 2]. Since the time reversed Q^* of Q is still Markov, Q admits a backward stochastic derivative $-\partial + \overleftarrow{L}^Q$ which is defined by

$$[-\partial_t + \overleftarrow{L}^Q_t]u(t,z) := \lim_{h\downarrow 0} h^{-1} E_Q\big(u(t-h, X_{t-h}) - u(t, X_t) \mid X_t = z\big)$$

for any measurable function $u : [0,1] \times \boldsymbol{X} \to \mathbb{R}$ in the set $\mathrm{Dom}\, \overleftarrow{L}^Q$ for which this limit exists Q_t-a.e. for all $0 < t \le 1$. When the function u only depends on the space variable: $z \mapsto u(z)$, one denotes for all $0 < t < 1$ and $z \in \boldsymbol{X}$,

$$\begin{aligned} \overrightarrow{L}_t u(z) &:= \lim_{h\downarrow 0} h^{-1} E_Q\big(u(X_{t+h}) - u(X_t) \mid X_t = z\big), \\ \overleftarrow{L}_t u(z) &:= \lim_{h\downarrow 0} h^{-1} E_Q\big(u(X_{t-h}) - u(X_t) \mid X_t = z\big). \end{aligned}$$

Notice that

$$\overleftarrow{L}^Q_t = \overrightarrow{L}^{Q^*}_{1-t}, \quad 0 < t \le 1.$$

It is proved in [25, §. 2] that these stochastic derivatives are extensions of the extended forward and backward generators of Q in the semimartingale sense, see [15]. In particular, they offer us a natural way for computing the generators.

Forward and backward dynamics of the entropic interpolation

The dynamics in both time directions of $[\mu_0, \mu_1]^R$ are specified by the stochastic derivatives $\overrightarrow{A} := \overrightarrow{L}^P$ and $\overleftarrow{A} := \overleftarrow{L}^P$ of the Markov measure $P \in \mathrm{P}(\Omega)$ which appears in Definition 6.1.3. For simplicity, we also denote $\overrightarrow{L}^R = \overrightarrow{L}$ and $\overleftarrow{L}^R = \overleftarrow{L}$ and assume that R is stationary, i.e. $\theta^h_\# R_{[a+h,b+h]} = R_{[0,b-a]}$ for all $0 \le a \le b \le b+h \le 1$, where θ^h is the time-shift defined for all t by $X_t \circ \theta^h = X_{t-h}$. This implies that

$$\overrightarrow{L}_t = \overrightarrow{L}, \quad \overleftarrow{L}_t = \overleftarrow{L}, \quad \forall 0 \le t \le 1,$$

meaning that the forward and backward transition mechanisms of R do not depend on t.

To derive the expressions of $\overrightarrow{A}$ and $\overleftarrow{A}$, we need to introduce the carré du champ Γ of both R and its time reversed R^*. The exact definition of the extended carré du champ

Γ and its domain is given in [25, Def. 4.10]. Restricting to functions u, v on $\mathbf{X}$ such that u, v and uv are in Dom $\overrightarrow{L}$, we recover the standard definition

$$\begin{cases} \overrightarrow{\Gamma}(u,v) & := & \overrightarrow{L}(uv) - u\overrightarrow{L}v - v\overrightarrow{L}u \\ \overleftarrow{\Gamma}(u,v) & := & \overleftarrow{L}(uv) - u\overleftarrow{L}v - v\overleftarrow{L}u \end{cases}$$

When R is a reversible path measure, we denote $L = \overrightarrow{L} = \overleftarrow{L}$ and $\Gamma = \overrightarrow{\Gamma} = \overleftarrow{\Gamma}$, dropping the useless time arrows.

Remark 6.1.8. *In the standard Bakry-Émery setting [4, 5], $\mathbf{X}$ is a Riemannian manifold and one considers the self-adjoint Markov diffusion generator on $L^2(\mathbf{X}, e^{-V}\,\mathrm{vol})$ given by $\widetilde{L} = \Delta - \nabla V \cdot \nabla$, with* vol *the volume measure and $V : \mathbf{X} \to \mathbb{R}$ a regular function. The usual definition of the carré du champ of $\widetilde{L}$ is $\widetilde{\Gamma}(u,v) := [\widetilde{L}(uv) - u\widetilde{L}v - v\widetilde{L}u]/2$. In the present paper, Γ is not divided by 2 and we consider $L = \widetilde{L}/2$, i.e.*

$$L = (-\nabla V \cdot \nabla + \Delta)/2, \tag{6.9}$$

which corresponds to an SDE driven by a standard Brownian motion. Consequently, $\Gamma(u,v) = \widetilde{\Gamma}(u,v) = \nabla u \cdot \nabla v$.

In general, even if R is a reversible path measure, the prescribed marginal constraints enforce a time-inhomogeneous transition mechanism: the forward and backward derivatives $(\partial_t + \overrightarrow{A}_t)_{0 \le t \le 1}$ and $(-\partial_t + \overleftarrow{A}_t)_{0 \le t \le 1}$ of P depend explicitly on t. It is known (see [25] for instance) that for any function $u : \mathbf{X} \to \mathbb{R}$ belonging to some class of regular functions to be made precise,

$$\begin{cases} \overrightarrow{A}_t u(z) & = & \overrightarrow{L}u(z) + \dfrac{\overrightarrow{\Gamma}(g_t, u)(z)}{g_t(z)}, & (t,z) \in [0,1) \times \mathbf{X}, \\ \overleftarrow{A}_t u(z) & = & \overleftarrow{L}u(z) + \dfrac{\overleftarrow{\Gamma}(f_t, u)(z)}{f_t(z)}, & (t,z) \in (0,1] \times \mathbf{X}, \end{cases} \tag{6.10}$$

where f and g are defined as in (6.4). Because of (6.6), for any t no division by zero occurs μ_t-a.e. For (6.10) to be meaningful, it is necessary that the functions f and g are regular enough for f_t and g_t to be in the domains of the carré du champ operators. In the remainder of the paper, we shall only be concerned with Brownian diffusion processes and random walks for which the class of regular functions will be specified, see Sections 6.3 and 6.4.

Going back to (6.4), we see that the processes $f_t(X_t)$ and $g_t(X_t)$ are respectively backward and forward local R-martingales. In terms of stochastic derivatives, this is equivalent to

$$\begin{cases} (-\partial_t + \overleftarrow{L})f(t,z) = 0, & 0 < t \le 1, \\ f_0, & t = 0, \end{cases} \qquad \begin{cases} (\partial_t + \overrightarrow{L})g(t,z) = 0, & 0 \le t < 1, \\ g_1, & t = 1, \end{cases} \tag{6.11}$$

where these identities hold μ_t-almost everywhere for almost all t.

Assumptions 6.1.9. *We assume that the kernels $\overrightarrow{r}, \overleftarrow{r}$ that appear in (6.4) are (a) positivity improving and (b) regularizing:*
(a) *For all $(t,z) \in (0,1) \times \mathbf{X}$, $\overrightarrow{r}(t,z;1,\cdot) \gg m$ and $\overleftarrow{r}(0,\cdot;t,z) \gg m$.*
(b) *The functions (f_0, g_1) and the kernels $\overrightarrow{r}, \overleftarrow{r}$ are such that $(t,z) \mapsto f(t,z), g(t,z)$ are twice t-differentiable* classical solutions *of the parabolic PDEs* (6.11).

Assumption (a) implies that for any $0 < t < 1$, f_t and g_t are positive everywhere. In particular, we see with (6.6) that $\mu_t \sim m$. Assumption (b) will be made precise later in specific settings. It will be used when computing time derivatives of $t \mapsto \mu_t$.

Mainly because we are going to study the relative entropy $H(\mu_t|m) = \int_{\mathbf{X}} \log(f_t g_t)\, d\mu_t$, it is sometimes worthwhile to express $\overrightarrow{A}$ and $\overleftarrow{A}$ in terms of the logarithms of f and g :

$$\begin{cases} \phi_t(z) := \log f_t(z) = \log E_R(f_0(X_0) \mid X_t = z), & (t,z) \in (0,1] \times \mathbf{X} \\ \psi_t(z) := \log g_t(z) = \log E_R(g_1(X_1) \mid X_t = z), & (t,z) \in [0,1) \times \mathbf{X}. \end{cases} \tag{6.12}$$

In analogy with the Kantorovich potentials which appear in the optimal transport theory, we call ϕ and ψ the *Schrödinger potentials*. Under Assumption 6.1.9-(b), they are classical solutions of the "second order" Hamilton-Jacobi-Bellman (HJB) equations

$$\begin{cases} (-\partial_t + \overleftarrow{B})\phi = 0, & 0 < t \le 1, \\ \phi_0 = \log f_0, & t = 0, \end{cases} \qquad \begin{cases} (\partial_t + \overrightarrow{B})\psi = 0, & 0 \le t < 1, \\ \psi_1 = \log g_1, & t = 1, \end{cases} \tag{6.13}$$

where the non-linear operators $\overrightarrow{B}$ and $\overleftarrow{B}$ are defined by

$$\begin{cases} \overrightarrow{B} u & := & e^{-u} \overrightarrow{L} e^u, \\ \overleftarrow{B} v & := & e^{-v} \overleftarrow{L} e^v, \end{cases}$$

for any functions u, v such that $e^u \in \operatorname{Dom} \overrightarrow{L}$ and $e^v \in \operatorname{Dom} \overleftarrow{L}$. Let us introduce the notation

$$\begin{cases} \overrightarrow{A}_\theta & := & \overrightarrow{L} + e^{-\theta} \overrightarrow{\Gamma}(e^\theta, \cdot) \\ \overleftarrow{A}_\theta & := & \overleftarrow{L} + e^{-\theta} \overleftarrow{\Gamma}(e^\theta, \cdot) \end{cases}$$

which allows us to rewrite (6.10) as $\overrightarrow{A}_t = \overrightarrow{A}_{\psi_t}$ and $\overleftarrow{A}_t = \overleftarrow{A}_{\phi_t}$, emphasizing their dependence on the Schrödinger potentials.

Forward-backward systems

The complete dynamics of $[\mu_0, \mu_1]^R$ is described by a forward-backward system. We see that

$$\begin{cases} (-\partial_t^- + \overrightarrow{A}_{\psi_t}^\star)\mu = 0, & 0 < t \le 1, \\ \mu_0, & t = 0, \end{cases} \quad \text{where} \quad \begin{cases} (\partial_t^+ + \overrightarrow{B})\psi = 0, & 0 \le t < 1, \\ \psi_1 = \log g_1, & t = 1, \end{cases} \tag{6.14}$$

and A^* is the algebraic adjoint of A : $\int_X Au\,d\mu =: \langle u, A^*\mu \rangle$. The evolution equation for μ is understood in the weak sense, in duality with respect to a large enough class of regular functions u. Similarly,

$$\begin{cases} (\partial_t^+ + \overleftarrow{A}^{\star}_{\phi_t})\mu = 0, & 0 \le t < 1, \\ \mu_1, & t = 1, \end{cases} \quad \text{where} \quad \begin{cases} (-\partial_t^- + \overleftarrow{B})\phi = 0, & 0 < t \le 1, \\ \phi_0 = \log f_0, & t = 0. \end{cases} \tag{6.15}$$

We have denoted ∂^+ and ∂^- the right and left time-derivatives but we will drop the superscripts + and - in the rest of the paper.
The boundary data for the systems (6.14) and (6.15) are respectively (μ_0, g_1) and (f_0, μ_1).

6.2 Second derivative of the entropy

The aim of this section is to provide basic formulas to study the convexity of the entropy

$$t \in [0,1] \mapsto H(t) := H(\mu_t|m) \in [0, \infty)$$

as a function of time, along the entropic interpolation $[\mu_0, \mu_1]^R$ associated with the (f, g)-transform P of R defined by (6.3). The interpolation $[\mu_0, \mu_1]^R$ is specified by its endpoint data (μ_0, μ_1) defined by (6.5). We have also seen in (6.14) and (6.15) that it is specified by either (μ_0, g_1) or (f_0, μ_1).

Basic rules of calculus and notation

We are going to use the subsequent rules of calculus and notation where we drop the subscript t as often as possible. Assumption 6.1.9 is used in a significant way: it allows us to work with everywhere defined derivatives.

- $\rho_t = f_t g_t\, m, \quad \phi := \log f, \quad \psi := \log g.$
  The first identity is Theorem 6.1.4. The others are notation changes well defined everywhere under Assumption 6.1.9.
- $\dot{u} := \partial_t u, \quad \dot{\mu} := \partial_t \mu, \quad \langle u, \eta \rangle := \int_X u\, d\eta.$
  These are simplifying notation.
- The first line of the following equalities is (6.10) and the other ones are definitions:
  $$\begin{cases} \overrightarrow{A} u &= \overrightarrow{L} u + \overrightarrow{\Gamma}(g, u)/g \\ \overrightarrow{B} u &:= e^{-u} \overrightarrow{L} e^u \\ \overrightarrow{C} u &:= \overrightarrow{B} u - \overrightarrow{L} u \end{cases}, \qquad \begin{cases} \overleftarrow{A} u &= \overleftarrow{L} u + \overleftarrow{\Gamma}(f, u)/f \\ \overleftarrow{B} u &:= e^{-u} \overleftarrow{L} e^u \\ \overleftarrow{C} u &:= \overleftarrow{B} u - \overleftarrow{L} u. \end{cases}$$
- $\langle u, \dot{\mu} \rangle = \langle \overrightarrow{A} u, \mu \rangle = \langle -\overleftarrow{A} u, \mu \rangle.$
  These identities hold since μ is the time-marginal flow of the Markov law P whose forward and backward derivatives are $\overrightarrow{A}$ and $\overleftarrow{A}$.

– A short way for writing (6.11) and (6.13) is:

$$\begin{cases} \dot{g} &= -\overrightarrow{L} g \\ \dot{\psi} &= -\overrightarrow{B} \psi \end{cases}, \qquad \begin{cases} \dot{f} &= \overleftarrow{L} f \\ \dot{\phi} &= \overleftarrow{B} \phi. \end{cases}$$

Entropy production

By Assumption 6.1.9, the evolution PDEs (6.11) and (6.13) are defined in the classical sense, and the functions $\overrightarrow{I}$ and $\overleftarrow{I}$ below are well defined.

Definition 6.2.1 (Forward and backward entropy production). *For each $0 \le t \le 1$, we define respectively the forward and backward entropy production at time t along the interpolation $[\mu_0, \mu_1]^R$ by*

$$\begin{cases} \overrightarrow{I}(t) &:= \int_{\mathrm{X}} \overrightarrow{\Theta} \psi_t \, d\mu_t \\ \overleftarrow{I}(t) &:= \int_{\mathrm{X}} \overleftarrow{\Theta} \phi_t \, d\mu_t \end{cases}$$

where for any regular enough function u,

$$\begin{cases} \overrightarrow{\Theta} u &:= e^{-u} \overrightarrow{\Gamma}(e^u, u) - \overrightarrow{C} u \\ \overleftarrow{\Theta} u &:= e^{-u} \overleftarrow{\Gamma}(e^u, u) - \overleftarrow{C} u. \end{cases} \tag{6.16}$$

Calling the functions $\overrightarrow{I}$ and $\overleftarrow{I}$ “entropy production” is justified by Corollary 6.2.4 below.

We shall see in Section 6.3 that in the reversible Brownian diffusion setting where R is associated with the generator (6.9) we have

$$\overrightarrow{\Theta} u = \overleftarrow{\Theta} u = \Gamma(u)/2$$

with no dependence on t and as usual, one simply writes $\Gamma(u) := \Gamma(u, u)$.

Proposition 6.2.2. *Suppose that the Assumptions 6.1.6 and 6.1.9 hold. The first derivative of $t \mapsto H(t)$ is*

$$\frac{d}{dt} H(\mu_t|m) = \overrightarrow{I}(t) - \overleftarrow{I}(t), \qquad 0 < t < 1.$$

Proof. With Theorem 6.1.4 and the relative entropy definition, we immediately see that $H(t) = \langle \log \rho, \mu \rangle = \langle \phi + \psi, \mu \rangle$. It follows from our basic rules of calculus that

$$H'(t) = \left\langle \dot{\phi} + \dot{\psi}, \mu \right\rangle + \langle \phi + \psi, \dot{\mu} \rangle = \langle \overleftarrow{B} \phi - \overrightarrow{B} \psi, \mu \rangle + \left\langle -\overleftarrow{A} \phi + \overrightarrow{A} \psi, \mu \right\rangle.$$

The point here is to apply the forward operators $\overrightarrow{A} = \overrightarrow{A}_\psi$, $\overrightarrow{B}$ to ψ and the backward operators $\overleftarrow{A} = \overleftarrow{A}_\phi$, $\overleftarrow{B}$ to ϕ. This is the desired result since $(\overrightarrow{A} - \overrightarrow{B})\psi = \overrightarrow{\Theta} \psi$ and $(\overleftarrow{A} - \overleftarrow{B})\phi = \overleftarrow{\Theta} \phi$. □

Notice that $\left\langle \dot\phi + \dot\psi, \mu \right\rangle = \langle \partial_t \log \rho, \mu\rangle = \langle \dot\rho/\rho, \mu\rangle = \langle \dot\rho, m\rangle = (d/dt)\,\langle \rho, m\rangle = 0$. Hence,

$$\left\langle \dot\phi + \dot\psi, \mu \right\rangle = \langle \overleftarrow{B}\phi - \overrightarrow{B}\psi, \mu\rangle = 0.$$

As an immediate consequence of this Proposition 6.2.2, we obtain the following Corollary 6.2.4 about the dynamics of heat flows.

Definitions 6.2.3 (Heat flows).

(a) *We call* forward heat flow *the time marginal flow of the (f, g)-transform described in Remark 6.1.2-(a). It corresponds to $f_0 = \rho_0$ and $g_1 = 1$, i.e. to $P = \rho_0(X_0)\,R$.*

(b) *We call* backward heat flow *the time marginal flow of the (f, g)-transform described in Remark 6.1.2-(b). It corresponds to $f_0 = 1$ and $g_1 = \rho_1$, i.e. to $P = \rho_1(X_1)\,R$.*

Corollary 6.2.4.

(a) *Along any forward heat flow $(\mu_t)_{0\le t\le 1}$, we have: $\frac{d}{dt}H(\mu_t|m) = -\overleftarrow{I}(t)$, $0 < t < 1$.*

(b) *Along any backward heat flow $(\mu_t)_{0\le t\le 1}$, we have: $\frac{d}{dt}H(\mu_t|m) = \overrightarrow{I}(t)$, $0 < t < 1$.*

Second derivative

The computations in this subsection are informal and the underlying regularity hypotheses are kept fuzzy. We assume that the Markov measure R is *nice enough* for the Schrödinger potentials ϕ and ψ to be in the domains of the compositions of the operators L, Γ, A, B and C which are going to appear below. To keep formulas to a reasonable size, we have assumed that R is m-stationary. The informal results that are stated below in Claims 1 and 2 will turn later in Sections 6.3 and 6.4 into formal statements in specific settings. We introduce

$$\begin{cases} \overrightarrow{\Theta}_2 u &:= \overrightarrow{L}\overrightarrow{\Theta} u + e^{-u}\overrightarrow{\Gamma}\left(e^u, \overrightarrow{\Theta} u\right) + e^{-u}\overrightarrow{\Gamma}(e^u, u)\overrightarrow{B} u - e^{-u}\overrightarrow{\Gamma}(e^u\overrightarrow{B} u, u), \\ \overleftarrow{\Theta}_2 u &:= \overleftarrow{L}\overleftarrow{\Theta} u + e^{-u}\overleftarrow{\Gamma}\left(e^u, \overleftarrow{\Theta} u\right) + e^{-u}\overleftarrow{\Gamma}(e^u, u)\overleftarrow{B} u - e^{-u}\overleftarrow{\Gamma}(e^u\overleftarrow{B} u, u), \end{cases} \tag{6.17}$$

provided that the function u is such that these expressions are well defined. It will be shown in Section 6.3, see (6.30), that in the Brownian diffusion setting where R is associated with the generators $\begin{cases} \overrightarrow{L} &= \overrightarrow{b}\cdot\nabla + \Delta/2 \\ \overleftarrow{L} &= \overleftarrow{b}\cdot\nabla + \Delta/2 \end{cases}$, we have $\begin{cases} \overrightarrow{\Theta}_2 &= \overrightarrow{\Gamma}_2/2 \\ \overleftarrow{\Theta}_2 &= \overleftarrow{\Gamma}_2/2 \end{cases}$, where

$$\begin{cases} \overrightarrow{\Gamma}_2(u) &:= \overrightarrow{L}\overrightarrow{\Gamma}(u) - 2\overrightarrow{\Gamma}(\overrightarrow{L} u, u), \\ \overleftarrow{\Gamma}_2(u) &:= \overleftarrow{L}\overleftarrow{\Gamma}(u) - 2\overleftarrow{\Gamma}(\overleftarrow{L} u, u), \end{cases} \tag{6.18}$$

are the forward and backward iterated carré du champ operators.

Remark 6.2.5. *We keep the notation of Remark 6.1.8. In the standard Bakry-Émery setting, the iterated carré du champ of $\widetilde{L} = \Delta - \nabla V\cdot\nabla$ is $\widetilde{\Gamma}_2(u) = [\widetilde{L}(\widetilde{\Gamma}(u)) - 2\widetilde{\Gamma}(u, \widetilde{L}u)]/2$.*

In the present paper, we consider $L = \widetilde{L}/2$ instead of $\widetilde{L}$ and we have already checked in Remark 6.1.8 that $\widetilde{\Gamma} = \Gamma$. Consequently, $\Gamma_2 = \widetilde{\Gamma}_2$.

Introducing these definitions is justified by the following claim.

Claim 1 (Informal result). *Assume that R is m-stationary. Then,*

$$\frac{d^2}{dt^2}H(\mu_t|m) = \left\langle \overleftarrow{\Theta}_2\phi_t + \overrightarrow{\Theta}_2\psi_t, \mu_t \right\rangle, \quad \forall 0 < t < 1.$$

Proof. Starting from $H(t) = \langle \rho \log \rho, m\rangle$ gives $H'(t) = \langle 1 + \log \rho, \dot{\mu}\rangle = \langle \log \rho, \dot{\mu}\rangle = \left\langle -\overleftarrow{A}\phi + \overrightarrow{A}\psi, \mu \right\rangle$ and $H''(t) = \frac{d}{dt}\left\langle -\overleftarrow{A}\phi, \mu \right\rangle + \frac{d}{dt}\left\langle \overrightarrow{A}\psi, \mu \right\rangle$ with

$$\begin{aligned}
\frac{d}{dt}\left\langle -\overleftarrow{A}\phi, \mu \right\rangle &= \left\langle \overleftarrow{A}^2\phi - \overleftarrow{A}\dot{\phi} - \dot{\overleftarrow{A}}\phi, \mu \right\rangle \\
&= \left\langle \overleftarrow{A}(\overleftarrow{A} - \overleftarrow{B})\phi - \left(\frac{\overleftarrow{\Gamma}(\dot{f}, \cdot)}{f} - \frac{\dot{f}}{f^2}\overleftarrow{\Gamma}(f, \cdot) \right)\phi, \mu \right\rangle \\
&= \left\langle \overleftarrow{A}(\overleftarrow{A} - \overleftarrow{B})\phi - \left(\frac{\overleftarrow{\Gamma}(\overleftarrow{L}f, \cdot)}{f} - \frac{\overleftarrow{L}f}{f^2}\overleftarrow{\Gamma}(f, \cdot) \right)\phi, \mu \right\rangle \\
&= \left\langle \overleftarrow{A}\left(\frac{\overleftarrow{\Gamma}(f, \cdot)}{f} - \overleftarrow{C} \right)\phi - \left(\frac{\overleftarrow{\Gamma}(\overleftarrow{L}f, \cdot)}{f} - \overleftarrow{B}\phi\frac{\overleftarrow{\Gamma}(f, \cdot)}{f} \right)\phi, \mu \right\rangle \\
&= \left\langle \overleftarrow{\Theta}_2\phi, \mu \right\rangle.
\end{aligned}$$

Similarly, we obtain

$$\begin{aligned}
\frac{d}{dt}\left\langle \overrightarrow{A}\psi, \mu \right\rangle &= \left\langle \overrightarrow{A}^2\psi + \overrightarrow{A}\dot{\psi} + \dot{\overrightarrow{A}}\psi, \mu \right\rangle \\
&= \left\langle \overrightarrow{A}(\overrightarrow{A} - \overrightarrow{B})\psi + \left(\frac{\overrightarrow{\Gamma}(\dot{g}, \cdot)}{g} - \frac{\dot{g}}{g^2}\overrightarrow{\Gamma}(g, \cdot) \right)\psi, \mu \right\rangle \\
&= \left\langle \overrightarrow{A}(\overrightarrow{A} - \overrightarrow{B})\psi - \left(\frac{\overrightarrow{\Gamma}(\overrightarrow{L}g, \cdot)}{g} - \frac{\overrightarrow{L}g}{g^2}\overrightarrow{\Gamma}(g, \cdot) \right)\psi, \mu \right\rangle \\
&= \left\langle \overrightarrow{\Theta}_2\psi, \mu \right\rangle, \qquad (6.19)
\end{aligned}$$

which completes the proof of the claim. □

Gathering Proposition 6.2.2 and Claim 1, we obtain the following

Claim 2 (Informal result). *When R is m-stationary, for all* $0 < t < 1$,

$$\begin{cases}
\dfrac{d}{dt}H(\mu_t|m) = \left\langle \overrightarrow{\Theta}\psi_t - \overleftarrow{\Theta}\phi_t, \mu_t \right\rangle, \\[2ex]
\dfrac{d^2}{dt^2}H(\mu_t|m) = \left\langle \overrightarrow{\Theta}_2\psi_t + \overleftarrow{\Theta}_2\phi_t, \mu_t \right\rangle.
\end{cases}$$

6.3 Brownian diffusion process

As a first step, we compute informally the operators Θ and Θ_2 associated with the Brownian diffusion process on $\boldsymbol{X} = \mathbb{R}^n$ whose forward and backward generators are given for all $0 \le t \le 1$ by

$$\overrightarrow{L} = \overrightarrow{b} \cdot \nabla + \Delta/2, \qquad \overleftarrow{L} = \overleftarrow{b} \cdot \nabla + \Delta/2. \tag{6.20}$$

Here, $z \mapsto \overrightarrow{b}(z), \overleftarrow{b}(z) \in \mathbb{R}^n$ are the forward and backward drift vector fields. The term "informally" means that we suppose that $\overrightarrow{b}$ and $\overleftarrow{b}$ satisfy some unstated growth and regularity properties which ensure the existence of R and also that Assumptions 6.1.9 are satisfied.

Then, we consider reversible Brownian diffusion processes on a compact manifold.

Dynamics of the entropic interpolations

The associated nonlinear operators are $\overrightarrow{B}u = \Delta u/2 + \overrightarrow{b} \cdot \nabla u + |\nabla u|^2/2$, $\overleftarrow{B}u = \Delta u/2 + \overleftarrow{b} \cdot \nabla u + |\nabla u|^2/2$ and $\overrightarrow{\Gamma}(u,v) = \overleftarrow{\Gamma}(u,v) = \nabla u \cdot \nabla v$ for any t and $u, v \in \mathcal{C}^2(\mathbb{R}^n)$. The expressions

$$\begin{cases} \overrightarrow{A}_t &=& \Delta/2 + (\overrightarrow{b} + \nabla\psi_t)\cdot\nabla, \\ \overleftarrow{A}_t &=& \Delta/2 + (\overleftarrow{b} + \nabla\varphi_t)\cdot\nabla, \end{cases}$$

of the forward and backward derivatives tell us that the density $\mu_t(z) := d\mu_t/dz$ solves the following forward-backward system of parabolic PDEs

$$\begin{cases} (\partial_t - \Delta/2)\mu_t(z) + \nabla\cdot(\mu_t(\overrightarrow{b} + \nabla\psi_t))(z) = 0, & (t,z) \in (0,1] \times \boldsymbol{X} \\ \mu_0, & t = 0, \end{cases}$$

where ψ solves the HJB equation

$$\begin{cases} (\partial_t + \Delta/2 + \overrightarrow{b}\cdot\nabla)\psi_t(z) + |\nabla\psi_t(z)|^2/2 = 0, & (t,z) \in [0,1) \times \boldsymbol{X} \\ \psi_1 = \log g_1, & t = 1. \end{cases} \tag{6.21}$$

In the reverse time direction of time, we obtain

$$\begin{cases} (-\partial_t - \Delta/2)\mu_t(z) + \nabla\cdot(\mu_t(\overleftarrow{b} + \nabla\varphi_t))(z) = 0, & (t,z) \in [0,1) \times \boldsymbol{X} \\ \mu_1, & t = 1, \end{cases}$$

where φ solves the HJB equation

$$\begin{cases} (-\partial_t + \Delta/2 + \overleftarrow{b}\cdot\nabla)\varphi_t(z) + |\nabla\varphi_t(z)|^2/2 = 0, & (t,z) \in (0,1] \times \boldsymbol{X} \\ \varphi_0 = \log f_0, & t = 0. \end{cases} \tag{6.22}$$

We fix $\log g_1 = -\infty$ on the set where g_1 vanishes, and the boundary condition at time $t = 1$ must be understood as $\lim_{t\uparrow 1}\psi_t = \log g_1$. Similarly, we have also $\lim_{t\downarrow 0}\varphi_t = \log f_0$ in $[-\infty, \infty)$.

Because of the stochastic representation formulas (6.4), the functions ϕ and ψ are the unique viscosity solutions of the above Hamilton-Jacobi-Bellman equations, see [20, Thm. II.5.1]. The existence of these solutions is ensured by the Assumptions 6.1.6.

Remarks 6.3.1.

(a) *In general, ϕ and ψ might not be classical solutions of their respective HJB equations; the gradients $\nabla\psi$ and $\nabla\phi$ are not defined in the usual sense. One has to consider the notion of P-extended gradient: $\widetilde{\nabla}^P$, that is introduced in [25]. The complete description of the Markov dynamics of P is a special case of [25, Thm. 5.4]: the forward and backward drift vector fields of the canonical process under P are respectively $\overrightarrow{b} + \widetilde{\nabla}^P\psi_t$ and $\overleftarrow{b} + \widetilde{\nabla}^P\phi_t$.*

(b) *Suppose that the stationary measure m associated with R is equivalent to the Lebesgue measure. Then, the forward and backward drift fields $\overrightarrow{b}$ and $\overleftarrow{b}$ are related to m as follows. Particularizing the evolution equations with $\mu_t = m$ and $\nabla\psi = \nabla\phi = 0$, we see that the requirement that R is m-stationary implies that*

$$\begin{cases} \nabla\cdot(m\,\{v^{\mathrm{os}} - \nabla\log\sqrt{m}\}) & = & 0 \\ \nabla\cdot(m\,v^{\mathrm{cu}}) & = & 0 \end{cases} \tag{6.23}$$

for all t, where $m(x) := dm/dx$, these identities are considered in the weak sense and

$$\begin{cases} v^{\mathrm{os}} & := & (\overrightarrow{b} + \overleftarrow{b})/2 \\ v^{\mathrm{cu}} & := & (\overrightarrow{b} - \overleftarrow{b})/2 \end{cases}$$

are respectively the osmotic and the (forward) current velocities of R that were introduced by E. Nelson in [37]. In fact, the first equation in (6.23) is satisfied in a stronger way, since the duality formula associated to time reversal [21] is

$$v^{\mathrm{os}} = \nabla\log\sqrt{m}. \tag{6.24}$$

An interesting situation is given by $\overrightarrow{b} = -\nabla V/2 + b_\perp$ where $V : \mathbb{R}^n \to \mathbb{R}$ is $\mathcal{C}^2$ and the drift vector field $b_\perp$ satisfies

$$\nabla\cdot(e^{-V}b_\perp) = 0. \tag{6.25}$$

It is easily seen that

$$m = e^{-V}\mathrm{Leb}$$

is the stationary measure of R. Moreover, with (6.24), we also obtain

$$\begin{cases} \overrightarrow{b} & = & -\nabla V/2 + b_\perp \\ \overleftarrow{b} & = & -\nabla V/2 - b_\perp \end{cases}, \qquad \begin{cases} v^{\mathrm{os}} & = & -\nabla V/2 \\ v^{\mathrm{cu}} & = & b_\perp \end{cases}. \tag{6.26}$$

In dimension 2, choosing $b_\perp = e^V(-\partial_y U, \partial_x U)$ with $U : \mathbb{R}^2 \to \mathbb{R}$ a $\mathcal{C}^2$-regular function, solves (6.25). Regardless of the dimension, $b_\perp = e^V S\nabla U$, where S is a constant skew-symmetric matrix and $U : \mathbb{R}^n \to \mathbb{R}$ a $\mathcal{C}^2$-regular function, is also a possible choice. In dimension 3 one can take $b_\perp = e^V \nabla \wedge A$ where $A : \mathbb{R}^3 \to \mathbb{R}^3$ is a $\mathcal{C}^2$-regular vector field.

Computing Θ and Θ_2

Our aim is to compute the operators $\overrightarrow{\Theta}$, $\overleftarrow{\Theta}$, $\overrightarrow{\Theta}_2$ and $\overleftarrow{\Theta}_2$ of the Brownian diffusion process R with the forward and backward derivatives given by (6.20) and (6.26):

$$\begin{cases} \overrightarrow{L} &= (-\nabla V \cdot \nabla + \Delta)/2 + b_\perp \cdot \nabla, \\ \overleftarrow{L} &= (-\nabla V \cdot \nabla + \Delta)/2 - b_\perp \cdot \nabla, \end{cases} \tag{6.27}$$

where $\nabla \cdot (e^V b_\perp) = 0$. For simplicity, we drop the arrows and the index t during the intermediate computations.

Computation of Θ

We have $\Gamma(u) = e^{-u}\Gamma(e^u, u) = |\nabla u|^2$ and $Cu = |\nabla u|^2/2$. Therefore, $\Theta u = Cu = \Gamma(u)/2 = |\nabla u|^2/2$. This gives

$$\begin{cases} \overrightarrow{\Theta}_t \psi_t &= |\nabla \psi_t|^2/2, \\ \overleftarrow{\Theta}_t \phi_t &= |\nabla \phi_t|^2/2, \end{cases} \tag{6.28}$$

and the entropy productions are written as follows, for all $0 < t < 1$,

$$\begin{cases} \overrightarrow{I}(t) &= \dfrac{1}{2} \displaystyle\int_X |\nabla \psi_t|^2 \, d\mu_t, \\ \overleftarrow{I}(t) &= \dfrac{1}{2} \displaystyle\int_X |\nabla \phi_t|^2 \, d\mu_t. \end{cases} \tag{6.29}$$

Recall that $f_t, g_t > 0$, μ_t-a.e. : for all $0 < t < 1$.

Computation of Θ_2

Since R is a diffusion, Γ is a derivation i.e. $\Gamma(uv, w) = u\Gamma(v, w) + v\Gamma(u, w)$, for any regular enough functions u, v and w. In particular, the two last terms in the expression of $\Theta_2 u$ simplify:

$$e^{-u}\Gamma(e^u Bu, u) - e^{-u}\Gamma(e^u, u)Bu = \Gamma(Bu, u) = \Gamma(Lu, u) + \Gamma(Cu, u),$$

and we get $\Theta_2 u = L\Gamma(u)/2 + \Gamma(Cu, u) - \Gamma(Lu, u) - \Gamma(Cu, u) = L\Gamma(u)/2 - \Gamma(Lu, u)$. This means that

$$\begin{cases} \overrightarrow{\Theta}_2 &= \overrightarrow{\Gamma}_2/2, \\ \overleftarrow{\Theta}_2 &= \overleftarrow{\Gamma}_2/2. \end{cases} \tag{6.30}$$

They are precisely half the iterated carré du champ operators $\overrightarrow{\Gamma}_2$ and $\overleftarrow{\Gamma}_2$ defined at (6.18). The iterated carré du champ Γ_2^o of $L^o = \Delta/2$ is

$$\Gamma_2^o(u) = \|\nabla^2 u\|_{\mathrm{HS}}^2 = \sum_{i,j} (\partial_{ij} u)^2$$

where $\nabla^2 u$ is the Hessian of u and $\|\nabla^2 u\|_{\mathrm{HS}}^2 = \mathrm{tr}((\nabla^2 u)^2)$ is its squared Hilbert-Schmidt norm. As $\Gamma_2(u) = \Gamma_2^o(u) - 2\nabla b(\nabla u, \nabla u)$ where $2\nabla b(\nabla u, \nabla u) = 2\sum_{i,j} \partial_i b_j\, \partial_i u \partial_j u = [\nabla + \nabla^*] b(\nabla u, \nabla u)$ with $\nabla^* b$ the adjoint of ∇b, it follows that

$$\begin{cases} \overrightarrow{\Gamma}_2(u) &= \|\nabla^2 u\|_{\mathrm{HS}}^2 + (\nabla^2 V - [\nabla + \nabla^*] b_\perp)(\nabla u, \nabla u), \\ \overleftarrow{\Gamma}_2(u) &= \|\nabla^2 u\|_{\mathrm{HS}}^2 + (\nabla^2 V + [\nabla + \nabla^*] b_\perp)(\nabla u, \nabla u). \end{cases} \tag{6.31}$$

Regularity problems

To make Claim 2 a rigorous statement, one needs to rely upon regularity hypotheses such as Assumptions 6.1.9. *If one knows that f_0, g_1, $\overrightarrow{b}$ and $\overleftarrow{b}$ are such that the linear parabolic equations* (6.11) *admit positive $\mathcal{C}^{2,2}((0,1) \times \boldsymbol{X})$-regular solutions then we are done.* Verifying this is a standard (rather difficult) problem which is solved under various hypotheses. Rather than giving details about this PDE problem which can be solved by means of Malliavin calculus, we present an example in next subsection.

Working with classical solutions is probably too demanding. The operators Θ and Θ_2 are functions of ∇u and the notion of gradient can be extended as in [25] in connection with the notion of extended stochastic derivatives, see Remark 6.3.1-(a). Furthermore, when working with integrals with respect to time, for instance when considering integrals of the entropy production (see Section 6.5), or in situations where $H(t)$ is known to be twice differentiable almost everywhere (e.g. $H(t)$ is the sum of a twice differentiable function and a convex function), it would be enough to consider dt-almost everywhere defined extended gradients. This program is not initiated in the present paper.

Reversible Brownian diffusion process on a Riemannian manifold $\boldsymbol{X}$

We give some detail about the standard Bakry-Émery setting that already appeared in Remark 6.1.8. It corresponds to the case where $b_\perp = 0$. Let us take

$$m = e^{-V}\,\mathrm{vol} \tag{6.32}$$

where $V \in \mathcal{C}^2$ satisfies $\int_{\boldsymbol{X}} e^{-V(x)}\,\mathrm{vol}(dx) < \infty$. The m-reversible Brownian diffusion process $R \in \mathrm{M}_+(\Omega)$ is the Markov measure with the initial (reversing) measure m and the semigroup generator

$$L = (-\nabla V \cdot \nabla + \Delta)/2.$$

The reversible Brownian motion $R^o \in \mathrm{M}_+(\Omega)$ corresponds to $V = 0$. Its stochastic derivatives are

$$\overrightarrow{L}^o = \overleftarrow{L}^o = \Delta/2 \tag{6.33}$$

and vol is its reversing measure. Since R^o is the unique solution to its own martingale problem, there is a unique R which is both absolutely continuous with respect to R^o and m-reversible.

Lemma 6.3.2. *In fact R is specified by*

$$R = \exp\left(-[V(X_0) + V(X_1)]/2 + \int_{[0,1]} \left[\Delta V(X_t)/4 - |\nabla V(X_t)|^2/8\right] dt \right) R^o.$$

Proof. To see this, let us define $\hat{R}$ by means of $d\hat{R}/dR^o := e^{-V(X_0)} Z_1$ where $Z_t = \exp\left(\int_0^t -\frac{\nabla V}{2}(X_s)\cdot dX_s - \frac{1}{2}\int_0^t |\frac{\nabla V}{2}(X_s)|^2\right)$ is a local positive forward R^o-martingale. Since $\int_{\boldsymbol{X}} e^{-V(z)}\,\mathrm{vol}(dz) < \infty$, it follows that $t \mapsto e^{-V(X_0)}Z_t$ is a forward R^o-supermartingale. In particular its expectation $t \mapsto E_{R^o}[e^{-V(X_0)}Z_t]$ is a decreasing function so that $\hat{R}(\Omega) = E_{R^o}(e^{-V(X_0)}Z_1) \le E_{R^o}(e^{-V(X_0)}Z_0) = \int_{\boldsymbol{X}} e^{-V(z)}\,\mathrm{vol}(dz) < \infty$.
On the other hand, since both R^o and $d\hat{R}/dR^o$ are invariant with respect to the time reversal X^*: $X_t^* := X_{1-t,\, t\in[0,1]}$, $\hat{R}$ is also invariant with respect to time reversal: $(X^*)_\#\hat{R} = \hat{R}$. In particular, its endpoint marginals are equal: $\hat{R}_0 = \hat{R}_1$. Consequently, $\hat{R}_{[0,t]}$ doesn't send mass to a cemetery point † outside $\boldsymbol{X}$ as time increases, for otherwise its terminal marginal $\hat{R}_1$ would give a positive mass to †, in contradiction with $\hat{R}_0(\dagger) = 0$ and $\hat{R}_0 = \hat{R}_1$. Hence, Z is a genuine forward R^o-martingale. With Itô's formula, we see that $dZ_t = -Z_t\nabla\frac{V}{2}(X_t)\cdot dX_t$, R^o-a.e. and by Girsanov's theory we know that $\hat{R}$ is a (unique) solution to the martingale problem associated with the generator $L = (-\nabla V\cdot\nabla + \Delta)/2$. Finally, we take $R = \hat{R}$ and it is easy to check that L is symmetric in $L^2(m)$, which implies that m is reversing. □

Remark 6.3.3. *When L is given by* (6.33), *for any non-zero nonnegative functions* $f_0, g_1 \in L^2(\mathrm{vol})$, *the smoothing effect of the heat kernels* $\overrightarrow{r}$ *and* $\overleftarrow{r}$ *in* (6.4) *allows us to define* classical *gradients* $\nabla\psi_t$ *and* $\nabla\phi_t$ *for all t in* $[0, 1)$ *and* $(0, 1]$ *respectively. We see that* $\nabla\psi_t$ *and* $\nabla\phi_t$ *are the forward and backward drift vector fields of the canonical process under P.*

The next result proposes a general setting where Assumptions 6.1.9 are satisfied. The manifold $\boldsymbol{X}$ is assumed to be compact to avoid integrability troubles.

Proposition 6.3.4. *Suppose that* $\boldsymbol{X}$ *is a compact Riemannian manifold without boundary and* $V : \boldsymbol{X} \to \mathbb{R}$ *is* $\mathcal{C}^4$*-regular. Then, for any* $\mathcal{C}^2$*-regular function* $u : \boldsymbol{X} \to \mathbb{R}$, *the function* $(t, x) \mapsto u(t, x) := E_R[u(X_t) \mid X_0 = x]$ *belongs to* $\mathcal{C}^{1,2}((0, 1) \times \boldsymbol{X})$.

In particular, if in addition $\boldsymbol{X}$ *is assumed to be connected and* f_0 *and* g_1 *are nonzero nonnegative* $\mathcal{C}^2$*-regular functions, the functions* f *and* g *defined in* (6.4) *and their logarithms* ϕ *and* ψ *are classical solutions of* (6.11) *and* (6.13).

Proof. For any path $\omega \in \Omega$ and any $0 \le t \le 1$, we denote

$$Z_t(\omega) := \exp\left(-[V(\omega_0)+V(\omega_t)]/2 + \int_0^t \left[\Delta V(\omega_s)/4 - |\nabla V(\omega_s)|^2/8\right] ds\right).$$

With Lemma 6.3.2 we see that for all $0 \le t \le 1$, $Z_t = \frac{dR_{[0,t]}}{dR^o_{[0,t]}}$ and

$$u(t,x) = \frac{E_{R^o}[u(X_t)Z_1 \mid X_0 = x]}{E_{R^o}[Z_1 \mid X_0 = x]} = e^{V(x)}\mathbb{E}[u(B^x_t)Z_t(B^x)]$$

where $(B^x_s)_{0 \le s \le 1}$ is a Brownian motion starting from x under some abstract probability measure whose expectation is denoted by $\mathbb{E}$. By means of parallel transport, it is possible to build on any small enough neighborhood U of x a coupling such that $x' \in U \mapsto B^{x'}$ is almost surely continuous with respect to the uniform topology on Ω. This coupling corresponds to $B^x = x + B^0$ in the Euclidean case. The announced x-regularity is a consequence of our assumptions which allow us to differentiate (in the usual deterministic sense) in the variable x under the expectation sign $\mathbb{E}$. On the other hand, the t-regularity is a consequence of stochastic differentiation: apply Itô's formula to $u(B^x_t)$ and take advantage of the martingale property of $Z_t(B^x)$.

With regard to the last statement, the connectivity assumption implies the positivity of f and g on $(0,1) \times \boldsymbol{X}$ if f_0 and g_1 are nonnegative and not vanishing everywhere. □

We gave detail on the proof of Proposition 6.3.4 because of its ease. Nonetheless, in view of Remark 6.3.3, the requirement that f_0 and g_1 are $\mathcal{C}^2$ is not optimal. However, this restriction will not be harmful when investigating convexity properties of the entropy along interpolations.

In view of Proposition 6.3.4, we remark that under its assumptions the functions f_t, g_t belong to the domain of the carré du champ operator. Hence, for any $0 < t < 1$, the stochastic derivatives of P are well-defined on $\mathcal{C}^2(\boldsymbol{X})$ and equal to

$$\begin{cases} \overrightarrow{A}_t &= \Delta/2 + \nabla(-V/2 + \psi_t)\cdot\nabla, \\ \overleftarrow{A}_t &= \Delta/2 + \nabla(-V/2 + \phi_t)\cdot\nabla. \end{cases}$$

Here, ψ and ϕ are respectively the classical solutions (compare with Remark 6.3.1-(a)) of the HJB equations (6.21) and (6.22) with $\overrightarrow{b} = \overleftarrow{b} = -\nabla V/2$.

Bochner's formula relates the iterated carré du champ Γ^o_2 of $L^o = \Delta/2$ and the Ricci curvature:

$$\Gamma^o_2(u) = \|\nabla^2 u\|^2_{\mathrm{HS}} + \mathrm{Ric}(\nabla u) \tag{6.34}$$

where $\nabla^2 u$ is the Hessian of u and $\|\nabla^2 u\|^2_{\mathrm{HS}} = \mathrm{tr}((\nabla^2 u)^2)$ is its squared Hilbert-Schmidt norm. As $\Gamma_2(u) = \Gamma_2^o(u) - [\nabla + \nabla^*]b(\nabla u)$, it follows that

$$\Gamma_2(u) = \|\nabla^2 u\|^2_{\mathrm{HS}} + (\mathrm{Ric} + \nabla^2 V)(\nabla u). \tag{6.35}$$

Theorem 6.3.5 (Reversible Brownian diffusion process). *Let the reference Markov measure R be associated with L given in (6.9) on a compact connected Riemannian manifold $\boldsymbol{X}$ without boundary. It is assumed that $V : \boldsymbol{X} \to \mathbb{R}$ is $\mathcal{C}^4$-regular and f_0, g_1 are non-zero nonnegative $\mathcal{C}^2$-regular functions. Then, along the entropic interpolation $[\mu_0, \mu_1]^R$ associated with R, f_0 and g_1, we have for all $0 < t < 1$,*

$$\begin{aligned}\frac{d}{dt}H(\mu_t|e^{-V}\,\mathrm{vol}) &= \left\langle \frac{1}{2}\left(|\nabla\psi_t|^2 - |\nabla\varphi_t|^2\right), \mu_t\right\rangle,\\ \frac{d^2}{dt^2}H(\mu_t|e^{-V}\,\mathrm{vol}) &= \left\langle \frac{1}{2}\left(\Gamma_2(\psi_t) + \Gamma_2(\varphi_t)\right), \mu_t\right\rangle\end{aligned}$$

where ψ and φ are the classical solutions of the HJB equations (6.21) *and* (6.22) *with $\overrightarrow{b} = \overleftarrow{b} = -\nabla V/2$, $\Gamma(u) = |\nabla u|^2$ and Γ_2 is given by* (6.35).

Proof. The assumptions imply that f_0 and g_1 satisfy (6.8), and they allow us to apply Proposition 6.3.4; hence, Assumptions 6.1.9 are satisfied and Claim 2 is a rigorous result. □

6.4 Random walk on a graph

Now we take as our reference measure R a continuous-time Markov process on a countable state space $\boldsymbol{X}$ with a graph structure $(\boldsymbol{X}, \sim)$. The set $\boldsymbol{X}$ of all vertices is equipped with the graph relation $x \sim y$ which signifies that x and y are adjacent, i.e. $\{x, y\}$ is a non-directed edge. The degree $n_x := \#\{y \in \boldsymbol{X}, x \sim y\}$ of each $x \in \boldsymbol{X}$ is the number of its neighbors. It is assumed that $(\boldsymbol{X}, \sim)$ is locally finite, i.e. $n_x < \infty$ for all $x \in \boldsymbol{X}$, and also that $(\boldsymbol{X}, \sim)$ is connected. This means that for any couple $(x, y) \in \boldsymbol{X}^2$ of different states, there exists a finite chain $(z_1, \dots, z_k)$ in $\boldsymbol{X}$ such that $x \sim z_1 \sim z_2 \cdots \sim z_k \sim y$. In particular $n_x \geq 1$, for all $x \in \boldsymbol{X}$.

Dynamics of the entropic interpolation

A (time homogeneous) random walk on $(\boldsymbol{X}, \sim)$ is a Markov measure with forward derivative $\partial + \overrightarrow{L}$ defined for any function $u \in \mathbb{R}^{\boldsymbol{X}}$ by

$$\overrightarrow{L}u(x) = \sum_{y:x\sim y}(u_y - u_x)\overrightarrow{J}_x(y) =: \int_{\boldsymbol{X}} D_x u\, d\overrightarrow{J}_x, \qquad 0 \leq t < 1, x \in \boldsymbol{X},$$

where $\overrightarrow{J}_x(y)$ is the instantaneous frequency of forward jumps from x to y,

$$D_X u(y) = Du(x,y) := u(y) - u(x)$$

is the discrete gradient and

$$\overrightarrow{J}_x = \sum_{y:x\sim y} \overrightarrow{J}_x(y)\,\delta_y \in \mathrm{M}_+(\boldsymbol{X}), \qquad 0 \le t < 1,\ x \in \boldsymbol{X}$$

is the forward jump kernel. Similarly, one denotes its backward derivative by

$$\overleftarrow{L}u(x) = \int_{\boldsymbol{X}} D_X u\, d\overleftarrow{J}_x, \qquad x \in \boldsymbol{X}, u \in \mathbb{R}^{\boldsymbol{X}}.$$

For the m-stationary Markov measure R with stochastic derivatives $\overrightarrow{L}$ and $\overleftarrow{L}$, the time-reversal duality formula is

$$m(x)\overrightarrow{J}_x(y) = m(y)\overleftarrow{J}_y(x), \quad \forall x \sim y \in \boldsymbol{X}.$$

The expressions $\overrightarrow{L}u$ and $\overleftarrow{L}u$ can be seen as the matrices

$$\begin{cases} \overrightarrow{L} = \left(\mathbf{1}_{\{x\sim y\}}\overrightarrow{J}_x(y) - \mathbf{1}_{\{x=y\}}\overrightarrow{J}_x(\boldsymbol{X})\right)_{x,y\in\boldsymbol{X}} \\ \overleftarrow{L} = \left(\mathbf{1}_{\{x\sim y\}}\overleftarrow{J}_x(y) - \mathbf{1}_{\{x=y\}}\overleftarrow{J}_x(\boldsymbol{X})\right)_{x,y\in\boldsymbol{X}} \end{cases} \tag{6.36}$$

acting on the column vector $u = [u_x]_{x\in\boldsymbol{X}}$. Therefore, the solutions f and g of (6.11) are

$$f(t) = e^{t\overleftarrow{L}} f_0, \qquad g(t) = e^{(1-t)\overrightarrow{L}} g_1,$$

where $f : t \in [0,1] \mapsto [f_x]_{x\in\boldsymbol{X}}(t) \in \mathbb{R}^{\boldsymbol{X}}$ and $g : t \in [0,1] \mapsto [g_x]_{x\in\boldsymbol{X}}(t) \in \mathbb{R}^{\boldsymbol{X}}$ are column vectors, whenever these exponential matrices are well-defined.

Let us compute the operators B, Γ and A. By a direct computation, we obtain for all $0 < t < 1$ and $x \in \boldsymbol{X}$,

$$\begin{cases} \overrightarrow{B}u(x) &= \int_{\boldsymbol{X}} (e^{D_X u} - 1)\, d\overrightarrow{J}_x, \\ \overleftarrow{B}u(x) &= \int_{\boldsymbol{X}} (e^{D_X u} - 1)\, d\overleftarrow{J}_x, \end{cases} \qquad \begin{cases} \overrightarrow{\Gamma}(u,v)(x) &= \int_{\boldsymbol{X}} D_X u D_X v\, d\overrightarrow{J}_x, \\ \overleftarrow{\Gamma}(u,v)(x) &= \int_{\boldsymbol{X}} D_X u D_X v\, d\overleftarrow{J}_x, \end{cases}$$

and

$$\begin{cases} \overrightarrow{A}_t u(x) &= \int_{\boldsymbol{X}} D_X u\, e^{D_X \psi_t}\, d\overrightarrow{J}_x &= \sum_{y:x\sim y} [u(y) - u(x)]\, \dfrac{g_t(y)}{g_t(x)}\, \overrightarrow{J}_x(y), \\ \overleftarrow{A}_t u(x) &= \int_{\boldsymbol{X}} D_X u\, e^{D_X \varphi_t}\, d\overleftarrow{J}_x &= \sum_{y:x\sim y} [u(y) - u(x)]\, \dfrac{f_t(y)}{f_t(x)}\, \overleftarrow{J}_x(y). \end{cases}$$

The matrix representation of these operators is

$$\begin{cases} \overrightarrow{A}_t = \overrightarrow{A}(g_t) = \left(\mathbf{1}_{\{x\sim y\}} \dfrac{g_t(y)}{g_t(x)} \overrightarrow{J}_x(y) - \mathbf{1}_{\{x=y\}} \sum_{z:z\sim x} \dfrac{g_t(z)}{g_t(x)} \overrightarrow{J}_x(z)\right)_{x,y\in\boldsymbol{X}} \\ \overleftarrow{A}_t = \overleftarrow{A}(f_t) = \left(\mathbf{1}_{\{x\sim y\}} \dfrac{f_t(y)}{f_t(x)} \overleftarrow{J}_x(y) - \mathbf{1}_{\{x=y\}} \sum_{z:z\sim x} \dfrac{f_t(z)}{f_t(x)} \overleftarrow{J}_x(z)\right)_{x,y\in\boldsymbol{X}} \end{cases}.$$

The forward-backward systems describing the evolution of $[\mu_0, \mu_1]^R$ are

$$\begin{aligned}\mu_t &= \mu_0 \overrightarrow{\exp}\left(\int_0^t \overrightarrow{A}\left(\overleftarrow{\exp}\left(\int_s^1 \overrightarrow{L}_r\,dr\right) g_1\right) ds\right)\\ &= \mu_1 \overleftarrow{\exp}\left(\int_t^1 \overleftarrow{A}\left(\overrightarrow{\exp}\left(\int_0^s \overleftarrow{L}_r\,dr\right) f_0\right) ds\right), \qquad 0 \le t \le 1\end{aligned}$$

where the measures μ_0, μ_1 are seen as row vectors and the functions f_0, g_1 as column vectors, and the ordered exponentials are defined by

$$\begin{aligned}\overrightarrow{\exp}\left(\int_0^s M_r\,dr\right) &:= \mathrm{Id} + \sum_{n\ge1} \int_{0\le r_1\le\cdots\le r_n\le s} M_{r_n}\cdots M_{r_1}\,dr_1\cdots dr_n,\\ \overleftarrow{\exp}\left(\int_s^1 M_r\,dr\right) &:= \mathrm{Id} + \sum_{n\ge1} \int_{s\le r_1\le\cdots\le r_n\le 1} M_{r_1}\cdots M_{r_n}\,dr_1\cdots dr_n.\end{aligned}$$

Derivatives of the entropy

In the discrete setting, no spatial regularity is required for a function to be in the domain of the operators $L, \Gamma, \ldots$ The only required regularity is that functions f and g defined by (6.4) are twice differentiable with respect to t on the open interval $(0, 1)$. This is ensured by the following lemma.

Lemma 6.4.1. *Consider f_0, g_1 as in Definition 6.1.1 and suppose that $f_0, g_1 \in L^1(m) \cap L^2(m)$ where the stationary measure m charges every point of $\boldsymbol{X}$. Then, under the assumption that*

$$\sup_{x\in\boldsymbol{X}}\{\overrightarrow{J}_x(\boldsymbol{X}) + \overleftarrow{J}_x(\boldsymbol{X})\} < \infty, \tag{6.37}$$

for each $x \in \boldsymbol{X}$, $t \mapsto f(t, x)$ and $t \mapsto g(t, x)$ are $\mathcal{C}^\infty$-regular on $(0, 1)$.

Proof. For any $0 \le t \le 1$, f_t and g_t are well-defined P_t-a.e., where P is given by (6.3). But, as R is an irreducible random walk, for each $0 < t < 1$, "P_t-a.e." is equivalent to "everywhere". Since $f_0 \in L^1(m)$ and R is m-stationary, we have $\int_{\boldsymbol{X}} f_t\,dm = E_R E_R[f_0(X_0) \mid X_t] = E_R(f_0(X_0)) = \int_{\boldsymbol{X}} f_0\,dm < \infty$, implying that $f_t \in L^1(m)$. As $\sup_x \overleftarrow{J}_x(\boldsymbol{X}) < \infty$, $\overleftarrow{L}$ is a bounded operator on $L^1(m)$. This implies that $t \in (0, 1) \mapsto f_t = \overrightarrow{\exp}\left(\int_0^s \overleftarrow{L}_r\,dr\right) f_0 \in L^1(m)$ is differentiable and $(d/dt)^k f_t = \overleftarrow{L}^k f_t$ for any k. As m charges every point, we also see that $t \in (0, 1) \mapsto f_t(x) \in \mathbb{R}$ is infinitely differentiable for every x. A similar proof works with g instead of f. □

As an important consequence of Lemma 6.4.1 for our purpose, we see that the statement of Claim 2 is rigorous in the present discrete setting.

Theorem 6.4.2. *Let R be an m-stationary random walk with jump measures $\overrightarrow{J}$ and $\overleftarrow{J}$ which satisfy (6.37). Along any entropic interpolation $[\mu_0, \mu_1]^R$ associated with a couple (f_0, g_1) as in Definition 6.1.1 and such that $f_0, g_1 \in L^1(m) \cap L^2(m)$, we have for all $0 < t < 1$,*

$$\begin{aligned}\frac{d}{dt}H(\mu_t|m) &= \sum_{x\in\mathcal{X}} \left(\overrightarrow{\Theta}\psi_t - \overleftarrow{\Theta}\phi_t\right)(x)\,\mu_t(x),\\ \frac{d^2}{dt^2}H(\mu_t|m) &= \sum_{x\in\mathcal{X}} \left(\overrightarrow{\Theta}_2\psi_t + \overleftarrow{\Theta}_2\phi_t\right)(x)\,\mu_t(x)\end{aligned}$$

where the expressions of $\overrightarrow{\Theta}\psi_t$ and $\overleftarrow{\Theta}\phi_t$ are given in (6.38) and the expressions of $\overrightarrow{\Theta}_2\psi_t$ and $\overleftarrow{\Theta}_2\phi_t$ are given in Proposition 6.4.4 below.

Computing Θ and Θ_2

Our aim now is to compute the operators $\overrightarrow{\Theta}$, $\overleftarrow{\Theta}$, $\overrightarrow{\Theta}_2$ and $\overleftarrow{\Theta}_2$ for a general random walk R. We use the shorthand notation $\int_{\mathcal{X}} a(x,y)\,J_x(dy) = [\int_{\mathcal{X}} a\,dJ](x)$ and drop the arrows and the index t during intermediate computations. The Hamilton-Jacobi operator is $Bu := e^{-u}Le^u = \int_{\mathcal{X}}(e^{Du}-1)\,dJ$ which gives

$$Cu := (B-L)u = \int_{\mathcal{X}} \theta(Du)\,dJ$$

where the function θ is defined by

$$\theta(a) := e^a - a - 1, \quad a \in \mathbb{R}.$$

Compare $Cu = |\nabla u|^2/2$, noting that $\theta(a) = a^2/2 + o_{a\to 0}(a^2)$. The convex conjugate θ^* of θ will be used in a moment. It is given by

$$\theta^*(b) = \begin{cases} (b+1)\log(b+1) - b, & b > -1,\\ 1, & b = -1,\\ \infty, & b < -1.\end{cases}$$

Computation of Θ

The carré du champ is $\Gamma(u,v) = \int_{\mathcal{X}} DuDv\,dJ$ so that $e^{-u}\Gamma(e^u, u) = \int_{\mathcal{X}} Du(e^{Du}-1)\,dJ$. Since $a(e^a - 1) - \theta(a) = ae^a - e^a + 1 = \theta^*(e^a - 1)$, with $\Theta u := e^{-u}\Gamma(e^u, u) - Cu$, we obtain

$$\begin{cases} \overrightarrow{\Theta}\psi_t(x) = \sum_{y:x\sim y} \theta^*\left(\frac{Dg_t(x,y)}{g_t(x)}\right)\overrightarrow{J}_x(y),\\ \overleftarrow{\Theta}\phi_t(x) = \sum_{y:x\sim y} \theta^*\left(\frac{Df_t(x,y)}{f_t(x)}\right)\overleftarrow{J}_x(y),\end{cases} \tag{6.38}$$

where we used $e^{D\psi_t(x,y)} - 1 = Dg_t(x,y)/g_t(x)$ and $e^{D\phi_t(x,y)} - 1 = Df_t(x,y)/f_t(x)$. The ratio Dg/g should be seen as the discrete logarithmic derivative of g. Recall that $g_t > 0$, μ_t-a.e. : we do not divide by zero. Compare with $\Theta\psi = |\nabla g/g|^2/2$, remarking that $\theta^*(b) = b^2/2 + o_{b\to 0}(b^2)$.

Entropy production

The entropy productions are

$$\begin{cases} \overrightarrow{I}(t) & = & \sum_{(x,y):x\sim y} \theta^* \left(\frac{Dg_t(x,y)}{g_t(x)}\right) \mu_t(x) \overrightarrow{J}_x(y), \\ \overleftarrow{I}(t) & = & \sum_{(x,y):x\sim y} \theta^* \left(\frac{Df_t(x,y)}{f_t(x)}\right) \mu_t(x) \overleftarrow{J}_x(y). \end{cases} \tag{6.39}$$

Computation of Θ_2

Unlike the diffusion case, no welcome cancellations occur. The carré du champ Γ is not a derivation anymore since $\Gamma(uv, w) - [u\Gamma(v,w) + v\Gamma(u,w)] = uvLw + \int_{\mathbf{X}} DuDvDw\, dJ$. Furthermore, we also loose the simplifying identity $C = \Theta$. To give a readable expression of Θ_2, it is necessary to introduce some simplifying shorthand notation:

$$\begin{cases} \sum_{x\to y} F(\varrho a) & = & \int_{\mathbf{X}} F(Du(x,y))\, J_x(dy) \\ \sum_{x\to y; x\to y'} F(\varrho a, \varrho a') & = & \int_{\mathbf{X}^2} F(Du(x,y), Du(x,y'))\, J_x(dy) J_x(dy') \\ \sum_{x\to y\to z} F(\varrho a, \mathbf{b}) & = & \int_{\mathbf{X}^2} F(Du(x,y), Du(y,z))\, J_x(dy) J_y(dz) \\ \sum_{x\to y\to z} F(\varrho a, \mathbf{c}) & = & \int_{\mathbf{X}^2} F(Du(x,y), Du(x,z))\, J_x(dy) J_y(dz). \end{cases}$$

where we have denoted $\begin{cases} \varrho a & = & Du(x,y) \\ \varrho a' & = & Du(x,y') \\ \mathbf{b} & = & Du(y,z) \\ \mathbf{c} & = & Du(x,z) \end{cases}$. Notice that $\sum_{x\to y\to z} F(\mathbf{c}) = \int_{\mathbf{X}} F(Du(x,z))\, J_x^2(dz)$ with $J_x^2(dz) := \int_{y\in\mathbf{X}} J_x(dy) J_y(dz)$ and $\varrho a + \mathbf{b} = \mathbf{c}$.

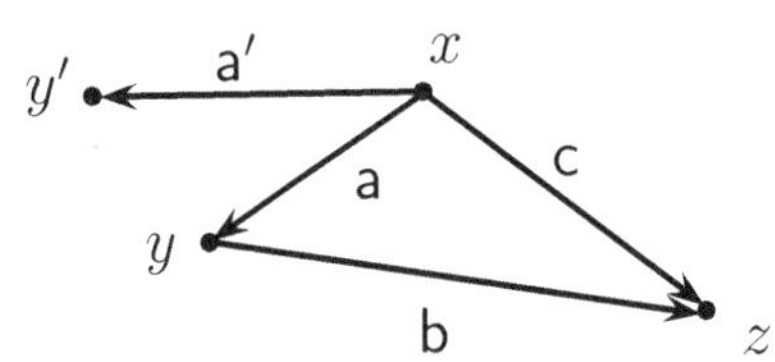

We also define the function

$$h(a) := \theta^*(e^a - 1) = ae^a - e^a + 1, \quad a \in \mathbb{R}. \tag{6.40}$$

Lemma 6.4.3. *For any function $u \in \mathbb{R}^{\mathbf{X}}$ and any $x \in \mathbf{X}$,*

$$\begin{aligned}
\Theta u(x) &= \sum_{x\to y} h(\varrho a),\\
\Theta_2 u(x) &= \Big(\sum_{x\to y}(e^{\varrho}a-1)\Big)^2 + \sum_{x\to y}[J_y(\mathbf{X})-J_x(\mathbf{X})]h(\varrho a)\\
&\quad + \sum_{x\to y\to z}[2e^{\varrho}ah(\boldsymbol{b})-h(\boldsymbol{c})].
\end{aligned}$$

Proof. The first identity is (6.38). Let us look at $\Theta_2 u(x)$. Rather than using formula (6.17), for an explicit formulation of Θ_2 in terms of J, it will be easier to go back to (6.19): $\overrightarrow{\Theta}_2\psi_t = \overrightarrow{A}_t^2\psi_t + \frac{d}{dt}(\overrightarrow{A}_t\psi_t)$ with $\overrightarrow{A}_t u = \int_{\mathbf{X}} Du\, e^{D\psi_t}\, d\overrightarrow{J}$ and $\dot{\psi} = -\overrightarrow{B}\psi = -\int_{\mathbf{X}}(e^{D\psi}-1)\, d\overrightarrow{J}$.
Making use of $A\psi(x) = \sum_{x\to y}\varrho a e^{\varrho}a$ and $\dot\psi(x) = \sum_{x\to y} -(e^{\varrho}a-1)$, we see that

$$\begin{aligned}
A^2\psi(x) &= \int_{\mathbf{X}}(A\psi_y - A\psi_x)e^{D\psi(x,y)}\,J_x(dy)\\
&= \sum_{x\to y\to z} e^{\varrho}a\boldsymbol{b}e^{\boldsymbol{b}} - \sum_{x\to y;x\to y'} e^{\varrho}a\varrho a' e^{\varrho a'}\\
\frac{d}{dt}(A\psi)(x) &= \int_{\mathbf{X}}(D\psi(x,y)e^{D\psi(x,y)}+e^{D\psi(x,y)})(\dot\psi_y-\dot\psi_x)\,J_x(dy)\\
&= -\sum_{x\to y\to z}(\varrho a e^{\varrho}a+\varrho a)(e^{\boldsymbol{b}}-1) + \sum_{x\to y;x\to y'}(\varrho a e^{\varrho}a+\varrho a)(e^{\varrho a'}-1)
\end{aligned}$$

where ϱa, $\varrho a'$ and $\boldsymbol{b}$ are taken with $u = \psi$. This shows that

$$\Theta_2 u(x) = \sum_{x\to y\to z} e^{\varrho}a\boldsymbol{b}e^{\boldsymbol{b}} - (\varrho a e^{\varrho}a+e^{\varrho}a)(e^{\boldsymbol{b}}-1) - \sum_{x\to y;x\to y'} e^{\varrho}a\varrho a' e^{\varrho a'} - (\varrho a e^{\varrho}a+e^{\varrho}a)(e^{\varrho a'}-1)$$

The functions defining the integrands rewrite as follows

$$\begin{aligned}
e^a b e^b - (ae^a+e^a)(e^b-1) &= h(a) + 2e^a h(b) - h(c) \quad \text{with} \quad c = a+b,\\
e^a a' e^{a'} - (ae^a+e^a)(e^{a'}-1) &= (e^a-1)h(a') - h(a)(e^{a'}-1)\\
&\quad + h(a') - (e^a-1)(e^{a'}-1).
\end{aligned}$$

Hence,

$$\begin{aligned}
&\sum_{x\to y;x\to y'} e^{\varrho}a\varrho a' e^{\varrho a'} - (\varrho a e^{\varrho}a + e^{\varrho}a)(e^{\varrho a'}-1)\\
&= \sum_{x\to y;x\to y'}(e^{\varrho}a-1)h(\varrho a') - \sum_{x\to y;x\to y'} h(\varrho a)(e^{\varrho a'}-1) + \sum_{x\to y;x\to y'} h(\varrho a')\\
&\quad - \sum_{x\to y;x\to y'}(e^{\varrho}a-1)(e^{\varrho a'}-1)
\end{aligned}$$

$$= J_x(\mathbf{X}) \sum_{x \to y} h(\varrho a) - \left(\sum_{x \to y} (e^{\varrho} a - 1) \right)^2$$

and the desired result follows. □

Lemma 6.4.3 leads us to the following evaluations.

Proposition 6.4.4. *Consider f_0, g_1 as in Definition 6.1.1 and suppose that $f_0, g_1 \in L^1(m) \cap L^2(m)$ where the stationary measure m charges every point of $\mathbf{X}$. Also assume that $\overrightarrow{J}, \overleftarrow{J}$ satisfy (6.37). Then,*

$$\begin{aligned} \overrightarrow{\Theta}_2 \psi_t(x) = & \left(\sum_{y:x\sim y} (e^{D\psi_t(x,y)} - 1) \overrightarrow{J}_x(y) \right)^2 \\ & + \sum_{y:x\sim y} [\overrightarrow{J}_y(\mathbf{X}) - \overrightarrow{J}_x(\mathbf{X})] \theta^* \left(e^{D\psi_t(x,y)} - 1 \right) \overrightarrow{J}_x(y) \\ & + \sum_{(y,z):x\sim y\sim z} \Big[2\theta^* \left(e^{D\psi_t(y,z)} - 1 \right) \overrightarrow{A}_{t,x}(y) \overrightarrow{J}_y(z) \\ & \qquad - \theta^* \left(e^{D\psi_t(x,z)} - 1 \right) \overrightarrow{J}_x(y) \overrightarrow{J}_y(z) \Big] \end{aligned} \tag{6.41}$$

and

$$\begin{aligned} \overleftarrow{\Theta}_2 \phi_t(x) = & \left(\sum_{y:x\sim y} (e^{D\phi_t(x,y)} - 1) \overleftarrow{J}_x(y) \right)^2 \\ & + \sum_{y:x\sim y} [\overleftarrow{J}_y(\mathbf{X}) - \overleftarrow{J}_x(\mathbf{X})] \theta^* \left(e^{D\phi_t(x,y)} - 1 \right) \overleftarrow{J}_x(y) \\ & + \sum_{(y,z):x\sim y\sim z} \Big[2\theta^* \left(e^{D\phi_t(y,z)} - 1 \right) \overleftarrow{A}_{t,x}(y) \overleftarrow{J}_y(z) \\ & \qquad - \theta^* \left(e^{D\phi_t(x,z)} - 1 \right) \overleftarrow{J}_x(y) \overleftarrow{J}_y(z) \Big] \end{aligned} \tag{6.42}$$

where

$$\overrightarrow{A}_{t,x}(y) = e^{D\psi_t(x,y)} \overrightarrow{J}_x(y) = \frac{g_t(y)}{g_t(x)} \overrightarrow{J}_x(y),$$

$$\overleftarrow{A}_{t,x}(y) = e^{D\phi_t(x,y)} \overleftarrow{J}_x(y) = \frac{f_t(y)}{f_t(x)} \overleftarrow{J}_x(y)$$

are the forward and backward jump frequencies of $P = f_0(X_0) g_1(X_1) R$.

Reversible random walks

Let us take a positive measure $m = \sum_{x \in \mathbf{X}} m_x \, \delta_x \in \mathrm{M}_+(\mathbf{X})$ with $m_x > 0$ for all $x \in \mathbf{X}$. It is easily checked with the detailed balance condition

$$m(dx) J_x(dy) = m(dy) J_y(dx) \tag{6.43}$$

which characterizes the reversibility of m that the jump kernel

$$J_x = \sum_{y:x\sim y} s(x,y)\sqrt{m_y/m_x}\,\delta_y, \tag{6.44}$$

where $s(x,y) = s(y,x) > 0$ for all $x \sim y$, admits m as a reversing measure. As $(\boldsymbol{X}, \sim)$ is assumed to be connected, the random walk is irreducible and the reversing measure is unique up to a multiplicative constant.

Examples 6.4.5. *Let us present the simplest examples of reversible random walks.*

(a) Reversible counting random walk. *The simplest example is provided by the counting jump kernel* $J_x = \sum_{y:x\sim y} \delta_y \in \mathrm{P}(\boldsymbol{X})$, $x \in \boldsymbol{X}$ *which admits the counting measure* $m = \sum_x \delta_x$ *as a reversing measure: take* $m_x = 1$ *and* $s(x,y) = 1$ *in (6.44).*

(b) Reversible simple random walk. *The simple jump kernel is* $J_x = \frac{1}{n_x}\sum_{y:x\sim y}\delta_y \in \mathrm{P}(\boldsymbol{X})$, $x \in \boldsymbol{X}$. *The measure* $m = \sum_x n_x\delta_x$ *is a reversing measure: take* $m_x = n_x$ *and* $s(x,y) = (n_x n_y)^{-1/2}$ *in (6.44).*

The dynamics of R is as follows. Once the walker is at x, one starts an exponential random clock with frequency $J_x(\boldsymbol{X})$. When the clock rings, the walker jumps at random onto a neighbor of x according to the probability law $J_x(\boldsymbol{X})^{-1}\sum_{y:x\sim y} J(x;y)\,\delta_y$. This procedure goes on and on. Since R is reversible, we have $\overrightarrow{J} = \overleftarrow{J} = J$ and the forward and backward frequencies of jumps of the (f,g)-transform $P = f_0(X_0)g_1(X_1)\,R$ are respectively

$$\overrightarrow{A}_x(y) = \frac{g_t(y)}{g_t(x)}J_x(y) \quad \text{and} \quad \overleftarrow{A}_x(y) = \frac{f_t(y)}{f_t(x)}J_x(y).$$

As a direct consequence of Theorem 6.4.2 and Proposition 6.4.4, we obtain the following result.

Theorem 6.4.6. *Let R be an m-reversible random walk with jump measure J which is given by (6.44) and satisfies*

$$\sup_{x\in\boldsymbol{X}} J_x(\boldsymbol{X}) < \infty.$$

Then, along any entropic interpolation $[\mu_0,\mu_1]^R$ associated with a couple (f_0, g_1) as in Definition 6.1.1 and such that $f_0, g_1 \in L^1(m)\cap L^2(m)$, we have for all $0 < t < 1$,

$$\begin{aligned}\frac{d}{dt}H(\mu_t|m) &= \sum_{x\in\boldsymbol{X}}[\Theta\psi_t - \Theta\phi_t](x)\,\mu_t(x),\\ \frac{d^2}{dt^2}H(\mu_t|m) &= \sum_{x\in\boldsymbol{X}}[\Theta_2\psi_t + \Theta_2\phi_t](x)\,\mu_t(x)\end{aligned}$$

where the expressions of $\Theta\psi_t$ and $\Theta\phi_t$ are given in (6.38) and the expressions of $\Theta_2\psi_t$ and $\Theta_2\phi_t$ are given in Proposition 6.4.4 (drop the useless time arrows).

Remarks 6.4.7.

(a) *For the above condition* $\sup_{x\in\mathbf{X}} J_x(\mathbf{X}) < \infty$ *to be satisfied, it suffices that for some* $0 < c, \sigma < \infty$, *we have for all* $x \sim y \in \mathbf{X}$,

$$\begin{cases} m_y/n_y \le c m_x/n_x, \\ 0 < s(x,y)\sqrt{n_x n_y} \le \sigma, \end{cases} \tag{6.45}$$

with the notation of (6.44).

(b) *In the special case where* $f_0 = \rho_0$ *and* $g_1 = 1$ *corresponding to the forward heat flow – see Definition 6.2.3 – we obtain*

$$\begin{aligned} \frac{d}{dt} H(\mu_t|m) &= -\sum_{x\sim y} D\rho_t(x,y) D(\log \rho_t)(x,y)\, m(x) J_x(y), \\ \frac{d^2}{dt^2} H(\mu_t|m) &= \sum_x \left\{ L\rho_t(x) L(\log \rho_t)(x) + \frac{(L\rho_t(x))^2}{\rho_t(x)} \right\} m(x). \end{aligned}$$

The first identity is the well-known entropy production formula and the second one was proved in [8].

6.5 Convergence to equilibrium

This section is dedicated to the detailed proofs of already known results such as the (modified) logarithmic Sobolev inequality in Theorem 6.5.8. Its motivation is two-pronged: (i) the proof of the convergence result Theorem 6.5.2 is usually omitted as part of the standard folklore, and (ii) it is interesting to look at heat flows as special entropic interpolations.

In this section R is an m-stationary Markov probability measure on the path space

$$\Omega = D([0,\infty), \mathbf{X})$$

built on the unbounded time interval ([0, ∞) and m is assumed to be a probability measure (and so is R). We are interested in the convergence as t tends to infinity of the forward heat flow

$$\mu_t := (X_t)_\#(\rho_0(X_0)\, R) \in \mathrm{P}(\mathbf{X}), \quad t \ge 0, \tag{6.46}$$

where $\rho_0 = d\mu_0/dm$ is the initial density. We are going to prove by implementing the “stochastic process strategy” of the present paper, that under some hypotheses, the following convergence to equilibrium

$$\lim_{t\to\infty} \mu_t = m$$

holds. As usual, the function $t \mapsto H(\mu_t|m)$ is an efficient Lyapunov function of the dynamics.

Stationary dynamics

Let us first make the basic assumptions precise.

Assumptions 6.5.1. *The stationary measure m is a probability measure and we assume the following.*

(a) Brownian diffusion. The forward derivative is

$$\overrightarrow{L} = \overrightarrow{b} \cdot \nabla + \Delta/2$$

on a connected Riemannian manifold $\boldsymbol{X}$ *without boundary. The initial density* $\rho_0 := d\mu_0/dm$ *is regular enough for* $\rho_t(z) = d\mu_t/dm(z)$ *to be positive and twice differentiable in t and z on* $(0, \infty) \times \boldsymbol{X}$.

(b) Random walk. The forward derivative is

$$\overrightarrow{L} u(x) = \sum_{y:x\sim y} [u(y) - u(x)] \, \overrightarrow{J}_x(y)$$

on the countable locally finite connected graph $(\boldsymbol{X}, \sim)$. *We assume that* (6.37) *holds:* $\sup_x \{\overrightarrow{J}_x(\boldsymbol{X}) + \overleftarrow{J}_x(\boldsymbol{X})\} < \infty$.

The point in the following Theorem 6.5.2 is that no reversibility is required. We define

$$\mathcal{I}(\mu|m) := \int_{\boldsymbol{X}} \overleftarrow{\Theta}(\log \rho) \, d\mu \in [0, \infty], \quad \mu = \rho \, m \in \mathrm{P}(\boldsymbol{X}).$$

We have seen in (6.28) and (6.38) that $\overleftarrow{\Theta} \geq 0$ (notice that $\theta^* \geq 0$), this shows that $\mathcal{I}(\mu|m) \geq 0$.

Theorem 6.5.2. *The Assumptions 6.5.1 are supposed to hold together with* $\Theta_2 \geq 0$. *Then,* $\lim_{t\to\infty} \mathcal{I}(\mu_t|m) = 0$.
Let us assume in addition that

$$\overleftarrow{\Theta}_2 \geq \kappa \overleftarrow{\Theta} \quad \text{for some constant } \kappa > 0 \tag{6.47}$$

and that the initial density $\rho_0 := d\mu_0/dm$ *is bounded:* $\sup_{\boldsymbol{X}} \rho_0 < \infty$.
Then, $\lim_{t\to\infty} H(\mu_t|m) = 0$.

Remark 6.5.3. *The total variation norm of a signed measure η on* $\boldsymbol{X}$ *is defined by* $\|\eta\|_{\mathrm{TV}} := |\eta|(\boldsymbol{X})$ *and the Csiszár-Kullback-Pinsker inequality is* $\frac{1}{2}\|\mu - m\|_{TV}^2 \leq H(\mu|m)$, *for any* $\mu \in \mathrm{P}(\boldsymbol{X})$. *Therefore we have* $\lim_{t\to\infty} \|\mu_t - m\|_{\mathrm{TV}} = 0$.

Proof. The main idea of this proof is to consider the forward heat flow $(\mu_t)_{t\geq 0}$ as an entropic interpolation, see Definition 6.2.3, and to apply Claim 2. It will be seen that the general regularity hypotheses of the theorem ensure that this claim becomes a rigorous statement.

Instead of restricting time to $[0,1]$, we allow it to be in $[0,T]$, $T>0$, and let T tend to infinity. We denote $\rho_t := d\mu_t/dm$. Since μ is the time-marginal flow of $P = \rho_0(X_0)\,R$, we have $f_0 = \rho_0$ and $g_T = 1$. This implies that $g_t = 1$ and $f_t = \rho_t$. Indeed, $g_t(z) = E_R[g_T(X_T) \mid X_t = z] = E_R(1 \mid X_t = z) = 1$ and $f_t(z) = E_R[f_0(X_0) \mid X_t = z] = E_R[\rho_0(X_0) \mid X_t = z] = E_R[dP/dR \mid X_t = z] = dP_t/dR_t(z) = dP_t/dm(z) =: \rho_t(z)$. Therefore, $\psi_t = 0$, $\phi_t = \log\rho_t$, for all $0 \le t \le T$. Our assumptions guarantee that for all $t>0$, ϕ_t is regular enough to use our previous results about "Θ_2-calculus". Denoting $H(t) = H(\mu_t|m)$, with Corollary 6.2.4, we see that

$$-H'(t) = \overleftarrow{I}(t) := \int_{\mathbf{X}} \overleftarrow{\Theta}(\log\rho_t)\,d\mu_t =: \mathcal{I}(\mu_t|m).$$

As $\mathcal{I}(\cdot|m) \ge 0$, H is decreasing. Of course, $H(0) - H(T) = \int_0^T \mathcal{I}(\mu_t|m)\,dt$. Now, we let T tend to infinity. As H is non-negative and decreasing, it admits a limit $H(\infty) := \lim_{T\to\infty} H(T)$ and the integral in

$$H(0) - H(\infty) = \int_0^\infty \mathcal{I}(\mu_t|m)\,dt \tag{6.48}$$

is convergent, implying with $\frac{d}{dt}\mathcal{I}(\mu_t|m) = -\Theta_2(\phi_t) \le 0$ that $\lim_{t\to\infty}\mathcal{I}(\mu_t|m) = 0$. This proves the first assertion of the theorem.

As ρ_0 is assumed to be bounded, for all t we have

$$\rho_t(X_t) = E_R(\rho_0 \mid X_t) \le \sup\rho_0 < \infty. \tag{6.49}$$

This will be used below.

Diffusion setting. By (6.29) with $\phi_t = \rho_t$, this limit implies that $\lim_{t\to\infty}\int_{\mathbf{X}} |\nabla\sqrt{\rho_t}|^2\,d\,\mathrm{vol} = 0$. Poincaré's inequality tells us that for any open bounded connected domain U with a smooth boundary, there exists a constant C_U such that $\|\sqrt{\rho_t} - \langle\sqrt{\rho_t}\rangle_U\|_{L^2(U)} \le C_U\|\nabla\sqrt{\rho_t}\|_{L^2(U)}$ for all t, where $\langle\sqrt{\rho_t}\rangle_U := \int_U \sqrt{\rho_t}\,d\,\mathrm{vol}\,/\,\mathrm{vol}(U)$. It follows that $\lim_{t\to\infty}\|\sqrt{\rho_t} - \langle\sqrt{\rho_t}\rangle_U\|_{L^2(U)} = 0$ which in turns implies that

$$\lim_{t\to\infty}\rho_t = c \tag{6.50}$$

everywhere on $\mathbf{X}$ for some constant $c \ge 0$.

Let us prove this last claim. By Bienaymé-Chebychev's inequality, a vanishing variance along a sequence of random variables implies convergence in measure towards a constant. On the other hand, the exponential rate of convergence ensured by our assumption (6.47) implies with the Borel-Cantelli lemma that this limit is almost sure on U. Finally, as ρ is continuous on $(0,\infty)\times\mathbf{X}$ and $\mathbf{X}$ is connected, we can extend this convergence to the whole space.

As $\sup_t H(\mu_t|m) \le H(\mu_0|m) < \infty$, the set $\{\mu_t, t \ge 0\}$ is uniformly tight and therefore relatively compact with respect to the narrow topology on $\mathrm{P}(\mathbf{X})$. It follows from (6.50) that $\lim_{t\to\infty} \rho_t = 1$ everywhere. Finally, the uniform bound (6.49) allows us to apply the Lebesgue dominated convergence theorem to assert $\lim_{t\to\infty} H(\mu_t|m) = 0$, which is the desired result.

Random walk setting. Since $\lim_{t\to\infty} \mathcal{I}(\mu_t|m) = 0$, we obtain

$$\lim_{t\to\infty} \sum_{(x,y):x\sim y} \rho_t(x)\theta^*\left(\rho_t(y)/\rho_t(x) - 1\right) m(x)\overleftarrow{J}_x(y) = 0. \tag{6.51}$$

For some constant $C > 0$, we have

$$a\theta^*(b/a - 1) \geq C\frac{(b-a)^2}{a}, \quad \text{if } b/a \leq 2.$$

We see that the convergence to zero of the left-hand side implies the convergence of $b - a$ to zero when on a domain such that a is bounded. This will be applied with $a = \rho_t(x)$ which is bounded in virtue of (6.49).
When $b/a \geq 2$ and $a \geq 2$, the left hand side cannot be close to zero. Finally, the remaining case when $b/a \geq 2$ and $a \leq 2$ is controlled by considering the symmetric term $b\theta^*(a/b - 1)$ which also appears in the series in the limit (6.51), reverting x and y; recall that $\overleftarrow{J}_x(y) > 0 \Leftrightarrow \overleftarrow{J}_y(x) > 0$. As $b\theta^*(a/b - 1) \geq b \geq 2a \geq 0$, the convergence of the left-hand term to zero implies the convergence of a and b to 0. Therefore, the limit (6.51) implies that for all adjacent x and $y \in \boldsymbol{X}$, $\lim_{t\to\infty}(\rho_t(x) - \rho_t(y)) = 0$.
It follows from (6.49) the countability of $\boldsymbol{X}$ and the connectedness of the graph, that there exists a sequence of times (t_k) tending to infinity and a non-negative number c such that $\lim_k \rho_{t_k}(x) = c$ for all $x \in \boldsymbol{X}$. The same argument as in the diffusion setting leads us to the relative compactness of $\{\mu_t; t \geq 0\}$ and in particular of (μ_{t_k}). It follows that $\lim_k \rho_{t_k}(x) = 1$ for all x. Finally, as $t \mapsto H(\mu_t|m) \geq 0$ is decreasing, it admits a limit as t tends to infinity and we obtain: $\lim_t H(\mu_t|m) = \lim_k H(\mu_{t_k}|m) = 0$ where the last identity follows as in the diffusion setting from the dominated convergence theorem. This completes the proof of the theorem. □

Theorem 6.5.4. *The Assumptions 6.5.1 are supposed to hold. Then, under the additional hypotheses that $H(\mu_0|m) < \infty$ and (6.47) i.e. $\overleftarrow{\Theta}_2 \geq \kappa\overleftarrow{\Theta}$ for some $\kappa > 0$, we have*

$$\mathcal{I}(\mu_t|m) \leq \mathcal{I}(\mu_0|m)e^{-\kappa t}, \quad t \geq 0 \tag{6.52}$$
$$H(\mu_t|m) \leq H(\mu_0|m)e^{-\kappa t}, \quad t \geq 0 \tag{6.53}$$

and the following functional inequality

$$H(\rho m|m) \leq \kappa^{-1}\mathcal{I}(\rho m|m), \tag{6.54}$$

holds for any nonnegative function $\rho : \boldsymbol{X} \to [0, \infty)$ such that $\int_{\boldsymbol{X}} \rho\, dm = 1$, which in the diffusion setting (a) is also assumed to be $\mathcal{C}^2$-regular.

Proof of Theorem 6.5.4. We already saw during the proof of Theorem 6.5.2 that our general assumptions allow us to apply Θ_2-calculus: Claim 2 is rigorous with $\phi_t = \log \rho_t$.

This gives: $\overleftarrow{I}(t) = -H'(t) = \int_{\mathbf{X}} \overleftarrow{\Theta}(\log \rho_t)\, d\mu_t$ and $\overleftarrow{I}'(t) = -H''(t) = -\int_{\mathbf{X}} \overleftarrow{\Theta}_2(\log \rho_t)\, d\mu_t$. The inequality (6.47) implies that $\overleftarrow{I}'(t) \le -\kappa \overleftarrow{I}(t)$, for all $t \ge 0$. Integrating leads to $\overleftarrow{I}(t) \le \overleftarrow{I}(0)e^{-\kappa t}$, $t \ge 0$, which is (6.52).

Let us assume for a while, in addition to (6.47), that $\sup_{\mathbf{X}} \rho_0 < \infty$. It is proved in Theorem 6.5.2 that under this restriction, $H(\infty) = 0$. With (6.48) and (6.52), we see that

$$H(\mu_0|m) = H(0) = \int_0^\infty \overleftarrow{I}(t)\, dt \le \overleftarrow{I}(0) \int_0^\infty e^{-\kappa t}\, dt = \overleftarrow{I}(0)/\kappa.$$

A standard approximation argument based on the sequence $\rho_n = (\rho \wedge n)/\int_{\mathbf{X}}(\rho \wedge n)\, dm$ and Fatou's lemma: $H(\rho m|m) \le \liminf_n H(\rho_n m|m)$, allows to extend this inequality to unbounded ρ. This proves (6.54).

Plugging μ_t instead of μ_0 into (6.54), one sees that $H(t) \le \overleftarrow{I}(t)/\kappa = -H'(t)/\kappa$. Integrating leads to $H(t) \le H(0)e^{-\kappa t}$, which is (6.53). □

Remark 6.5.5 (About time reversal)**.** *The backward arrows in* (6.47) *suggest that time reversal is tightly related to this convergence to stationarity. Let us propose an informal interpretation of this phenomenon. The forward entropy production* $\overrightarrow{I}(t)$ *vanishes along the forward heat flow* $(\mu_t)_{t\ge 0}$ *since* $\overrightarrow{I}(t) = \int_{\mathbf{X}} \overrightarrow{\Theta}\psi_t\, d\mu_t$ *and* $\psi = 0$*. Similarly* $\int_{\mathbf{X}} \overrightarrow{\Theta}_2\psi_t\, d\mu_t = 0$*. No work is needed to drift along the heat flow. In order to evaluate the rate of convergence, one must measure the strength of the drift toward equilibrium. To do so "one has to face the wind" and measure the work needed to reach* μ_0 *when starting from m, reversing time.*

Remarks 6.5.6 (About $\overleftarrow{\Theta}_2 \ge \kappa \overleftarrow{\Theta}$)**.**

(a) *In the diffusion setting,* (6.47) *can be written*

$$\overleftarrow{\Gamma}_2 \ge \kappa \Gamma.$$

When $\overrightarrow{L}$ *is the reversible forward derivative* (6.9)*, this is the standard Bakry-Émery curvature condition* CD(κ, ∞)*, see [4, 5]. Further detail is given below at Theorem 6.5.8.*

(b) *In the random walk setting, we see with Lemma 6.4.3 that* (6.47) *can be written*

$$\begin{aligned} \Big(\sum_{y:x\sim y} (e^{\varrho a_y} - 1)\overleftarrow{J}_x(y)\Big)^2 &+ \sum_{y:x\sim y} [\overleftarrow{J}_y(\mathbf{X}) - \overleftarrow{J}_x(\mathbf{X})]h(\varrho a_y)\overleftarrow{J}_x(y) \\ &+ \sum_{(y,z):x\sim y\sim z} [2e^{\varrho a_y} h(\mathbf{c}_z - \varrho a_y) - h(\mathbf{c}_z)]\overleftarrow{J}_x(y)\overleftarrow{J}_y(z) \\ &\ge \kappa \sum_{y:x\sim y} h(\varrho a_y)\overleftarrow{J}_x(y) \end{aligned} \tag{6.55}$$

for all $x \in \mathbf{X}$ *and any numbers* $\varrho a_y, \mathbf{c}_z \in \mathbb{R}$, $(y,z) : x \sim y \sim z$*, where* $h(a) := \theta^*(e^a - 1) = ae^a - e^a + 1$, $a \in \mathbb{R}$*, see* (6.40)*.*

(c) *An inspection of the proof of Theorem 6.5.4 shows that the integrated version of (6.47):*

$$\int_X \overleftarrow{\Theta}_2(\rho)\, d\mu \geq \kappa \int_X \overleftarrow{\Theta}(\rho)\, d\mu, \quad \mu = \rho m \in \mathrm{P}(\boldsymbol{X})$$

is sufficient.

In the diffusion setting (6.27) on $\boldsymbol{X} = \mathbb{R}^n$, we know through (6.31) that (6.47) becomes

$$\|\nabla^2 u\|_{\mathrm{HS}}^2 + \left(\nabla^2 V + [\nabla + \nabla^*] b_\perp\right)(\nabla u) \geq \kappa |\nabla u|^2$$

for any sufficiently regular function u. This Γ_2-criterion was obtained by Arnold, Carlen and Ju [3]; see also the paper [24] by Hwang, Hwang-Ma and Sheu for a related result. These results are recovered in the recent paper [22] by Fontbona and Jourdain who implement a stochastic process approach, slightly different from the present article's one but where time reversal also plays a crucial role, see Remark 6.5.5.

Reversible dynamics

More precisely, we are concerned with the following already encountered m-reversible generators L.

(a) On a Riemannian manifold $\boldsymbol{X}$, see (6.9):

$$L = (-\nabla V \cdot \nabla + \Delta)/2 \tag{6.56}$$

with the reversing measure $m = e^{-V}\,\mathrm{vol}$.

(b) On a graph $(\boldsymbol{X}, \sim)$, see (6.44):

$$Lu(x) = \sum_{y:x\sim y} [u(y) - u(x)] s(x, y) \sqrt{m_y/m_x} \tag{6.57}$$

with $s(x, y) = s(y, x) > 0$ for all $x \sim y$.

Let us recall our hypotheses.

Assumptions 6.5.7. *It is required that the reversing measure m is a probability measure.*

The assumptions (a) and (b) below allow us to apply respectively Theorems 6.3.5 and 6.4.6.

(a) *The Riemannian manifold $\boldsymbol{X}$ is compact connected without boundary and L is given by (6.56). We assume that $V : \boldsymbol{X} \to \mathbb{R}$ is $\mathcal{C}^4$-regular and without loss of generality that V is normalized by an additive constant such that $m = e^{-V}$ vol is a probability measure.*

(b) *The countable graph* $(\mathbf{X}, \sim)$ *is assumed to be locally finite and connected. The generator* L *is given by* (6.57), *and the equilibrium probability measure* m *and the function* s *satisfy* (6.45).

Reversibility allows for a simplified expression of $\mathcal{I}(\cdot|m)$. Indeed, as $\rho\Theta(\log\rho) = L(\rho\log\rho - \rho) - \log\rho\, L\rho$, we obtain

$$\int_{\mathbf{X}} \Theta(\log\rho)\rho\, dm = \int_{\mathbf{X}} L(\rho\log\rho - \rho)\, dm - \int_{\mathbf{X}} \log\rho\, L\rho\, dm = \frac{1}{2}\int_{\mathbf{X}} \Gamma(\rho, \log\rho)\, dm$$

where the last equality follows from the symmetry of L in $L^2(m)$ (a consequence of the m-reversibility of R), whenever $(\rho, \log\rho) \in \operatorname{Dom}\Gamma$. Therefore,

$$\mathcal{I}(\mu|m) = I(\mu|m) := \frac{1}{2}\int_{\mathbf{X}} \Gamma(\rho, \log\rho)\, dm. \tag{6.58}$$

It is the Fisher information of μ with respect to m. A direct computation shows that

(a) in the diffusion setting (a),

$$I(\mu|m) = \frac{1}{2}\int_{\mathbf{X}} |\nabla\log\rho|^2\, d\mu = \frac{1}{2}\int_{\mathbf{X}} \frac{|\nabla\rho|^2}{\rho}\, dm = 2\int_{\mathbf{X}} |\nabla\sqrt{\rho}|^2\, dm;$$

(b) in the random walk setting (b),

$$I(\mu|m) = \frac{1}{2}\sum_{(x,y):x\sim y} [\rho(y) - \rho(x)][\log\rho(y) - \log\rho(x)]\, m(x)J_x(y).$$

Theorem 6.5.4 becomes

Theorem 6.5.8. *The Assumptions 6.5.7 are supposed to hold. Then, under the additional hypotheses that* $H(\mu_0|m) < \infty$ *and* (6.47): $\overleftarrow{\Theta}_2 \geq \kappa\overleftarrow{\Theta}$ *for some* $\kappa > 0$, *we have*

$$\begin{aligned} I(\mu_t|m) &\leq I(\mu_0|m)e^{-\kappa t}, \quad t \geq 0 \\ H(\mu_t|m) &\leq H(\mu_0|m)e^{-\kappa t}, \quad t \geq 0 \end{aligned}$$

and the following (modified) logarithmic Sobolev inequality

$$H(\mu|m) \leq \kappa^{-1} I(\mu|m), \tag{6.59}$$

for any $\mu \in \mathrm{P}(\mathbf{X})$ *which in the diffusion setting (a) is also restricted to be such that* $d\mu/dm$ *is* $\mathcal{C}^2$*-regular.*

In the diffusion setting (a), Theorem 6.5.8 is covered by the well-known result by Bakry and Émery [5]. Inequality (6.59) is the standard logarithmic Sobolev inequality and

(6.47): $\overleftarrow{\Theta}_2 \geq \kappa \overleftarrow{\Theta}$ for some $\kappa > 0$, is the usual Γ_2-criterion: $\Gamma_2 \geq \kappa\Gamma$. In the random walk setting (b), inequality (6.59) is a modified logarithmic Sobolev inequality which was introduced by Bobkov and Tetali [6] in a general setting, and was already employed by Dai Pra, Paganoni and Posta [13] in the context of Gibbs measures on $\mathbb{Z}^d$. Later Caputo, Dai Pra and Posta [8, 9, 14] have derived explicit criteria for (6.59) to be satisfied in specific settings related to Gibbs measures of particle systems. These criteria are close in spirit to (6.47).

6.6 Some open questions about curvature and entropic interpolations

We have seen in Section 6.5 that convergence to equilibrium is obtained by considering heat flows. The latter are the simplest entropic interpolations since $f = 1$ or $g = 1$. In the present section, general entropic interpolations are needed to explore curvature properties of Markov generators and their underlying state space $\boldsymbol{X}$ by means of the main result of the article:

$$H'(t) = \int_{\boldsymbol{X}} \frac{1}{2}(\overrightarrow{\Theta}\psi_t - \overleftarrow{\Theta}\phi_t)\, d\mu_t, \quad H''(t) = \int_{\boldsymbol{X}} \frac{1}{2}(\overrightarrow{\Theta}_2\psi_t + \overleftarrow{\Theta}_2\phi_t)\, d\mu_t.$$

Convexity properties of the entropy along displacement interpolations are fundamental for the Lott-Sturm-Villani (LSV) theory of lower bounded curvature of metric measure spaces [31, 41, 42, 44]. An alternative approach would be to replace displacement interpolations with entropic interpolations, taking advantage of the analogy between these two types of interpolations. Although this program is interesting in itself, it can be further motivated by the following remarks.

1. Entropic interpolations are more regular than displacement interpolations.
2. Displacement interpolations are (semiclassical) limits of entropic interpolations, see [26, 29].
3. Entropic interpolations work equally well in continuous and discrete settings while LSV theory fails in the discrete setting (see below).

With respect to (1), note that the dynamics of an entropic interpolation share the regularity of the solutions f and g of the backward and forward “heat equations” (6.11). Their logarithms, the Schrödinger potentials ϕ and ψ, solve *second order* Hamilton-Jacobi-Bellman equations. This is in contrast with a displacement interpolation whose dynamics are driven by the Kantorovich potentials (analogues of ϕ and ψ) which are solutions of *first order* Hamilton-Jacobi equations. Entropic interpolations are linked to regularizing dissipative PDEs, while displacement interpolations are linked to transport PDEs. More detail about these relations is given in [28].

In practice, once the regularity of the solutions of the dissipative linear PDEs (6.11) is ensured, the rules of calculus that are displayed at the beginning of Section 6.2 are efficient. Again, this is in contrast with Otto's calculus which only yields heuristics about displacement interpolations.

With respect to (3), recall that we have seen in Theorems 6.5.4 and 6.5.8 unified proofs of entropy-entropy production inequalities, available in the diffusion and random walk settings.

In order to provide a better understanding of the analogy between displacement and entropic interpolations, we start describing two thought experiments. Then, keeping the LSV strategy as a guideline, we raise a few open questions about the use of entropic interpolations in connection with curvature problems on Riemannian manifolds and graphs.

Thought experiments

Let us describe two thought experiments.

- *Schrödinger's hot gas experiment.* The dynamical Schrödinger problem (see (S_{dyn}) in Section 6.1) is a formalization of Schrödinger's thought experiment which was introduced in [39]. Ask a perfect gas of particles living in a Riemannian manifold $\boldsymbol{X}$ which are in contact with a *thermal bath* to start from a configuration profile $\mu_0 \in \mathrm{P}(\boldsymbol{X})$ at time $t = 0$ and to end up at the unlikely profile $\mu_1 \in \mathrm{P}(\boldsymbol{X})$ at time $t = 1$. For instance the gas may be constituted of undistinguishable mesoscopic particles (e.g. grains of pollen) with a random motion communicated by numerous shocks with the microscopic particles of the thermal bath (e.g. molecules of hot water). Large deviation considerations lead to the following conclusion. In the (thermodynamical) limit of infinitely many mesoscopic particles, the most likely trajectory of the whole mesoscopic particle system from μ_0 to μ_1 is the R-entropic interpolation $[\mu_0, \mu_1]^R$ where R describes the random motion of the non-interacting (the gas is assumed to be perfect) mesoscopic particles. Typically, R describes a Brownian motion as in Schrödinger's original papers [39, 40]. For more detail see [28, §. 6] for instance.
- *Cold gas experiment.* The same thought experiment is also described in [44, pp. 445-446] in a slightly different setting where it is called the *lazy gas experiment.* The only difference with Schrödinger's thought experiment is that no thermal bath enters the game. The perfect gas is cold so that the paths of the particles are Riemannian geodesics in order to minimize the average kinetic action of the whole system: entropy minimization is replaced by quadratic optimal transport.

Connection between entropic and displacement interpolations

It is shown in [26, 29] that displacement interpolations are limits of entropic interpolations when the frequency of the random elementary motions encrypted by R tends down to zero. With the above thought experiment in mind, this corresponds to a thermal bath whose density tends down to zero. Replacing the hot water by an increasingly rarefied gas, at the zero-temperature limit, the entropic interpolation $[\mu_0, \mu_1]^R$ is replaced by McCann's displacement interpolation $[\mu_0, \mu_1]^{\mathrm{disp}}$.

Cold gas experiment

As the gas is (absolutely) cold, the particle sample paths are regular and deterministic. The trajectory of the mass distribution of the cold gas is the displacement interpolation $[\mu_0, \mu_1]^{\mathrm{disp}}$ which is the time marginal flow of a probability measure P on Ω concentrated on geodesics:

$$P = \int_{\boldsymbol{X}^2} \delta_{\gamma^{xy}}\, \pi(dxdy) \in \mathrm{P}(\Omega) \tag{6.60}$$

where $\pi \in \mathrm{P}(\boldsymbol{X}^2)$ is an optimal transport plan between μ_0 and μ_1, see [28]. Therefore,

$$\mu_t = P_t = \int_{\boldsymbol{X}^2} \delta_{\gamma_t^{xy}}\, \pi(dxdy) \in \mathrm{P}(\boldsymbol{X}), \quad 0 \le t \le 1, \tag{6.61}$$

where for simplicity it is assumed that there is a unique minimizing geodesic γ^{xy} between x and y (otherwise, one is allowed to replace $\delta_{\gamma^{xy}}$ with any probability concentrated on the set of all minimizing geodesics from x to y).

In a positively curved manifold, several geodesics starting from the same point have a tendency to approach each other. Therefore, it is necessary that the gas initially spreads out to thwart its tendency to concentrate. Otherwise, it would not be possible to reach a largely spread target distribution. The left side of Figure 6.1 (taken from [44]) depicts this phenomenon. This is also the case for the right side as one can verify by reversing time. The central part is obtained by interpolating between the initial and final segments of the geodesics.

On the other hand, in a negatively curved manifold, several geodesics starting from the same point have a tendency to depart form each other. Therefore, it is necessary that the gas initially concentrates to thwart its tendency to spread out. Otherwise, it would not be possible to reach a condensed target distribution.

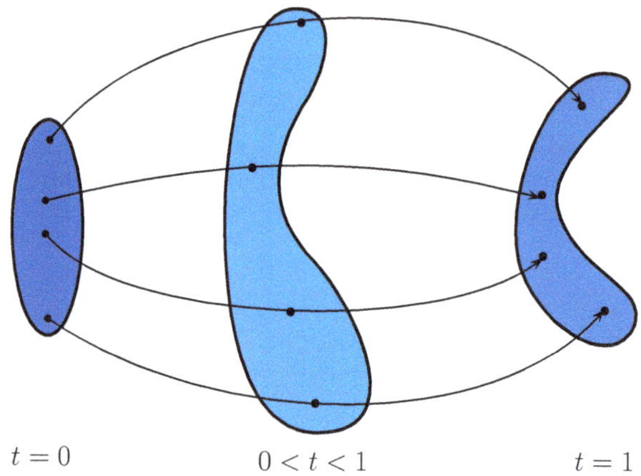

Fig. 6.1. The cold gas experiment. Positive curvature

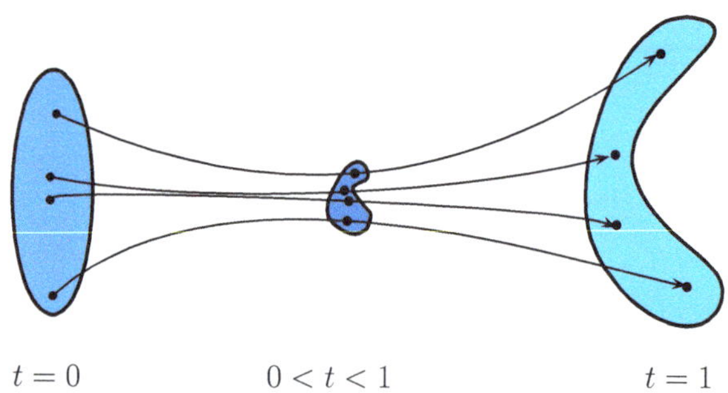

Fig. 6.2. The cold gas experiment. Negative curvature

Schrödinger's hot gas experiment

As the gas is hot, the particle sample paths are irregular and random. The trajectory of the mass distribution of the hot gas is the entropic interpolation $[\mu_0, \mu_1]^R$. An R-entropic interpolation is the time marginal flow of a probability measure P on Ω which is a mixture of bridges of R:

$$P = \int_{\boldsymbol{X}^2} R^{xy}\, \pi(dxdy) \in \mathrm{P}(\Omega) \tag{6.62}$$

where $\pi \in \mathrm{P}(\boldsymbol{X}^2)$ is the unique minimizer of $\eta \mapsto H(\eta|R_{01})$ among all $\eta \in \mathrm{P}(\boldsymbol{X}^2)$ with prescribed marginals μ_0 and μ_1, see [28]. Therefore,

$$\mu_t = P_t = \int_{\boldsymbol{X}^2} R_t^{xy}\, \pi(dxdy) \in \mathrm{P}(\boldsymbol{X}), \quad 0 \le t \le 1. \tag{6.63}$$

Comparing (6.60) and (6.62), one sees that the deterministic evolution $\delta_{\gamma^{xy}}$ is replaced by the random evolution R^{xy}. The grey leaf in Figure 6.3 is a symbol for the (say) 95%-support of the bridge R^{xy}. It spreads around the geodesic γ^{xy} and in general the support of R^{xy}_t for each intermediate time $0 < t < 1$ is the whole space $\boldsymbol{X}$.

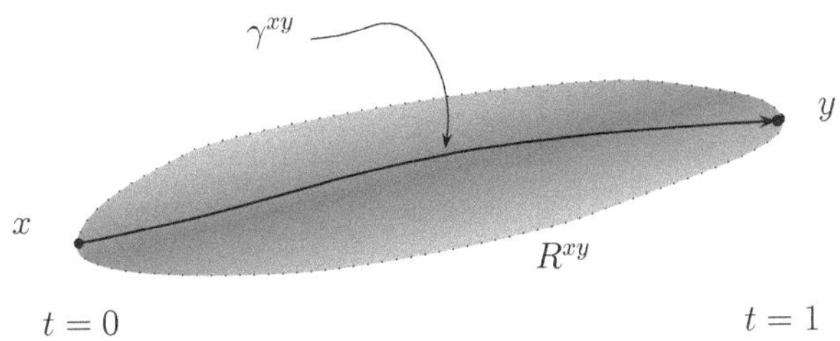

Fig. 6.3. Hot gas experiment with $\mu_0 = \delta_x$ and $\mu_1 = \delta_y$

At a lower temperature, R^{xy} is closer to $\delta_{\gamma^{xy}}$, see Figure 6.4.

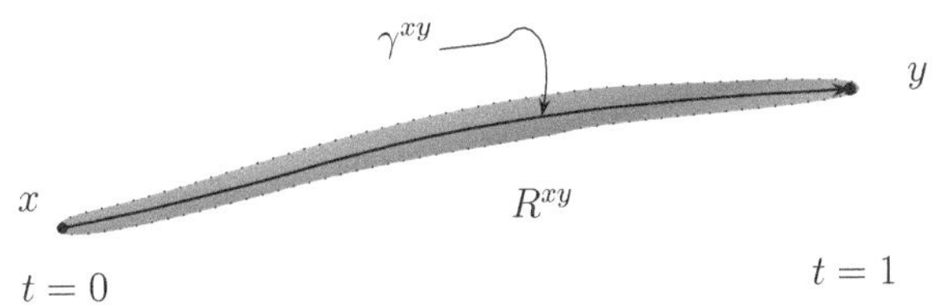

Fig. 6.4. Hot gas experiment with $\mu_0 = \delta_x$ and $\mu_1 = \delta_y$. At a lower temperature

And eventually, at zero temperature R^{xy} must be replaced by its deterministic limit $\delta_{\gamma^{xy}}$. Figure 6.5 is a superposition of Figures 6.2 and 6.3. It describes the hot gas experiment in a negative curvature manifold.

Figures 6.3, 6.4 and 6.5 are only suggestive pictures describing most visited space areas.

Open questions

As far as one is interested in the curvature of the metric measure space $(\boldsymbol{X}, d, m)$, rather than in the curvature of some Markov generator $\overrightarrow{L}$ such as in Section 6.5, it seems natural to restrict our attention to a reversible reference path measure R. For instance when looking at a Riemannian manifold $\boldsymbol{X}$ considered as a metric measure space $(\boldsymbol{X}, d, m)$ with the Riemannian metric d and the weighted measure $m = e^{-V}\,\mathrm{vol}$, the efficient choice with respect to LSV theory is to take R as the reversible Markov

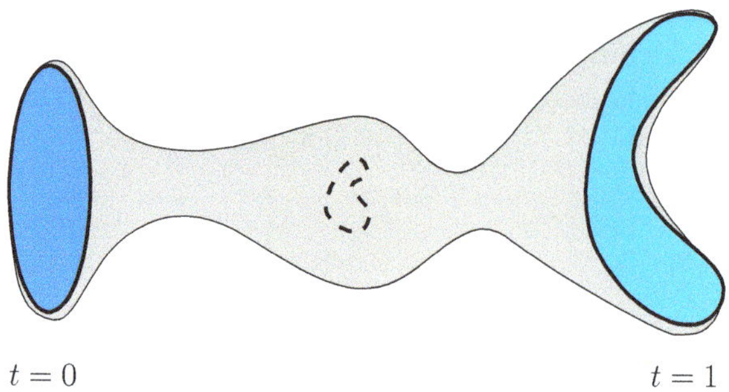

Fig. 6.5. The not so lazy gas experiment in a negative curvature manifold

measure with generator $L = (\Delta - \nabla V \cdot \nabla)/2$ (or any positive multiple) and initial measure m.

A representative result in the case where the reference measure m is the volume measure ($V = 0$) is that relative entropy is convex along any displacement interpolation in a Riemannian manifold with a nonnegative Ricci curvature. Based on McCann's seminal work [33, 34] and Otto's heuristic calculus, Otto and Villani have conjectured this result in [38]. This was proved by Cordero-Erausquin, McCann and Schmuckenschläger in [12]. The converse is also true, as was shown later by Sturm and von Renesse [43].

In the same spirit, an immediate consequence of Theorem 6.3.5 is the following

Corollary 6.6.1. *Let R^o be the reversible Brownian motion on a connected Riemannian manifold $\boldsymbol{X}$ without boundary.*
Suppose that the Ricci curvature is non-negative. Then, along any entropic interpolation $[\mu_0, \mu_1]^{R^o}$ associated with R^o and $f_0, g_1 \in L^2(\mathrm{vol})$ such that (6.8) holds, the entropy $t \in [0,1] \mapsto H(\mu_t|\,\mathrm{vol}) \in \mathbb{R}$ is a convex function.

Notice that in Theorem 6.3.5, $\boldsymbol{X}$ is assumed to be compact and f_0, g_1 must be $\mathcal{C}^2$. This was assumed to ensure the regularity of the interpolation (Proposition 6.3.4). However, in the present setting this regularity follows from the regularity of the heat kernel.

Open questions 6.6.2.

(a) *Is the converse of Corollary 6.6.1 true?*

(b) *Is there a way to define a notion of κ-convexity on $\mathrm{P}(\boldsymbol{X})$ along entropic interpolations: $H(\mu_t) \le (1-t)H(\mu_0) + tH(\mu_1) - \kappa C(\mu_0, \mu_1)t(1-t)/2$, $0 \le t \le 1$? What term $C(\mu_0, \mu_1)$ should replace the quadratic transport cost $W_2^2(\mu_0, \mu_1)$ of the LSV and AGS (Ambrosio, Gigli and Savaré [1, 2]) theories?*

(c) *Is there a notion of "entropic gradient flow" related to entropic interpolations that would be similar to a gradient flow and would allow interpreting heat flows as "en-*

tropic gradient flows" of the entropy? This question might be related to the previous one: the entropic cost $C(\mu_0, \mu_1)$ could play the role of a squared distance on $\mathrm{P}(\boldsymbol{X})$.

The LSV theory requires the metric space $(\boldsymbol{X}, d)$ to be geodesic. Consequently, discrete metric graphs are ruled out. Alternative approaches are necessary to develop a theory of lower bounded curvature on discrete metric graphs.

1. Bonciocat and Sturm [7] have obtained precise results for a large class of planar graphs by introducing the notion of h-approximate t-midpoint interpolations: $d(x_0, x_t) \le td(x_0, x_1)+h$, $d(x_t, x_1) \le (1-t)d(x_0, x_1)+h$ where typically h is of order 1.
2. Erbar and Maas [18, 32] and independently Mielke [35, 36] have shown that the evolution of a reversible random walk on a graph is the gradient flow of an entropy for some distance on $\mathrm{P}(\boldsymbol{X})$, and derived curvature bounds for discrete graphs by following closely the gradient flow strategy developed by Ambrosio, Gigli and Savaré [1] in the setting of the LSV theory.
3. Recently, Gozlan, Roberto, Samson and Tetali [23] obtained curvature bounds by studying the convexity of entropy along binomial interpolations. On a geodesic $(x := z_0, z_1, \dots, z_{d(x,y)} =: y)$ for the standard graph distance d, the binomial interpolation is defined by $\mu_t(z_k) = \mathcal{B}(d(x, y), t)(k)$, $0 \le k \le d(x, y)$, $0 \le t \le 1$, where $\mathcal{B}(n, p)(k) = \binom{n}{k} p^k(1 - p)^{n-k}$ is the usual binomial weight.

   We know by Lemma 6.4.3 that

$$\begin{aligned}\Theta_2 u(x) = \left(Bu(x)\right)^2 &+ \sum_{y:x\sim y} [J_y(\boldsymbol{X}) - J_x(\boldsymbol{X})]h\big(Du(x, y)\big)\, J_x(y)\\ &+ \sum_{(y,z):x\sim y\sim z} \left[2e^{Du(x,y)}h\big(Du(y, z)\big) - h\big(Du(x, z)\big)\right] J_x(y)J_y(z)\end{aligned} \tag{6.64}$$

where $Bu(x) = \sum_{y:x\sim y}(e^{Du(x,y)} - 1)\, J_x(y)$ and $h(a) := \theta^*(e^a - 1) = ae^a - e^a + 1$, $a \in \mathbb{R}$. In the diffusion setting where $L = (\Delta - \nabla V \cdot \nabla)/2$, we have seen in Section 6.3 that $\Theta_2 = \Gamma_2/2$ where Γ_2 is given by the Bochner formula

$$\Gamma_2(u) = \|\nabla^2 u\|_{\mathrm{HS}}^2 + \nabla^2 V(\nabla u) + \mathrm{Ric}(\nabla u). \tag{6.65}$$

Open questions 6.6.3.

1. *In view of* (6.64) *and* (6.65), *is* (6.64) *a Bochner formula? Where is the curvature?*
2. *Are the following definitions relevant?*
   (a) *The m-reversible random walk generator L has curvature bounded below by* $\kappa \in \mathbb{R}$ *if*

$$\sum_x \Theta_2(u)(x)m(x) \ge \kappa \sum_x \Theta(u)(x)m(x), \quad \forall u.$$

(b) *The infimum of the curvature at $x \in \boldsymbol{X}$ of the Markov generator L is*

$$\mathrm{curv}_L(x) := \inf_u \frac{\Theta_2(u)}{\Theta(u)}(x).$$

A definition of curvature should be justified by its usefulness in terms of rate of convergence to equilibrium, concentration of measure or isoperimetric behavior.

Let us indicate that with the discrete Laplacian $Lu(x) = [u(x+1)-2u(x)+u(x-1)]/2$ on $\mathbb{Z}$, a rather tedious computation leads to the desired flatness result: $\mathrm{curv}_L(x) = 0$, for all $x \in \mathbb{Z}$.

A. Basic definitions

This appendix section is part of articles [27, 30]. We repeat it here for the reader's convenience.

Markov measures

We slightly extend the notion of Markov property to unbounded positive measures on Ω.

Definitions .0.4. *Any positive measure on Ω is called a* path measure. *Let Q be a path measure.*

(a) *It is said to be a* conditionable path measure *if for any $0 \le t \le 1$, Q_t is a σ-finite measure on $\boldsymbol{X}$.*

(b) *It is said to be a Markov measure if it is conditionable and for all $0 \le t \le 1$ and $B \in \sigma(X_{[t,1]})$, we have: $Q(B \mid X_{[0,t]}) = Q(B \mid X_t)$.*

When Q has an infinite mass, the notion of conditional expectation must be handled with care. We refer to the following definition.

Definition .0.5 (Conditional expectation). *Let $\varphi : \Omega \to Y$ be a measurable map. Suppose that $\varphi_\# Q$ is a σ-finite measure on Y. Then, for any $f \in L^p(Q)$ with $p = 1, 2$ or ∞, the conditional expectation of f knowing φ is the unique (up to Q-a.e. equality) function $E_Q(f \mid \varphi) := \theta_f(\varphi) \in L^p(Q)$ such that for all measurable function h on Y in $L^{p_*}(m)$ where $1/p + 1/p_* = 1$, we have $\int_\Omega h(\varphi) f \, dQ = \int_\Omega h(\varphi)\theta_f(\varphi) \, dQ$.*

It is essential in this definition that the measure $\varphi_\# Q$ is σ-finite on Y; otherwise, we can't invoke the Radon-Nikodym theorem when proving the existence of θ_f. Inspecting the above definition of a Markov measure, one sees that it is necessary that $Q_{[0,t]}$ and Q_t are σ-finite for all $t \in [0, 1]$ for the corresponding conditional expectations to be defined. Nonetheless, this is warranted by the requirement that Q is conditionable, as shown by the following result.

Lemma .0.6. *Let Q be a path measure and $\mathcal{T} \subset [0,1]$. For $Q_{\mathcal{T}}$ to be σ-finite, it is enough that Q_{t_o} is σ-finite for some $t_o \in \mathcal{T}$.*

Proof. Let $t_o \in \mathcal{T}$ be such that Q_{t_o} is σ-finite with $(\boldsymbol{X}_n)_{n\geq 1}$ an increasing sequence of measurable subsets of $\boldsymbol{X}$ such that $Q_{t_o}(\boldsymbol{X}_n) < \infty$ and $\cup_{n\geq 1}\boldsymbol{X}_n = \boldsymbol{X}$. Then, $Q_{\mathcal{T}}$ is also σ-finite since $Q_{\mathcal{T}}(X_{t_o} \in \boldsymbol{X}_n) = Q_{t_o}(\boldsymbol{X}_n)$ for all $n \geq 1$ and $\cup_{n\geq 1}[X_{\mathcal{T}}(\Omega) \cap \{X_{t_o} \in \boldsymbol{X}_n\}] = X_{\mathcal{T}}(\Omega)$. □

Consequently, for any Markov measure Q and any $\mathcal{T} \subset [0,1]$, the conditional expectation $E_Q(\cdot \mid X_{\mathcal{T}})$ is well-defined; since Ω is a Polish space, there exists a regular conditional *probability* kernel $Q(\cdot \mid X_{\mathcal{T}}) : \Omega \to \mathrm{P}(\Omega)$ such that for all $f \in L^1(Q)$, $E_Q(f \mid X_{\mathcal{T}}) = \int_\Omega f \, dQ(\cdot \mid X_{\mathcal{T}})$, see [17, Thm. 10.2.2].

Stationary path measures

Let us make precise a couple of other known notions.

Definitions .0.7. *Let Q be a path measure.*

(a) It is said to be stationary *if for all $0 \leq t \leq 1$, $Q_t = m$, for some $m \in \mathrm{M}(\boldsymbol{X})$. One says that Q is m-stationary.*

(b) It is said to be reversible *if for any subinterval $[a,b] \subset [0,1]$, we have $(\mathrm{rev}^{[a,b]})_\# Q_{[a,b]} = Q_{[a,b]}$ where $\mathrm{rev}^{[a,b]}$ is the time reversal on $[a,b]$ which is defined by $\mathrm{rev}^{[a,b]}[\eta](t) = \eta([a+b-t]^+)$ for any $\eta \in D([a,b],\boldsymbol{X})$, $t \in [a,b]$.*

A reversible path measure is stationary. When Q is reversible, one sometimes says that $m = Q_0 = Q_1$ is a reversing measure for the forward kernel $(Q(\cdot \mid X_0 = x); x \in \boldsymbol{X})$, or the backward kernel $(Q(\cdot \mid X_1 = y); y \in \boldsymbol{X})$ or shortly for Q. One also says that Q is m-reversible to emphasize the role of the reversing measure.

Relative entropy

This subsection is a short part of [27, §. 2] which we refer to for more detail. Let r be some σ-finite positive measure on some space Y. The relative entropy of the probability measure p with respect to r is loosely defined by

$$H(p|r) := \int_Y \log(dp/dr)\, dp \in (-\infty, \infty], \qquad p \in \mathrm{P}(Y) \tag{6.66}$$

if $p \ll r$, and $H(p|r) = \infty$ otherwise. More precisely, when r is a probability measure, we have

$$H(p|r) = \int_Y h(dp/dr)\, dr \in [0, \infty], \qquad p, r \in \mathrm{P}(Y)$$

with $h(a) = a \log a - a + 1 \geq 0$ for all $a \geq 0$, (take $h(0) = 1$). Hence, the definition (6.66) is meaningful. If r is unbounded, one must restrict the definition of $H(\cdot|r)$ to some subset of $\mathrm{P}(Y)$ as follows. As r is assumed to be σ-finite, there exist measurable functions $W : Y \to [1, \infty)$ such that

$$z_W := \int_Y e^{-W}\, dr < \infty. \tag{6.67}$$

Define the probability measure $r_W := z_W^{-1} e^{-W} r$ so that $\log(dp/dr) = \log(dp/dr_W) - W - \log z_W$. It follows that for any $p \in \mathrm{P}(Y)$ satisfying $\int_Y W\, dp < \infty$, the formula

$$H(p|r) := H(p|r_W) - \int_Y W\, dp - \log z_W \in (-\infty, \infty]$$

is a meaningful definition of the relative entropy which is coherent in the following sense. If $\int_Y W'\, dp < \infty$ for another measurable function $W' : Y \to [0, \infty)$ such that $z_{W'} < \infty$, then $H(p|r_W) - \int_Y W\, dp - \log z_W = H(p|r_{W'}) - \int_Y W'\, dp - \log z_{W'} \in (-\infty, \infty]$. Therefore, $H(p|r)$ is well-defined for any $p \in \mathrm{P}(Y)$ such that $\int_Y W\, dp < \infty$ for some measurable nonnegative function W verifying (6.67).

Bibliography

[1] L. Ambrosio, N. Gigli, and G. Savaré: *Gradient flows in metric spaces and in the space of probability measures*. Lectures in Mathematics ETH Zürich. Birkhäuser, Basel, second edition, 2008.

[2] L. Ambrosio, N. Gigli, and G. Savaré: Metric measure spaces with Riemannian Ricci curvature. *Duke Math. J.*, 163:1405–1490, 2014.

[3] A. Arnold, E. Carlen, and Q. Ju: Large-time behavior of non-symmetric Fokker-Planck type equations. *Commun. Stoch. Anal.*, 2(1):153–175, 2008.

[4] D. Bakry: L'hypercontractivité et son utilisation en théorie des semigroupes. In *Ecole d'Eté de Probabilités de Saint-Flour (1992)*, number 1581 in Lecture Notes in Mathematics. Springer-Verlag, 1994.

[5] D. Bakry and M. Émery: Diffusions hypercontractives. In *Séminaire de Probabilités*, number 1123 in Lecture Notes in Mathematics. Springer-Verlag, 1985.

[6] S. Bobkov and P. Tetali: Modified logarithmic Sobolev inequalities in discrete settings. *J. Theoret. Probab.*, 19(2):289–336, 2006.

[7] A.-I. Bonciocat and K.T. Sturm: Mass transportation and rough curvature bounds for discrete spaces. *J. Funct. Anal.*, 256:2944–2966, 2009.

[8] P. Caputo, P. Dai Pra, and G. Posta: Convex entropy decay via the Bochner-Bakry-Emery approach. *Ann. Inst. Henri Poincaré Probab. Stat.*, 45(3):734–753, 2009.

[9] P. Caputo and G. Posta: Entropy dissipation estimates in a zero-range dynamics. *Probab. Theory Related Fields*, 139:65–87, 2007.
[10] P. Cattiaux: A pathwise approach of some classical inequalities. *Potential Anal.*, 20(4):361–394, 2004.
[11] L. Chaumont and G. Uribe-Bravo: Markovian bridges: weak continuity and pathwise constructions. *Ann. Probab.*, 39:609–647, 2011.
[12] D. Cordero-Erausquin, R. McCann, and M. Schmuckenschläger: A Riemannian interpolation inequality à la Borell, Brascamp and Lieb. *Invent. Math.*, 146(2):219–257, 2001.
[13] P. Dai Pra, A.M. Paganoni, and G. Posta: Entropy inequalities for unbounded spin systems. *Ann. Probab.*, 30(4):1959–1976, 2002.
[14] P. Dai Pra and G. Posta: Entropy decay for interacting systems via the Bochner–Bakry–Emery approach. *Electron. J. Probab.*, 18(52), 2013.
[15] C. Dellacherie and P.A. Meyer: *Probabilités et Potentiel. Ch. XII à XVI. Théorie du potentiel associée à une résolvante, théorie des processus de Markov.* Hermann. Paris, 1987.
[16] J.L. Doob: Conditional Brownian motion and the boundary limits of harmonic functions. *Bull. Soc. Math. France*, 85:431–458, 1957.
[17] R. Dudley: *Real Analysis and Probability*, volume 74 of *Cambridge Studies in Advanced Mathematics*. Cambridge University Press, Cambridge, 2002. Revised reprint of the 1989 original.
[18] M. Erbar and J. Maas: Ricci curvature of finite Markov chains via convexity of the entropy. *Arch. Ration. Mech. Anal.*, 206(3):997–1038, 2012.
[19] M. Fathi and J. Maas: Entropic Ricci curvature bounds for discrete interacting systems. To appear in Ann. Appl. Probab., arXiv:1501.00562.
[20] W.H. Fleming and H.M. Soner: *Controlled Markov Processes and Viscosity Solutions*, volume 25 of *Applications of Mathematics*. Springer, 1993.
[21] H. Föllmer: Time reversal on Wiener space. In *Stochastic Processes - Mathematic and Physics*, volume 1158 of *Lecture Notes in Math.*, pages 119–129. Springer, Berlin, 1986.
[22] J. Fontbona and B. Jourdain: A trajectorial interpretation of the dissipations of entropy and Fisher information for stochastic differential equations. *Ann. Probab.*, 44(1):131–170, 2016.
[23] N. Gozlan, C. Roberto, P.-M. Samson, and P. Tetali: Displacement convexity of entropy and related inequalities on graphs. *Probab. Theory Related Fields*, 160(1-2):47–94, 2014.
[24] C.-R. Hwang, S.-Y. Hwang-Ma, and S.-J. Sheu: Accelerating diffusions. *Ann. Appl. Probab.*, 15(2):1433–1444, 2005.
[25] C. Léonard: Stochastic derivatives and generalized h-transforms of Markov processes. Preprint, arXiv:1102.3172.
[26] C. Léonard: From the Schrödinger problem to the Monge-Kantorovich problem. *J. Funct. Anal.*, 262(1879-1920), 2012.
[27] C. Léonard: Some properties of path measures. In *Séminaire de probabilités de Strasbourg, vol. 46.*, pages 207–230. Lecture Notes in Mathematics 2123. Springer., 2014.
[28] C. Léonard: A survey of the Schrödinger problem and some of its connections with optimal transport. *Discrete Contin. Dyn. Syst. A*, 34(4):1533–1574, 2014.
[29] C. Léonard: Lazy random walks and optimal transport on graphs. *Ann. Probab.*, 44(3):1864–1915, 2016.
[30] C. Léonard, S. Rœlly, and J-C. Zambrini: Reciprocal processes. A measure-theoretical point of view. *Probab. Surv.*, 11:237–269, 2014.
[31] J. Lott and C. Villani: Ricci curvature for metric-measure spaces via optimal transport. *Ann. of Math.*, 169(3):903–991, 2009.
[32] J. Maas: Gradient flows of the entropy for finite Markov chains. *J. Funct. Anal*, 261:2250–2292, 2011.

[33] R. McCann: *A convexity theory for interacting gases and equilibrium crystals*. PhD thesis, Princeton Univ., 1994.
[34] R. McCann: A convexity principle for interacting gases. *Adv. Math.*, 128:153–179, 1997.
[35] A. Mielke: A gradient structure for reaction-diffusion systems and for energy-drift-diffusion systems. *Nonlinearity*, 24(4):1329–1346, 2011.
[36] A. Mielke: Geodesic convexity of the relative entropy in reversible Markov chains. *Calc. Var. Part. Diff. Equ.*, 48(1):1–31, 2013.
[37] E. Nelson: *Dynamical theories of Brownian motion*. Princeton University Press, 1967.
[38] F. Otto and C. Villani: Generalization of an inequality by Talagrand and links with the logarithmic Sobolev inequality. *J. Funct. Anal.*, 173(2):361–400, 2000.
[39] E. Schrödinger: Über die Umkehrung der Naturgesetze. *Sitzungsberichte Preuss. Akad. Wiss. Berlin. Phys. Math.*, 144:144–153, 1931.
[40] E. Schrödinger: Sur la théorie relativiste de l'électron et l'interprétation de la mécanique quantique. *Ann. Inst. H. Poincaré*, 2:269–310, 1932.
[41] K-T. Sturm: On the geometry of metric measure spaces, I. *Acta Math*, 196:65–131, 2006.
[42] K-T. Sturm: On the geometry of metric measure spaces, II. *Acta Math*, 196:133–177, 2006.
[43] K-T. Sturm and M-K. von Renesse: Transport inequalities, gradient estimates, entropy, and Ricci curvature. *Comm. Pure Appl. Math.*, 58(7):923–940, 2005.
[44] C. Villani: Optimal Transport. Old and New, volume 338 of *Grundlehren der mathematischen Wissenschaften*. Springer, 2009.

Nageswari Shanmugalingam

Brief survey on ∞-Poincaré inequality and existence of ∞-harmonic functions

7.1 Introduction

Recent work on analysis in metric measure spaces saw the development of Sobolev-type function theory and associated potential theory in the non-smooth setting of metric measure spaces. A significant part of this development is based on the theory of upper gradients first proposed by Heinonen and Koskela in [21], see [5] for potential theory based on this notion, and [22] for the corresponding study of metric space-valued Sobolev-type function theory in non-smooth settings.

In much of the above-mentioned analysis the basic requirements on the metric measure space under study are that the measure should be doubling and support a p-Poincaré inequality for some fixed p with $1 \le p < \infty$. The weakest of all the p-Poincaré inequalities is the ∞-Poincaré inequality. The paper [9] proved an analog of Rademacher's theorem for metric measure spaces whose measure is doubling and supports a p-Poincaré inequality for some $1 \le p < \infty$ and showed that for such metric measure spaces, the minimal p-weak upper gradient of a Lipschitz function f on the metric space *is* its pointwise Lipschitz constant function $\operatorname{Lip} f$. The conclusions of [9] did not depend on the precise value of p. Furthermore, it is known that spaces X whose measure is doubling and supports a p-Poincaré inequality are *quasiconvex*, that is, there is a constant $C \ge 1$ such that whenever $x, y \in X$ there is a rectifiable curve in X with end points x, y with length $\ell(\gamma) \le C\, d(x, y)$. Again, this consequence does not depend on the precise value of p.

It was therefore natural to ask whether one could obtain the results of [9] for metric measure spaces whose measure is doubling and supports an ∞-Poincaré inequality. The series of papers [12–16, 18] studied metric measure spaces equipped with a doubling measure supporting an ∞-Poincaré inequality, with this goal in mind. The purpose of this present article is to give an overview of the results obtained in these papers.

Throughout this article we will assume that the metric space X is complete and that the measure μ on X is a Radon measure and is doubling, that is, there is a constant $C_d \ge 1$ such that whenever $x \in X$ and $r > 0$, we have

$$0 < \mu(B(x, 2r)) \le C_d\, \mu(B(x, r)) < \infty.$$

Nageswari Shanmugalingam: Department of Mathematical Sciences, P.O.Box 210025, University of Cincinnati, Cincinnati, OH 45221–0025, U.S.A., E-mail: shanmun@uc.edu

The structure of this paper is as follows. In Section 2 we describe the basic notions needed to study first order calculus in the metric setting. In Section 3 we describe properties related to spaces supporting an ∞-Poincaré inequality, and in Section 4 we describe results from [15] demonstrating that, unlike in [26], one needs only ∞-Poincaré inequality in order to know existence and uniqueness of ∞-harmonic functions with prescribed Lipschitz boundary data. The final section, Section 5, will give examples demonstrating sharpness of the results in the previous section.

7.2 Background

In the non-smooth metric setting (including weighted Euclidean spaces and sub-Riemannian manifolds) the derivative of a function is not available. However, there are some possible notions that take on the role of magnitude of the derivative. Of these possible notions, the notion of (weak) upper gradients, developed in [21], has the so-called strong locality property.

Definition 7.2.1. *Given a function $f : X \to \mathbb{R}$, we say that a non-negative Borel-measurable function $g : X \to [0, \infty]$ is an upper gradient of f if whenever γ is a non-constant compact rectifiable curve in X,*

$$|f(y) - f(x)| \le \int_\gamma g\, ds, \tag{7.1}$$

where x and y denote the two end points of γ.

See [2] or [22] for more on path integrals, and [5, 19–22] for more on the notion of upper gradients.

Definition 7.2.2. *Let $1 \le p \le \infty$. A non-negative Borel-measurable function $g : X \to [0, \infty]$ is a p-weak upper gradient of f if the collection Γ of all non-constant compact rectifiable curves γ for which (7.1) fails has p-modulus zero, that is, there is a non-negative Borel measurable function $\rho \in L^p(X)$ such that for each $\gamma \in \Gamma$ we have $\int_\gamma \rho\, ds = \infty$.*

It was shown in [12] that a family Γ of non-constant compact rectifiable curves in X has ∞-modulus zero if and only if there is a non-negative Borel measurable function ρ on X with $\rho = 0$ μ-a.e. in X such that $\int_\gamma \rho\, ds = \infty$ whenever $\gamma \in \Gamma$. For $1 \le p < \infty$ such a strong result does not hold, but in this case Γ has p-modulus zero if and only if there is a non-negative Borel measurable function $\rho \in L^p(X)$ such that $\int_\gamma \rho\, ds = \infty$ for each $\gamma \in \Gamma$, see [29].

For $1 \le p \le \infty$ we set $D_p(f)$ to be the collection of all p-weak upper gradients of f that also belong to $L^p(X)$. For $1 < p < \infty$, the uniform convexity of $L^p(X)$ together with the lattice properties and convexity property of $D_p(f)$ imply the existence of a "minimal" function $g_f \in D_p(f)$ such that whenever $g \in D_p(f)$, we have $g_f \le g$ μ-a.e. in X.

Remark 7.2.3. It turns out that the existence of such $g_f \in D_p(f)$ holds for all $1 \le p \le \infty$, see [31, Theorem 4.6]. We can say even more for the case $p = \infty$. From the work of [12] mentioned above, we can modify every $g \in D_\infty(f)$ on a set of μ-measure zero such that the modified function is an *upper gradient* of f. Thus, without loss of generality, we may assume that every function in $D_\infty(f)$ is an upper gradient of f.

It was shown in [34] and [5] that if f is a measurable function with a p-weak upper gradient $g \in D_p(f)$, and if f is constant on an open set $U \subset X$, then $g\chi_{X\setminus U} \in D_p(f)$. This property, called the *strong locality property* of p-weak upper gradients, is highly useful in the development of potential theory in the metric setting, see [5].

Definition 7.2.4. *For $1 \le p < \infty$ we say that X supports a p-Poincaré inequality if there are constants $C > 0$ and $\lambda \ge 1$ such that whenever $f \in L^p(X)$ and $g \in D_p(X)$, for all balls $B = B(x, r) \subset X$ we have*

$$\fint_B |f - f_B|\, d\mu \le C\, r \left(\fint_{\lambda B} g^p\, d\mu \right)^{1/p}.$$

Here $f_B := \mu(B)^{-1} \int_B f\, d\mu =: \fint_B f\, d\mu$ is the average of f on the ball B, and $\lambda B := B(x, \lambda r)$. We say that X supports an ∞-Poincaré inequality if there are constants $C > 0$ and $\lambda \ge 1$ such that whenever $f \in L^\infty(X)$ and $g \in D_\infty(X)$, for all balls $B = B(x, r) \subset X$ we have

$$\fint_B |f - f_B|\, d\mu \le C\, r\, \|g\|_{L^\infty(\lambda B)}.$$

The Sobolev-type function spaces under consideration here are the Newton-Sobolev spaces, first developed in [34].

Definition 7.2.5. *The set $\widetilde{N^{1,p}}(X)$ is the collection of all functions $f : X \to \mathbb{R}$ such that $f \in L^p(X)$ and $D_p(f)$ is non-empty. For functions $f \in \widetilde{N^{1,p}}(X)$ we set*

$$\|f\|_{N^{1,p}(X)} := \|f\|_{L^p(X)} + \inf_{g \in D_p(f)} \|g\|_{L^p(X)}.$$

From the above discussion it is clear that when $f \in \widetilde{N^{1,p}}(X)$,

$$\|f\|_{N^{1,p}(X)} = \|f\|_{L^p(X)} + \|g_f\|_{L^p(X)}.$$

We say that $f_1 \sim f_2$ if $f_1, f_2 \in \widetilde{N^{1,p}}(X)$ and $\|f_1 - f_2\|_{N^{1,p}(X)} = 0$. It was shown in [34] that if $f_1, f_2 \in \widetilde{N^{1,p}}(X)$, then $f_1 \sim f_2$ if and only if $f_1 = f_2$ μ-a.e. in X and that $\sim$ is an equivalence relation on $\widetilde{N^{1,p}}(X)$.

Definition 7.2.6. *The Newton-Sobolev space $N^{1,p}(X)$ is the collection of all equivalence classes of functions from $\widetilde{N^{1,p}}(X)$ from the equivalence relation $\sim$.*

It was shown in [34] and [12] that $N^{1,p}(X)$ is a Banach space when equipped with the norm $\|\cdot\|_{N^{1,p}(X)}$.

A function $f : X \to \mathbb{R}$ is said to be L-Lipschitz on X if for all $x, y \in X$ we have $|f(x) - f(y)| \le L\, d(x, y)$. For such functions f and $x \in X$, we set

$$\operatorname{Lip} f(x) = \lim_{r\to 0} \sup_{x \neq y \in B(x,r)} \frac{|f(x) - f(y)|}{d(x, y)}$$

and call this the pointwise Lipschitz constant function of f. For $A \subset X$ we set

$$\operatorname{LIP}(f, A) = \sup_{x,y\in A, x\neq y} \frac{|f(x) - f(y)|}{d(x, y)}.$$

Note that $\operatorname{Lip} f$ is an upper gradient of the Lipschitz function f, see for example [20]. The Sobolev-type spaces studied in [9] are also based on the notion of upper gradients, combined with the idea of relaxation. It was shown in [34] that when $1 < p < \infty$ the Sobolev-type space of [9] agrees with the above $N^{1,p}(X)$. Combining the results of [9] with [34] it is seen that if X supports a p-Poincaré inequality for some $1 \le p < \infty$ and f is Lipschitz continuous on X, then $g_f = \operatorname{Lip} f$ μ-a.e. in X. Thus g_f in this case becomes independent of the choice of p. More specifically, $\operatorname{Lip} f$ is the minimal q-weak upper gradient of f in the class $D_q(f)$ for all $q \ge p$.

If X supports a p-Poincaré inequality for some $1 \le p < \infty$, then by Hölder's inequality it follows that X supports a q-Poincaré inequality for all $p \le q \le \infty$. It is a highly non-trivial result of Keith and Zhong [27] that if X is complete (recall that we also assume the measure to be doubling in our paper), and X supports a p-Poincaré inequality for some $1 < p < \infty$, then there is some $q \in [1, p)$ such that X supports a q-Poincaré inequality. The exposition of the proof of this result, given in [22], shows that q depends only on p and the doubling and p-Poincaré constants of μ and X. Such Gehring-type self-improvement is highly useful in regularity theory for p-harmonic functions in the metric setting, see for example [28], [5], and [22].

Definition 7.2.7. *Let $\Omega \subset X$ be a bounded domain such that $X\setminus\Omega$ has positive measure, and let $f : X \to \mathbb{R}$ be in $N^{1,p}(X)$. For $1 < p < \infty$ we say that a function $u \in N^{1,p}(X)$ with $u = f$ on $X \setminus \Omega$ is p-harmonic in Ω with boundary data f if whenever $\phi \in N^{1,p}(X)$ with $\phi = f$ on $X \setminus \Omega$, we have*

$$\|g_u\|_{L^p(\Omega)} \le \|g_\phi\|_{L^p(\Omega)}. \tag{7.2}$$

Because of the strongly local nature of p-weak upper gradients (that is, if two functions in $N^{1,p}(X)$ agree on a Borel set, then their minimal p-weak upper gradients agree almost everywhere on that set, see [34] or [5, Lemma 2.19], *and* the local nature of integrals, we see that

$$\int_X g_\phi^p \, d\mu = \int_{\operatorname{supt}(\phi-u)} g_\phi^p \, d\mu + \int_{X\setminus\operatorname{supt}(\phi-u)} g_u^p \, d\mu.$$

Therefore, u is a p-harmonic function on Ω with boundary data f if and only if $u = f$ on $X \setminus \Omega$ and whenever $V \subset \Omega$ is an open set and $\phi \in N^{1,p}(X)$ such that $\phi = u$ on $X \setminus V$, we have

$$\|g_u\|_{L^p(V)} \le \|g_\phi\|_{L^p(V)}.$$

While the notion of p-harmonicity from (7.2) will not yield a local notion of ∞-harmonicity when $p \to \infty$, the above notion does. Thus we have the following definition of ∞-harmonicity.

Definition 7.2.8. *We say that $u \in N^{1,\infty}(X)$ is ∞-harmonic in Ω with boundary data $f \in N^{1,\infty}(X)$ if $u = f$ on $X \setminus \Omega$ and whenever $V \subset \Omega$ is an open set and $\phi \in N^{1,\infty}(X)$ such that $\phi = u$ on $X \setminus V$, we have*

$$\|g_u\|_{L^\infty(V)} \le \|g_\phi\|_{L^\infty(V)}.$$

With the above definition, a function u that is ∞-harmonic in Ω is also ∞-harmonic in every subdomain U of Ω, that is, ∞-harmonicity is a local property.

Note that a function f is in $N^{1,\infty}(X)$ if and only if $\|f\|_{L^\infty(X)} = \operatorname{esssup}_{x\in X}|f(x)|$ is finite and if it has an ∞-weak upper gradient g such that $\|g\|_{L^\infty(X)}$ is also finite. We do not claim that in general such functions are Lipschitz continuous on X; the paper [12] has examples of functions in $N^{1,\infty}(X)$ that fail to be Lipschitz continuous on X. As we will see in the next section, if X in addition supports an ∞-Poincaré inequality, then indeed such functions must be Lipschitz continuous on X.

Unlike ∞-harmonicity, the notion of minimal Lipschitz extension is not a local property. A Lipschitz function $u : \overline{\Omega} \to \mathbb{R}$ is said to be a minimal Lipschitz extension of $f = u|_{\partial\Omega}$ if

$$\operatorname{LIP}(u, \Omega) \le \operatorname{LIP}(u, \partial\Omega) = \operatorname{LIP}(f, \partial\Omega). \tag{7.3}$$

For every Lipschitz function $w : \overline{\Omega} \to \mathbb{R}$, we always have that

$$\operatorname{LIP}(w, \Omega) \ge \operatorname{LIP}(w, \partial\Omega),$$

and hence the minimality of u in the above definition. Every Lipschitz function $f : \partial\Omega \to \mathbb{R}$ has a minimal Lipschitz extension to Ω, as demonstrated by McShane [32]. In fact, the proof given in [32] also shows the non-uniqueness of such extension, for both the following two extensions are minimal Lipschitz extensions to Ω:

$$u^+(x) = \inf\{f(y) + L\, d(x, y) \,:\, y \in \partial\Omega\},$$
$$u^-(x) = \sup\{f(y) - L\, d(x, y) \,:\, y \in \partial\Omega\}.$$

Here $L = \mathrm{LIP}(f, \partial\Omega)$. In the seminal paper [3] Aronsson sought the optimal extensions that are minimal Lipschitz extensions locally as well; that is, (7.3) is satisfied not only for Ω but also for every non-empty open subset U of Ω (by replacing Ω with U in (7.3)). Functions satisfying this condition are called *absolute minimal Lipschitz extensions*, or AMLEs for short.

It was shown in [3] that AMLEs F in Euclidean domains satisfy $\Delta_\infty F = 0$, that is, they are ∞-harmonic. Here,

$$\Delta_\infty F = \sum_{i,j=1}^{n} \frac{\partial F}{\partial x_i}\frac{\partial F}{\partial x_j}\frac{\partial^2 F}{\partial x_i \partial x_j}$$

is the ∞-Laplacian of F. Indeed, a function on an Euclidean domain is an AMLE if and only if it is ∞-harmonic, see for example [11] or [4, Theorem 4.13].

There are at least two ways of constructing AMLEs in the Euclidean setting, see [3] or [4] for a general overview of the topic. The first of the two methods employed in [3] uses a Perron method (which, in [4] is also called *comparison with cones*) and requires only the knowledge of the metric, see [25] for the extension of this method to metric spaces that are length spaces. The second method employed in [3] was to construct an ∞-harmonic extension, and since ∞-harmonicity and the AMLE property concide in the Euclidean setting, this construction will also yield an AMLE. This latter method used the non-linear potential theory to construct p-harmonic extensions u_p of the Lipschitz boundary data $f : \partial\Omega \to \mathbb{R}$ for $1 < p < \infty$, and showed that there exists a sequence $p_k \to \infty$ for which u_{p_k} converges uniformly to a limiting Lipschitz function u_∞, which was then shown to be ∞-harmonic. This method was extended to the setting of doubling metric measure spaces supporting a p-Poincaré inequality for some finite p in [26] to construct ∞-harmonic extensions of the boundary data f. It was shown in [26] that if the underlying metric measure space satisfies a p-weak Fubini property (see Section 4), then ∞-harmonic functions are AMLEs. In Section 4 we will explain how to construct ∞-harmonic functions and describe some connections between AMLEs and ∞-harmonic functions in the setting of metric measure spaces that support an ∞-Poincaré inequality but might not support any p-Poincaré inequality for any finite $p \geq 1$. In Section 5 we will describe examples of such metric measure spaces (such as the Sierpinski Carpet).

7.3 Characterizations of ∞-Poincaré inequality

The notion of p-modulus zero family of curves, as described in Definition 7.2.2, is extended to the notion of p-modulus as an outer measure on the collection of all rectifiable curves in X as follows.

Definition 7.3.1. *Let Γ be a family of rectifiable curves in X, and let $\mathcal{A}(\Gamma)$ denote the collection of all non-negative Borel measurable functions ρ on X such that $\int_\gamma \rho\, ds \geq 1$*

for each $\gamma \in \Gamma$. *For* $1 \le p < \infty$, *it is traditional to set*

$$\mathrm{Mod}_p(\Gamma) = \inf_{\rho \in \mathcal{A}(\Gamma)} \int_X \rho^p \, d\mu,$$

see [20] for example. For $p = \infty$ *we extend the above notion by considering the* p*-th root of* Mod_p *and letting* $p \to \infty$:

$$\mathrm{Mod}_\infty(\Gamma) = \inf_{\rho \in \mathcal{A}(\Gamma)} \|\rho\|_{L^\infty(X)}.$$

Recall that X is a complete metric space equipped with a doubling measure μ. The following characterizations of ∞-Poincaré inequality were established in [13] and [14]. In what follows, we will automatically assume that every rectifiable path (and these are the only paths we will consider in this article) is arc-length parametrized.

Theorem 7.3.2. ([14, Theorem 3.1], [13, Theorem 4.7]) *Let* X *be complete, connected, and* μ *be doubling. Then the following are equivalent:*

1. X *supports an ∞-Poincaré inequality.*
2. *There exist constants* $C, \lambda \ge 1$ *such that if* $f \in L^\infty(X)$ *with an upper gradient* $g \in L^\infty(X)$, *then* f *is* $C\|g\|_{L^\infty(X)}$*-Lipschitz continuous on* X *and is* $C\|g\|_{L^\infty(\lambda B)}$*-Lipschitz continuous on each ball* $B \subset X$.
3. *There is a constant* $C \ge 1$ *such that for all* $N \subset X$ *with* $\mu(N) = 0$ *and* $x, y \in X$ *with* $x \neq y$, *there is a rectifiable curve* γ *with end points* x, y *such that* $\ell(\gamma) \le C d(x, y)$ *and* $\mathcal{H}^1(\gamma^{-1}(N)) = 0$.
4. *There is a constant* $C \ge 1$ *such that whenever* $x, y \in X$ *are two distinct points, setting* $\Gamma(x, y, C)$ *to be the collection of all rectifiable curves* γ *in* X *with end points* x, y *such that* $\ell(\gamma) \le C\, d(x, y)$, *we have*

$$\mathrm{Mod}_\infty(\Gamma(x, y, C)) > 0.$$

5. *There is a constant* $C \ge 1$ *such that whenever* $x, y \in X$ *are two distinct points, with* $\Gamma(x, y, C)$ *as above we have*

$$\frac{1}{C\, d(x, y)} \le \mathrm{Mod}_\infty(\Gamma(x, y, C)) \le \frac{C}{d(x, y)}.$$

Note that while Property (4) gives only a qualitative control of the ∞-modulus of the family $\Gamma(x, y, C)$, Property (5) gives quantitative control. Property (3) on the other hand is a purely geometric measure-theoretic property; thus it is clear that ∞-Poincaré inequality is a geometric measure-theoretic notion.

Remark 7.3.3. As a consequence of the above theorem, more specifically Property (2) of the theorem, we know that if $f : X \to \mathbb{R}$ has an ∞-weak upper gradient g in X and that $g \le L$ μ-a.e. in X, then f is CL-Lipschitz continuous on X. This is of particular use to us in Section 4.

Observe also that if X does *not* support any ∞-Poincaré inequality, then for each positive integer n we can find two distinct points $x_n, y_n \in X$ and a set $N_n \subset X$ with $\mu(N_n) = 0$ such that N_n separates x_n from y_n, that is, every rectifiable curve γ in X with end points x_n, y_n and with $\ell(\gamma) \le n\, d(x_n, y_n)$ must see N_n as a large set ($\mathcal{H}^1(\gamma^{-1}(N_n)) > 0$). One can always choose N_n to be independent of n by replacing N_n with $N := \bigcup_{k\in\mathbb{N}} N_k$. It would be interesting to know whether one can choose x_n, y_n to be also independent of n. In Section 5 we will give an example (Sierpiński Carpet) of a complete doubling metric measure space X for which there is a set $N \subset X$ with $\mu(N) = 0$ such that *every* pair of distinct points $x, y \in X$ plays the role of x_n, y_n in the above discussion.

We now compare the above result regarding ∞-Poincaré inequality to analogous results concerning p-Poincaré inequalities for finite $p \ge 1$. For the sake of brevity, we focus on metric measure spaces whose measure μ is Ahlfors Q-regular for some $Q \ge 1$, that is, there is a constant $C \ge 1$ such that whenever $x \in X$ and $0 < r < 2\mathrm{diam}\,(\mathrm{X})$,

$$\frac{r^Q}{C} \le \mu(B(x,r)) \le C\, r^Q.$$

A metric measure space supports a 1-Poincaré inequality if and only if it supports a *relative isoperimetric inequality*, that is, with $P(E, A)$ denoting the perimeter measure of the set $E \subset X$ inside an open set $A \subset X$ (see for example [1]),

$$\min\{\mu(B \cap E), \mu(B \setminus E)\} \le C\, \mathrm{rad}(B)\, P(E, \lambda B)$$

for all balls $B \subset X$. The perimeter measure $P(E, B)$ is comparable to the co-dimension 1 Hausdorff measure of the part of the measure-theoretic boundary of E that is contained in B when E is of finite perimeter in the sense of [1], that is,

$$P(E,B) \approx \liminf_{r\to 0^+} \left\{ \sum_{i\in I} \frac{\mu(B_i)}{\mathrm{rad}(B_i)} \;:\; I \subset \mathbb{N},\ \text{each } B_i \text{ is a ball, } E\cap B \subset \bigcup_{i\in I} B_i,\ \mathrm{rad}(B_i) \le r \right\}.$$

An Ahlfors Q-regular metric measure space supports a Q-Poincaré inequality if and only if it is a *Q-Loewner space* in the sense of Heinonen and Koskela [21]. A space is Q-Loewner if there is a decreasing homeomorphism $\phi : (0, \infty) \to (0, \infty)$ such that whenever $F, K \subset X$ are two compact connected sets with at least two points each such that $F \cap K$ is empty, then the p-modulus of the family of all rectifiable curves in X with one end point in F and the other in K is at least $\phi(\mathrm{d}(F, K)/\mathrm{diam}\,(\mathrm{F}) \wedge \mathrm{diam}\,(\mathrm{K}))$. Compared to the above characterization of 1-Poincaré inequality, the characterizaton of Q-Poincaré inequality is more intimately connected with the number of rectifiable curves in X, and is more similar to Property (5) of Theorem 7.3.2 above. A refinement of the argument found in [21] would allow us to refine the notion of Q-Loewner property by letting us to restrict attention to C-quasiconvex curves (as in the sense of Properties (4) and (5) above) connecting F to K. In the Q-Loewner property one needs the quantitative lower bound for the p-modulus of the relevant family of curves; surprisingly, from

Property (4) of Theorem 7.3.2 we know that a characterization of ∞-Poincaré inequality only requires that the ∞-modulus of the relevant family of curves be positive.

This connection between the amount of rectifiable curves in X and the support of p-Poincaré inequality becomes stronger when $p > Q$.

Proposition 7.3.4 ([14, Theorem 5.1(3)]). *Suppose that X is Ahlfors Q-regular and that $p > Q$. Then X supports a p-Poincaré inequality if and only if there is a constant $C \geq 1$ such that whenever $x, y \in X$ are two distinct points, then*

$$\operatorname{Mod}_p(\Gamma(x, y, C)) \geq \frac{1}{C\, d(x, y)^{p-Q}}.$$

Note that the lower bound above for $\operatorname{Mod}_p(\Gamma(x, y, C))^{1/p}$ tends to the lower bound found in Property (4) of Theorem 7.3.2. This gives us hope that other geometric properties related to p-Poincaré inequalities persist also for ∞-Poincaré inequality. Unfortunately this is not the case. Properties such as persistence of p-Poincaré inequalities under pointed measured Gromov-Hausdorff limits, self-improvement of p-Poincaré inequality to q-Poincaré inequality for some $q < p$ when $p > 1$, Rademacher-type differentiability of Lipschitz functions in the sense of Cheeger [9] all fail for spaces that support ∞-Poincaré inequality but no p-Poincaré inequality for any finite $p \geq 1$. We will describe some examples regarding this in Section 5. These examples are from [18] and [14].

Now we revert back to our standing assumptions that X is complete and μ is doubling. If X supports a p-Poincaré inequality and $\Omega \subset X$ is a *uniform domain*, then Ω supports a p-Poincaré inequality, see [6]. A domain Ω is a uniform domain if there is a constant $C \geq 1$ such that whenever $x, y \in \Omega$ there is a curve γ, called a *uniform curve*, with end points x, y such that $\ell(\gamma) \leq C\, d(x, y)$ and whenever z is a point in γ, and $\gamma_{x,z}$, $\gamma_{y,z}$ are two subcurves of γ with end points x, z and y, z respectively, we have

$$\min\{\ell(\gamma_{x,z}), \ell(\gamma_{y,z})\} \leq C \operatorname{dist}(z, X \setminus \Omega).$$

No geometric characterization is known for domains in X that would inherit the property of supporting a p-Poincaré inequality for $1 \leq p < \infty$. However, we have the following geometric characterization for inheritance of ∞-Poincaré inequality.

Lemma 7.3.5. *Suppose that X supports an ∞-Poincaré inequality. Let $\Omega \subset X$ be a domain such that the restriction of μ to Ω is doubling. Then Ω, equipped with the restriction of the measure μ and the metric d to Ω, supports an ∞-Poincaré inequality if and only if it is quasiconvex.*

Recall that a set $A \subset X$ is quasiconvex if there is a constant $C \geq 1$ such that whenever $x, y \in A$ there is a rectifiable curve γ in A with end points x, y such that $\ell(\gamma) \leq C\, d(x, y)$. Since this characterization has not appeared in any other current literature, we provide its proof here. The proof relies heavily on the characterization (3) of Theorem 7.3.2.

As stated, this theorem requires X to be complete. However, it does remain valid when X is locally complete as well, as demonstrated in [14]. If X is complete and Ω is a domain in X, then Ω is necessarily locally complete.

Proof. Let $N \subset \Omega$ such that $\mu(N) = 0$, and $x, y \in \Omega$. By assumption, there is a quasiconvex curve γ in Ω connecting x to y. For each point $z \in \gamma$, there exists $r_z > 0$ such that $B(z, 2Cr_z) \subset \Omega$, where C is the constant for X from Theorem 7.3.2(3) (which exists because X supports an ∞-Poincaré inequality). The collection of balls $B(z, r_z)$ forms a cover of the compact set γ, and so there is a finite subcover, say $B_i = B(z_i, r_i)$, $i = 1, \cdots, k$. Let a_i be the location at which γ first enters B_i, and b_i be the last time γ exits B_i. Note that

$$\sum_{i=1}^{k} d(a_i, b_i) \le 2\ell(\gamma) \le 2Cd(x, y).$$

Now we use the fact that X supports an ∞-Poincaré inequality (see Theorem 7.3.2(3)) to find a quasiconvex curve β_i (with $\ell(\beta_i) \le Cd(a_i, b_i)$) connecting a_i to b_i such that $\beta_i \subset 2CB_i \subset \Omega$ and $\mathcal{H}^1(\beta_i^{-1}(N)) = 0$. Let γ_0 be the concatenation of the curves β_i, $i = 1, \cdots, k$. Then $\ell(\gamma_0) \le 2C^2 d(x, y)$ and γ_0 lies in Ω and connects x to y, with $\mathcal{H}^1(\gamma_0^{-1}(N)) = 0$. By Theorem 7.3.2(3), the support of an ∞-Poincaré inequality follows.

Conversely, if Ω supports an ∞-Poincaré inequality, then by Theorem 7.3.2 we know that Ω has to be quasiconvex. This completes the proof. □

7.4 Existence of ∞-harmonic extensions of Lipschitz functions

Throughout this section we will assume in addition to the doubling property of μ that μ also supports an ∞-Poincaré inequality.

We fix a bounded domain $\Omega \subset X$ such that $\mu(X \setminus \Omega) > 0$, and an L-Lipschitz function $f : X \to \mathbb{R}$. In this section we seek to find a function $u \in N^{1,\infty}(X)$ that is ∞-harmonic in Ω and $u = f$ on $X \setminus \Omega$.

The definition of ∞-harmonic functions can be found in Definition 7.2.8 above. From Remark 7.2.3 we know that the minimal ∞-weak upper gradient g_f of a function f with $g_f \in L^\infty(X)$ can be modified on a set of measure zero such that the modified function is an upper gradient of f. Thus from now on g_f will denote such a *minimal upper gradient* of f.

Definition 7.4.1. *The space $N_0^{1,\infty}(\Omega)$ consists of all the functions $v \in N^{1,\infty}(X)$ that satisfy $v = 0$ on $X \setminus \Omega$.*

It follows from Definition 7.2.8 that a function $u \in N^{1,\infty}(X)$ is ∞-harmonic in Ω if and only if for each $v \in N_0^{1,\infty}(X)$, we have

$$\|g_u\|_{L^\infty(\mathrm{supt}(v)\cap\Omega)} \le \|g_{u+v}\|_{L^\infty(\mathrm{supt}(v)\cap\Omega)}. \tag{7.4}$$

Note that Lipschitz functions belong to $N^{1,p}_{loc}(X)$ for each $1 \le p \le \infty$, and hence if X supports a p-Poincaré inequality, for $\infty > q \ge p$ one can find a Hölder continuous q-harmonic function on Ω that agrees with f in $X \setminus \Omega$, with Hölder continuity constant dependent solely on q, the doubling constant and the constants related to the Poincaré inequality, and the bound on f and the Lipschitz constant of f. The uniform limit of a subsequence of the sequence of q-harmonic functions, as $q \to \infty$, will yield an ∞-harmonic function that solves the above-stated problem, see for example [26]. There are many complete doubling metric measure spaces that support an ∞-Poincaré inequality but support no p-Poincaré inequality for any finite $p \ge 1$. For such spaces this approach might fail to give an ∞-harmonic function. In [15] one modifies the above approach by considering the following rather than q-harmonic functions.

From now on, L will denote the essential supremum $\|g_f\|_{L^\infty(X)}$ of the minimal ∞-weak upper gradient of f.

Definition 7.4.2. *We fix $L > 0$ as above, and set $N^{1,\infty}_L(X)$ to be the collection of all functions u on X that have an upper gradient g with $\|g\|_{L^\infty(X)} \le L$. For $u \in N^{1,\infty}_L(X)$ the set $D_L(u)$ is the collection of all upper gradients g of u such that $\|g\|_{L^\infty(X)} \le L$.*

Functions in $N^{1,\infty}_L(X)$ might not be L-Lipschitz, but given that X supports an ∞-Poincaré inequality, they are CL-Lipschitz where C is the constant given by the ∞-Poincaré inequality on X.

Definition 7.4.3. *Fix $1 < p < \infty$. For $u \in N^{1,\infty}(X)$ we set $I^p_L(u) := \inf_{g \in D_L(u)} \int_\Omega g^p\, d\mu$, and let*

$$J^p_f = \inf_{u \in N^{1,\infty}(X)\,:\, u=f \text{ on } X\setminus\Omega} I^p_L(u).$$

In the above, if $u \in N^{1,\infty}(X)$ but $D_L(u)$ is empty, then $I^p_L(u) = \infty$.

Note that $J^p_f \le I_L(f)^p \le L^p\mu(\Omega) < \infty$, and so we can find a sequence $u_k \in N^{1,\infty}_L(X)$ with $u_k = f$ on $X\setminus\Omega$ such that $\lim_k I^p_L(u_k) = J^p_f$. Since each u_k is CL-Lipschitz, the family $\{u_k\}_k$ is equicontinuous on X, and since $u_k = f$ on $X\setminus\Omega$ with Ω bounded, it follows that the family is also equibounded on X. Thus an application of the Arzelà-Ascoli theorem allows us to, by passing to a subsequence if necessary, find a CL-Lipschitz function u_p on X such that $u_k \to u_p$ uniformly on X.

It was shown in [15] that $J^p_f = I^p_L(u_p)$ and that $u_p \in N^{1,\infty}_L(X)$ with $u_p = f$ on $X \setminus \Omega$. It was also shown there that such u_p is unique given f. This uniqueness result was used to show that solutions u_p satisfy a weak comparison principle: if $F \in N^{1,\infty}_L(X)$ such that $f \le F$ on $X \setminus \Omega$, then the solution U_p associated with the boundary data F satisfies $u_p \le U_p$ on Ω. In the proof of this comparison principle the local nature of the L^p-integral was a key tool.

The next step was to fix a monotone increasing sequence $\{p_k\}_k$ with $1 < p_k < \infty$ and for each $k \in \mathbb{N}$ consider u_{p_k} as above. By passing to a subsequence if necessary, it was possible to obtain a uniform limit ϕ of the equicontinuous equibounded sequence

$\{u_{p_k}\}_k$ of CL-Lipschitz functions, and show that ϕ is ∞-harmonic in Ω. Thus we have the following theorem (recall the standing assumption for this section that X supports an ∞-Poincaré inequality).

Theorem 7.4.4 ([15]). *The function ϕ is ∞-harmonic in Ω with $\phi \in N^{1,\infty}(X)$ and $\phi = f$ on $X \setminus \Omega$.*

The next natural question to ask is whether ∞-harmonic functions are necessarily AMLEs. It was shown in [15] that this is not the case, see the next section for a description of this example. In [26] it was shown that if the metric measure space satisfies a p-weak Fubini property, then ∞-harmonic functions are AMLEs and that AMLEs are also ∞-harmonic.

Definition 7.4.5. *Let $1 \le p \le \infty$. A metric measure space is said to satisfy a p-weak Fubini property if there are positive constants C and τ_0 such that whenever $0 < \tau < \tau_0$ and $B_1, B_2 \subset X$ are measurable sets with positive measure such that $\mathrm{d}(B_1, B_2) > \tau \max\{\mathrm{diam}\,(B_1), \mathrm{diam}\,(B_2)\}$, then $\mathrm{Mod}_p(\Gamma(B_1, B_2, \tau) > 0$ where $\Gamma(B_1, B_2, \tau)$ is the collection of all rectifiable curves γ in X with one end point in B_1, the other in B_2, and $\ell(\gamma) < \mathrm{d}(B_1, B_2) + C\tau$.*

A space satisfying a p-weak Fubini property will necessarily satisfy a q-weak Fubini property for each $p \le q \le \infty$. The following simple geometric characterization of ∞-weak Fubini property, akin to that of Theorem 7.3.2(3), holds. No such characterization is known to hold for p-weak Fubini property for finite $p \ge 1$. Note also by Theorem 7.3.2 that if X satisfies an ∞-weak Fubini property, then X supports an ∞-Poincaré inequality.

Lemma 7.4.6 ([15]). *X satisfies an ∞-weak Fubini property if and only if for every set $N \subset X$ with $\mu(N) = 0$, every $\varepsilon > 0$, and every pair of distinct points $x, y \in X$, there is a rectifiable curve γ with end points x, y such that $\ell(\gamma) \le d(x, y) + \varepsilon$ and $\mathcal{H}^1(\gamma^{-1}(N)) = 0$.*

Theorem 7.4.7 ([15]). *If X satisfies an ∞-weak Fubini property and $f : X \to \mathbb{R}$ belongs to $N^{1,\infty}(X)$, then every ∞-harmonic extension of f to Ω is also an AMLE. Furthermore, every AMLE of f to Ω is necessarily ∞-harmonic in Ω.*

The ∞-weak Fubini property is not an unreasonable property to consider, as the following proposition shows. Given a set $N \subset X$ with $\mu(N) = 0$ we set the function $d_N : X \times X \to [0, \infty]$ as follows:

$$d_N(x, y) = \inf\{\ell(\gamma) \;:\; \gamma \text{ is rectifiable with end points } x, y \text{ and } \mathcal{H}^1(\gamma^{-1}(N)) = 0\}.$$

If X supports an ∞-Poincaré inequality, then by Property (3) of Theorem 7.3.2 we see that d_N is a metric on X and that $d(x, y) \le d_N(x, y) \le C\, d(x, y)$. We set

$$d_\mu(x, y) := \sup\{d_N(x, y) \;:\; N \subset X \text{ with } \mu(N) = 0\}.$$

By above, if X supports an ∞-Poincaré inequality, then d_μ is biLipschitz equivalent to the original metric d.

Proposition 7.4.8 ([15]). *If X supports an ∞-Poincaré inequality, then the metric measure space (X, d_μ, μ) satisfies an ∞-weak Fubini property.*

Thus if X supports an ∞-Poincaré inequality, then the class of AMLEs and the class of ∞-harmonic functions with respect to the metric d_μ are the same class and this class is non-empty. It was shown in [24] and [33] that AMLEs of Lipschitz boundary functions in a geodesic space are unique. It follows then that if X supports an ∞-weak Fubini property, then $d_\mu = d$ and so ∞-harmonic extensions of Lipschitz functions are unique.

Going one step further, it was shown in [15] that the class of ∞-harmonic functions with respect to the metric d and the class of ∞-harmonic functions with respect to the metric d_μ coincide if X supports an ∞-Poincaré inequality. Hence even if X does not satisfy an ∞-weak Fubini property, uniqueness of ∞-harmonic functions in (X, d, μ) follows under the uniqueness of ∞-harmonic functions in (X, d_μ, μ). We will demonstrate in the next section that if the ∞-weak Fubini property fails, then ∞-harmonic extensions need not be unique even if the metric space is a geodesic space. We do not know whether if X is a *geodesic space* supporting an ∞-Poincaré inequality then $d_\mu = d$.

7.5 Examples

In this section we give some examples that show the optimality of the results presented in this paper. We begin with an example of a complete metric measure space whose measure is doubling and is quasiconvex, but does not support any ∞-Poincaré inequality.

Example 7.5.1. Let X be the Sierpiński carpet, equipped with the Euclidean metric and the natural Hausdorff measure $\mu = \mathcal{H}^{\log(8)/\log(3)}$. Then μ is doubling, and indeed is Ahlfors $\log(8)/\log(3)$-regular. Furthermore, X is $\sqrt{2}$-quasiconvex. However, by the results of [7] X cannot support an ∞-Poincaré inequality. Indeed, the projection of the measure μ to $[0, 1]$ via the first coordinate projection map yields a measure μ_0 that is singular with respect to the 1-dimensional Lebesgue measure $\mathcal{L}^1$ on $[0, 1]$ (see [7]). Thus there is a set $N_1 \subset [0, 1]$ with $\mu_0(N_1) = 0$ but $\mathcal{L}^1(N_1) = 1$, and since both μ_0 and $\mathcal{L}^1$ are Radon measures, we can even choose N to be a Borel set. Denoting by $\Pi_1 : X \to [0, 1]$ the first coordinate projection map, we define a function $f : X \to \mathbb{R}$ by

$$f(x) := \int_0^{\Pi_1(x)} \chi_{N_1}(t)\, dt.$$

It is easy to see that f is $\sqrt{2}$-Lipschitz continuous on X and that $g = \chi_{N_1} \circ \Pi_1$ is an upper gradient of f, see for example [13, Lemma 4.13]. Therefore $f \in N^{1,\infty}(X)$, and we note that f is non-constant on X (since N_1 has full measure in $[0, 1]$ with respect to $\mathcal{L}^1$, the measure with respect to which the above integral was taken). On the other hand, $X = B((0, 0), 2) \cap X =: B$, and we have that $\fint_B |f - f_B|\, d\mu > 0$ but $\|g\|_{L^\infty(\lambda B)} = 0$; hence X cannot support any ∞-Poincaré inequality.

The above example shows that one needs the full strength of Property (3) of Theorem 7.3.2 in order to characterize ∞-Poincaré inequality.

The next two examples explore the Sierpinski carpet further in the context of ∞-harmonic functions and AMLEs, see [15] for more details.

Example 7.5.2. If X satisfies the stronger requirement of ∞-weak Fubini property, it is directly seen that $d_\mu = d$ on X. As explained above, the Sierpiński carpet does not support any ∞-Poincaré inequality and hence cannot satisfy any ∞-weak Fubini property. Since the length metric on this carpet is biLipschitz equivalent to the Euclidean metric, it follows that the above statement holds also when the carpet is equipped with the length metric. Recall the set N_1 from Example 7.5.1 above, and let $N = (\Pi_1^{-1}(N_1) \cup \Pi_2^{-1}(N_1)) \cap X$ where X is the carpet. Here Π_1 and Π_2 are the first coordinate and the second coordinate projection maps from the carpet to $[0, 1]$. Note that $\mu(N) = 0$, but given any curve γ in the carpet with end points x, y such that $(x_1, x_2) = x \neq y = (y_1, y_2)$, we must have

$$\mathcal{H}^1(\gamma^{-1}(N)) \geq \max\{\mathcal{H}^1(\Pi_1 \circ \gamma(\gamma^{-1}(N))), \mathcal{H}^1(\Pi_2 \circ \gamma(\gamma^{-1}(N)))\} \geq \max\{|x_1 - y_1|, |x_2 - y_2|\} > 0.$$

It follows that $d_N(x, y) = \infty$, and so $d_\mu \neq d$ in the carpet.

Example 7.5.3. The Sierpiński carpet also gives a situation where an ∞-harmonic function is not necessarily an AMLE. To construct such a function, set $g := \chi_N$, where N is the set given in Example 7.5.2. Let $E := \{0\} \times [0, 1]$, and for points $x = (x_1, x_2)$ in the carpet we define

$$f(x) := \inf_\gamma \int_\gamma g\, ds,$$

where the infimum is over all rectifiable curves γ in the carpet with one end point at x and the other at E. Note that f is zero on E, but for $x \notin E$ we have $f(x) \geq x_1 > 0$. Hence f is non-constant. From [15] we know that f is $\sqrt{2}$-Lipschitz on the carpet with respect to the Euclidean metric, and is 1-Lipschitz with respect to the length metric on X. Its minimal ∞-weak upper gradient g_f satisfies $g_f \leq g$ μ-almost everywhere, and so by the fact that $\mu(N) = 0$, we have $\|g_f\|_{L^\infty(X)} \leq \|g\|_{L^\infty(X)} = 0$, and therefore f is automatically ∞-harmonic in the carpet. However, $\mathrm{LIP}(f, X) > 0$ because f is non-constant. Let Ω be the domain $X \setminus E$; then f is not AMLE in Ω since the only AMLE extension of the zero function on $E = \partial\Omega$ is the zero extension.

We next turn our attention to examples related to spaces supporting an ∞-Poincaré inequality.

Example 7.5.4. In this example we describe a metric measure space that is complete, doubling, supports an ∞-Poincaré inequality, but does *not* support any p-Poincaré inequality for any finite $p \geq 1$. The details regarding this example can be found in [18, Example 3.7].

Let $Q_1 = [0,1]^2$, and let Q_2 be the set obtained from Q_1 in the first step of the construction of the Sierpiński carpet, that is, Q_2 is obtained from Q_1 by first dividing Q_1 into 9 equal squares, each of side length $1/3$, and then removing the middle *open* square $(1/3, 2/3)^2$. Note that Q_2 is the union of 8 squares, each of side length $1/3$. Q_3 is obtained from Q_2 by repeating the above process for each of the 8 squares that make up Q_2 to obtain 8^2 squares, each of side length $1/3^2$. Proceeding inductively, for each positive integer n we have a union of 8^n *closed* squares, each of side length $1/3^n$, making up the set Q_n. Note that the Sierpinski carpet is the set $\bigcap_{n\in\mathbb{N}} Q_n$. In this example we are not interested in this carpet, as we saw in the previous examples that the carpet does not support any ∞-Poincaré inequality. Instead we consider the complete set $X \subset [0,\infty) \times [0,1] \subset \mathbb{R}^2$ given by

$$X = \bigcup_{n\in\mathbb{N}} Q_n + (n-1, 0).$$

Thus X is obtained from the strip $[0,\infty) \times [0,1]$ by removing the first middle-third square from the first unit square $[0,1]^2$ in the strip, removing the first and second steps of the construction of the carpet from the second unit square $[1,2] \times [0,1]$ and so on. The metric on X is the Euclidean metric, but the measure μ is not the Lebesgue measure $\mathcal{L}^2$ restricted to X (since this would fail to be doubling on X at large scales), but the following measure: for each $n \in \mathbb{N}$ we set

$$\mu_n = \left(\frac{9}{8}\right)^{n-1} \mathcal{L}^2|_{Q_n+(n-1,0)},$$

and set

$$\mu = \sum_{n\in\mathbb{N}} \mu_n.$$

Then X, equipped with the Euclidean metric and the measure μ, is doubling and is complete. It is directly verifiable by the use of Fubini theorem that X satisfies Property (3) of Theorem 7.3.2 (with constant $C = \sqrt{2}$), and so X supports an ∞-Poincaré inequality.

Suppose that X supports a p-Poincaré inequality for some finite $p \geq 1$. The domains $Q_n + (n-1, 0)$ are uniform domains in X with uniformity constant 2; hence, it support a p-Poincaré inequality with constants that do not depend on n, see [6]. Thus the spaces $X_n := Q_n$, equipped with the Euclidean metric and the measure

$\nu_n := (\frac{9}{8})^{n-1}\mathcal{L}^2|_{Q_n}$ will be doubling and support a p-Poincaré inequality, with the relevant constants independent of n. Note that the Sierpiński carpet is the pointed (at the point $(0,0) \in X$) measured Gromov-Hausdorff limit of the sequence $(X_n, d_{Euc}, \nu_n, p_n)$ with $p_n = (0,0)$, and so by a result of Cheeger [9] (see also [22, Chapter 11]) the Sierpinski carpet should support a p-Poincaré inequality as well and hence should support an ∞-Poincaré inequality. Since this is not possible (see Example 7.5.1 above), it follows that X does not support any p-Poincaré inequality for any finite $p \geq 1$ either.

The above example serves two purposes. It shows that the self-improvement property of Keith and Zhong [27] does not hold for $p = \infty$, and it also shows that the support of an ∞-Poincaré inequality does *not* persist under pointed measured Gromov-Hausdorff limits. Indeed, the sequence of spaces (X_n, ν_n) each support an ∞-Poincaré inequality with associated constants independent of n, but this sequence converges under Gromov-Hausdorff limit to the Sierpinski carpet equipped with its natural Hausdorff measure, which in turn does *not* support an ∞-Poincaré inequality.

The technique of sphericalization, studied in [8, 16, 17, 23, 30], applied to the above-constructed X yields a *compact* doubling metric measure space supporting an ∞-Poincaré inequality (see [16]) but *no* p-Poincaré inequality for any finite $p \geq 1$. The failure of supporting a finite p-Poincaré inequality happens locally and asymptotically at one point in this compact space, and so the results of [26] do not apply in this space when the domain of interest contains this bad point in it.

The final example of this section deals with the lack of Rademacher-type differentiability for spaces that support ∞-Poincaré inequality but no better.

Example 7.5.5. Since the Euclidean space $\mathbb{R}$, equipped with the Lebesgue measure $\mathcal{L}^1$ and the Euclidean metric, supports a 1-Poincaré inequality, it clearly also supports an ∞-Poincaré inequality. Let ν be a singular doubling measure on $\mathbb{R}$ (for example, the Riesz product measure, see [36], [18, Section 4], or [35, page 40, Section 8.8(a)]). Then the measure $\mu = \mathcal{L}^1 + \nu$ is a doubling measure which, by Theorem 7.3.2(3) and by the fact that null sets for μ are necessarily null sets for $\mathcal{L}^1$, also supports an ∞-Poincaré inequality.

Since ν is singular to $\mathcal{L}^1$, there is a set $N \subset \mathbb{R}$ with $\mathcal{L}^1(N) = 0$ such that $\nu(N) > 0$. By a result of Choquet [10], there is a Lipschitz function f on $\mathbb{R}$ that is not Euclidean differentiable anywhere in N. Suppose that $X = \mathbb{R}$, equipped with the Euclidean metric and the measure μ, supports a Cheeger type differentiable structure as in [9]. Then, the Cheeger differential Df exists μ-almost everywhere in $\mathbb{R}$, and hence as shown in [14, Example 4.7], (note that N must have infinitely many points as ν cannot have finite support) f has to be Euclidean differentiable at ν-almost every point in N, which violates the choice of f.

Acknowledgment: The author's research was partially supported from grant # DMS-1500440 of NSF (U.S.A.). The author wishes to thank Estibalitz Durand-Cartagena and

Lukáš Malý for reading the manuscript and giving valuable suggestions that improved the exposition. If there are any errors and typos still found in this paper, they are the fault of the author alone.

Bibliography

[1] L. Ambrosio, M. Miranda Jr., D. Pallara: Special functions of bounded variation in doubling metric measure spaces. Calculus of variations: topics from the mathematical heritage of E. De Giorgi, 1–45, Quad. Mat., 14, Dept. Math., Seconda Univ. Napoli, Caserta, 2004.
[2] L. Ambrosio, P. Tilli: Topics on analysis in metric spaces, Oxford Lecture Series in Mathematics and its Applications 25 (2003) viii+133.
[3] G. Aronsson: Extension of functions satisfying Lipschitz conditions, Ark. Mat. 6 (1967), 551–561.
[4] G. Aronsson, M. G. Crandall, P. Juutinen: A tour of the theory of absolutely minimizing functions, Bull. Amer. Math. Soc. (N.S.) 41 (2004), 439–505.
[5] A. Björn, J. Björn: Nonlinear potential theory on metric spaces, phEMS Tracts Mathematics, vol. 17. European Math. Soc., Zurich, November 2011.
[6] J. Björn, N. Shanmugalingam: Poincaré inequalities, uniform domains, and extension properties for Newton-Sobolev spaces in metric spaces. J. Math. Anal. Appl. 332, no. 1 (2007) 190–208.
[7] M. Bourdon, H. Pajot: Quasi-conformal geometry and hyperbolic geometry. Rigidity in dynamics and geometry (Cambridge, 2000), 1–17, Springer, Berlin, 2002.
[8] S. Buckley, D. Herron and X. Xie: Metric space inversions, quasihyperbolic distance, and uniform spaces, Indiana Univ. Math. J. 57 (2008), 837–890.
[9] J. Cheeger: Differentiability of Lipschitz Functions on metric measure spaces. Geom. Funct. Anal. 9 (1999), 428–517.
[10] G. Choquet: Applications des propriétés descriptives de la fonction contingent à la théorie des fonctions de variable réelle et à la géometrie différentielle des variétés cartésiennes. J. Math. Pures Appl. 26 (1947), 115–226.
[11] M. G. Crandall, L. C. Evans, and R. F. Gariepy: Optimal Lipschitz extensions and the infinity Laplacian. Calc. Var. Partial Differential Equations 13 (2001), 123–139.
[12] E. Durand-Cartagena, J. A. Jaramillo: Pointwise Lipschitz functions on metric spaces. J. Math. Anal. Appl. 363 (2010), 525–548.
[13] E. Durand-Cartagena, J. A. Jaramillo, N. Shanmugalingam: The ∞-Poincaré inequality in metric measure spaces. Michigan Math. Journal. 60 (2011).
[14] E. Durand-Cartagena, J. A. Jaramillo, N. Shanmugalingam: Geometric characterizations of p-Poincaré inequalities, Pub. Mat. 60 (2016) 81–111.
[15] E. Durand-Cartagena, J. A. Jaramillo, N. Shanmugalingam: Existence of ∞-harmonic functions under assumption of ∞-Poincaré inequality, In preparation.
[16] E. Durand-Cartagena, X. Li: *Preservation of bounded geometry under sphericalization and flattening: quasiconvexity and ∞-Poincaré inequality,* preprint in http://cvgmt.sns.it/paper/2571/.
[17] E. Durand-Cartagena, X. Li: *Preservation of p-Poincaré inequality for large p under sphericalization and flattening,* preprint in http://cvgmt.sns.it/paper/2791/.
[18] E. Durand-Cartagena, N. Shanmugalingam, A. Williams: p-Poincaré inequality vs. ∞-Poincaré inequality; some counter-examples. Math. Z. 271 (2012) 447–467.
[19] P. Hajłasz, P. Koskela: Sobolev met Poincaré. phMem. Amer. Math. Soc. 145 (2000), no. 688, x+101 pp.

[20] J. Heinonen: Lectures on Analysis on Metric Spaces. Springer (2001).
[21] J. Heinonen, P. Koskela: Quasiconformal maps in metric spaces with controlled geometry. Acta Math. 181 (1998), 1–61.
[22] J. Heinonen, P. Koskela, N. Shanmugalingam, J. Tyson: Sobolev Spaces on Metric Measure Spaces: An Approach Based on Upper Gradients. Cambridge University Press, New Mathematical Monographs series: February 2015.
[23] D. Herron, N. Shanmugalingam, X. Xie: Uniformity from Gromov hyperbolicity. Illinois J. Math. 52 (2008), no. 4, 1065–1109.
[24] R. Jensen: Uniqueness of Lipschitz extensions: minimizing the sup norm of the gradient. Arch. Rational Mech. Anal. 123 (1993), no. 1, 51–74.
[25] P. Juutinen: Absolutely minimizing Lipschitz extensions on a metric space. Ann. Acad. Sci. Fenn. Math. 27 (2002), no. 1, 57–67.
[26] P. Juutinen, N. Shanmugalingam: Equivalence of AMLE, strong AMLE, and comparison with cones in metric measure spaces. Math. Nachr., 279 (2006), no. 9-10, 1083–1098.
[27] S. Keith, X. Zhong: The Poincaré inequality is an open ended condition. Ann. Math. 167 (2) (2008), 575–599.
[28] J. Kinnunen, N. Shanmugalingam: Regularity of quasi-minimizers on metric spaces. Manuscripta Math. 105 (2001), no. 3, 401–423.
[29] P. Koskela and P. McManus: Quasiconformal mappings and Sobolev spaces. Studia Math. 131 (1) (1998), 1–17.
[30] X. Li, N. Shanmugalingam: Preservation of bounded geometry under sphericalization and flattening. Indiana Univ. Math. J. 64 (2015), no. 5, 1303–1341.
[31] L. Malý: Minimal weak upper gradients in Newtonian spaces based on quasi-Banach function lattices. Ann. Acad. Sci. Fenn. Math., 38 (2013), no. 2, 727–745.
[32] E. J. McShane: Extension of range of functions. Bull. Amer. Math. Soc. 40 (1934), no. 12, 837–842.
[33] Y. Peres, O. Schramm, S. Sheffield, D. B. Wilson: Tug-of-war and the infinity Laplacian. J. Amer. Math. Soc. 22 (2009), no. 1, 167–210.
[34] N. Shanmugalingam: Newtonian Spaces: An extension of Sobolev spaces to Metric Measure Spaces. Rev. Mat. Iberoamericana, 16 (2000), 243–279.
[35] E. M. Stein: Harmonic analysis: real-variable methods, orthogonality, and oscillatory integrals. With the assistance of Timothy S. Murphy. Princeton Mathematical Series, 43. Princeton University Press, Princeton, NJ, 1993.
[36] A. Zygmund: Trigonometric series. 2nd ed. Vols. I, II. Cambridge University Press, New York 1959 Vol. I. xii+383 pp.; Vol. II. vii+354 pp.

Takashi Shioya

Metric measure limits of spheres and complex projective spaces

8.1 Introduction

Gromov [11]*S. $3\frac{1}{2}$ developed the metric measure geometry based on the idea of concentration of measure phenomenon due to Lévy and Milman (see [14–17]). This is particularly useful to study a family of spaces with unbounded dimensions. A *metric measure space* (or *mm-space* for short) is a triple (X, d_X, μ_X), where (X, d_X) is a complete separable metric space and μ_X a Borel probability measure on X. Gromov defined the *observable distance*, say $d_{\mathrm{conc}}(X, Y)$, between two mm-spaces X and Y by the difference between 1-Lipschitz functions on X and those on Y, and studied the geometry of the space of mm-spaces, say $\mathcal{X}$, with metric d_{conc}. (In [11], the observable distance function is denoted by $\underline{H}_1\mathcal{L}\iota_1$.) The observable distance is much more useful than the Gromov-Hausdorff distance to study a sequence of spaces with unbounded dimensions. We say that a sequence of mm-spaces X_n, $n = 1, 2, \dots$, *concentrates* to an mm-space X if X_n d_{conc}-converges to X as $n \to \infty$.

We have a specific natural compactification, say Π, of $(\mathcal{X}, d_{\mathrm{conc}})$. The space Π consists of pyramids, where a *pyramid* is a directed subfamily of $\mathcal{X}$ with respect to a natural order relation $\prec$, called the *Lipschitz order*. $X \prec Y$ holds if there exists a 1-Lipschitz continuous map from Y to X that pushes μ_Y forward to μ_X. For a given mm-space $X \in \mathcal{X}$, the set of $X' \in \mathcal{X}$ with $X' \prec X$ is a pyramid, denoted by $\mathcal{P}_X$. We call $\mathcal{P}_X$ the *pyramid associated with* X. The space Π has a natural compact topology such that the map

$$\iota : \mathcal{X} \ni X \longmapsto \mathcal{P}_X \in \Pi$$

is a topological embedding map. The image $\iota(\mathcal{X})$ is dense in Π and so Π is a compactification of $\mathcal{X}$.

It follows that $X \prec Y$ if and only if $\mathcal{P}_X \subset \mathcal{P}_Y$, namely the Lipschitz order $\prec$ on $\mathcal{X}$ extends to the inclusion relation on Π. $\mathcal{X}$ itself is a maximal element of Π, and a one-point mm-space corresponds to a minimal element of Π. A sequence of mm-spaces is a *Lévy family* if and only if it concentrates to the one-point mm-space. A sequence of mm-spaces *infinitely dissipates* if and only if the sequence of the associated pyramids converges to the maximal element $\mathcal{X}$.

It is interesting to study concrete examples of nontrivial sequences of mm-spaces and their limits (in Π and in $\mathcal{X}$), where 'nontrivial' means that it neither is a Lévy

Takashi Shioya: Mathematical Institute, Tohoku University, Sendai 980-8578, Japan, E-mail: shioya@math.tohoku.ac.jp

family, dissipates, nor $\square$-converges, where $\square$ denotes the box distance function on $\mathcal{X}$, which is an elementary metric on $\mathcal{X}$ and satisfies $d_{\mathrm{conc}} \le \square$. We remark that there are very few nontrivial examples that are studied in detail before. All such known examples are of the type of product spaces (see [11]*S. $3\frac{1}{2}$.49,56). In this paper, we study two examples of the non-product type, spheres and complex projective spaces with unbounded dimensions. Those are also the first discovered examples of sequence with the property that the limit space is drastically different from the spaces in the sequence.

We present some definitions needed to state our main theorems. The precise definitions are given in S. 8.4. Let γ^∞ be the infinite-dimensional standard Gaussian measure on $\mathbb{R}^\infty$. We call $\Gamma^\infty := (\mathbb{R}^\infty, \|\cdot\|_2, \gamma^\infty)$ the *infinite-dimensional standard Gaussian space*, where $\|\cdot\|_2$ denotes the l_2 norm on $\mathbb{R}^\infty$ (which takes values in $[0, +\infty]$). Note that γ^∞ is not a Borel measure with respect to the l_2 norm (cf. [2]*S. 2.3) and that Γ^∞ is not an mm-space. Nevertheless we have the associated pyramid $\mathcal{P}_{\Gamma^\infty}$. We call $\mathcal{P}_{\Gamma^\infty}$ the *virtual infinite-dimensional standard Gaussian space*. In the same way, we consider the infinite-dimensional centered Gaussian measure $\gamma^\infty_{\lambda^2}$ with variance λ^2, $\lambda > 0$, and define the virtual infinite-dimensional Gaussian space $\mathcal{P}_{\Gamma^\infty_{\lambda^2}}$ as a pyramid. $\mathcal{P}_{\Gamma^\infty_{\lambda^2}}$ coincides with the scale change of $\mathcal{P}_{\Gamma^\infty}$ of factor λ. We consider the Hopf action on $\Gamma^\infty_{\lambda^2}$ by identifying $\mathbb{R}^\infty$ with $\mathbb{C}^\infty$. The Hopf action is isometric with respect to the l_2 norm and also preserves $\gamma^\infty_{\lambda^2}$. The quotient space $\Gamma^\infty_{\lambda^2}/S^1$ has a natural measure and a metric. We also have the associated pyramid $\mathcal{P}_{\Gamma^\infty_{\lambda^2}/S^1}$. Let $S^n(r)$ be the n-dimensional sphere in $\mathbb{R}^{n+1}$ centered at the origin and of radius $r > 0$. We equip the Riemannian distance function or the restriction of the Euclidean distance function with $S^n(r)$. We also equip the normalized volume measure with $S^n(r)$. Then $S^n(r)$ is an mm-space. We consider the Hopf quotient

$$\mathbb{C}P^n(r) := S^{2n+1}(r)/S^1$$

that has a natural mm-structure induced from that of $S^{2n+1}(r)$ (see S. 8.2.5). This is topologically an n-dimensional complex projective space. Note that, if the distance function on $S^{2n+1}(r)$ is assumed to be Riemannian, then the distance function on $\mathbb{C}P^n(r)$ coincides with that induced from the Fubini-Study metric scaled with factor r.

One of our main theorems in this paper is stated as follows.

Theorem 8.1.1. *Let $\{r_n\}_{n=1}^\infty$ be a given sequence of positive real numbers, and let $\lambda_n := r_n/\sqrt{n}$ (resp. $\lambda_n := r_n/\sqrt{2n+1}$). Then we have the following* (1), (2), *and* (3).

1. *$\lambda_n \to 0$ as $n \to \infty$ if and only if $\{S^n(r_n)\}_{n=1}^\infty$ (resp. $\{\mathbb{C}P^n(r_n)\}_{n=1}^\infty$) is a Lévy family.*
2. *$\lambda_n \to +\infty$ as $n \to \infty$ if and only if $\{S^n(r_n)\}_{n=1}^\infty$ (resp. $\{\mathbb{C}P^n(r_n)\}_{n=1}^\infty$) infinitely dissipates.*
3. *Assume that $\lambda_n \to \lambda$ as $n \to \infty$ for a positive real number λ. Then, as $n \to \infty$, $\mathcal{P}_{S^n(r_n)}$ (resp. $\mathcal{P}_{\mathbb{C}P^n(r_n)}$) converges to $\mathcal{P}_{\Gamma^\infty_{\lambda^2}}$ (resp. $\mathcal{P}_{\Gamma^\infty_{\lambda^2}/S^1}$).*

(1) and (2) both hold for the Riemannian metric (resp. the scaled Fubini-Study metric) and also for the Euclidean distance function (resp. the distance induced from the Euclidean). (3) holds only for for the Euclidean distance function (resp. the distance induced from the Euclidean).

Theorem 8.1.1 is analogous to phase transition phenomena in statistical mechanics.

If r_n is bounded away from zero, then $\{S^n(r_n)\}_{n=1}^{\infty}$ and $\{\mathbb{C}P^n(r_n)\}_{n=1}^{\infty}$ both have no $\square$-convergent subsequence (see Proposition 8.7.4). The theorem also holds for any subsequence of $\{n\}$. We have the same statement as in Theorem 8.1.1 also for real and quaternionic projective spaces in the same way.

(1) of Theorem 8.1.1 follows essentially from the works of Lévy and Milman (see [10]*S. 1.1 for the proof of (1) for spheres). For (1), we give some fine estimates of the observable diameter (Corollaries 8.5.7 and 8.5.10) by using the normal law à la Lévy (Theorem 8.5.2).

(2) follows from a discussion using the Maxwell-Boltzmann distribution law (Proposition 8.5.1).

(3) is the most important part of the theorem. It follows from the Maxwell-Boltzmann distribution law that the limit pyramid of (the Hopf quotient of) spheres contains the (Hopf quotient of) virtual Gaussian space. For the proof of the reverse inclusion, we define a metric ρ on the space Π of pyramids compatible with the topology on Π (see Definition 8.3.4) that satisfies the following theorem.

Theorem 8.1.2. *For any two mm-spaces X and Y, we have*

$$\rho(\mathcal{P}_X, \mathcal{P}_Y) \le d_{\mathrm{conc}}(X, Y),$$

i.e., the embedding map $\iota : \mathcal{X} \ni X \mapsto \mathcal{P}_X \in \Pi$ is 1-Lipschitz continuous with respect to d_{conc} and ρ.

Applying this theorem, we prove the reverse inclusion. We give an outline of the proof of the reverse inclusion for spheres with $r_n = \sqrt{n}$. The key point of the proof is to show that the $(n+1)$-dimensional standard Gaussian measure of the Euclidean ball, say B_θ^{n+1}, centered at the origin and with radius $\theta\sqrt{n}$ converges to zero as $n \to \infty$ for any fixed $0 < \theta < 1$ (see Lemma 8.6.1). Since the measure of B_θ^{n+1} is almost equal to zero for large n, we ignore this part of Γ^n and consider the central projection from $\mathbb{R}^{n+1} \setminus B_\theta^{n+1}$ to the boundary $S^n(\theta\sqrt{n})$ of B_θ^{n+1}, which proves $S^n(\theta\sqrt{n}) \prec \Gamma^{n+1} \setminus B_\theta^{n+1}$ and so $\mathcal{P}_{S^n(\theta\sqrt{n})} \subset \mathcal{P}_{\Gamma^{n+1}\setminus B_\theta^{n+1}}$. The box distance between Γ^{n+1} and $\Gamma^{n+1} \setminus B_\theta^{n+1}$ can be estimated by the measure $\gamma^{n+1}(B_\theta^{n+1})$. Thus, applying Theorem 8.1.2 proves

$$\lim_{n\to\infty} \mathcal{P}_{S^n(\theta\sqrt{n})} \subset \lim_{n\to\infty} \mathcal{P}_{\Gamma^{n+1}\setminus B_\theta^{n+1}} = \lim_{n\to\infty} \mathcal{P}_{\Gamma^{n+1}} = \mathcal{P}_{\Gamma^\infty}.$$

As $\theta \to 1$, we obtain the reverse inclusion.

Note that Gromov [11]*S. $3\frac{1}{2}$ gave only the notion of convergence of pyramids and did not define the topology on Π. Note also that there exists no metric on Π (strongly) equivalent to d_{conc}. In fact we have the following as a consequence of Theorem 8.1.1.

Theorem 8.1.3. *There exist mm-spaces X_n and Y_n, $n = 1, 2, \ldots$, such that*
1. *$d_{\mathrm{conc}}(X_n, Y_n)$ is bounded away from zero;*
2. *the associated pyramids $\mathcal{P}_{X_n}$ and $\mathcal{P}_{Y_n}$ both converge to a common pyramid as $n \to \infty$.*

After this paper had been completed, we have the following progress. In [19], we study the phase transition phenomenon appeared in Theorem 8.1.1 for a general sequence of mm-spaces. Using a criterion of the phase transition in the observable diameter, we investigate several sequences of manifolds. For the phase transition property of the product spaces, we refer to [20]. In [22], we determine the limits of Stiefel manifolds and flag manifolds as the dimension diverges to infinity. A main part of this paper is also explained in [21].

8.2 Preliminaries

In this section, we give the definitions and the facts stated in [11]*S. $3\frac{1}{2}$. In [11]*S. $3\frac{1}{2}$, many details are omitted. We refer to [8, 9, 18, 21] for such details. The reader is expected to be familiar with basic measure theory and metric geometry (cf. [1, 3, 4, 12]).

8.2.1 mm-Isomorphism and Lipschitz order

Definition 8.2.1 (mm-Space). *Let (X, d_X) be a complete separable metric space and μ_X a Borel probability measure on X. We call the triple (X, d_X, μ_X) an* mm-space. *We sometimes say that X is an mm-space, in which case the metric and the measure of X are respectively indicated by d_X and μ_X.*

Definition 8.2.2 (mm-Isomorphism). *Two mm-spaces X and Y are said to be* mm-isomorphic *to each other if there exists an isometry $f : \operatorname{supp} \mu_X \to \operatorname{supp} \mu_Y$ such that $f_*\mu_X = \mu_Y$, where $f_*\mu_X$ is the push-forward of μ_X by f. Such an isometry f is called an* mm-isomorphism. *Denote by $\mathcal{X}$ the set of mm-isomorphism classes of mm-spaces.*

Any mm-isomorphism between mm-spaces is automatically surjective, even if we do not assume it.

Note that X is mm-isomorphic to $(\operatorname{supp} \mu_X, d_X, \mu_X)$. *We assume that an mm-space X satisfies*

$$X = \operatorname{supp} \mu_X$$

unless otherwise stated.

Definition 8.2.3 (Lipschitz order). *Let X and Y be two mm-spaces. We say that X (Lipschitz)* dominates *Y and write $Y \prec X$ if there exists a 1-Lipschitz map $f : X \to Y$ satisfying*

$$f_* \mu_X = \mu_Y.$$

We call the relation $\prec$ on $\mathcal{X}$ the Lipschitz order.

The Lipschitz order $\prec$ is a partial order relation on $\mathcal{X}$.

8.2.2 Observable diameter

The observable diameter is one of the most fundamental invariants of an mm-space.

Definition 8.2.4 (Partial and observable diameter). *Let X be an mm-space. For a real number α, we define the* partial diameter $\operatorname{diam}(X; \alpha) = \operatorname{diam}(\mu_X; \alpha)$ *of X to be the infimum of* $\operatorname{diam} A$, *where $A \subset X$ runs over all Borel subsets with $\mu_X(A) \geq \alpha$ and* $\operatorname{diam} A$ *denotes the diameter of A. For a real number $\kappa > 0$, we define the* observable diameter *of X to be*

$$\operatorname{ObsDiam}(X; -\kappa) := \sup\{ \operatorname{diam}(f_* \mu_X; 1-\kappa) \mid f : X \to \mathbb{R} \text{ is 1-Lipschitz continuous} \}.$$

Definition 8.2.5 (Lévy family). *A sequence of mm-spaces X_n, $n = 1, 2, \dots$, is called a* Lévy family *if*

$$\lim_{n\to\infty} \operatorname{ObsDiam}(X_n; -\kappa) = 0$$

for any $\kappa > 0$.

Proposition 8.2.6. *If $X \prec Y$, then*

$$\operatorname{ObsDiam}(X; -\kappa) \leq \operatorname{ObsDiam}(Y; -\kappa)$$

for any $\kappa > 0$.

8.2.3 Separation distance

Definition 8.2.7 (Separation distance). *Let X be an mm-space. For any real numbers $\kappa_0, \kappa_1, \cdots, \kappa_N > 0$ with $N \geq 1$, we define the* separation distance

$$\operatorname{Sep}(X; \kappa_0, \kappa_1, \cdots, \kappa_N)$$

of X as the supremum of $\min_{i \neq j} d_X(A_i, A_j)$ over all sequences of $N + 1$ Borel subsets $A_0, A_2, \cdots, A_N \subset X$ satisfying that $\mu_X(A_i) \geq \kappa_i$ for all $i = 0, 1, \cdots, N$, where $d_X(A_i, A_j) := \inf_{x \in A_i, y \in A_j} d_X(x, y)$. If there exists no sequence $A_0, \ldots, A_N \subset X$ with $\mu_X(A_i) \geq \kappa_i$, $i = 0, 1, \cdots, N$, then we define

$$\mathrm{Sep}(X; \kappa_0, \kappa_1, \cdots, \kappa_N) := 0.$$

Lemma 8.2.8. *Let X and Y be two mm-spaces. If X is dominated by Y, then we have, for any real numbers $\kappa_0, \ldots, \kappa_N > 0$,*

$$\mathrm{Sep}(X; \kappa_0, \ldots, \kappa_N) \leq \mathrm{Sep}(Y; \kappa_0, \ldots, \kappa_N).$$

Proposition 8.2.9. *For any mm-space X and any real numbers κ and κ' with $\kappa > \kappa' > 0$, we have*

$$\mathrm{ObsDiam}(X; -2\kappa) \leq \mathrm{Sep}(X; \kappa, \kappa), \tag{1}$$

$$\mathrm{Sep}(X; \kappa, \kappa) \leq \mathrm{ObsDiam}(X; -\kappa'). \tag{2}$$

8.2.4 Box distance and observable distance

Definition 8.2.10 (Prokhorov distance). *The* Prokhorov distance $d_P(\mu, \nu)$ between two Borel probability measures μ and ν on a metric space X *is defined to be the infimum of $\varepsilon > 0$ satisfying*

$$\mu(U_\varepsilon(A)) \geq \nu(A) - \varepsilon \tag{8.1}$$

for any Borel subset $A \subset X$, where

$$U_\varepsilon(A) := \{ x \in X \mid d_X(x, A) < \varepsilon \}.$$

The Prokhorov metric is a metrization of weak convergence of Borel probability measures on X provided that X is a separable metric space.

Definition 8.2.11 (me). *Let (X, μ) be a measure space and Y a metric space. For two μ-measurable maps $f, g : X \to Y$, we define* $\mathrm{me}_\mu(f, g)$ *to be the infimum of $\varepsilon \geq 0$ satisfying*

$$\mu(\{ x \in X \mid d_Y(f(x), g(x)) > \varepsilon \}) \leq \varepsilon. \tag{8.2}$$

We sometimes write $\mathrm{me}(f, g)$ *by omitting μ.*

me_μ is a metric on the set of μ-measurable maps from X to Y by identifying two maps if they are equal μ-a.e.

Lemma 8.2.12. *Let X be a topological space with a Borel probability measure μ and Y a metric space. For any two μ-measurable maps $f, g : X \to Y$, we have*

$$d_P(f_*\mu, g_*\mu) \leq \mathrm{me}_\mu(f, g).$$

Definition 8.2.13 (Parameter). *Let $I := [0, 1)$ and let X be an mm-space. A map $\phi : I \to X$ is called a* parameter *of X if ϕ is a Borel measurable map such that*

$$\phi_*\mathcal{L}^1 = \mu_X,$$

where $\mathcal{L}^1$ denotes the one-dimensional Lebesgue measure on I.

Any mm-space has a parameter.

Definition 8.2.14 (Box distance). *We define the* box distance $\square(X, Y)$ between two mm-spaces X and Y *to be the infimum of $\varepsilon \geq 0$ satisfying that there exist parameters $\phi : I \to X$, $\psi : I \to Y$, and a Borel subset $I_0 \subset I$ such that*

$$|\phi^* d_X(s, t) - \psi^* d_Y(s, t)| \leq \varepsilon \quad \text{for any } s, t \in I_0; \tag{1}$$

$$\mathcal{L}^1(I_0) \geq 1 - \varepsilon, \tag{2}$$

where $\phi^* d_X(s, t) := d_X(\phi(s), \phi(t))$ for $s, t \in I$.

The box distance function $\square$ is a complete separable metric on $\mathcal{X}$.

Proposition 8.2.15. *Let X be a complete separable metric space. For any two Borel probability measures μ and ν on X, we have*

$$\square((X, \mu), (X, \nu)) \leq 2\, d_P(\mu, \nu).$$

Definition 8.2.16 (Observable distance $d_{\mathrm{conc}}(X, Y)$). *Denote by* $\mathrm{Lip}_1(X)$ *the set of* 1-*Lipschitz continuous functions on an mm-space X. For any parameter ϕ of X, we set*

$$\phi^* \mathrm{Lip}_1(X) := \{ f \circ \phi \mid f \in \mathrm{Lip}_1(X) \}.$$

We define the observable distance $d_{\mathrm{conc}}(X, Y)$ between two mm-spaces X and Y *by*

$$d_{\mathrm{conc}}(X, Y) := \inf_{\phi, \psi} d_H(\phi^* \mathrm{Lip}_1(d_X), \psi^* \mathrm{Lip}_1(d_Y)),$$

where $\phi : I \to X$ and $\psi : I \to Y$ run over all parameters of X and Y, respectively, and where d_H is the Hausdorff distance with respect to the metric $\mathrm{me}_{\mathcal{L}^1}$. We say that a sequence of mm-spaces X_n, $n = 1, 2, \ldots$, concentrates *to an mm-space X if X_n d_{conc}-converges to X as $n \to \infty$.*

Proposition 8.2.17. *Let $\{X_n\}_{n=1}^\infty$ be a sequence of mm-spaces. Then, $\{X_n\}$ is a Lévy family if and only if X_n concentrates to a one-point mm-space as $n \to \infty$.*

Proposition 8.2.18. *For any two mm-spaces X and Y we have*

$$d_{\mathrm{conc}}(X, Y) \leq \square(X, Y).$$

8.2.5 Quotient space

Let X be a metric space and G a group acting on X isometrically. We define a pseudo-metric on the quotient space X/G by

$$d_{X/G}([x], [y]) := \inf_{x' \in [x], y' \in [y]} d_X(x', y'), \qquad [x], [y] \in X/G.$$

We call $d_{X/G}$ the *pseudo-metric on X/G induced from d_X*. If every orbit in X of G is closed, then $d_{X/G}$ is a metric.

Assume that we have a Borel measure μ_X on X. Then, we call the measure $\mu_{X/G} := \pi_* \mu_X$ the *measure on X/G induced from μ_X*, where $\pi : X \to X/G$ is the natural projection.

8.2.6 Pyramid

Definition 8.2.19 (Pyramid). *A subset $\mathcal{P} \subset \mathcal{X}$ is called a* pyramid *if it satisfies the following (1), (2), and (3).*

1. *If $X \in \mathcal{P}$ and if $Y \prec X$, then $Y \in \mathcal{P}$.*
2. *For any two mm-spaces $X, X' \in \mathcal{P}$, there exists an mm-space $Y \in \mathcal{P}$ such that $X \prec Y$ and $X' \prec Y$.*
3. *$\mathcal{P}$ is nonempty and $\square$-closed.*

We denote the set of pyramids by Π.

For an mm-space X we define

$$\mathcal{P}_X := \{\, X' \in \mathcal{X} \mid X' \prec X \,\}.$$

We call $\mathcal{P}_X$ the pyramid associated with X.

It is trivial that $\mathcal{X}$ is a pyramid.

In Gromov's book [11], the definition of a pyramid is only by (1) and (2) of Definition 8.2.19. We here put (3) as an additional condition for the Hausdorff property of Π.

Definition 8.2.20 (Weak convergence). *Let $\mathcal{P}_n, \mathcal{P} \in \Pi$, $n = 1, 2, \ldots$. We say that $\mathcal{P}_n$* converges weakly to $\mathcal{P}$ *as $n \to \infty$ if the following (1) and (2) are both satisfied.*

1. *For any mm-space $X \in \mathcal{P}$, we have*

$$\lim_{n\to\infty} \square(X, \mathcal{P}_n) = 0.$$

2. *For any mm-space $X \in \mathcal{X} \setminus \mathcal{P}$, we have*

$$\liminf_{n\to\infty} \square(X, \mathcal{P}_n) > 0.$$

Theorem 8.2.21. *The set Π of pyramids is sequentially compact, i.e., any sequence of pyramids has a subsequence that converges weakly to a pyramid.*

Proposition 8.2.22. *For given mm-spaces X and X_n, $n = 1, 2, \ldots$, the following* (1) *and* (2) *are equivalent to each other.*

1. *X_n concentrates to X as $n \to \infty$.*
2. *$\mathcal{P}_{X_n}$ converges weakly to $\mathcal{P}_X$ as $n \to \infty$.*

Dissipation is an opposite notion to concentration.

Definition 8.2.23 (Infinite dissipation). *Let X_n, $n = 1, 2, \ldots$, be mm-spaces. We say that $\{X_n\}$* infinitely dissipates *if for any real numbers $\kappa_0, \kappa_1, \ldots, \kappa_N > 0$ with $\sum_{i=0}^{N} \kappa_i < 1$, the separation distance*

$$\mathrm{Sep}(X_n; \kappa_0, \kappa_1, \ldots, \kappa_N)$$

diverges to infinity as $n \to \infty$.

Proposition 8.2.24. *Let X_n, $n = 1, 2, \ldots$, be mm-spaces. Then, $\{X_n\}$ infinitely dissipates if and only if $\mathcal{P}_{X_n}$ converges to $\mathcal{X}$ as $n \to \infty$.*

Let X be an mm-space. Denote by $\mathcal{L}_1(X)$ the quotient of $\mathrm{Lip}_1(X)$ by the $\mathbb{R}$-action:

$$\mathbb{R} \times \mathrm{Lip}_1(X) \ni (t, f) \mapsto t + f \in \mathrm{Lip}_1(X).$$

The $\mathbb{R}$-action on $\mathrm{Lip}_1(X)$ is isometric with respect to the metric me. We denote also by 'me' the induced metric on $\mathcal{L}_1(X)$ from 'me'. We see that $\mathrm{me}([f], [g]) = \inf_{t\in\mathbb{R}} \mathrm{me}(f + t, g) = \mathrm{me}(f + t_0, g)$ for some $t_0 \in \mathbb{R}$.

Lemma 8.2.25. *For any two mm-spaces X and Y we have*

$$d_{GH}(\mathcal{L}_1(X), \mathcal{L}_1(Y)) \le d_{\mathrm{conc}}(X, Y),$$

where d_{GH} denotes the Gromov-Hausdorff distance function.

Definition 8.2.26. *We say that a pyramid $\mathcal{P}$ is* concentrated *if $\{\mathcal{L}_1(X)\}_{X\in\mathcal{P}}$ is d_{GH}-precompact.*

Lemma 8.2.27. *Let $\mathcal{P}$ be a pyramid. Then the following* (1) *and* (2) *are equivalent to each other.*

1. *$\mathcal{P}$ is concentrated.*
2. *$\mathcal{P}$ is the weak limit of $\{\mathcal{P}_{X_n}\}$ for some d_{conc}-Cauchy sequence $\{X_n\}$ of mm-spaces.*

8.3 Metric on the space of pyramids

The purpose of this section is to define a metric ρ on Π compatible with weak convergence such that the embedding map

$$\iota : \mathcal{X} \ni X \longmapsto \mathcal{P}_X \in \Pi$$

is a 1-Lipschitz continuous with respect to d_{conc} on $\mathcal{X}$.

Definition 8.3.1 ($\mathcal{M}(N)$, $\mathcal{M}(N,R)$, $\mathcal{X}(N,R)$). *Let N be a natural number and R a nonnegative real number. Denote by $\mathcal{M}(N)$ the set of Borel probability measures on $\mathbb{R}^N$ equipped with the Prokhorov metric d_P, and set*

$$\mathcal{M}(N,R) := \{\, \mu \in \mathcal{M}(N) \mid \operatorname{supp}\mu \subset B_R^N \,\},$$

where $B_R^N := \{\, x \in \mathbb{R}^N \mid \|x\|_\infty \le R \,\}$ and $\|\cdot\|_\infty$ denotes the l_∞ norm on $\mathbb{R}^N$. We define

$$\mathcal{X}(N,R) := \{\, (B_R^N, \|\cdot\|_\infty, \mu) \mid \mu \in \mathcal{M}(N,R) \,\}.$$

Note that $\mathcal{M}(N,R)$ and $\mathcal{X}(N,R)$ are compact with respect to d_P and $\square$, respectively.

Definition 8.3.2 (N-Measurement). *Let X be an mm-space, N a natural number, and R a nonnegative real number. We define*

$$\mathcal{M}(X;N) := \{\, \Phi_*\mu_X \mid \Phi : X \to (\mathbb{R}^N, \|\cdot\|_\infty) \text{ is 1-Lipschitz} \,\},$$
$$\mathcal{M}(X;N,R) := \{\, \mu \in \mathcal{M}(X;N) \mid \operatorname{supp}\mu \subset B_R^N \,\}.$$

We call $\mathcal{M}(X;N)$ (resp. $\mathcal{M}(X;N,R)$) the N-measurement *(resp.* (N,R)-measurement*) of X.*

The N-measurement $\mathcal{M}(X;N)$ is a closed subset of $\mathcal{M}(N)$ and the (N,R)-measurement $\mathcal{M}(X;N,R)$ is a compact subset of $\mathcal{M}(N)$.

The following lemma is claimed in [11]*S. $3\frac{1}{2}$ without proof. Since the lemma is important for the definition of ρ, we give a sketch of proof (the detailed proof is lengthy and contained in the book [21]).

Lemma 8.3.3 ([11]*S. $3\frac{1}{2}$). *For given pyramids $\mathcal{P}$ and $\mathcal{P}_n$, $n = 1, 2, \dots$, the following* (1) *and* (2) *are equivalent to each other.*

1. *$\mathcal{P}_n$ converges weakly to $\mathcal{P}$ as $n \to \infty$.*
2. *For any natural number k, the set $\mathcal{P}_n \cap \mathcal{X}(k,k)$ Hausdorff converges to $\mathcal{P} \cap \mathcal{X}(k,k)$ as $n \to \infty$, where the Hausdorff distance is induced from the box metric.*

Sketch of proof. We prove '(1) $\Longrightarrow$ (2)'. Suppose that $\mathcal{P}_n$ converges weakly to $\mathcal{P}$, but $\mathcal{P}_n \cap \mathcal{X}(k,k)$ does not Hausdorff converge to $\mathcal{P} \cap \mathcal{X}(k,k)$ for some k. We then find a subsequence $\{\mathcal{P}_{n_i}\}$ of $\{\mathcal{P}_n\}$ in such a way that $\liminf_{n\to\infty} d_H(\mathcal{P}_n \cap \mathcal{X}(k,k), \mathcal{P} \cap \mathcal{X}(k,k)) >$

0. Since $\mathcal{X}(k, k)$ is $\square$-compact and by replacing $\{\mathcal{P}_{n_i}\}$ with a subsequence, $\mathcal{P}_{n_i} \cap \mathcal{X}(k, k)$ Hausdorff converges to some compact subset $\mathcal{P}_\infty \subset \mathcal{X}(k, k)$ different from $\mathcal{P} \cap \mathcal{X}(k, k)$. Since any mm-space $X \in \mathcal{P}_\infty$ is the limit of some $X_i \in \mathcal{P}_{n_i} \cap \mathcal{X}(k, k)$, $i = 1, 2, \dots$, the set $\mathcal{P}_\infty$ is contained in $\mathcal{P}$, so that $\mathcal{P}_\infty \subset \mathcal{P} \cap \mathcal{X}(k, k)$. For any mm-space $X \in \mathcal{P} \cap \mathcal{X}(k, k)$, there is a sequence of mm-spaces $X_i \in \mathcal{P}_{n_i}$ $\square$-converging to X as $i \to \infty$. We are able to find a sequence of mm-spaces $X_i' \in \mathcal{X}(k, k)$ with $X_i' \prec X_i$ that $\square$-converges to X. Since $X_i' \in \mathcal{P}_{n_i} \cap \mathcal{X}(k, k)$, the space X belongs to $\mathcal{P}_\infty$. Thus we have $\mathcal{P}_\infty = \mathcal{P} \cap \mathcal{X}(k, k)$. This is a contradiction.

We prove '(2) $\implies$ (1)'. We assume (2). Let $\underline{\mathcal{P}}_\infty$ be the set of the limits of convergent sequences of mm-spaces $X_n \in \mathcal{P}_n$, and $\overline{\mathcal{P}}_\infty$ the set of the limits of convergent *sub*sequences of mm-spaces $X_n \in \mathcal{P}_n$. We have $\underline{\mathcal{P}}_\infty \subset \overline{\mathcal{P}}_\infty$ in general. We shall prove $\underline{\mathcal{P}}_\infty = \overline{\mathcal{P}}_\infty = \mathcal{P}$.

To prove $\mathcal{P} \subset \underline{\mathcal{P}}_\infty$, we take any mm-space $X \in \mathcal{P}$. Since $\mathcal{P} \cap \bigcup_{N=1}^\infty \mathcal{X}(N, N)$ is $\square$-dense in $\mathcal{P}$, there is a sequence of mm-spaces $X_i \in \mathcal{P} \cap \bigcup_{N=1}^\infty \mathcal{X}(N, N)$ that $\square$-converges to X. For each i we find a natural number N_i with $X_i \in \mathcal{X}(N_i, N_i)$. By (2), there is a sequence of mm-spaces $X_{in} \in \mathcal{P}_n \cap \mathcal{X}(N_i, N_i)$, $n = 1, 2, \dots$, that $\square$-converges to X_i for each i. There is a sequence $i_n \to \infty$ such that $X_{i_n n}$ $\square$-converges to X, so that X belongs to $\underline{\mathcal{P}}_\infty$. We obtain $\mathcal{P} \subset \underline{\mathcal{P}}_\infty$.

To prove $\overline{\mathcal{P}}_\infty \subset \mathcal{P}$, we take any mm-space $X \in \overline{\mathcal{P}}_\infty$. X is approximated by some $\underline{X}_N = (\mathbb{R}^N, \|\cdot\|_\infty, \underline{\mu}_N)$, $\underline{\mu}_N \in \mathcal{M}(X; N)$. It is easy to see that for any $R > 0$ there is a unique nearest point projection $\pi_R : \mathbb{R}^N \to B_R^N$ with respect to the l_∞ norm. π_R is 1-Lipschitz continuous with respect to the l_∞ norm. Since $(\pi_R)_*\underline{\mu}_N \to \underline{\mu}_N$ weakly as $R \to +\infty$, X is approximated by some $X' \in \mathcal{X}(N, R)$ with $X' \prec X$. By the $\square$-closedness of $\mathcal{P}$, it suffices to prove that X' belongs to $\mathcal{P}$. It follows from $X \in \overline{\mathcal{P}}_\infty$ that there are sequences $n_i \to \infty$ and $X_i \in \mathcal{P}_{n_i}$, $i = 1, 2, \dots$, such that X_i $\square$-converges to X. We find a sequence of mm-spaces X_i' with $X_i' \prec X_i$ that $\square$-converges to X'. We are also able to find a sequence $X_i'' \in \mathcal{X}(N, R)$ such that $X_i'' \prec X_i'$ for any i and X_i'' converges to X' as $i \to \infty$. Since $\mathcal{P}_{n_i}$ is a pyramid, X_i'' belongs to $\mathcal{P}_{n_i}$. By (2), X' is an element of $\mathcal{P}$. We thus obtain $\mathcal{P} = \underline{\mathcal{P}}_\infty = \overline{\mathcal{P}}_\infty$.

We prove the weak convergence $\mathcal{P}_n \to \mathcal{P}$. Let us verify the first condition of Definition 8.2.20. Take any mm-space $X \in \mathcal{P}$. Since $X \in \underline{\mathcal{P}}_\infty$, there is a sequence of mm-spaces $X_n \in \mathcal{P}_n$, $n = 1, 2, \dots$, that $\square$-converges to X. Therefore,

$$\limsup_{n\to\infty} \square(X, \mathcal{P}_n) \le \lim_{n\to\infty} \square(X, X_n) = 0.$$

Let us verify the second condition of Definition 8.2.20. Suppose that $\liminf_{n\to\infty} \square(X, \mathcal{P}_n) = 0$ for an mm-space X. It suffices to prove that X belongs to $\mathcal{P}$. We find a subsequence $\{\mathcal{P}_{n_i}\}$ of $\{\mathcal{P}_n\}$ in such a way that $\lim_{i\to\infty} \square(X, \mathcal{P}_{n_i}) = 0$. There is an mm-space $X_i \in \mathcal{P}_{n_i}$ for each i such that X_i $\square$-converges to X as $i \to \infty$. Therefore, X belongs to $\overline{\mathcal{P}}_\infty = \mathcal{P}$.

This completes the proof. $\square$

Definition 8.3.4 (Metric on the space of pyramids). *Define, for a natural number k and for two pyramids $\mathcal{P}$ and $\mathcal{P}'$,*

$$\rho_k(\mathcal{P}, \mathcal{P}') := \frac{1}{4k} d_H(\mathcal{P} \cap \chi(k,k), \mathcal{P}' \cap \chi(k,k)),$$

$$\rho(\mathcal{P}, \mathcal{P}') := \sum_{k=1}^{\infty} 2^{-k} \rho_k(\mathcal{P}, \mathcal{P}').$$

Note that $1/(4k)$ in the definition of ρ_k is necessary for the proof of Theorem 8.1.2.

Proposition 8.3.5. *ρ is a metric on the space Π of pyramids that is compatible with weak convergence. Π is compact with respect to ρ.*

Proof. We first prove that ρ is a metric. Since $\square \le 1$, we have $\rho_k \le 1/(4k)$ for each k and then $\rho \le 1/4$. Each ρ_k is a pseudo-metric on Π and so is ρ. If $\rho(\mathcal{P}, \mathcal{P}') = 0$ for two pyramids $\mathcal{P}$ and $\mathcal{P}'$, then $\rho_k(\mathcal{P}, \mathcal{P}') = 0$ for any k, which implies $\mathcal{P} = \mathcal{P}'$ (see Lemma 8.3.3). Thus, ρ is a metric on Π.

We next prove the compatibility of the metric ρ with weak convergence in Π. It follows from Lemma 8.3.3 that a sequence of pyramids $\mathcal{P}_n$, $n = 1, 2, \dots$, converges weakly to a pyramid $\mathcal{P}$ if and only if $\lim_{n\to\infty} \rho_k(\mathcal{P}_n, \mathcal{P}) = 0$ for any k, which is also equivalent to $\lim_{n\to\infty} \rho(\mathcal{P}_n, \mathcal{P}) = 0$.

Since Π is sequentially compact (see Theorem 8.2.21), it is compact with respect to ρ. This completes the proof. □

The rest of this section is devoted to the proof of Theorem 8.1.2.

Lemma 8.3.6. *Let X and Y be two mm-spaces. For any natural number N we have*

$$d_H(\mathcal{M}(X;N), \mathcal{M}(Y;N)) \le N \cdot d_{\mathrm{conc}}(X, Y),$$

where the Hausdorff distance d_H is defined with respect to the Prokhorov metric d_P.

Proof. Assume that $d_{\mathrm{conc}}(X, Y) < \varepsilon$ for a real number ε. There are two parameters $\phi : I \to X$ and $\psi : I \to Y$ such that

$$d_H(\phi^* \operatorname{Lip}_1(X), \psi^* \operatorname{Lip}_1(Y)) < \varepsilon. \tag{8.3}$$

Let us prove that $\mathcal{M}(X;N) \subset B_{N\varepsilon}(\mathcal{M}(Y;N))$. Take any $F_*\mu_X \in \mathcal{M}(X;N)$, where $F : X \to (\mathbb{R}^N, \|\cdot\|_\infty)$ is a 1-Lipschitz map. Setting $(f_1, \dots, f_N) := F$ we have $f_i \in \operatorname{Lip}_1(X)$ and so $f_i \circ \phi \in \phi^* \operatorname{Lip}_1(X)$. By (8.3), there is a function $g_i \in \operatorname{Lip}_1(Y)$ such that $\mathrm{me}(f_i \circ \phi, g_i \circ \psi) < \varepsilon$. Since $G := (g_1, \dots, g_N) : Y \to (\mathbb{R}^N, \|\cdot\|_\infty)$ is 1-Lipschitz, we have $G_*\mu_Y \in \mathcal{M}(Y;N)$. We prove $d_P(F_*\mu_X, G_*\mu_Y) \le N\varepsilon$ in the following. For this, it suffices to prove $F_*\mu_X(B_\varepsilon(A)) \ge G_*\mu_Y(A) - N\varepsilon$ for any Borel subset $A \subset \mathbb{R}^N$. Since $F_*\mu_X = (F \circ \phi)_*\mathcal{L}^1$ and $G_*\mu_Y = (G \circ \psi)_*\mathcal{L}^1$, we have

$$F_*\mu_X(B_\varepsilon(A)) = \mathcal{L}^1((F \circ \phi)^{-1}(B_\varepsilon(A))), \quad G_*\mu_Y(A) = \mathcal{L}^1((G \circ \psi)^{-1}(A)).$$

It is sufficient to prove

$$\mathcal{L}^1((G \circ \psi)^{-1}(A) \setminus (F \circ \phi)^{-1}(B_\varepsilon(A))) \le N\varepsilon.$$

If we take $s \in (G \circ \psi)^{-1}(A) \setminus (F \circ \phi)^{-1}(B_\varepsilon(A))$, then $G \circ \psi(s) \in A$ and $F \circ \phi(s) \notin B_\varepsilon(A)$ together imply

$$\|F \circ \phi(s) - G \circ \psi(s)\|_\infty > \varepsilon$$

and therefore

$$\begin{aligned}
\mathcal{L}^1((G \circ \psi)^{-1}(A) \setminus (F \circ \phi)^{-1}(B_\varepsilon(A))) &\le \mathcal{L}^1(\{\, s \in I \mid \|F \circ \phi(s) - G \circ \psi(s)\|_\infty > \varepsilon \,\}) \\
&= \mathcal{L}^1\left(\bigcup_{i=1}^{N} \{\, s \in I \mid |f_i \circ \phi(s) - g_i \circ \psi(s)| > \varepsilon \,\}\right) \\
&\le \sum_{i=1}^{N} \mathcal{L}^1(\{\, s \in I \mid |f_i \circ \phi(s) - g_i \circ \psi(s)| > \varepsilon \,\}) \\
&\le N\varepsilon,
\end{aligned}$$

where the last inequality follows from $\mathrm{me}(f_i \circ \phi, g_i \circ \psi) < \varepsilon$. We thus obtain $d_P(F_*\mu_X, G_*\mu_Y) \le N\varepsilon$, so that $\mathcal{M}(X; N) \subset B_{N\varepsilon}(\mathcal{M}(Y; N))$. Since this also holds if we exchange X and Y, we have

$$d_H(\mathcal{M}(X; N), \mathcal{M}(Y; N)) \le N\varepsilon.$$

This completes the proof. □

Lemma 8.3.7. *Let X and Y be two mm-spaces. Then, for any natural number N and nonnegative real number R we have*

$$d_H(\mathcal{M}(X; N, R), \mathcal{M}(Y; N, R)) \le 2\, d_H(\mathcal{M}(X; N), \mathcal{M}(Y; N)),$$

where the Hausdorff distance d_H is defined with respect to the Prokhorov metric d_P on $\mathcal{M}(N)$

Proof. Let $\varepsilon := d_H(\mathcal{M}(X; N), \mathcal{M}(Y; N))$. For any measure $\mu \in \mathcal{M}(X; N, R)$ there is a measure $\nu \in \mathcal{M}(Y; N)$ such that $d_P(\mu, \nu) \le \varepsilon$. This implies

$$\nu(B_\varepsilon(B_R^N)) \ge \mu(B_R^N) - \varepsilon = 1 - \varepsilon. \tag{8.4}$$

Let $\pi = \pi_R : \mathbb{R}^N \to B_R^N$ be the nearest point projection. This is 1-Lipschitz and satisfies $\pi|_{B_R^N} = \mathrm{Id}_{B_R^N}$. We have $\pi_*\nu \in \mathcal{M}(Y; N, R)$. By Lemma 8.2.12 and (8.4), we see $d_P(\pi_*\nu, \nu) \le \mathrm{me}_\nu(\pi, \mathrm{Id}_{\mathbb{R}^N}) \le \varepsilon$ and hence

$$d_P(\mu, \pi_*\nu) \le d_P(\mu, \nu) + d_P(\nu, \pi_*\nu) \le 2\varepsilon,$$

so that $\mathcal{M}(X;N,R) \subset B_{2\varepsilon}(\mathcal{M}(Y;N,R))$. Exchanging X and Y yields $\mathcal{M}(Y;N,R) \subset B_{2\varepsilon}(\mathcal{M}(X;N,R))$. We thus obtain

$$d_H(\mathcal{M}(X;N,R), \mathcal{M}(Y;N,R)) \le 2\varepsilon.$$

This completes the proof. □

Lemma 8.3.8. *Let X and Y be two mm-spaces, N a natural number, and R a nonnegative real number. Then we have*

$$d_H(\mathcal{P}_X \cap \mathcal{X}(N,R), \mathcal{P}_Y \cap \mathcal{X}(N,R)) \le 2\, d_H(\mathcal{M}(X;N,R), \mathcal{M}(Y;N,R)).$$

Proof. The lemma follows from Proposition 8.2.15. □

Proof of Theorem 8.1.2. By Lemmas 8.3.8, 8.3.6, and 8.3.7,

$$\begin{aligned} &d_H(\mathcal{P}_X \cap \mathcal{X}(N,R), \mathcal{P}_Y \cap \mathcal{X}(N,R)) \\ &\le 2 d_H(\mathcal{M}(X;N,R), \mathcal{M}(Y;N,R)) \\ &\le 4 d_H(\mathcal{M}(X;N), \mathcal{M}(Y;N)) \le 4N\, d_{\mathrm{conc}}(X,Y), \end{aligned}$$

so that $\rho_k(\mathcal{P}_X, \mathcal{P}_Y) \le d_{\mathrm{conc}}(X,Y)$ for any k. This completes the proof of the theorem. □

8.4 Gaussian space and Hopf quotient

In this section we present the precise definitions of the spaces which appeared in Theorem 8.1.1.

For $\lambda > 0$, let $\gamma^n_{\lambda^2}$ denote the *n-dimensional centered Gaussian measure on* $\mathbb{R}^n$ *with variance* λ^2, i.e.,

$$\gamma^n_{\lambda^2}(A) := \frac{1}{(2\pi\lambda^2)^{\frac{n}{2}}} \int_A e^{-\frac{1}{2\lambda^2}\|x\|_2^2}\, dx$$

for a Lebesgue measurable subset $A \subset \mathbb{R}^n$, where dx is the Lebesgue measure on $\mathbb{R}^n$ and $\|\cdot\|_2$ the l_2 (or Euclidean) norm on $\mathbb{R}^n$. We put $\gamma^n := \gamma^n_1$, which is the *n-dimensional standard Gaussian measure on* $\mathbb{R}^n$. Note that the n-th product measure of $\gamma^1_{\lambda^2}$ coincides with $\gamma^n_{\lambda^2}$. We call the mm-space $\Gamma^n_{\lambda^2} := (\mathbb{R}^n, \|\cdot\|_2, \gamma^n_{\lambda^2})$ the *n-dimensional Gaussian space with variance* λ^2. Call $\Gamma^n := \Gamma^n_1$ the *n-dimensional standard Gaussian space*. For $k \le n$, we denote by $\pi^n_k : \mathbb{R}^n \to \mathbb{R}^k$ the natural projection, i.e.,

$$\pi^n_k(x_1, x_2, \dots, x_n) := (x_1, x_2, \dots, x_k), \quad (x_1, x_2, \dots, x_n) \in \mathbb{R}^n.$$

Since the projection $\pi^n_{n-1} : \Gamma^n_{\lambda^2} \to \Gamma^{n-1}_{\lambda^2}$ is 1-Lipschitz continuous and measure-preserving for any $n \ge 2$, the Gaussian space $\Gamma^n_{\lambda^2}$ is monotone increasing in n with respect to the Lipschitz order, so that, as $n \to \infty$, the associated pyramid $\mathcal{P}_{\Gamma^n_{\lambda^2}}$ converges to the $\square$-closure of $\bigcup_{n=1}^{\infty} \mathcal{P}_{\Gamma^n_{\lambda^2}}$, denoted by $\mathcal{P}_{\Gamma^\infty_{\lambda^2}}$. We call $\mathcal{P}_{\Gamma^\infty_{\lambda^2}}$ the *virtual infinite-dimensional Gaussian space with variance* λ^2. Call $\mathcal{P}_{\Gamma^\infty} := \mathcal{P}_{\Gamma^\infty_1}$ the *virtual infinite-dimensional standard Gaussian space.*

Recall that the *Hopf action* is the following S^1-action on $\mathbb{C}^n$:

$$S^1 \times \mathbb{C}^n \ni (e^{\sqrt{-1}t}, z) \longmapsto e^{\sqrt{-1}t} z \in \mathbb{C}^n,$$

where S^1 is the group of unit complex numbers under multiplication. Since the projection $\pi^{2n}_{2k} : \mathbb{C}^n \to \mathbb{C}^k$, $k \le n$, is S^1-equivariant, i.e., $\pi^{2n}_{2k}(e^{\sqrt{-1}t} z) = e^{\sqrt{-1}t}\pi^{2n}_{2k}(z)$ for any $e^{\sqrt{-1}t} \in S^1$ and $z \in \mathbb{C}^n$, there exists a unique map $\bar{\pi}^{2n}_{2k} : \mathbb{C}^n/S^1 \to \mathbb{C}^k/S^1$ such that the following diagram commutes:

$$\begin{array}{ccc} \mathbb{C}^n & \xrightarrow{S^1} & \mathbb{C}^n/S^1 \\ {\scriptstyle \pi^{2n}_{2k}}\downarrow & & \downarrow{\scriptstyle \bar{\pi}^{2n}_{2k}} \\ \mathbb{C}^k & \xrightarrow{S^1} & \mathbb{C}^k/S^1 \end{array}$$

We consider the Hopf action on $\Gamma^{2n}_{\lambda^2}$ by identifying $\mathbb{R}^{2n}$ with $\mathbb{C}^n$. The Hopf action is isometric with respect to the Euclidean distance and also preserves the Gaussian measure $\gamma^{2n}_{\lambda^2}$. Let

$$\Gamma^{2n}_{\lambda^2}/S^1 = (\mathbb{C}^n/S^1, d_{\mathbb{C}^n/S^1}, \bar{\gamma}^{2n}_{\lambda^2})$$

be the quotient space with the induced mm-structure (see S. 8.2.5). Note that this is isometric to the Euclidean cone (cf. [4]) over a complex projective space of complex dimension $n-1$ with the Fubini-Study metric. Since the map $\bar{\pi}^{2n}_{2(n-1)} : \Gamma^{2n}_{\lambda^2}/S^1 \to \Gamma^{2(n-1)}_{\lambda^2}/S^1$ is 1-Lipschitz continuous and pushes $\bar{\gamma}^{2n}_{\lambda^2}$ forward to $\bar{\gamma}^{2(n-1)}_{\lambda^2}$, the quotient space $\Gamma^{2n}_{\lambda^2}/S^1$ is monotone increasing in n with respect to the Lipschitz order. The associated pyramid $\mathcal{P}_{\Gamma^{2n}_{\lambda^2}/S^1}$ converges to the $\square$-closure of $\bigcup_{n=1}^{\infty} \mathcal{P}_{\Gamma^{2n}_{\lambda^2}/S^1}$, which we denote by $\mathcal{P}_{\Gamma^\infty_{\lambda^2}/S^1}$. We put $\mathcal{P}_{\Gamma^\infty/S^1} := \mathcal{P}_{\Gamma^\infty_1/S^1}$.

8.5 Estimate of observable diameter

In this section, we give some estimates of the observable diameters of spheres and complex projective spaces, which are little extensions of known results (see [11, 13]).

Let σ^n denotes the normalized volume measure on the sphere

$$S^n(r) := \{\ x \in \mathbb{R}^{n+1} \mid \|x\|_2 = r\ \}$$

of radius $r > 0$. For $k \le n$, we consider the restriction of the projection $\pi^{n+1}_k : S^n(r) \subset \mathbb{R}^{n+1} \to \mathbb{R}^k$, which is 1-Lipschitz continuous with respect to the geodesic distance and also to the restriction of the Euclidean distance to $S^n(r)$.

Recall that *weak* (resp. *vague*) *convergence of measures* means weak-⋆ convergence in the dual of the space of bounded continuous functions (resp. continuous functions with compact support).

The following is well-known and the proof is elementary.

Proposition 8.5.1 (Maxwell-Boltzmann distribution law[8.1]). *For any natural number k we have*

$$\frac{d(\pi_k^{n+1})_*\sigma^n}{d\mathcal{L}^k} \to \frac{d\gamma^k}{d\mathcal{L}^k} \qquad \text{as } n \to \infty,$$

where σ^n is the normalized volume measure on $S^n(\sqrt{n})$ and $\frac{d}{d\mathcal{L}^k}$ means the Radon-Nikodym derivative with respect to the k-dimensional Lebesgue measure. In particular,

$$(\pi_k^{n+1})_*\sigma^n \to \gamma^k \text{ weakly} \qquad \text{as } n \to \infty.$$

We prove the following theorem, since we find no proof in any literature.

Theorem 8.5.2 (Normal law à la Lévy; Gromov [11]*S. $3\frac{1}{2}$).
Let $f_n : S^n(\sqrt{n}) \to \mathbb{R}$, $n = 1, 2, \ldots$, be 1-Lipschitz continuous functions with respect to the geodesic distance on $S^n(\sqrt{n})$. Assume that, for a subsequence $\{f_{n_i}\}$ of $\{f_n\}$, the push-forward $(f_{n_i})_*\sigma^{n_i}$ converges vaguely to a Borel measure σ_∞ on $\mathbb{R}$, and that σ_∞ is not identically equal to zero. Then, σ_∞ is a probability measure and

$$(\mathbb{R}, |\cdot|, \sigma_\infty) \prec (\mathbb{R}, |\cdot|, \gamma^1),$$

i.e., there exists a 1-Lipschitz continuous function $\alpha : \mathbb{R} \to \mathbb{R}$ such that $\alpha_*\gamma^1 = \sigma_\infty$.

Note that there always exists a subsequence $\{f_{n_i}\}$ such that $(f_{n_i})_*\sigma^{n_i}$ converges vaguely to some finite Borel measure on $\mathbb{R}$.

We need some claims for the proof of Theorem 8.5.2. The following theorem is well-known.

Theorem 8.5.3 (Lévy's isoperimetric inequality; [6, 14]). *For any closed subset $\Omega \subset S^n(1)$, we take a metric ball B_Ω of $S^n(1)$ with $\sigma^n(B_\Omega) = \sigma^n(\Omega)$. Then we have*

$$\sigma^n(U_r(\Omega)) \geq \sigma^n(U_r(B_\Omega))$$

for any $r > 0$, where $U_r(\Omega)$ denotes the open r-neighborhood of Ω with respect to the geodesic distance on $S^n(\sqrt{n})$.

We assume the condition of Theorem 8.5.2. Consider a natural compactification $\bar{\mathbb{R}} := \mathbb{R} \cup \{-\infty, +\infty\}$ of $\mathbb{R}$. Then, by replacing $\{f_{n_i}\}$ with a subsequence, $\{(f_{n_i})_*\sigma^{n_i}\}$ converges weakly to a probability measure $\bar{\sigma}_\infty$ on $\bar{\mathbb{R}}$. We have $\bar{\sigma}_\infty|_{\mathbb{R}} = \sigma_\infty$ and $\sigma_\infty(\mathbb{R}) + \bar{\sigma}_\infty\{-\infty, +\infty\} = \bar{\sigma}_\infty(\bar{\mathbb{R}}) = 1$. We prove the following

8.1 This is also called the Poincaré limit theorem in many literature. However, there is no evidence that Poincaré proved this (see [5]*S. 6.1).

Lemma 8.5.4. *Let x and x' be two given real numbers. If $\gamma^1(-\infty, x] = \bar{\sigma}_\infty[-\infty, x']$ and if $\sigma_\infty\{x'\} = 0$, then*

$$\sigma_\infty[x' - \varepsilon_1, x' + \varepsilon_2] \geq \gamma^1[x - \varepsilon_1, x + \varepsilon_2]$$

for all real numbers $\varepsilon_1, \varepsilon_2 \geq 0$. In particular, σ_∞ is a probability measure on $\mathbb{R}$ and $\bar{\sigma}_\infty\{-\infty, +\infty\} = 0$.

Proof. We set $\Omega_+ := \{f_{n_i} \geq x'\}$ and $\Omega_- := \{f_{n_i} \leq x'\}$. We have $\Omega_+ \cup \Omega_- = S^{n_i}(\sqrt{n_i})$. Let us prove

$$U_{\varepsilon_1}(\Omega_+) \cap U_{\varepsilon_2}(\Omega_-) \subset \{x' - \varepsilon_1 < f_{n_i} < x' + \varepsilon_2\}. \tag{8.5}$$

In fact, for any point $\xi \in U_{\varepsilon_1}(\Omega_+)$, there is a point $\xi' \in \Omega_+$ such that the geodesic distance between ξ and ξ' is less than ε_1. The 1-Lipschitz continuity of f_{n_i} proves that $f_{n_i}(\xi) > f_{n_i}(\xi') - \varepsilon_1 \geq x' - \varepsilon_1$. Thus we have $U_{\varepsilon_1}(\Omega_+) \subset \{x' - \varepsilon_1 < f_{n_i}\}$ and, in the same way, $U_{\varepsilon_2}(\Omega_-) \subset \{f_{n_i} < x' + \varepsilon_2\}$. Combining these two inclusions implies (8.5).

It follows from (8.5) and $U_{\varepsilon_1}(\Omega_+) \cup U_{\varepsilon_2}(\Omega_-) = S^{n_i}(\sqrt{n_i})$ that

$$\begin{aligned}
&(f_{n_i})_*\sigma^{n_i}[x' - \varepsilon_1, x' + \varepsilon_2] \\
&= \sigma^{n_i}(x' - \varepsilon_1 \leq f_{n_i} \leq x' + \varepsilon_2) \geq \sigma^{n_i}(U_{\varepsilon_1}(\Omega_+) \cap U_{\varepsilon_2}(\Omega_-)) \\
&= \sigma^{n_i}(U_{\varepsilon_1}(\Omega_+)) + \sigma^{n_i}(U_{\varepsilon_2}(\Omega_-)) - 1,
\end{aligned}$$

where $\sigma^{n_i}(P)$ means the σ^{n_i}-measure of the set of points satisfying a conditional formula P. The Lévy's isoperimetric inequality (Theorem 8.5.3) implies $\sigma^{n_i}(U_{\varepsilon_1}(\Omega_+)) \geq \sigma^{n_i}(U_{\varepsilon_1}(B_{\Omega_+}))$ and $\sigma^{n_i}(U_{\varepsilon_2}(\Omega_-)) \geq \sigma^{n_i}(U_{\varepsilon_2}(B_{\Omega_-}))$, so that

$$(f_{n_i})_*\sigma^{n_i}[x' - \varepsilon_1, x' + \varepsilon_2] \geq \sigma^{n_i}(U_{\varepsilon_1}(B_{\Omega_+})) + \sigma^{n_i}(U_{\varepsilon_2}(B_{\Omega_-})) - 1.$$

It follows from $\sigma_\infty\{x'\} = 0$ that $\sigma^{n_i}(\Omega_+)$ converges to $\bar{\sigma}_\infty(x', +\infty]$ as $n \to \infty$. We besides have $\bar{\sigma}_\infty(x', +\infty] = \gamma^1[x, +\infty) \neq 0, 1$. Therefore, there is a number i_0 such that $\sigma^{n_i}(\Omega_+) \neq 0, 1$ for all $i \geq i_0$. For each $i \geq i_0$ we have a unique real number a_i satisfying $\gamma^1[a_i, +\infty) = \sigma^{n_i}(\Omega_+)$. The number a_i converges to x as $i \to \infty$. By the Maxwell-Boltzmann distribution law (Proposition 8.5.1),

$$\lim_{i\to\infty} \sigma^{n_i}(U_{\varepsilon_1}(B_{\Omega_+})) = \gamma^1[x - \varepsilon_1, +\infty)$$

as well as

$$\lim_{i\to\infty} \sigma^{n_i}(U_{\varepsilon_2}(B_{\Omega_-})) = \gamma^1(-\infty, x + \varepsilon_2].$$

Therefore,

$$\begin{aligned}
\sigma_\infty[x' - \varepsilon_1, x' + \varepsilon_2] &\geq \liminf_{i\to\infty}(f_{n_i})_*\sigma^{n_i}[x' - \varepsilon_1, x' + \varepsilon_2] \\
&\geq \gamma^1[x - \varepsilon_1, +\infty) + \gamma^1(-\infty, x + \varepsilon_2] - 1
\end{aligned}$$

$$= \gamma^1[\,x - \varepsilon_1, x + \varepsilon_2\,].$$

The first part of the lemma is obtained.

The rest of the proof is to show that $\sigma_\infty(\mathbb{R}) = 1$. Suppose $\sigma_\infty(\mathbb{R}) < 1$. Then, since $\sigma_\infty(\mathbb{R}) > 0$, there is a non-atomic point x' of σ_∞ such that $0 < \bar{\sigma}_\infty[-\infty, x') < 1$. We find a real number x in such a way that $\gamma^1(-\infty, x] = \bar{\sigma}_\infty[-\infty, x']$. The first part of the lemma implies $\sigma_\infty[x' - \varepsilon_1, x' + \varepsilon_2] \geq \gamma^1[x - \varepsilon_1, x + \varepsilon_2]$ for all $\varepsilon_1, \varepsilon_2 \geq 0$. Taking the limit as $\varepsilon_1, \varepsilon_2 \to +\infty$, we obtain $\sigma_\infty(\mathbb{R}) = 1$. This completes the proof. □

Lemma 8.5.5. supp σ_∞ *is a closed interval.*

Proof. supp σ_∞ is a closed set by the definition of the support of a measure. It then suffices to prove the connectivity of supp σ_∞. Suppose not. Then, there are numbers x' and $\varepsilon > 0$ such that $\sigma_\infty(-\infty, x' - \varepsilon) > 0$, $\sigma_\infty[x' - \varepsilon, x' + \varepsilon] = 0$, and $\sigma_\infty(x' + \varepsilon, +\infty) > 0$. There is a number x such that $\gamma^1(-\infty, x] = \sigma_\infty(-\infty, x']$. Lemma 8.5.4 shows that $\sigma_\infty[x' - \varepsilon, x' + \varepsilon] \geq \gamma^1[x - \varepsilon, x + \varepsilon] > 0$, which is a contradiction. The lemma has been proved. □

Proof of Theorem 8.5.2. For any given real number x, there exists a smallest number x' satisfying $\gamma^1(-\infty, x] \leq \sigma_\infty(-\infty, x']$. The existence of x' follows from the right-continuity and the monotonicity of $y \mapsto \sigma_\infty(-\infty, y]$. Setting $\alpha(x) := x'$ we have a function $\alpha : \mathbb{R} \to \mathbb{R}$, which is monotone nondecreasing.

We first prove the continuity of α in the following. Take any two numbers x_1 and x_2 with $x_1 < x_2$. We have $\gamma^1(-\infty, x_1] \leq \sigma_\infty(-\infty, \alpha(x_1)]$ and $\gamma^1(-\infty, x_2] \geq \sigma_\infty(-\infty, \alpha(x_2))$, which imply

$$\gamma^1[\,x_1, x_2\,] \geq \sigma_\infty(\,\alpha(x_1), \alpha(x_2)\,). \tag{8.6}$$

This shows that, as $x_1 \to a - 0$ and $x_2 \to a + 0$ for a number a, we have $\sigma_\infty(\alpha(x_1), \alpha(x_2)) \to 0$, which together with Lemma 8.5.5 implies $\alpha(x_2) - \alpha(x_1) \to 0$. Thus, α is continuous on $\mathbb{R}$.

Let us next prove the 1-Lipschitz continuity of α. We take two numbers x and $\varepsilon > 0$ and fix them. It suffices to prove that

$$\Delta\alpha := \alpha(x + \varepsilon) - \alpha(x) \leq \varepsilon.$$

Claim 1. *If* $\sigma_\infty\{\alpha(x)\} = 0$, *then* $\Delta\alpha \leq \varepsilon$.

Proof. The claim is trivial if $\Delta\alpha = 0$. We thus assume $\Delta\alpha > 0$. Since $\sigma_\infty\{\alpha(x)\} = 0$, we have $\gamma^1(-\infty, x] = \sigma_\infty(-\infty, \alpha(x)]$, so that Lemma 8.5.4 implies that

$$\sigma_\infty[\,\alpha(x), \alpha(x) + \delta\,] \geq \gamma^1[\,x, x + \delta\,] \tag{8.7}$$

for all $\delta \geq 0$. By (8.6) and (8.7),

$$\gamma^1[\,x, x + \varepsilon\,] \geq \sigma_\infty(\,\alpha(x), \alpha(x + \varepsilon)\,)$$

$$
\begin{aligned}
&= \sigma_\infty[\,\alpha(x), \alpha(x) + \Delta\alpha\,) \\
&= \lim_{\delta \to \Delta\alpha - 0} \sigma_\infty[\,\alpha(x), \alpha(x) + \delta\,] \\
&\geq \lim_{\delta \to \Delta\alpha - 0} \gamma^1[\,x, x + \delta\,] \\
&= \gamma^1[\,x, x + \Delta\alpha\,],
\end{aligned}
$$

which implies $\Delta\alpha \leq \varepsilon$. □

We next prove that $\Delta\alpha \leq \varepsilon$ in the case where $\sigma_\infty\{\alpha(x)\} > 0$. We may assume that $\Delta\alpha > 0$. Let $x_+ := \sup \alpha^{-1}(\alpha(x))$. It follows from $\alpha(x) < \alpha(x+\varepsilon)$ that $x_+ < x + \varepsilon$. The continuity of α implies that $\alpha(x_+) = \alpha(x)$. There is a sequence of positive numbers $\varepsilon_i \to 0$ such that $\sigma_\infty\{\alpha(x_+ + \varepsilon_i)\} = 0$. By applying the claim above,

$$
\alpha(x_+ + \varepsilon_i + \varepsilon) - \alpha(x_+ + \varepsilon_i) \leq \varepsilon.
$$

Moreover we have $\alpha(x + \varepsilon) \leq \alpha(x_+ + \varepsilon_i + \varepsilon)$ and $\alpha(x_+ + \varepsilon_i) \to \alpha(x_+) = \alpha(x)$ as $i \to \infty$. Thus,

$$
\alpha(x + \varepsilon) - \alpha(x) \leq \varepsilon
$$

and so α is 1-Lipschitz continuous.

The rest is to prove that $\alpha_*\gamma^1 = \sigma_\infty$. Take any number $x' \in \alpha(\mathbb{R})$ and fix it. Set $x := \sup \alpha^{-1}(x')$ ($\leq +\infty$). We then have $\alpha(x) = x'$ provided $x < +\infty$. Since x is the largest number to satisfy $\gamma^1(-\infty, x\,] \leq \sigma_\infty(-\infty, x'\,]$, we have $\gamma^1(-\infty, x\,] = \sigma_\infty(-\infty, x'\,]$, where we agree $\gamma^1(-\infty, +\infty\,] = 1$. By the monotonicity of α, we obtain

$$
\alpha_*\gamma^1(-\infty, x'\,] = \gamma^1(\alpha^{-1}(-\infty, x'\,]) = \gamma^1(-\infty, x\,] = \sigma_\infty(-\infty, x'\,],
$$

which implies that $\alpha_*\gamma^1 = \sigma_\infty$ because σ_∞ is a Borel probability measure. This completes the proof. □

Proposition 8.5.6. *Let X be an mm-space. Then, for any real number $t > 0$ and $\kappa \geq 0$, we have*

$$
\operatorname{ObsDiam}(tX; -\kappa) = t \operatorname{ObsDiam}(X; -\kappa).
$$

Proof. We have

$$
\begin{aligned}
\operatorname{ObsDiam}(tX; -\kappa) &= \sup\{\, \operatorname{diam}(f_*\mu_X; 1 - \kappa) \mid f : tX \to \mathbb{R} \text{ 1-Lipschitz} \,\} \\
&= \sup\{\, \operatorname{diam}(f_*\mu_X; 1 - \kappa) \mid t^{-1}f : X \to \mathbb{R} \text{ 1-Lipschitz} \,\} \\
&= \sup\{\, \operatorname{diam}((tg)_*\mu_X; 1 - \kappa) \mid g : X \to \mathbb{R} \text{ 1-Lipschitz} \,\} \\
&= t \operatorname{ObsDiam}(X; -\kappa).
\end{aligned}
$$

This completes the proof. □

Corollary 8.5.7. *Let $\{r_n\}_{n=1}^{\infty}$ be a sequence of positive real numbers. If $r_n/\sqrt{n} \to \lambda$ as $n \to \infty$ for a real number λ, then*

$$\lim_{n\to\infty} \mathrm{ObsDiam}(S^n(r_n); -\kappa) = \mathrm{diam}(\gamma^1_{\lambda^2}; 1-\kappa) = 2\lambda I^{-1}((1-\kappa)/2),$$

for any real number κ with $0 < \kappa < 1$, where $I(r) := \gamma^1[0, r]$. This holds for the geodesic distance function on $S^n(r_n)$ and also for the restriction to $S^n(r_n)$ of the Euclidean distance function on $\mathbb{R}^n$.

Proof. It suffices to prove the corollary in the case where $r_n = \sqrt{n}$, because of Proposition 8.5.6. We assume that $r_n = \sqrt{n}$.

The corollary for the geodesic distance function follows from the normal law à la Lévy (Theorem 8.5.2) and the Maxwell-Boltzmann distribution law (Proposition 8.5.1).

Assume that the distance function on $S^n(\sqrt{n})$ to be Euclidean. Since the Euclidean distance is not greater than the geodesic distance, the normal law à la Lévy still holds. Clearly, the projection $\pi^{n+1}_k : S^n(\sqrt{n}) \to \mathbb{R}^k$ is 1-Lipschitz with respect to the Euclidean distance function. Thus, the corollary follows from the normal law à la Lévy and the Maxwell-Boltzmann distribution law. This completes the proof. □

Lemma 8.5.8. *Let X be a metric space and G a group acting on X isometrically. Then, for any two Borel probability measures μ and ν on X, we have*

$$d_P(\bar{\mu}, \bar{\nu}) \le d_P(\mu, \nu),$$

where $\bar{\mu}$ denotes the push-forward of μ by the projection $X \to X/G$.

Proof. Take any Borel subset $\bar{A} \subset X/G$ and set $A := \pi^{-1}(\bar{A})$, where $\pi : X \to X/G$ is the projection. Assume that $d_P(\mu, \nu) < \varepsilon$ for a real number ε. Since $\pi^{-1}(B_\varepsilon(\bar{A})) \supset B_\varepsilon(A)$, we have

$$\bar{\mu}(B_\varepsilon(\bar{A})) = \mu(\pi^{-1}(B_\varepsilon(\bar{A}))) \ge \mu(B_\varepsilon(A)) \ge \nu(A) - \varepsilon = \bar{\nu}(\bar{A}) - \varepsilon$$

and therefore $d_P(\bar{\mu}, \bar{\nu}) \le \varepsilon$. This completes the proof. □

Denote by ζ^n the normalized volume measure on $\mathbb{C}P^n(r)$ with respect to the Fubini-Study metric. Note that $\zeta_n = \bar{\sigma}^{2n+1}$, where $\bar{\sigma}^{2n+1}$ denotes the push-forward of σ^{2n+1} by the Hopf fibration $S^{2n+1}(r) \to \mathbb{C}P^n(r)$.

Proposition 8.5.9. *Let $\{r_n\}_{i=1}^{\infty}$ be a sequence of positive real numbers such that $r_n/\sqrt{2n+1} \to \lambda$ as $n \to \infty$ for a real number λ. Then, for any natural number k we have*

$$(\bar{\pi}^{2n+2}_{2k})_* \zeta^n \to \bar{\gamma}^{2k}_{\lambda^2} \text{ weakly} \quad \text{as } n \to \infty,$$

where $\bar{\pi}^{2n+2}_{2k} : \mathbb{C}P^n(r_n) \subset \mathbb{C}^{n+1}/S^1 \to \mathbb{C}^k/S^1$ is as in S. 8.4 and $\bar{\gamma}^{2k}_{\lambda^2}$ the push-forward of $\gamma^{2k}_{\lambda^2}$ by the projection $\mathbb{C}^k \to \mathbb{C}^k/S^1$.

Proof. By Proposition 8.5.1, $(\pi^{2n+2}_{2k})_*\sigma^{2n}$ converges weakly to $\gamma^{2k}_{\lambda^2}$ as $n \to \infty$. In the following, a measure with upper bar means the push-forward by the projection to the Hopf quotient. Since $(\bar\pi^{2n+2}_{2k})_*\zeta^n = (\bar\pi^{2n+2}_{2k})_*\bar\sigma^{2n} = \overline{(\pi^{2n+2}_{2k})_*\sigma^{2n}}$, and by Lemma 8.5.8, we have

$$d_P((\bar\pi^{2n+2}_{2k})_*\zeta^n, \bar\gamma^{2k}_{\lambda^2}) \le d_P((\pi^{2n+2}_{2k})_*\sigma^{2n}, \gamma^{2k}_{\lambda^2}) \to 0 \quad \text{as } n \to \infty.$$

This completes the proof. □

Corollary 8.5.10. *Let $\{r_n\}_{i=1}^\infty$ be a sequence of positive real numbers. If $r_n/\sqrt{2n+1} \to \lambda$ as $n \to \infty$ for a real number λ, then*

$$\limsup_{n\to\infty} \mathrm{ObsDiam}(\mathbb{C}P^n(r_n); -\kappa) \le 2\lambda I^{-1}((1-\kappa)/2),$$

$$\liminf_{n\to\infty} \mathrm{ObsDiam}(\mathbb{C}P^n(r_n); -\kappa) \ge \lambda \mathrm{diam}(([0, +\infty), re^{-\frac{r^2}{2}}\,dr); 1-\kappa).$$

These inequalities both hold for the scaled Fubini-Study metric and also for the induced distance function from the Euclidean distance on $S^{2n+1}(r_n)$.

Proof. The upper estimate follows from $\mathbb{C}P^n(r_n) \prec S^{2n}(r_n)$ and Corollary 8.5.7.

Since $\bar\pi^{2n+2}_2 : \mathbb{C}P^n(r_n) \subset \mathbb{C}^{n+1}/S^1 \to \mathbb{C}/S^1$ is 1-Lipschitz, Proposition 8.5.9 proves

$$\begin{aligned}\liminf_{n\to\infty} \mathrm{ObsDiam}(\mathbb{C}P^n(r_n); -\kappa) &\ge \mathrm{diam}((\mathbb{C}/S^1, \bar\gamma^2_{\lambda^2}); 1-\kappa)\\ &= \lambda \mathrm{diam}(([0, +\infty), re^{-\frac{r^2}{2}}\,dr); 1-\kappa).\end{aligned}$$

This completes the proof. □

We conjecture that

$$\lim_{n\to\infty} \mathrm{ObsDiam}(\mathbb{C}P^n(r_n); -\kappa) = \lambda \mathrm{diam}(([0, +\infty), re^{-\frac{r^2}{2}}\,dr); 1-\kappa)$$

if $r_n/\sqrt{2n+1} \to \lambda$ as $n \to \infty$.

8.6 Limits of spheres and complex projective spaces

In this section, we prove Theorem 8.1.1 by using several claims proved before. We need some more lemmas.

Lemma 8.6.1. *For any real number θ with $0 < \theta < 1$, we have*

$$\lim_{n\to\infty} \gamma^{n+1}\{\, x \in \mathbb{R}^{n+1} \mid \theta\sqrt{n} \le \|x\|_2 \le \theta^{-1}\sqrt{n} \,\} = 1.$$

Proof. Considering the polar coordinates on $\mathbb{R}^n$, we see that

$$\gamma^{n+1}\{\, x \in \mathbb{R}^{n+1} \mid \|x\|_2 \le r \,\} = \frac{\int_0^r t^n e^{-t^2/2}\,dt}{\int_0^\infty t^n e^{-t^2/2}\,dt}.$$

Integrating both sides of $(\log(t^n e^{-t^2/2}))'' = -n/t^2 - 1 \le -1$ over $[\,t, \sqrt{n}\,]$ with $0 < t \le \sqrt{n}$ yields

$$-(\log(t^n e^{-t^2/2}))' = (\log(t^n e^{-t^2/2}))'|_{t=\sqrt{n}} - (\log(t^n e^{-t^2/2}))' \le t - \sqrt{n}.$$

Integrating this again over $[\,t, \sqrt{n}\,]$ implies

$$\log(t^n e^{-t^2/2}) - \log(n^{n/2} e^{-n/2}) \le -\frac{(t-\sqrt{n})^2}{2},$$

so that $t^n e^{-t^2/2} \le n^{n/2} e^{-n/2} e^{-(t-\sqrt{n})^2/2}$ and then, for any r with $0 \le r \le \sqrt{n}$,

$$\int_0^{\sqrt{n}-r} t^n e^{-t^2/2}\,dt \le n^{n/2} e^{-n/2} \int_r^{\sqrt{n}} e^{-t^2/2}\,dt \le n^{n/2} e^{-n/2} e^{-r^2/2}.$$

Stirling's approximation implies

$$\int_0^{\infty} t^n e^{-t^2/2}\,dt = 2^{\frac{n-1}{2}} \int_0^{\infty} s^{\frac{n-1}{2}} e^{-s}\,ds \approx \sqrt{\pi}(n-1)^{\frac{n}{2}} e^{-\frac{n-1}{2}}.$$

Therefore,

$$\lim_{n\to\infty} \gamma^{n+1}\{\, x \in \mathbb{R}^{n+1} \mid \|x\|_2 \le \theta\sqrt{n}\,\} \le \lim_{n\to\infty} \frac{n^{n/2} e^{-n/2} e^{-(1-\theta)^2 n/2}}{\sqrt{\pi}(n-1)^{\frac{n}{2}} e^{-\frac{n-1}{2}}} = 0.$$

The same calculation leads us to obtain

$$\lim_{n\to\infty} \gamma^{n+1}\{\, x \in \mathbb{R}^{n+1} \mid \|x\|_2 \ge \theta^{-1}\sqrt{n}\,\} = 0.$$

This completes the proof. □

For a positive real number t and a pyramid $\mathcal{P}$, we put

$$t\mathcal{P} := \{\, tX \mid X \in \mathcal{P}\,\},$$

where $tX := (X, t\,d_X, \mu_X)$. The following lemma is obvious and the proof is omitted.

Lemma 8.6.2. *Assume that a sequence of pyramids $\mathcal{P}_n$, $n = 1, 2, \dots,$ converges weakly to a pyramid $\mathcal{P}$, and that a sequence of positive real numbers t_n, $n = 1, 2, \dots,$ converges to a positive real number t. Then, $t_n\mathcal{P}_n$ converges weakly to $t\mathcal{P}$ as $n \to \infty$.*

Proof of Theorem 8.1.1. (1) follows from Corollaries 8.5.7 and 8.5.10.

We prove (2) for $S^n(r_n)$. We first assume that $r_n/\sqrt{n} \to +\infty$ as $n \to \infty$. Take any finitely many positive real numbers $\kappa_0, \kappa_1, \dots, \kappa_N$ with $\sum_{i=0}^N \kappa_i < 1$, and fix them. We find positive real numbers $\kappa_0', \kappa_1', \dots, \kappa_N'$ in such a way that $\kappa_i < \kappa_i'$ for any i and

$\sum_{i=0}^{N} \kappa_i' < 1$. For any $\varepsilon > 0$, there are Borel subsets $A_0, A_1, \dots, A_N \subset \mathbb{R}$ such that $\gamma^1(A_i) \geq \kappa_i'$ for any i and

$$\min_{i \neq j} d_{\mathbb{R}}(A_i, A_j) > \operatorname{Sep}((\mathbb{R}, \gamma^1); \kappa_0', \dots, \kappa_N') - \varepsilon.$$

Without loss of generality we may assume that all A_i's are open. Let $f_n := \pi_1^n|_{S^n(\sqrt{n})} : S^n(\sqrt{n}) \to \mathbb{R}$. Proposition 8.5.1 tells us that $(f_n)_* \sigma^n$ converges weakly to γ^1 as $n \to \infty$. Since A_i is open,

$$\liminf_{n\to\infty} \sigma^n((f_n)^{-1}(A_i)) \geq \gamma^1(A_i) \geq \kappa_i' > \kappa_i.$$

There is a natural number n_0 such that $\sigma^n((f_n)^{-1}(A_i)) \geq \kappa_i$ for any i and $n \geq n_0$. Since f_n is 1-Lipschitz continuous for the Riemannian metric and also for the Euclidean distance on $S^n(\sqrt{n})$, we have

$$d_{S^n(\sqrt{n})}((f_n)^{-1}(A_i), (f_n)^{-1}(A_j)) \geq d_{\mathbb{R}}(A_i, A_j).$$

Therefore, for any $n \geq n_0$,

$$\operatorname{Sep}(S^n(\sqrt{n}); \kappa_0, \dots, \kappa_N) > \operatorname{Sep}((\mathbb{R}, \gamma^1); \kappa_0', \dots, \kappa_N') - \varepsilon,$$

which proves

$$\liminf_{n\to\infty} \operatorname{Sep}(S^n(\sqrt{n}); \kappa_0, \dots, \kappa_N) \geq \operatorname{Sep}((\mathbb{R}, \gamma^1); \kappa_0', \dots, \kappa_N') > 0.$$

Since $r_n/\sqrt{n} \to \infty$ as $n \to \infty$,

$$\operatorname{Sep}(S^n(r_n); \kappa_0, \dots, \kappa_N) = \frac{r_n}{\sqrt{n}} \operatorname{Sep}(S^n(\sqrt{n}); \kappa_0, \dots, \kappa_N) \tag{8.8}$$

is divergent to infinity and so $\{S^n(r_n)\}_{n=1}^{\infty}$ infinitely dissipates.

We next prove the converse. Assume that $\{S^n(r_n)\}$ infinitely dissipates and $r_n/\sqrt{n}$ is not divergent to infinity. Then, there is a subsequence $\{r_{n(j)}\}$ of $\{r_n\}$ such that $r_{n(j)}/\sqrt{n(j)}$ is bounded for all j. By (8.8), $\{S^{n(j)}(\sqrt{n(j)})\}_j$ infinitely dissipates. However, for each fixed κ with $0 < \kappa < 1/2$, $\operatorname{ObsDiam}(S^n(\sqrt{n}); -\kappa)$ is bounded for all n by Corollary 8.5.7, and, by Proposition 8.2.9, so is $\operatorname{Sep}(S^n(\sqrt{n}); \kappa, \kappa)$, which contradicts that $\{S^{n(j)}(\sqrt{n(j)})\}$ infinitely dissipates. This completes the proof of (2).

(2) for $\mathbb{C}P^n(r_n)$ is proved in the same way as for $S^n(r_n)$ by using Proposition 8.5.9 and Corollary 8.5.10 instead of Proposition 8.5.1 and Corollary 8.5.7.

We prove (3) for $S^n(r_n)$. By Lemma 8.6.2 and by $\Gamma_{\lambda^2}^n = \lambda \Gamma^n$, it suffices to prove it in the case of $r_n = \sqrt{n}$. We assume that $r_n = \sqrt{n}$. Suppose that $\mathcal{P}_{S^n(\sqrt{n})}$ does not converge weakly to $\mathcal{P}_{\Gamma^\infty}$ as $n \to \infty$. Then, by the compactness of Π, there is a subsequence $\{\mathcal{P}_{S^{n_i}(\sqrt{n_i})}\}$ of $\{\mathcal{P}_{S^n(\sqrt{n})}\}$ that converges weakly to a pyramid $\mathcal{P}$ with $\mathcal{P} \neq \mathcal{P}_{\Gamma^\infty}$. It follows from the Maxwell-Boltzmann distribution law (Proposition 8.5.1) that Γ^k belongs to $\mathcal{P}$ for any k, so that $\mathcal{P}_{\Gamma^\infty} \subset \mathcal{P}$. We take any real number θ with $0 < \theta < 1$ and fix it. It follows from Lemma 8.6.2 that $\mathcal{P}_{S^{n_i}(\theta\sqrt{n_i})}$ converges weakly to $\theta\mathcal{P}$ as $i \to \infty$. Define a function $f_{\theta,n} : \mathbb{R}^{n+1} \to \mathbb{R}^{n+1}$ by

$$f_{\theta,n}(x) := \begin{cases} \frac{\theta\sqrt{n}}{\|x\|_2}x & \text{if } \|x\|_2 > \theta\sqrt{n}, \\ x & \text{if } \|x\|_2 \le \theta\sqrt{n}, \end{cases}$$

for $x \in \mathbb{R}^{n+1}$. $f_{\theta,n}$ is 1-Lipschitz continuous with respect to the Euclidean distance. Let σ_θ^n be the normalized volume measure on $S^n(\theta\sqrt{n})$. We consider σ_θ^n as a measure on $\mathbb{R}^{n+1}$ via the natural embedding $S^n(\theta\sqrt{n}) \subset \mathbb{R}^{n+1}$. From Lemma 8.6.1, we have

$$d_P((f_{\theta,n})_*\gamma^{n+1}, \sigma_\theta^n) \le \gamma^{n+1}\{\, x \in \mathbb{R}^{n+1} \mid \|x\|_2 < \theta\sqrt{n}\,\} \to 0 \text{ as } n \to \infty,$$

so that the box distance between $S_{\theta,n} := (\mathbb{R}^{n+1}, \|\cdot\|_2, (f_{\theta,n})_*\gamma^{n+1})$ and $S^n(\theta\sqrt{n})$ converges to zero as $n \to \infty$. By Proposition 8.2.18 and Theorem 8.1.2, we have $\rho(\mathcal{P}_{S_{\theta,n}}, \mathcal{P}_{S^n(\theta\sqrt{n})}) \to 0$ as $n \to \infty$. Therefore, $\mathcal{P}_{S_{\theta,n_i}}$ converges weakly to $\theta\mathcal{P}$ as $i \to \infty$. Since $S_{\theta,n} \prec (\mathbb{R}^{n+1}, \|\cdot\|_2, \gamma^{n+1})$, we have $\mathcal{P}_{S_{\theta,n}} \subset \mathcal{P}_{\Gamma^{n+1}} \subset \mathcal{P}_{\Gamma^\infty}$. We thus obtain $\theta\mathcal{P} \subset \mathcal{P}_{\Gamma^\infty} \subset \mathcal{P}$ for any θ with $0 < \theta < 1$ and so $\mathcal{P} = \mathcal{P}_{\Gamma^\infty}$, which is a contradiction. This completes the proof of (3) for $S^n(r_n)$.

We prove (3) for $\mathbb{C}P^n(r_n)$. The proof is similar to that for $S^n(r_n)$. We may assume that $r_n = \sqrt{2n+1}$. Suppose that $\mathcal{P}_{\mathbb{C}P^n(\sqrt{2n+1})}$ does not converge weakly to $\mathcal{P}_{\Gamma^\infty/S^1}$ as $n \to \infty$. Then, by the compactness of Π, there is a subsequence $\{\mathbb{C}P^{n(j)}(\sqrt{2n(j)+1})\}_j$ of $\{\mathbb{C}P^n(\sqrt{2n+1})\}_n$ that converges weakly to a pyramid $\mathcal{P}$ with $\mathcal{P} \neq \mathcal{P}_{\Gamma^\infty/S^1}$. Proposition 8.5.9 proves that $\Gamma^{2k}/S^1 \subset \mathcal{P}$ for any k and so $\mathcal{P}_{\Gamma^\infty/S^1} \subset \mathcal{P}$. We take any real number θ with $0 < \theta < 1$ and fix it. By Lemma 8.6.2, $\mathcal{P}_{\mathbb{C}P^{n(j)}(\theta\sqrt{2n(j)+1})}$ converges weakly to $\theta\mathcal{P}$ as $j \to \infty$. Let $f_{\theta,2n+1} : \mathbb{C}^{n+1} \to \mathbb{C}^{n+1}$ be as above by identifying $\mathbb{C}^{n+1}$ with $\mathbb{R}^{2n+2}$. $f_{\theta,2n+1}$ is S^1-equivariant and induces a map $\bar{f}_{\theta,n} : \mathbb{C}^{n+1}/S^1 \to \mathbb{C}^{n+1}/S^1$, which is 1-Lipschitz continuous. Let σ_θ^{2n+1} be the normalized volume measure on $S^{2n+1}(\theta\sqrt{2n+1})$. Since $(f_{\theta,2n+1})_*\gamma^{2n+2}|_{S^{2n+1}(\theta\sqrt{2n+1})}$ is a constant multiple of σ_θ^{2n+1}, the measure $\overline{(f_{\theta,2n+1})_*\gamma^{2n+2}}|_{\mathbb{C}P^n(\theta\sqrt{2n+1})}$ is also a constant multiple of $\bar{\sigma}_\theta^{2n+1}$, where the upper bar means the push-forward of a measure by the projection to the Hopf quotient space. We see that

$$\overline{(f_{\theta,2n+1})_*\gamma^{2n+2}} = (\bar{f}_{\theta,2n+2})_*\bar{\gamma}^{2n+2}.$$

By Lemmas 8.5.8 and 8.6.1, we have

$$d_P((\bar{f}_{\theta,2n+2})_*\bar{\gamma}^{2n+2}, \bar{\sigma}_\theta^{2n+1}) \le d_P((f_{\theta,2n+1})_*\gamma^{2n+2}, \sigma_\theta^{2n+1}) \to 0 \quad \text{as } n \to \infty,$$

so that the box distance between $Y_{\theta,n} := (\mathbb{C}^{n+1}, \|\cdot\|_2, (\bar{f}_{\theta,n})_*\bar{\gamma}^{2n+2})$ and $\mathbb{C}P^n\{\theta\sqrt{2n+1}\}$ converges to zero as $n \to \infty$. The rest of the proof is the same as before. This completes the proof of the theorem. □

8.7 d_{conc}-Cauchy property and box convergence

In this section, we prove Theorem 8.1.3 and prove the non-convergence property for spheres and complex projective spaces with respect to the box distance.

Proposition 8.7.1. *The virtual infinite-dimensional standard Gaussian space $\mathcal{P}_{\Gamma^\infty}$ is not concentrated.*

Proof. Let $\phi_{n,i} : \mathbb{R}^n \to \mathbb{R}$, $i = 1, 2, \dots, n$, be the functions defined by

$$\phi_{n,i}(x_1, x_2, \dots, x_n) := x_i, \qquad (x_1, x_2, \dots, x_n) \in \mathbb{R}^n.$$

Each $\phi_{n,i}$ is 1-Lipschitz continuous and we have the elements $[\phi_{n,i}]$ of $\mathcal{L}_1(\Gamma^n)$. For any different i and j,

$$\mathrm{me}_{\gamma^n}([\phi_{n,i}], [\phi_{n,j}]) = \mathrm{me}_{\gamma^2}([\phi_{2,1}], [\phi_{2,2}]) = \mathrm{me}_{\gamma^2}(\phi_{2,1}, \phi_{2,2} + t)$$

for some real number t. If $\mathrm{me}_{\gamma^2}(\phi_{2,1}, \phi_{2,2} + t) = 0$ were to hold, then $\phi_{2,1} = \phi_{2,2} + t$ almost everywhere, which is a contradiction. Thus, $\mathrm{me}_{\gamma^n}([\phi_{n,i}], [\phi_{n,j}])$ is a positive constant independent of n, i, and j with $i \neq j$. This implies that $\{\mathcal{L}_1(\Gamma^n)\}_{n=1}^\infty$ is not d_{GH}-precompact. Since $\Gamma^n \in \mathcal{P}_{\Gamma^\infty}$, the pyramid $\mathcal{P}_{\Gamma^\infty}$ is not concentrated. This completes the proof. □

As a direct consequence of Proposition 8.7.1, Lemma 8.2.27, and Theorem 8.1.1, we have the following.

Corollary 8.7.2. *1. $\{\Gamma^n\}_{n=1}^\infty$ has no d_{conc}-Cauchy subsequence.*
2. If $\{r_n\}$ is a sequence of positive real numbers with $r_n/\sqrt{n} \to 1$ as $n \to \infty$, then $\{S^n(r_n)\}$ has no d_{conc}-Cauchy subsequence.

Proof. (1) follows from Proposition 8.7.1 and Lemma 8.2.27.

Theorem 8.1.1(3) implies that $\mathcal{P}_{S^n(r_n)}$ converges weakly to $\mathcal{P}_{\Gamma^\infty}$ as $n \to \infty$, which together with Proposition 8.7.1 and Lemma 8.2.27 proves (2) of the corollary. This completes the proof. □

Proof of Theorem 8.1.3. Since $\{S^n(\sqrt{n})\}$ has no d_{conc}-Cauchy subsequence, there exist two subsequences $\{X_n\}$ and $\{Y_n\}$ of $\{S^n(\sqrt{n})\}$ such that $\inf_n d_{\mathrm{conc}}(X_n, Y_n) > 0$. Theorem 8.1.1(3) implies that $\mathcal{P}_{X_n}$ and $\mathcal{P}_{Y_n}$ both converge weakly to $\mathcal{P}_{\Gamma^\infty}$. This completes the proof. □

Lemma 8.7.3. *Let X_n and Y_n, $n = 1, 2, \dots$, be mm-spaces such that $X_n \prec Y_n$ for any n. If $\{Y_n\}$ is $\square$-precompact, then so is $\{X_n\}$. In particular, if $\{Y_n\}$ is $\square$-precompact and if X_n concentrates to an mm-space X, then X_n $\square$-converges to X.*

Proof. Recall (see [11]*S. $3\frac{1}{2}$.14 and [21]*S. 4) that $\{Y_n\}$ is $\square$-precompact if and only if for any $\varepsilon > 0$ there exists a number $\Delta(\varepsilon) > 0$ such that we have Borel subsets $K_{n1}, K_{n2}, \dots, K_{nN} \subset Y_n$ for each n with the property that

(i) $N \le \Delta(\varepsilon)$;
(ii) $\operatorname{diam} K_{ni} \le \varepsilon$ for any $i = 1, 2, \dots, N$;
(iii) $\operatorname{diam} \bigcup_{i=1}^{n} K_{ni} \le \Delta(\varepsilon)$;
(iv) $\mu_{Y_n}(\bigcup_{i=1}^{n} K_{ni}) \ge 1 - \varepsilon$.

We assume that $\{Y_n\}$ is $\square$-precompact, and then have Borel subsets $K_{ni} \subset Y_n$ satisfying (i)–(iv). Without loss of generality we may assume that all K_{ni} are compact, since each μ_{Y_n} is inner regular. By $X_n \prec Y_n$, we find a 1-Lipschitz continuous map $f_n : Y_n \to X_n$ with $(f_n)_*\mu_{Y_n} = \mu_{X_n}$. The sets $K'_{ni} := f_n(K_{ni})$ are compact and satisfy (i)–(iv), so that $\{X_n\}$ is $\square$-precompact. The first part of the lemma has been proved.

We prove the second part. Assume that $\{Y_n\}$ is $\square$-precompact and that X_n concentrates to an mm-space X. If X_n does not concentrate to X, then the $\square$-precompactness of $\{X_n\}$ proves that it has a $\square$-convergent subsequence whose limit is different from X. This contradicts that X_n concentrates to X as $n \to \infty$ (see Proposition 8.2.18). The proof of the lemma is completed. □

Proposition 8.7.4. *Let $\{r_n\}_{i=1}^{\infty}$ be a sequence of positive real numbers. If r_n is bounded away from zero, then $\{S^n(r_n)\}_{n=1}^{\infty}$ and $\{\mathbb{C}P^n(r_n)\}_{n=1}^{\infty}$ both have no $\square$-convergent subsequence.*

Proof. Assume that $r_n \ge c > 0$ for any natural number n and for a constant c. We have $S^n(c) \prec S^n(r_n)$ and $\mathbb{C}P^n(c) \prec \mathbb{C}P^n(r_n)$. According to [7], the two sequences $\{S^n(c)\}$ and $\{\mathbb{C}P^n(c)\}$ both have no $\square$-convergent subsequence. By Lemma 8.7.3, $\{S^n(r_n)\}$ and $\{\mathbb{C}P^n(r_n)\}$ also have no $\square$-convergent subsequence. This completes the proof. □

Acknowledgment: The author would like to thank Professors Takefumi Kondo and Asuka Takatsu for valuable discussions. He also thanks to Professors Tomohiro Fukaya and Ayato Mitsuishi for useful comments.

The author is partially supported by a Grant-in-Aid for Scientific Research from the Japan Society for the Promotion of Science.

Bibliography

[1] P. Billingsley: Convergence of probability measures, 2nd ed., Wiley Series in Probability and Statistics: Probability and Statistics, John Wiley & Sons Inc., New York, 1999. A Wiley-Interscience Publication.
[2] V. I. Bogachev: Gaussian measures, Mathematical Surveys and Monographs, vol. 62, American Mathematical Society, Providence, RI, 1998.

[3] V. I. Bogachev: Measure theory. Vol. I, II, Springer-Verlag, Berlin, 2007.
[4] D. Burago, Y. Burago, and S. Ivanov: A course in metric geometry, Graduate Studies in Mathematics, vol. 33, American Mathematical Society, Providence, RI, 2001.
[5] P. Diaconis and D. Freedman: A dozen de Finetti-style results in search of a theory, Ann. Inst. H. Poincare Probab. Statist. 23 (1987), no. 2, suppl., 397–423 (English, with French summary).
[6] T. Figiel, J. Lindenstrauss, and V. D. Milman: The dimension of almost spherical sections of convex bodies, Acta Math. 139 (1977), no. 1-2, 53–94.
[7] K. Funano: Estimates of Gromov's box distance, Proc. Amer. Math. Soc. 136 (2008), no. 8, 2911–2920.
[8] K. Funano: Asymptotic behavior of mm-spaces. Doctoral Thesis, Tohoku University, 2009.
[9] K. Funano and T. Shioya: Concentration, Ricci curvature, and eigenvalues of Laplacian, Geom. Funct. Anal. 23 (2013), Issue 3, 888–936.
[10] M. Gromov and V. D. Milman: A topological application of the isoperimetric inequality, Amer. J. Math. 105 (1983), no. 4, 843–854.
[11] M. Gromov: Metric structures for Riemannian and non-Riemannian spaces, Reprint of the 2001 English edition, Modern Birkhäuser Classics, Birkhäuser Boston Inc., Boston, MA, 2007. Based on the 1981 French original; With appendices by M. Katz, P. Pansu and S. Semmes; Translated from the French by Sean Michael Bates.
[12] A. S. Kechris: Classical descriptive set theory, Graduate Texts in Mathematics, vol. 156, Springer-Verlag, New York, 1995.
[13] M. Ledoux: The concentration of measure phenomenon, Mathematical Surveys and Monographs, vol. 89, American Mathematical Society, Providence, RI, 2001.
[14] P. Lévy: Problèmes concrets d'analyse fonctionnelle. Avec un complément sur les fonctionnelles analytiques par F. Pellegrino, Gauthier-Villars, Paris, 1951 (French), 2d ed.
[15] V. D. Milman: The heritage of P. Lévy in geometrical functional analysis, Astérisque 157-158 (1988), 273–301. Colloque Paul Lévy sur les Processus Stochastiques (Palaiseau, 1987).
[16] V. D. Milman: A certain property of functions defined on infinite-dimensional mani-folds, Dokl. Akad. Nauk SSSR 200 (1971), 781–784 (Russian).
[17] V. D. Milman: Asymptotic properties of functions of several variables that are defined on homogeneous spaces, Dokl. Akad. Nauk SSSR 199 (1971), 1247–1250 (Russian); English transl., Soviet Math. Dokl. 12 (1971), 1277–1281.
[18] Y. Ollivier: Diamètre observable des sous-variétés de Sn et CPn (1999). mémoire de DEA, université d'Orsay.
[19] R. Ozawa and T. Shioya: Limit formulas for metric measure invariants and phase transition property, Math. Z. 280 (2015), no. 3-4, 759–782, DOI 10.1007/s00209-015-1447-2.
[20] R. Ozawa and T. Shioya: Estimate of observable diameter of lp-product spaces, Manuscripta Math. 147 (2015), no. 3-4, 501–509, DOI 10.1007/s00229-015-0730-1.
[21] T. Shioya: Metric measure geometry–Gromov's theory of convergence and concentration of metrics and measures, IRMA Lectures in Mathematics and Theoretical Physics, vol. 25, European Mathematical Society.
[22] T. Shioya and A. Takatsu: High-dimensional metric-measure limit of Stiefel and flag manifolds. preprint.

Christina Sormani

Scalar Curvature and Intrinsic Flat Convergence

9.1 Introduction

Gromov proved that sequences of Riemannian manifolds with nonnegative sectional curvature have subsequences which converge in the Gromov-Hausdorff sense to Alexandrov spaces with nonnegative Alexandrov curvature [21]. Burago-Gromov-Perelman proved that such spaces are rectifiable [11]. Building upon Gromov's Compactness Theorem, Cheeger-Colding proved that sequences of Riemannian manifolds with nonnegative Ricci curvature have subsequences which converge in the metric measure sense to metric measure spaces with generalized nonnegative Ricci curvature which are also rectifiable [14] and [21].

Sequences of manifolds with nonnegative scalar curvature need not have subsequences which converge in the Gromov-Hausdorff or metric measure sense. Gromov has suggested that perhaps under the right conditions a subsequence will converge in the intrinsic flat sense to a metric space with generalized nonnegative scalar curvature [24]. This is an open question: the notion of generalized scalar curvature has not yet been defined.

Intrinsic flat convergence was first defined by the author and Wenger in [61]. The limits obtained under this convergence are countably $\mathcal{H}^m$ rectifiable metric spaces called integral current spaces. We review the definitions of these notions within this chapter along with various continuity and compactness theorems by the author, Perales, Portegies, Matveev, Munn [42, 45–47, 49]. We also review applications of intrinsic flat convergence to study sequences of manifolds with nonnegative scalar curvature that arise in General Relativity by the author, Huang, Jauregui, Lee, LeFloch, and Stavrov [28, 31, 39, 40, 57]. We present many examples and state a number of open problems concerning limits of manifolds with nonnegative scalar curvature.

Recall that a Riemannian manifold, M^m, is endowed with a metric tensor, $g : TM \times TM \to \mathbb{R}$. One can then define lengths of curves and distances,

$$L(C) = \int_0^1 g(C', C')^{1/2}\, dt, \quad d(p, q) = \inf\{L(C) : C(0) = p,\ C(1) = q\}. \tag{9.1}$$

If M is compact the distances are achieved as the lengths of curves called geodesics. Given any $p \in M$ and any vector $V \in T_p(M)$ there is a geodesic,

$$\gamma(t) = \exp_p(tV) \text{ such that } \gamma(0) = p \text{ and } \gamma'(0) = V. \tag{9.2}$$

Christina Sormani: CUNY GC and Lehman College

Taking $e_1...e_m \in TM$ such that $g(e_i, e_j) = \delta_{i,j}$ one defines Scalar curvature to be the trace of the Ricci curvature, and Ricci to be the trace of the Sectional curvature:

$$\mathrm{Scal}_p = \sum_{i=1}^{m} \mathrm{Ric}_p(e_i, e_i) \text{ where } \mathrm{Ric}_p(e_i, e_i) = \sum_{j \neq i} \mathrm{Sect}_p(e_i \wedge e_j), \tag{9.3}$$

$$\mathrm{Sect}_p(e_i \wedge e_j) = \lim_{t \to 0} 6 \left(\frac{tg(e_i, e_j) - d(\exp_p(te_i), \exp_p(te_j))}{t^3} \right). \tag{9.4}$$

Scalar curvature can also be computed using volumes of balls:

$$\mathrm{Scal}_p = \lim_{r \to 0} 6(m+2) \left(\frac{\omega_m r^m - \mathrm{vol}(B(p,r))}{\omega_m r^{m+2}} \right). \tag{9.5}$$

In particular,

$$\mathrm{Scal}_p > 00 \iff \exists r_p > 0 \ s.t. \ \forall r < r_p \ \ \mathrm{vol}(B(p,r)) < \omega_m r^m. \tag{9.6}$$

This control on volume is too local to apply to prove any global results. All properties of manifolds with lower bounds on their scalar curvature are built using the fact that curvature is defined using tensors as in (9.3). While it may be tempting to define generalized positive scalar curvature on a limit space using (9.6) it is unlikely to lead to any consequences because the limit spaces are not smooth and have no tensors. We need a stronger definition which implies (9.6) and other properties.

Schoen and Yau applied the three dimensional version of (9.3) to study minimal surfaces in manifolds with positive scalar curvature. They proved that a strictly stable closed minimal surface in a manifold with Scal ≥ 0 is diffeomorphic to a sphere in [50]. In [51], they applied minimal surface techniques to prove the Positive Mass Theorem: if M^3 is an asymptotically flat Riemannian manifold with nonnegative scalar curvature then $m_{ADM}(M^3) \geq 0$. They also proved the following Positive Mass Rigidity Theorem:

$$\mathrm{Scal} \geq 0 \text{ and } m_{ADM}(M^3) = 0 \ \Rightarrow \ M^3 \text{ is isometric to } \mathbb{E}^3. \tag{9.7}$$

Here $\mathbb{E}^3$ is Euclidean space and the ADM mass is the limit of the Hawking masses of asymptotically expanding spheres $m_{ADM}(M) = \lim_{r\to\infty} m_H(\Sigma_r)$ where

$$m_H(\Sigma) = \sqrt{\frac{\mathrm{Area}(\Sigma)}{16\pi}} \left(1 - \frac{1}{16\pi} \int_{\Sigma} H^2 \, d\sigma \right). \tag{9.8}$$

Geroch proved that if $N_t : \mathbb{S}^2 \to M^3$ evolves by inverse mean curvature flow and M^3 has Scal ≥ 0 then the Hawking mass, $m_H(N_t)$, is nondecreasing. Huisken-Ilmanen introduced weak inverse mean curvature flow, proving it also satisfies Geroch monotonicity and $\lim_{t\to\infty} m_H(N_t) = m_{ADM}(M)$. They applied this to prove the Penrose Inequality:

$$m_{ADM}(M^3) \geq m_H(\partial M^3) = \sqrt{\frac{\mathrm{Area}(\Sigma)}{16\pi}} \tag{9.9}$$

when M^3 is asymptotically flat with a connected *outermost minimizing boundary* (e.g. ∂M is a minimal surface and there are no other closed minimal surfaces in M). Bray extended their result to have boundaries with more than one connected component in [9]. In addition, there is the Penrose Rigidity Theorem:

$$m_{ADM}(M^3) = m_H(\partial M^3) \quad \Rightarrow \quad M^3 \text{ is isometric to } M_{Sch,m} \tag{9.10}$$

where $M_{Sch,m}$ is the Riemannian Schwarschild space with mass $m = m_{ADM}(M^3)$.

In addition to Hawking mass, there are other quasilocal masses defined on manifolds with Scal ≥ 0 including the Brown-York mass (which has nice properties proven by Shi-Tam in [53]) and the Bartnik mass [4]. It is not a simple task to define and apply these quasilocal masses on limit spaces because they all involve the mean curvatures of surfaces. Perhaps more promising is Huisken's new isoperimetric quasilocal mass of a region $\Omega \subset M^3$,

$$m_{ISO}(\Omega) = \frac{2}{\text{Area}(\partial\Omega)}\left(\text{vol}(\Omega) - \frac{\text{Area}(\partial\Omega)^{3/2}}{6\sqrt{\pi}}\right), \tag{9.11}$$

and $m_{ISO}(M) = \limsup_{r\to\infty} m_{ISO}(\Omega_r)$ introduced in [29]. Miao has proven that $m_{ISO}(M) = m_{ADM}(M)$ using volume estimates of Fan-Shi-Tam in [18]. See also work of Jauregui, Lee, Carlotto, Chodosh, and Eichmair [32] [12].

Gromov-Lawson applied the Lichnerowicz formula to prove many things (cf. [37]) including the Scalar Torus Rigidity Theorem [25] in all dimensions:

$$\text{Scal} \geq 0 \text{ and } M^n \text{ diffeom to a torus} \quad \Rightarrow \quad M^n \text{ is isom to a flat torus.} \tag{9.12}$$

Witten applied it to prove the Positive Mass Theorem for Spin manifolds [65].

Hamilton's Ricci flow leads to a precise control on the scalar curvature as well as the areas of minimal surfaces in the evolving manifolds (cf. [27]). It was applied to prove the following rigidity theorems about minimal surfaces. Let

$$\text{MinA}(M^3) = \inf\{\text{Area}(\Sigma^2) : \Sigma^2 \text{ is a closed min surf in } M^3\} \text{ and} \tag{9.13}$$

$$\text{MinA}_1(M^3) = \inf\{\text{Area}(\Sigma^2) : \Sigma^2 \text{ is a min noncontr sphere in } M^3\}. \tag{9.14}$$

Note that these invariants are infinite if there are no qualifying minimal surfaces in M^3. Bray, Brendle, and Neves proved the Cover Splitting Rigidity Theorem:

$$\text{Scal} \geq 2 \text{ and } \text{MinA}_1(M^3) = 4\pi \quad \Rightarrow \quad \tilde{M}^3 \text{ is isom to } \mathbb{S}^2 \times \mathbb{R} \tag{9.15}$$

where $\tilde{M}^3$ is the universal cover of M^3 [8]. Bray, Brendle, Eichmair and Neves proved the $\mathbb{RP}^3$ Rigidity Theorem in [7]: if M^3 is diffeomorphic to $\mathbb{RP}^3$ then

$$\text{Scal} \geq 6 \text{ and } \text{MinA}(M^3) = 2\pi \quad \Rightarrow \quad M^3 \text{ is isom to } \mathbb{RP}^3. \tag{9.16}$$

They have in fact proven theorems which imply these two more simply stated theorems (9.15)- (9.16) as corollaries.

Recall that a rigidity theorem has a statement in the following form:

$$M \text{ satisfies a hypothesis} \quad \Rightarrow \quad M \text{ isometric to } M_0. \tag{9.17}$$

The corresponding almost rigidity theorem (if it exists) would then be:

$$M \text{ almost satisfies a hypothesis} \quad \Rightarrow \quad M \text{ is close to } M_0. \tag{9.18}$$

The almost rigidity theorem can also be stated as follows:

$$M_j \text{ closer and closer to satisfying a hypothesis} \quad \Rightarrow \quad M_j \to M_0. \tag{9.19}$$

Within we describe conjectured almost rigidity theorems for each of the rigidity theorems described above. All of those conjectures remain open although some have been proven under additional hypotheses.

First one needs to define closeness for pairs of Riemannian manifolds and convergence of sequences of Riemannian manifolds. A pair of compact Riemannian manifolds, M_1 and M_2, may be mapped into a common metric space, Z, via distance preserving maps $\phi_i : M_i \to Z$ which satisfy

$$d_Z(\phi_j(x), \phi_j(y)) = d_{M_j}(x, y) \qquad x, y \in M_j. \tag{9.20}$$

Once they lie in a common metric space, Z, then one may use the Hausdorff distance or the flat distance to measure the distance between the images with respect to the extrinsic space, Z. We review these extrinsic distances which depend on both Z and the location of the M_i within Z in Section 9.3.

However, an intrinsic notion of distance between M_1 and M_2 can only depend on intrinsic data about these spaces and not on how they may be embedded into some extrinsic Z. Thus Gromov defined his "intrinsic Hausdorff distance" in [21], now known as the Gromov-Hausdorff distance, by taking the infimum over all distance preserving maps into arbitrary compact metric spaces, Z, of the Hausdorff distance, d_H^Z, between the images:

$$d_{GH}(M_1, M_2) = \inf_{Z,\phi_i} \left\{ d_H^Z(\phi_1(M_1), \phi_2(M_2)) | \, \phi_i : M_i \to Z \right\}. \tag{9.21}$$

Many almost rigidity theorems have been proven for manifolds with nonnegative Ricci curvature using the Gromov-Hausdorff distance (cf. [15], [13] and [55]).

For manifolds with nonnegative scalar curvature, one does not obtain Gromov-Hausdorff closeness in the almost rigidity theorems. Counterexamples will be described in Section 9.2. Gromov has suggested that intrinsic flat convergence may be more well suited towards proving an Almost Rigidity for the Torus Rigidity Theorem in [23]. Indeed some progress has been made by the author, Lee, LeFloch, Huang, and Stavrov proving special cases of Almost Rigidity for the Positive Mass Theorem in [39][40] [28][57].

The intrinsic flat distance between compact oriented Riemannian manifolds was defined by the author with Wenger in [61] with an infimum over all distance preserving maps into arbitrary complete metric spaces, Z, of the flat distance, d_F^Z, between the images:

$$d_{\mathcal{F}}(M_1, M_2) = \inf_{Z,\phi_i} \left\{ d_F^Z(\phi_{1\#}[M_1], \phi_{2\#}[M_2]) : \phi_i : M_i \to Z \right\}. \tag{9.22}$$

Intuitively this distance is measuring the filling volume between the two spaces. One may also consider the intrinsic volume flat distance:

$$d_{\mathrm{vol}\,\mathcal{F}}(M_1, M_2) = d_{\mathcal{F}}(M_1, M_2) + \left| \mathrm{vol}(M_1) - \mathrm{vol}(M_2) \right|. \tag{9.23}$$

Full details about the intrinsic flat distance and limits obtained under intrinsic flat convergence are provided in Section 9.4 after a review of Ambrosio-Kirchheim theory in Section 9.3.

There are a few methods that can be applied to prove almost rigidity theorems. To apply the *explicit control method* one provides enough controls on the M in (9.18) so that one can explicitly construct an embedding of M and of M_0 into a common metric space and explicitly estimate the distance between them. This technique was applied to prove GH almost rigidity theorems by Colding in [15] and by Cheeger-Colding in [13]. It was also applied to prove the $\mathcal{F}$ almost rigidity of the Positive Mass Theorem under additional hypotheses in joint work with Lee [39] and in joint work with Stavrov [57]. Lakzian and the author have proven a theorem which provides such a construction and estimate if one can show M and M_0 are close on large regions in [36]. See Section 9.5.

A second technique used to prove almost rigidity theorems is the *compactness and weak rigidity method*. One first provides enough controls on M_j in (9.19) so that a subsequence converges to a limit space M_∞. Then one proves the limit space satisfies the hypothesis in some weak sense. Finally one proves the rigidity theorem in that weak setting. This technique was applied by the author to prove a GH almost rigidity theorem in [55] using Gromov's Compactness Theorem, which states that

$$Ric_j \geq -(n-1) \text{ and } \mathrm{diam}\,(\mathrm{M_j}) \leq \mathrm{D} \quad \Rightarrow \quad \mathrm{M_{j_k}} \xrightarrow{\mathrm{GH}} \mathrm{M_\infty}. \tag{9.24}$$

Wenger's Compactness Theorem [63] states that

$$\mathrm{diam}\,(\mathrm{M_j}) \leq \mathrm{D},\ \ \mathrm{vol}(\mathrm{M_j}) \leq \mathrm{V},\ \ \mathrm{vol}(\partial \mathrm{M_j}) \leq \mathrm{A} \quad \Rightarrow \quad \mathrm{M_{j_k}} \xrightarrow{\mathcal{F}} \mathrm{M_\infty}. \tag{9.25}$$

Huang, Lee and the author proved Almost Rigidity of the Positive Mass Theorem for graph manifolds using Wenger's Compactness Theorem combined with an Arzela-Ascoli Theorem, and a number of other theorems concerning intrinsic flat convergence in [28]. We will review $\mathcal{F}$ compactness and Arzela-Ascoli theorems in Section 9.6.

We begin with Section 9.2 surveying examples of sequences of manifolds with nonnegative scalar curvature. These examples reveal that one cannot simply use intrinsic flat convergence to handle all the problems that arise when trying to prove almost rigidity theorems involving nonnegative scalar curvature. There is a phenomenon called *bubbling*. One may also have tiny tunnels and construct sequences of manifolds through a process called *sewing* developed by the author with Basilio in [6], which lead to limit spaces that do not even satisfy (9.5). These examples with bubbling and sewing have $\mathrm{MinA}(M_j) \to 0$.

We next present the general theory of Intrinsic Flat convergence and Integral Current Spaces, and survey the key theorems proven in this area. We begin with Section 9.3 by reviewing work of Federer-Flemming and Ambrosio-Kirchheim on integral currents in Euclidean space and metric spaces. In Section 9.4 we rigorously define Intrinsic Flat Convergence and Integral Current Spaces and survey known compactness theorems and proposed compactness theorems. In Section 9.5 we present various methods that may be used to estimate the intrinsic flat distance between two spaces and describe how these estimates have been used to prove almost rigidity theorems using the explicit control method. In Section 9.6 we present theorems about intrinsic flat convergence including theorems about disappearing and converging points, converging balls, semicontinuity theorems, Arzela-Ascoli Theorems and Intrinsc Flat Volume Convergence and mention how these results have been applied to prove almost rigidity theorems using the compactness and weak rigidity method.

We close with Section 9.7 which includes statements of conjectures, surveys of partial solutions to the conjectures, and recommended related problems. We discuss the Almost Rigidity of the Positive Mass Theorem, the Bartnik Conjecture, the Almost Rigidity of the Scalar Torus Theorem, the Almost Rigidity of Rigidity Theorems proven using Ricci Flow, Gromov's Prism Conjecture and the Regularity of Limit Spaces. Throughout one hopes to devise a generalized notion of nonnegative scalar curvature on limit spaces. Conjectures and problems are interspersed throughout the paper. If a reader is interested in studying any of these questions, please contact the author. More details can be provided and the author can coordinate the research of those working on these problems.

The author must apologize up front that there is no possible way to mention all the fundementally important papers that have been written concerning manifolds with scalar curvature bounds. The results mentioned here have been selected because the author has read them and developed some idea as to how they may applied to study the intrinsic flat convergence of manifolds with nonnegative scalar curvature.

9.2 Examples with Positive Scalar Curvature

In this section we survey examples of sequences of three dimensional Riemannian manifolds, M_j, with lower scalar curvature bounds, and describe their intrinsic flat limits. Many of these examples were found by mathematicians interested in applications to General Relativity. Manifolds with positive scalar curvature can be viewed as time symmetric spacelike slices of spacetime satisfying the positive energy condition. Such manifolds are curved by matter and can have gravity wells and/or black holes with horizons that are minimal surfaces.

We do not provide the explicit details or the proofs for these examples but instead provide references. We also propose new examples as open problems that could be written up and published by an interested reader. We present these examples before presenting intrinsic flat convergence because they provide some intuitive understanding of intrinsic flat convergence, and what may occur when one has a sequence of manifolds with nonnegative scalar curvature.

9.2.1 Examples with Wells

Arbitrarily thin arbitrarily deep wells can be constructed with positive scalar curvature. By the Positive Mass Theorem, one cannot attach such wells smoothly to Euclidean space. However they may be glued to spheres of constant positive sectional curvature. In fact, the Ilmanen Example, which initially inspired the definition of intrinsic flat convergence, consists of a sequence of spheres with increasingly many increasingly thin wells as in Figure 9.1. This sequence converges in the intrinsic flat sense to a standard sphere because a sphere with many thin holes and a standard sphere can be mapped into a common metric space, and the flat distance between them, which is intuitively a filling volume between them, will be very small.

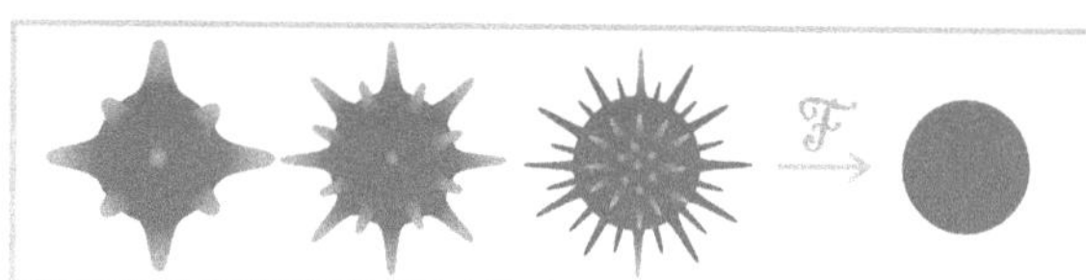

Fig. 9.1. The Ilmanen Example. Image owned by the author.

Example 9.2.1. *Lakzian has explicitly constructed sequences of spheres with one increasingly thin well in [35]. He has also explicitly constructed the Ilmanen Example of sequences of spheres with increasingly many increasingly thin wells in the same paper. He proves both sequences $\mathcal{F}$ converge to the standard sphere. He proves the sequence with one increasingly thin well converges in the GH sense to a sphere with a line segment attached, and the Ilmanen example has no GH limit.*

Example 9.2.2. *In joint work with Lee, the author has constructed examples of asymptotically flat rotationally symmetric manifolds of positive scalar curvature with arbitrarily thin and arbitrarily deep wells and $m_{ADM}(M_j) \to 0$ [39]. They have no smooth limits and the the pointed GH limits of such examples are Euclidean spaces with line segments of arbitrary length attached. Lee and the author have also constructed sequences which* *are not rotationally symmetric that have $m_{ADM}(M_j) \to 0$ and increasingly many increasingly thin wells [38]. Such sequences have no GH converging subsequences because they have increasingly many disjoint balls [21]. Thus almost rigidity of the positive mass theorem cannot be proven with GH or smooth convergence, only $\mathcal{F}$ convergence.*

Example 9.2.3. *It is possible to construct M_j with $Scalar \geq -1/j$ that are diffeomorphic to tori and contain balls of radius $1/2$ that are isometric to balls in rescaled standard spheres. This will appear in work of the author with Basilio [6]. One may then attach an increasingly thin well of arbitrary depth to such M_j that have positive scalar curvature. These examples would $\mathcal{F}$ converge to a standard flat torus and would GH converge to a standard flat torus with a line segment attached. One may also attach increasingly many increasingly thin wells of arbitrary depth to the M_j and still $\mathcal{F}$ converge to a standard flat torus but there will be no GH limit of such a sequence. Thus one must use intrinsic flat convergence to prove almost rigidity for the Scalar Torus Theorem.*

9.2.2 Tunnels and Bubbling

Gromov-Lawson and Schoen-Yau constructed tunnels diffeomoerphic to $\mathbb{S}^2\times[0, 1]$ with positive scalar curvature which attach smoothly on either end to the standard spheres [26] [52]. These tunnels may be arbitrarily thin and long or thin and short. At the center of the tunnel, there is a closed minimal surface diffeomorphic to a sphere. Sometimes these tunnels are called necks.

Example 9.2.4. *Using these tunnels one may construct sequences of M_j which consist of a pair of standard spheres joined by increasingly thin tunnels of length L_j. If $L_j \to 0$, then the GH and $\mathcal{F}$ limit can be shown to be a pair of standard spheres joined at a point as in Figure 9.2. This effect is called* bubbling. *Note that in this example,* $\mathrm{MinA}(M_j) \to 0$.

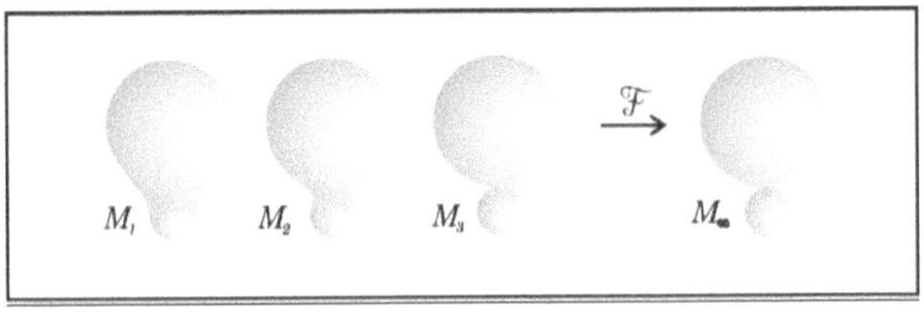

Fig. 9.2. Bubbling. Image owned by the author.

Example 9.2.5. *If $L_j \to L_\infty > 0$, then the GH limit is a pair of standard spheres joined by a line segment of length L and the $\mathcal{F}$ limit is just the pair of spheres without the line segment with the restricted distance from the GH limit. Examples similar to these are described by Wenger and the author in the [61]. Notice that the $\mathcal{F}$ limit is not geodesic.*

Example 9.2.6. *In fact one may add increasingly many bubbles with increasingly short and thin tunnels, and the sequence will have no GH limit and no $\mathcal{F}$ limit. This does not contradict Wenger's Compactness Theorem as in (9.25) because the volume is diverging to infinity even though the diameter is bounded and there is no boundary. Examples similar to these with many bubbles of various sizes and tunnels of various lengths converging to rectifiable limit spaces appear in [61].*

Example 9.2.7. *One may also have bubbling in asymptotically flat sequences of M_j with $m_{ADM}(M_j) \to 0$ obtaining limits which are Euclidean planes with spheres attached. This can be done by attaching bubbles instead of wells to the sequences in Example 9.2.2. Such an example demonstrates that any almost rigidity theorem for the Positive Mass Theorem must somehow avoid bubbling. In joint work with Lee, we require that the manifolds have outward minimizing boundaries just as in the Penrose Inequality [39]. This is effectively cutting off the bubbles. One could alternately eliminate bubbling by requiring the sequence to have a uniform lower bound on the area of the smallest closed minimal surface,* $\mathrm{MinA}(M_j) \geq A_0 > 0$.

Example 9.2.8. *One may add a bubble to each M_j of Example 9.2.3 with* $\mathrm{Scal}_j \geq -1/j$ *that are diffeomorphic to tori and contain balls of constant sectional curvature isometric to balls in rescaled standard spheres. Such sequences would $\mathcal{F}$ converge to standard flat tori with a sphere of arbitrary radius attached at a point. This will appear in work of the author with Basilio [6]. One could eliminate such examples by requiring the sequence have a uniform lower bound on the area of the smallest closed minimal surface,* $\mathrm{MinA}(M_j) \geq A_0 > 0$. *One cannot require that there are no closed minimal surfaces here since the manifolds are diffeomorphic to tori.*

9.2.3 Cancellation and Doubling

The next two examples are described by the author and Wenger in [60] [61]. Intrinsic flat limit spaces may be the **0** or a rectifiable space with integer weight.

Example 9.2.9. *There are sequences of manifolds M_j^3 with positive scalar curvature which have a $\mathcal{F}$ limit which is the* **0** *space, while converging in the GH sense to a standard three sphere. This cancelling sequence can be constructed with positive scalar curvature by taking a pair of standard three spheres and connecting them by increasingly dense increasingly small tunnels. These sequences converge to the* **0** *space because their filling volumes converge to 0. In fact they are the totally geodesic boundaries of four dimensional manifolds whose volume converges to 0.*

Example 9.2.10. *If in the previous example all the tunnels are cut and glued back together with reversed orientation, then the GH limit is still a standard three sphere and the $\mathcal{F}$ limit is a sphere with weight two everywhere.*

9.2.4 Sewing Manifolds

In joint doctoral work of Basilio with the author and Dodziuk, the notion of sewing Riemannian manifolds is introduced [5, 6]. One starts with a three dimensional manifold, M, that contains a curve, $C : [0,1] \to M$, such that a tubular neighborhood around the curve has constant positive sectional curvature. One then creates a sequence of manifolds sewn along this curve. That is, short thin tunnels are attached along the curve pulling the points on the curve closer together. The GH and $\mathcal{F}$ limit of such a sequence is then the original manifold with a *pulled thread* along C. That is, all the points in the image of C have been identified. One can also sew entire regions with constant positive sectional curvature to obtain sequences converging to the original manifold with the entire region identified as a single point. If the original manifold has positive scalar curvature, then so does the sequence. In addition one may consider sequences of M_j and sew along curves or in regions of those M_j. Using this construction, Basilio and the author construct the following examples.

Example 9.2.11. *If one takes the M_j of Example 9.2.3, one may sew along curves lying in the balls of radius $1/2$ that have constant sectional curvature to obtain a sequence of manifolds, M_j' with $\mathrm{Scal}_j \geq -1/j$ that are no longer tori but converge to a limit which is the standard flat torus with either a contractible circle, a contractible sphere or ball of radius $1/2$ each pulled to a point. These examples demonstrate that limits of manifolds with $\mathrm{Scal}_j \geq -1/j$ may fail to have generalized nonnegative scalar curvature in the sense that limit in (9.5) fails to be nonnegative. These limits can be biLipschitz to tori and still not be isometric to a flat torus. Like all examples created with this sewing construction,* $\mathrm{MinA}(M_j) \to 0$.

9.3 Integral Currents on a Metric Space

Before we can rigorously define intrinsic flat convergence and describe the limit spaces obtained under intrinsic flat convergence, we need to review Ambrosio and Kirchheim's notion of currents and convergence of currents on a complete metric space [1]. Note that like Federer-Fleming's earlier work on the flat and weak convergence of submanifolds viewed as currents in Euclidean space, the flat and weak convergence of Ambrosio-Kirchheim's currents are extrinsic notions of convergence, depending very much on the way in which the submanifold or current lies within an extrinsic space.

9.3.1 Federer-Fleming currents on Euclidean Space

In [19] Federer and Fleming first introduced the notion of a current on Euclidean space as a generalization of the notion of an oriented submanifold, $\phi : M \to \mathbb{E}^N$, which views M as a linear functional, $T = [M]$, on differential forms:

$$T(\omega) = [M]\omega = \int_M \phi^* \omega. \tag{9.26}$$

In particular

$$T(f\, d\pi_1 \wedge \cdots \wedge d\pi_m) = \int_M (f \circ \phi_i)\, d(\pi_1 \circ \phi_1) \wedge \cdots \wedge d(\pi_m \circ \phi_m). \tag{9.27}$$

Observe that this is perfectly well defined when $\phi : M \to \mathbb{E}^N$ is only Lipschitz. This linear functional captures the notion of boundary,

$$\partial T(\omega) = \int_{\partial M} \omega = \int_M d\omega = T(d\omega). \tag{9.28}$$

So that

$$\partial T(f\, d\pi_1 \wedge \cdots \wedge d\pi_{m-1}) = T(1\, df \wedge d\pi_1 \wedge \cdots \wedge d\pi_{m-1}). \tag{9.29}$$

Federer and Fleming then studied sequences of submanifolds by considering the weak limits of their corresponding linear functionals. They applied this to study the Plateau Problem: searching for the submanifold of smallest area with a given boundary. They proved sequences of submanifolds approaching the smallest area converge in the weak sense to a limit which they called an integral current.

9.3.2 Ambrosio-Kirchheim Integer Rectifiable Currents

In [1], Ambrosio and Kirchheim defined currents on Euclidean space to integral currents on any complete metric space, Z. In Federer-Fleming, currents were defined as linear functionals on differential forms [19]. Since there are no differential forms on a metric space, Ambrosio and Kirchheim's currents are multilinear functionals which act on DiGeorgi's $m+1$ tuples [17]. A *tuple* $(f, \pi_1, \dots, \pi_m)$ is in $\mathcal{D}^m(Z)$ iff $f : Z \to \mathbb{R}$ is a bounded Lipschitz function and $\pi_i : Z \to \mathbb{R}$ are Lipschitz. These tuples have no antisymmetry properties.

In [1] Ambrosio and Kirchheim began their work by defining currents. As we do not need the notion of a current in this paper, we jump directly to their notion of an integer rectifiable current applying their Theorems 9.1 and 9.5 as an explanation rather than using their definition.

A linear functional $T : \mathcal{D}^m(Z) \to \mathbb{R}$ is an m dimensional *integer rectifiable current*, denoted $T \in \mathcal{I}_m(Z)$ if and only if it can be parametrized as follows

$$T(f, \pi_1, ..., \pi_m) = \sum_{i=1}^{\infty} \left(\theta_i \int_{A_i} (f \circ \phi_i)\, d(\pi_1 \circ \phi_i) \wedge \cdots \wedge d(\pi_m \circ \phi_i) \right) \tag{9.30}$$

where $\theta_i \in \mathbb{Z}$ and $\phi_i : A_i \to \phi_i(A_i) \subset Z$ are biLipschitz maps defined on precompact Borel measurable sets, $A_i \subset \mathbb{R}^m$, with pairwise disjoint images such that

$$\sum_{i=1}^{\infty} |\theta_i|\, \mathcal{H}_m(\phi_i(A_i)) < \infty \quad \text{where } \mathcal{H}_m \text{ is the Hausdorff measure.} \tag{9.31}$$

A 0 dimensional integer rectifiable current can be parametrized by a finite collection of distinct weighted points

$$T(f) = \sum_{i=1}^{N} \theta_i f(p_i) \text{ where } \theta_i \in \mathbb{Z} \text{ and } p_i \in Z. \tag{9.32}$$

Observe that we then have the following antisymmetry property,

$$T(f, \pi_1, ..., \pi_m) = \operatorname{sgn}(\sigma)\, T(f, \pi_{\sigma(1)}, ..., \pi_{\sigma(m)}) \tag{9.33}$$

for any permutation $\sigma : \{1, ..., m\} \to \{1, ..., m\}$. In addition, $T(f, \pi_1, ..., \pi_m) = 0$ if f is the zero function or one of the π_i is constant. So while the tuples do not have the properties of differential forms, the action of the integer rectifiable currents on the tuples has these properties.

Ambrosio-Kirchheim's *mass measure* $\|T\|$ of a current T, is the smallest Borel measure, μ, such that

$$\left| T(f, \pi) \right| \le \int_X |f|\, d\mu \qquad \forall\, (f, \pi) \text{ where } \operatorname{Lip}(\pi_i) \le 1. \tag{9.34}$$

In Theorem 9.5 of [1], the mass measure is explicitly computed. For the purposes of this paper we need only the following consequence of their theorem:

$$m^{-m/2} H_T(A) \le ||T||(A) \le \frac{2^m}{\omega_m} H_T(A) \tag{9.35}$$

where

$$H_T(A) = \sum_{i=1}^{\infty} |\theta_i| \mathcal{H}_m(\phi_i(A_i) \cap A). \tag{9.36}$$

Furthermore the mass measure of a 0 dimensional integer rectifiable current satisfies

$$||T||(A) = \sum_{p_i \in A} |\theta_i|. \tag{9.37}$$

The Ambrosio-Kirchheim *mass* of T is defined

$$\mathbf{M}(T) = ||T||(Z). \tag{9.38}$$

By the definition of the Ambrosio-Kirchheim mass we have

$$T(f, \pi_1, \ldots, \pi_m) \leq \sup|f| \prod_{i=1}^{m} \mathrm{Lip}(\pi_i)\,\mathbf{M}(T). \tag{9.39}$$

The *restriction* of a current T by a $k+1$ tuple $\omega = (g, \tau_1, \ldots\tau_k) \in \mathcal{D}^k(Z)$ with $k < m$ is defined by

$$(T \llcorner \omega)(f, \pi_1, \ldots\pi_m) := T(f \cdot g, \tau_1, \ldots\tau_k, \pi_1, \ldots\pi_m). \tag{9.40}$$

Given a Borel set, A,

$$T \llcorner A(f, \pi_1, \ldots\pi_m) := T(1_A \cdot f, \pi_1, \ldots\pi_m) \tag{9.41}$$

where 1_A is the indicator function of the set. Observe that $T \llcorner \omega$ is an integer rectifiable current of dimension $m-k$ and that

$$\mathbf{M}(T \llcorner \omega) = ||T||(A). \tag{9.42}$$

Given a Lipschitz map, $\phi : Z \to Z'$, the *push forward* of a current T on Z to a current $\phi_\# T$ on Z' is given by

$$\phi_\# T(f, \pi_1, \ldots\pi_m) := T(f \circ \phi, \pi_1 \circ \phi, \ldots\pi_m \circ \phi) \tag{9.43}$$

which is clearly still an integer rectifiable current. Observe that

$$(\phi_\# T) \llcorner (f, \pi_1, \ldots\pi_k)) = \phi_\#(T \llcorner (f \circ \phi, \pi_1 \circ \phi, \ldots\pi_k \circ \phi)) \tag{9.44}$$

and

$$(\phi_\# T) \llcorner A = (\phi_\# T) \llcorner (1_A) = \phi_\#(T \llcorner (1_A \circ \phi)) = \phi_\#(T \llcorner \phi^{-1}(A)). \tag{9.45}$$

In (2.4) of [1], Ambrosio and Kirchheim show that

$$||\phi_\# T|| \leq [\mathrm{Lip}(\phi)]^m \phi_\# ||T||, \tag{9.46}$$

so that when ϕ is an isometric embedding

$$||\phi_\# T|| = \phi_\# ||T|| \text{ and } \mathbf{M}(T) = \mathbf{M}(\phi_\# T). \tag{9.47}$$

In [1][Theorem 4.6] Ambrosio and Kirchheim define the (canonical) set of a current, T, as the collection of points in Z with positive lower density:

$$\mathrm{Set}(T) = \{p \in Z : \Theta_{\star m}(||T||, p) > 0\}, \tag{9.48}$$

where the definition of lower density is

$$\Theta_{*m}(\mu, p) = \liminf_{r \to 0} \frac{\mu(B_p(r))}{\omega_m r^m}. \tag{9.49}$$

When T is an integer rectifiable current then $\mathrm{Set}(T)$ is countably $\mathcal{H}^m$ rectifiable, which means there exists a collection of biLipschitz maps, $\phi_i : A_i' \to \mathrm{Set}(T) \subset Z$, defined on Borel sets $A_i' \in \mathbb{R}^m$ such that

$$\mathcal{H}_m \left(\mathrm{Set}(T) \setminus \bigcup_{i=1}^{\infty} \phi_i(A_i) \right) = 0. \tag{9.50}$$

These ϕ_i can be taken from the parametrization of T with $A_i' \subset A_i \subset \bar{A}_i$.

9.3.3 Ambrosio-Kirchheim Integral Currents

The *boundary* of T is defined

$$\partial T(f, \pi_1, \ldots \pi_{m-1}) := T(1, f, \pi_1, \ldots \pi_{m-1}). \tag{9.51}$$

Note that $\phi_\#(\partial T) = \partial(\phi_\# T)$ and it can easily be shown that $\partial\partial T = 0$. The boundary of an integer rectifiable current is not necessarily an integer rectifiable current.

An integer rectifiable current $T \in \mathcal{I}_m(Z)$ is an *integral current*, denoted $T \in \mathbf{I}_m(Z)$, if ∂T is an integer rectifiable current. This includes the zero current

$$0(f, \pi_1, \ldots, \pi_m) := 0 \text{ with } \partial 0(f, \pi_1, \ldots, \pi_{m-1}) = 0(1, f, \pi_1, \ldots, \pi_{m-1}) = 0. \tag{9.52}$$

Note that Ambrosio and Kirchheim define an integral current as an integer rectifiable current whose boundary has finite mass and the more easily applied statement we have here is their Theorem 8.6 in [1].

Given an oriented Riemannian manifold with boundary, M^m, such that $\mathrm{vol}_m(M) < \infty$ and $\mathrm{vol}_{m-1}(\partial M) < \infty$, and given a Lipschitz map $\phi : M \to Z$, we can define an integral current $\phi_\#[M] \in \mathbf{I}_m(Z)$ as follows

$$\phi_\#[M](f, \pi_1, \ldots, \pi_m) = \int_M (f \circ \phi)\, d(\pi_1 \circ \phi) \wedge \cdots \wedge d(\pi_m \circ \phi). \tag{9.53}$$

Note that $\partial \phi_\#[M] = \phi_\#[\partial M]$ where ∂M is the boundary of M and

$$\mathbf{M}(\phi_\#[M]) = \mathrm{vol}_m(\phi(M)). \tag{9.54}$$

If $\mathrm{vol}_m(M) < \infty$ and $\mathrm{vol}_{m-1}(\partial M) = \infty$, then $[M]$ is only integer rectifiable and not integral.

Whenever T is an integral current, $\partial\partial T = 0$, and

$$\partial : \mathbf{I}_m(Z) \to \mathbf{I}_{m-1}(Z). \tag{9.55}$$

In addition, if $\phi : Z_1 \to Z_2$ is Lipschitz, then by (9.43)

$$\phi_\# : \mathbf{I}_m(Z_1) \to \mathbf{I}_m(Z_2). \tag{9.56}$$

The restriction of an integral current defined in (9.40) need not be an integral current. However, the Ambrosio-Kirchheim Slicing Theorem implies that

$$T \llcorner B(p, r) \text{ is an integral current for almost every } r > 0 \tag{9.57}$$

where $B(p, r) = \{x : d(x, p) < r\}$.

9.3.4 Convergence of Currents in a Metric Space

In Definition 3.6 of [1], Ambrosio and Kirchheim state that a sequence of integral currents $T_j \in \mathbf{I}_m(Z)$ lying in a complete metric space, Z, is said to converge weakly to a current T, denoted $T_j \to T$, iff the pointwise limits satisfy

$$\lim_{j\to\infty} T_j(f, \pi_1, \dots \pi_m) = T(f, \pi_1, \dots \pi_m) \tag{9.58}$$

for all bounded Lipschitz $f : Z \to \mathbb{R}$ and Lipschitz $\pi_i : Z \to \mathbb{R}$. Ambrosio and Kirchheim next observe that if T_j converges weakly to T, then the boundaries converge

$$\partial T_j \to \partial T, \tag{9.59}$$

and the mass is lower semicontinuous

$$\liminf_{j\to\infty} \mathbf{M}(T_j) \geq \mathbf{M}(T). \tag{9.60}$$

Thus, the weak limit of a sequence of integer rectifiable currents with a uniform upper bound on mass is an integer rectifiable current:

$$T_j \in \mathfrak{I}_m(Z), \quad \mathbf{M}(T_j) \leq V_0 \text{ and } T_j \to T \quad \Rightarrow \quad T \in \mathfrak{I}_m(Z). \tag{9.61}$$

Similarly for integral currents we have

$$T_j \in \mathbf{I}_m(Z), \quad \mathbf{M}(T_j) \leq V_0, \quad \mathbf{M}(\partial T_j) \leq A_0 \text{ and } T_j \to T \quad \Rightarrow \quad T \in \mathbf{I}_m(Z). \tag{9.62}$$

For any open set, $A \subset Z$, if $T_j \to T$ then

$$\liminf_{j\to\infty} ||T_j||(A) \geq ||T||(A). \tag{9.63}$$

However, $T_j \llcorner A$ need not converge weakly to $T \llcorner A$ (cf. Example 2.21 of [56]).

Ambrosio and Kirchheim prove the following compactness theorem:

Theorem 9.3.1. *[1] consider any complete metric space Z, a compact set $K \subset Z$ and $A_0, V_0 > 0$. Given any sequence of integral currents $T_j \in \mathbf{I}_m(Z)$ satisfying*

$$\mathbf{M}(T_j) \le V_0, \mathbf{M}(\partial T_j) \le A_0 \text{ and } \mathrm{Set}\,(T_j) \subset K, \tag{9.64}$$

there exists a subsequence, T_{j_i}, which converges weakly to $T \in \mathbf{I}_m(Z)$.

It is possible that the limit obtained in this theorem is the 0 integral current. Observe that whenever the sequence of currents is *collapsing*,

$$\mathbf{M}(T_j) \to 0, \tag{9.65}$$

then by (9.39) we have

$$|T_j(f, \pi_1, \ldots, \pi_m)| \le \sup|f| \prod_{i=1}^{m} \mathrm{Lip}(\pi_i)\,\mathbf{M}(T_j) \to 0 \tag{9.66}$$

and so T_j converges weakly to 0.

It is also possible for T_j to converge weakly to 0 without collapsing. This can occur due to *cancellation*, when the T_j fold over on themselves as in Example 9.3.2. We include this example in detail because it inspires the notion of flat convergence and will be refered to repeatedly in this paper.

Example 9.3.2. *Let $T_j = \phi_{j\#}[M] \subset \mathbf{I}_2(\mathbb{E}^3)$ where*

$$\phi_j(s,t) = \left(s,\ t/j,\ |t|b_j/j\right) \text{ where } b_j = \sqrt{j^2 - 1} \tag{9.67}$$

on $M = \{(s,t) : s \in [-1,1],\ t \in [-1,1]\}$. Since $\mathbf{M}(\phi_{j\#}[M]) = \mathrm{vol}(M)$ does not converge to 0, this sequence is not collapsing. Observe that $T_j = A_j + \partial B_j$ where

$$B_j = [\{(x,y,z) : |x| \le 1,\ |y| \le 1/j,\ z \in [|y|b_j, b_j/j]\}] \in \mathbf{I}_3(\mathbb{E}^3) \text{ and} \tag{9.68}$$

$$A_j = A_j^- + A_j^+ + A_j^0 \in \mathbf{I}_2(\mathbb{E}^3) \text{ where} \tag{9.69}$$

$$A_j^- = -[\{(-1,y,z) : y \in [-1/j, 1/j],\ z \in [|y|b_j, b_j/j]\}] \tag{9.70}$$

$$A_j^+ = [\{(+1,y,z) : y \in [-1/j, 1/j],\ z \in [|y|b_j, b_j/j]\}] \tag{9.71}$$

$$A_j^0 = [\{(x,y,b_j/j) : x \in [-1,1],\ y \in [-1/j, 1/j]\}]. \tag{9.72}$$

Since

$$\mathbf{M}(B_j) \le (4/j) \text{ and } \mathbf{M}(A_j) \le (2/j) + (2/j) + 4/j, \tag{9.73}$$

we have $B_j \to 0$ and $A_j \to 0$. By (9.59) we have $\partial B_j \to \partial 0 = 0$, and thus

$$T_j = A_j + \partial B_j \to 0. \tag{9.74}$$

Sometimes part of a sequence disappears under weak convergence and part remains. This happens in the following example:

Example 9.3.3. *Let* $T_j = \phi_{j\#}[D^2] \subset \mathbf{I}_2(\mathbb{E}^3)$ *with* $D^2 = \{(x,y) : x^2+y^2 \le 1\}$ *and*

$$\phi_{j\#}(x,y) = (x, y, f_j(\sqrt{x^2+y^2})), \tag{9.75}$$

where $f_j : [0,1] \to [0,1]$ *is a smooth cutoff function such that* $f_j(r) = 1$ *near* $r = 0$ *and* $f_j(r) = 0$ *for* $r \ge 1/j$. *Then* $\partial T_j = \phi_{j\#}[S^1]$ *is constant and so the sequence does not disappear. In fact* T_j *converges weakly to* $T_\infty = \phi_{\infty\#}[D^2]$ *where*

$$\phi_{j\#}(x,y) = (x,y,0) \tag{9.76}$$

since $T_j - T_\infty = \partial B_j$ *where*

$$B_j = [\{(x,y,z) : x^2+y^2 \le 1,\ 0 \le z \le f_j(\sqrt{x^2+y^2})\}]. \tag{9.77}$$

Since

$$\mathbf{M}(B_j) \le \pi(1/j)^2 \to 0 \tag{9.78}$$

we have $B_j \to 0$ *and thus* $T_j - T_\infty \to 0$ *and* $T_j \to T_\infty$.

9.3.5 The Flat distance vs the Hausdorff distance

In [62], Wenger defines the flat distance between two integral currents, $T_1, T_2 \in \mathbf{I}_m(Z)$, lying in a common complete metric space, Z, to be

$$d_F^Z(T_1,T_2) = \inf\left\{\mathbf{M}(A) + \mathbf{M}(B) :\ A + \partial B = T_1 - T_2\right\} \tag{9.79}$$

where the infimum is taken over all $A \in \mathbf{I}_m(Z)$ and $B \in \mathbf{I}_{m+1}(Z)$ such that $A + \partial B = T_1 - T_2$. This is the same definition given by Federer and Fleming in [19] building on work of Whitney [64] for the flat distance in Euclidean space, where it is a norm, $|T_1 - T_2|_\flat$. The lack of scaling in (9.80) is a result of setting the flat distance to be a norm on Euclidean space. A scalable version of the flat distance might be defined for Z with a finite diameter diam $(Z) = \mathrm{D}$ as follows (cf. [40]):

$$d_{DF}^Z(T_1,T_2) = \inf\left\{D\mathbf{M}(A) + \mathbf{M}(B) :\ A + \partial B = T_1 - T_2\right\} \tag{9.80}$$

Observe that if two oriented hypersurfaces share a boundary, then the flat distance between them is bounded above by the volume between them. In Example 9.3.3, we have

$$d_F^{\mathbb{E}^3}(T_j, T_\infty) \to 0 \tag{9.81}$$

by taking B_j as in (9.77) so that $T_j - T_\infty = \partial B_j$. Then $d_F^Z(T_j, 0) \le \mathbf{M}(B_j)$ which converges to 0 as $j \to \infty$ by (9.78). In Example 9.3.2, we produce a sequence of integral currents T_j in Euclidean space such that

$$d_F^{\mathbb{E}^3}(T_j, 0) \to 0 \tag{9.82}$$

by taking B_j as in (9.68) and A_j as in (9.69) so that $T_j - 0 = \partial B_j + A_j$. Then $d_F^Z(T_j, 0) \leq \mathbf{M}(A_j) + \mathbf{M}(B_j)$ which converges to 0 as $j \to \infty$ by (9.73).

In [62], Wenger proves that when

$$\mathbf{M}(T_j) \leq V_0 \text{ and } \mathbf{M}(\partial T) \leq A_0 \tag{9.83}$$

then weak and flat convergence are equivalent:

$$T_j \to T \text{ if and only if } d_F^Z(T_j, T) \to 0. \tag{9.84}$$

One should contrast the flat distance between submanifolds viewed as integral currents with the Hausdorff distance between submanifolds viewed as subsets, $X_i = \phi_i(M_i)$. Note that the Hausdorff distance is defined to be

$$d_H^Z(X_1, X_2) = \inf\{r > 0 : X_1 \subset T_r(X_2),\ X_2 \subset T_r(X_1)\} \tag{9.85}$$

where $T_r(X) = \{z : \exists x \in X s.t.\ d(x,z) < r\} \subset Z$. There is no notion of a disappearing Hausdorff limit. The Hausdorff limit of a collapsing sequence of sets like $[0, 1/j] \times [0, 1] \subset \mathbb{E}^2$ is easily seen to be $\{0\} \times [0, 1] \subset \mathbb{E}^2$, which is simply a lower dimensional set. The Hausdorff limit of the sequence of cancelling submanifolds, $\phi_j(M)$, in Example 9.3.2 is easily seen to be the set $[-1, 1] \times \{0\} \times [0, 1]$. No points in a Hausdorff limit can disappear. In Example 9.3.3, the Hausdorff limit of $\phi_j(D^2)$ is a disk with a line segment attached: $(D^2 \times \{0\}) \cup (\{0, 0\} \times [0, 1]) \subset \mathbb{E}^3$.

One reason Federer and Fleming introduced integral currents and flat convergence was to solve the Plateau Problem of finding a minimal surface with a given boundary, Γ. Suppose for example that

$$\Gamma = \{(\cos(t), \sin(t), 0) : t \in \mathbb{S}^1\} \subset \mathbb{E}^3. \tag{9.86}$$

One must find the surface $\phi(D^2)$ such that $\partial(\phi(D^2)) = \Gamma$ of smallest area. One may try to find this minimal surface by taking a sequence of such surfaces, $\phi_j(D^2)$, with area decreasing to the infimum of these areas, and look for a limit. In Example 9.3.3, we have such a sequence with a thinner and thinner spine so that the Hausdorff limit is a disk with a line segment attached, not a minimal surface. Even worse, one may have a sequence of $\phi_j(D^2) = Y_j$ with increasingly many increasingly dense spines as in Figure 9.3 so that the Hausdorff limit is

$$Y = \{(x, y, z) : x^2 + y^2 \leq 1,\ z \in [0, 1]\}. \tag{9.87}$$

This Hausdorff limit has no notion of boundary and is no longer two dimensional. There is no smooth or even C_0 limit of such ϕ_j.

On the other hand the flat limit of a sequence, $\phi_j(D^2)$, with area decreasing to the infimum of these areas does exist and is the standard disk

$$\phi_\infty : \{(x, y) : x^2 + y^2 \leq 1\} \to \mathbb{E}^3 \text{ with } \phi_\infty(x, y) = (x, y, 0). \tag{9.88}$$

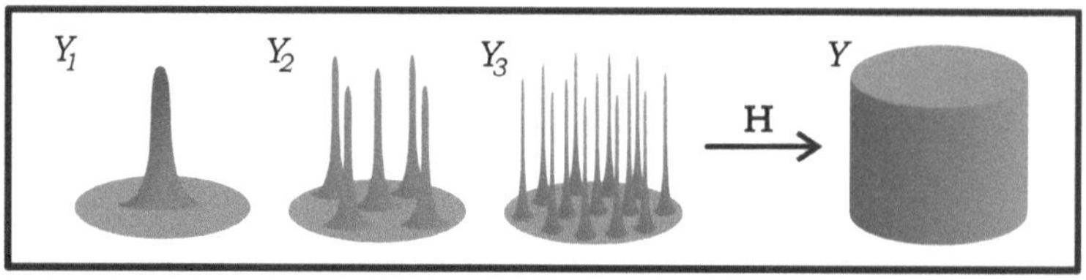

Fig. 9.3. A troublesome Hausdorff limit. Image owned by the author.

This was proven in the case with one spine in Example 9.3.3. This can be seen in Figure 9.3 because the volume between the $\phi_j(M_j)$ and $\phi_\infty(M_\infty)$ converges to 0.

Even on a compact metric space, Z, flat convergence is well suited to the Plateau Problem, where one is given $\Gamma \in \mathbf{I}_{m-1}(Z)$ and asked to find an integral current $T \in \mathbf{I}_m(Z)$, such that $\partial T = \Gamma$ and

$$\mathbf{M}(T) = M_0 = \inf\{\mathbf{M}(T) : \partial T = \Gamma\}. \tag{9.89}$$

One takes $T_j \in \mathbf{I}_m(Z)$ such that $\partial T_j = \Gamma$, and $\mathbf{M}(T_j) \to M_0$. By Ambrosio Kirchheim's Compactness Theorem [Theorem 9.3.1 above] and (9.84), a subsequence converges in the weak and flat sense to some $T_\infty \in \mathbf{I}_m(Z)$. Since $\partial T_\infty = \Gamma$ by (9.59) and $\mathbf{M}(T_\infty) = M_0$ by (9.60), we have a desired solution to the Plateau Problem.

9.4 Integral Current Spaces and Intrinsic Flat Convergence

In this section we provide the rigorous definition for the intrinsic flat convergence of a sequence of oriented Riemannian manifolds or, more generally, a sequence of integral current spaces [61]. It is crucial to remember that the manifolds in the sequence are not submanifolds of any common Euclidean space. This is an intrinsic notion about the intrinsic geometry of the Riemannian manifolds.

As described in the introduction, the intrinsic flat distance is defined much like the Gromov-Hausdorff distance, by taking an infimum over all distance preserving maps, $\phi_j : M_j^m \to Z$ into any common complete metric space, Z:

$$d_{\mathcal{F}}(M_1, M_2) = \inf_{Z,\phi_j} \left\{ d_F^Z\left(\phi_{1\#}[M_1], \phi_{2\#}[M_2]\right)\right\} \tag{9.90}$$

where this is now rigorously defined using (9.53) and (9.80). In fact we can define the intrinsic flat distance between M_1 and an abstract $\mathbf{0}$ space as well:

$$d_{\mathcal{F}}(M_1, \mathbf{0}) = \inf_{Z,\phi_j} \left\{ d_F^Z\left(\phi_{1\#}[M_1], 0\right)\right\}. \tag{9.91}$$

Keep in mind that by the Kuratowski Embedding Theorem any pair of separable metric spaces can be isometrically embedded into a Banach space, Z, so these infima are always finite (cf. [61]). In [61], the author and Wenger prove that for M_j compact, we have $d_{\mathcal{F}}(M_1, M_2) = 0$ if and only if there is an orientation preserving isometry between them.

It is essential to remember that the ϕ_j are distance preserving maps or isometric embeddings in the sense of Gromov as in (9.20). They are not Riemannian isometric embeddings which only preserve lengths of curves. For example, the Riemannian isometry from the standard circle, $\mathbb{S}^1$, to the boundary of the closed Euclidean disk, D^2, is not a distance preserving map. The Riemannian isometry from the standard circle, $\mathbb{S}^1$, to the boundary of the hemisphere, $\mathbb{S}^2_+$, is a distance preserving map. In Example 9.3.2 we have a single flat square, $M = [-1, 1] \times [-1, 1]$, with a sequence of $\phi_j : M \to \mathbb{E}^3$ which preserve lengths of curves, and yet the flat limit of the images is 0 due to cancellation. If the intrinsic flat distance were defined using such maps, then $d_{\mathcal{F}}(M_1, \mathbf{0}) = 0$, and similarly the intrinsic flat distance between any pair of oriented manifolds would be 0.

In this section we introduce a larger class of spaces, integral current spaces, which are metric spaces with an additional structure. These spaces include oriented Riemannian manifolds with boundary and their intrinsic flat limits. We then define the intrinsic flat distance between this larger class of spaces and review fundamental theorems about intrinsic flat convergence.

9.4.1 Integral Current Spaces

Unlike the Gromov-Hausdorff distance, the intrinsic flat distance cannot be defined between an arbitrary pair of metric spaces, $M_j = (X_j, d_j)$. One needs an additional structure which guarantees that the isometric embeddings of the M_j into Z may be viewed as integral currents. Thus the author and Wenger introduced the following notion in [61]:

Definition 9.4.1. *An m dimensional metric space $M = (X, d, T)$ is called an integral current space if it has an integral current structure $T \in \mathbf{I}_m(\bar{X})$ where $\bar{X}$ is the metric completion of X and* $\mathrm{Set}(T) = X$. *Also included in the m dimensional integral current spaces is the* **0** *space, denoted $\mathbf{0} = (\emptyset, 0, 0)$. We say two such spaces are equal, $M_1 = M_2$, if there is a current preserving isometry, $F : M_1 \to M_2$:*

$$d_2(F(x), F(t)) = d_1(x, y) \text{ and } F_\# T_1 = T_2. \tag{9.92}$$

The mass of the integral current space is, $\mathbf{M}(M) = \mathbf{M}(T)$. The diameter is $\mathrm{diam}(\mathrm{M}) = \sup\{\mathrm{d}(\mathrm{x}, \mathrm{y}) : \mathrm{x}, \mathrm{y} \in \mathrm{X}\}$ *if $M \neq 0$ and* $\mathrm{diam}(\mathbf{0}) = 0$.

Any m dimensional integral current space (X, d, T) is countably $\mathcal{H}^m$ rectifiable in the sense that there exists biLipschitz charts $\phi_i : A_i \to X$ where $A_i \subset \mathbb{R}^m$ are Borel and

$$\mathcal{H}^m\left(X \setminus \bigcup_{i=1}^{\infty} \phi_i(A_i)\right) = 0. \tag{9.93}$$

In fact these charts can be viewed as oriented with weights $\theta_i \in \mathbb{N}$ as in (9.30). Any 0 dimensional integral current space, (X, d, T) is a finite collection of points, $X = \{p_1, \ldots, p_N\}$, with a metric d and with weights $\theta_i \in \mathbb{Z}$ so that $T(f) = \sum_{i=1}^{N} \theta_i f(p_i)$.

A compact oriented Riemannian manifold with boundary, (M^m, g), is an integral current space which $X = M^m$, where $d = d_g$ is the standard metric on M,

$$d_g(p, q) = \inf\{L_g(C) : \; C(0) = p, \; C(1) = q\} \tag{9.94}$$

where

$$L_g(C) = \int_0^1 g(C'(t), C'(t))^{1/2}\, dt, \tag{9.95}$$

and where $T = [M]$ is integration over M,

$$T(f, \pi_1, \ldots, \pi_m) = \int_M f\, d\pi_1 \wedge \cdots \wedge d\pi_m. \tag{9.96}$$

In this setting $\mathbf{M}(M) = \mathrm{vol}(M)$. Note that if (M_1, g_1) and (M_2, g_2) are diffeomorphic then they have the same integral current structure up to a sign. In fact they need only be biLipschitz equivalent. They do not have the same mass unless there is a volume preserving diffeomorphism between them. They are not viewed as the same integral current space unless there is an orientation preserving isometry between them.

If M is a precompact oriented Riemannian manifold with boundary, (M^m, g), then we can define an integral current space (X, d, T), by taking the metric completion $\bar{X} = \bar{M}$, defining d_g as the continuous extension to $\bar{X}$ of (9.94), and defining $T \in \mathbf{I}_m(\bar{X})$ exactly as in (9.96). Then $X = \mathrm{Set}(T) \subset \bar{X}$. This set is called the *settled completion* of M, and is denoted, M'. In particular

$$M' = \{x \in \bar{M} : \; \liminf_{r \to 0} \mathrm{vol}_m(B(x, r) \cap M)/r^m > 0\}. \tag{9.97}$$

Thus if M is a manifold with a singular point removed, that point is always included in the metric completion $\bar{M}$ but if it is a cusp singularity it is not included in the settled completion.

The boundary of an integral current space, (X, d, T), is the integral current space:

$$\partial\,(X, d_X, T) := \left(\mathrm{Set}\,(\partial T)\,, d_{\bar{X}}, \partial T\right) \tag{9.98}$$

where the distance on the boundary is $d_{\bar{X}}$ which is restricted from the distance on the metric completion $\bar{X}$. If $\partial T = 0$ then one says (X, d, T) is an integral current without boundary. The $\mathbf{0}$ space has no boundary.

Note that the boundary of $(M, d_g, [M])$ when M is an oriented Riemannian manifold with boundary is $(\partial M, d_g, \partial[M])$ endowed with the restricted distance, d_g, defined on M as in (9.94). It is only a geodesic space if ∂M is totally geodesic in M. For example, when $M = D^2 \subset \mathbb{E}^2$, then $\partial M = (\mathbb{S}^1, d_{\mathbb{E}^2}, [\mathbb{S}^1])$ is not a geodesic space because there are no curves whose length is equal to the distance between the points,

$$d_{\mathbb{E}^2}(p,q) = |p-q| < d_{\mathbb{S}^1}(p,q) = \cos^{-1}(1 - |p-q|^2/2). \tag{9.99}$$

The boundary of the upper hemisphere, $M = \left(\mathbb{S}^2_+, d_{\mathbb{S}^2_+}, [\mathbb{S}^2_+]\right)$, is a geodesic integral current space, $\partial M = \left(\mathbb{S}^1, d_{\mathbb{S}^1}, [\mathbb{S}^1]\right)$.

In [56] the author proves that a ball in an integral current space, $M = (X, d, T)$, with the current restricted from the current structure of M is an integral current space itself,

$$S(p,r) := \left(\mathrm{Set}(T \llcorner B(p,r)), d, T \llcorner B(p,r)\right) \tag{9.100}$$

for almost every $r > 0$. Furthermore,

$$B(p,r) \subset \mathrm{Set}(S(p,r)) \subset \bar{B}(p,r) \subset X. \tag{9.101}$$

Note that the outside of the ball, $(M \setminus B(p,r), d, T - S(p,r))$, and the sphere,

$$\partial S(p,r) := \left(\mathrm{Set}(\partial(T \llcorner B(p,r))), d, \partial(T \llcorner B(p,r))\right), \tag{9.102}$$

are integral current spaces for the same values of $r > 0$. If $\partial M = 0$ then

$$\mathrm{Set}(\partial(T \llcorner B(p,r))) \subset \{x : d(x,p) = r\}. \tag{9.103}$$

In [49] the author investigate the notion of the filling volume

$$\mathrm{FillVol}(\partial M) = \inf\{\mathbf{M}(N) : \partial N = \partial M\} \tag{9.104}$$

where the infimum is over all integral current spaces, N, such that there is a current preserving isometry from ∂N to ∂M. This notion of filling volume does not quite agree with Gromov's notion of Filling Volume in [22] because in our notion there is a larger collection of candidates, N, for filling the manifold. This is because we require a current preserving isometry on the boundary, and because we use the Ambrosio-Kirchheim mass. With our notion, we immediately have

$$\mathbf{M}(M) \geq \mathrm{FillVol}(\partial M). \tag{9.105}$$

Applying the filling volume to balls, we have for almost every $r > 0$ that

$$||T||(B(p,r)) = \mathbf{M}(T \llcorner B(p,r)) = \mathbf{M}(S(p,r)) \geq \mathrm{FillVol}(\partial S(p,r)). \tag{9.106}$$

In particular, if a point $x \in \bar{X}$ satisfies

$$\liminf_{r \to 0} \mathrm{FillVol}(\partial S(p,r))/r^m > 0 \tag{9.107}$$

then $x \in \mathrm{Set}(T) = X$. This idea was first applied jointly with Wenger in [60] before the notion of integral current space was precisely defined in [61]. Further exploration of filling volume and a new notion called the sliced filling volume appeared in [49] with Portegies.

9.4.2 Intrinsic Flat Convergence

The intrinsic flat distance between integral current spaces was first defined by the author and Wenger in [61]:

Definition 9.4.2. *For $M_1 = (X_1, d_1, T_1)$ and $M_2 = (X_2, d_2, T_2) \in \mathcal{M}^m$ let the intrinsic flat distance be defined:*

$$d_{\mathcal{F}}(M_1, M_2) := \inf d_F^Z(\phi_{1\#}T_1, \phi_{2\#}T_2), \tag{9.108}$$

where the infimum is taken over all complete metric spaces (Z, d) and distance preserving maps $\phi_j : (\bar{X}_j, d_j) \to (Z, d)$.

When M_j are precompact integral current spaces we prove the infimum in this definition is obtained [61][Thm 3.23] and consequently $d_{\mathcal{F}}$ is a distance [61][Thm3.27] on the class of precompact integral current spaces up to current preserving isometries, as in (9.92). In particular, it is a distance on the class of oriented compact manifolds with boundary of a given dimension.

We say

$$M_j \xrightarrow{\mathcal{F}} M_\infty \text{ iff } d_{\mathcal{F}}(M_j, M_\infty) \to 0. \tag{9.109}$$

By the definition of intrinsic flat convergence, $M_j \xrightarrow{\mathcal{F}} M_\infty$ if and only if there exists distance preserving maps to complete metric spaces, $\phi_j : M_j \to Z_j$ and $\phi_j' : M_\infty \to Z_j$, and integral currents, $B_j \in \mathbf{I}_{m+1}(Z_j)$ and $A_j \in \mathbf{I}_m(Z_j)$, such that

$$\phi_{j\#}T_j - \phi'_{j\#}T_\infty = \partial B_j + A_j \tag{9.110}$$

and

$$d_{\mathcal{F}}(M_j, M_\infty) \le d_F^{Z_j}\left(\phi_{j\#}T_j, \phi'_{j\#}T_\infty\right) \le \mathbf{M}(B_j) + \mathbf{M}(A_j) \to 0. \tag{9.111}$$

We could then replace Z_j with Z_j' that are closures of countably $\mathcal{H}^{m+1}$ rectifiable spaces by taking

$$Z_j' = Cl(\mathrm{Set}(B_j) \cup \mathrm{Set}(A_j)) \subset Z_j. \tag{9.112}$$

So in fact the Z in the infimum of the Definition 9.4.2 may be chosen in this class.

Note that if $M_j \xrightarrow{\mathcal{F}} M_\infty$ then using the same distance preserving maps we have

$$\phi_{j\#}\partial T_j - \phi'_{j\#}\partial T_\infty = \partial A_j \tag{9.113}$$

and

$$d_{\mathcal{F}}(\partial M_j, \partial M_\infty) \le d_F^{Z_j}\left(\phi_{j\#}\partial T_j, \phi'_{j\#}\partial T_\infty\right) \le \mathbf{M}(A_j) \to 0. \tag{9.114}$$

So $\partial M_j \xrightarrow{\mathcal{F}} \partial M_\infty$.

The following theorem in [61] is an immediate consequence of Gromov's Embedding Theorem and Ambrosio-Kirchheim's Compactness Theorems:

Theorem 9.4.3. *Given a sequence of precompact m dimensional integral current spaces $M_j = (X_j, d_j, T_j)$ such that*

$$\left(\bar{X}_j, d_j\right) \xrightarrow{GH} (Y, d_Y), \quad \mathbf{M}(M_j) \le V_0 \quad \text{and} \quad \mathbf{M}(\partial M_j) \le A_0 \tag{9.115}$$

then a subsequence converges in the intrinsic flat sense

$$\left(X_{j_i}, d_{j_i}, T_{j_i}\right) \xrightarrow{\mathcal{F}} (X, d_X, T) \tag{9.116}$$

*where either (X, d_X, T) is the **0** integral current space or (X, d_X, T) is an m dimensional integral current space such that $X \subset Y$ with the restricted metric $d_X = d_Y$.*

Immediately one notes that if Y has Hausdorff dimension less than m, then $(X, d, T) =$ **0**. In Section 9.4.3 we survey theorems in which it is proven under additional hypotheses that the intrinsic flat and GH limits agree. There are many examples with nonnegative scalar curvature where these limits do not agree presented in [61]. In fact one may not even have a GH converging subsequence for a sequence with an intrinsic flat limit (see Section 9.2.1).

Gromov's Embedding Theorem, which is applied to prove Theorem 9.4.3, states that if $\left(X_j, d_j\right) \xrightarrow{\text{GH}} (X_\infty, d_\infty)$ then there is a compact metric space Z and a collection of isometric embeddings $\phi_j : X_j \to Z$ such that

$$d_H^Z(\phi_j(X_j), \phi_\infty(X_\infty)) \to 0. \tag{9.117}$$

Note that without his embedding theorem one needs different Z_j for each term in the sequence, and then one would not be able to apply the Ambrosio-Kirchheim Compactness Theorem (cf. Theorem 9.3.1) to complete the proof of Theorem 9.4.3.

In [61][Thms 4.2-4.3], the author and Wenger prove similar embedding theorems for sequences which converge in the intrinsic flat sense: if

$$M_j = \left(X_j, d_j, T_j\right) \xrightarrow{\mathcal{F}} M_0 = (X_\infty, d_\infty, T_\infty), \tag{9.118}$$

then there is a common separable complete metric space, Z, and distance preserving maps $\phi_j : X_j \to Z$ such that

$$d_F^Z(\phi_{j\#}T_j, \phi_{\infty\#}T_\infty) \to 0. \tag{9.119}$$

In the case where $M_0 = \mathbf{0}$ then we have (9.119) as well with $\phi_{\infty\#}T_\infty = 0$, and can find $z \in Z$ and $x_j \in X_j$ such that $\phi_j(x_j) = z$. In fact this Z can be chosen to be the closure of a countably $\mathcal{H}^{m+1}$ rectifiable metric space and is glued together from the Z'_j in (9.112).

These embedding theorems do not require uniform bounds on the masses or volumes of the M_j and ∂M_j. Combining them with Ambrosio-Kirchheim's lower semicontinuity of mass (9.60) we see that

$$M_j \xrightarrow{\mathcal{F}} M_\infty \quad \Rightarrow \quad \liminf_{i\to\infty} \mathbf{M}(M_i) \geq \mathbf{M}(M_\infty). \tag{9.120}$$

In [56] the author proves lower semicontinuity of the diameter as well:

$$M_j \xrightarrow{\mathcal{F}} M_\infty \quad \Rightarrow \quad \liminf_{i\to\infty} \operatorname{diam}(\mathrm{M_i}) \geq \operatorname{diam}(\mathrm{M_\infty}), \tag{9.121}$$

In [49], the author and Portegies prove that

$$\partial M_j \xrightarrow{\mathcal{F}} \partial M_\infty \quad \Rightarrow \quad \operatorname{FillVol}(\partial M_j) \to \operatorname{FillVol}(\partial M_\infty). \tag{9.122}$$

This idea was first observed in joint work of the author with Wenger [60]. Since

$$\mathbf{M}(M) \geq \operatorname{FillVol}(\partial M) \tag{9.123}$$

one can use filling volumes to provide a lower bound on the mass of the limit

$$\mathbf{M}(M_\infty) \geq \operatorname{FillVol}(\partial M_\infty) = \lim_{j\to\infty} \operatorname{FillVol}(\partial M_j). \tag{9.124}$$

Portegies and the author also introduce the notion of a sliced filling volume [49]. They prove that it is continuous with respect to intrinsic flat convergence and that it provides a lower bound for mass.

9.4.3 Compactness Theorems for Intrinsic Flat Convergence

The first compactness theorem for intrinsic flat convergence is stated by the author with Wenger in [61]. It is a combination of Gromov's Compactness and Embedding Theorems with Ambrosio-Kirchheim Compactness; it states that if $M_j = (X_j, d_j, T_j)$ satisfy the hypothesis of Gromov's Compactness Theorem and of Ambrosio-Kirchheim's Compactness theorem, then a subsequence converges in the GH sense and the $\mathcal{F}$ sense where the $\mathcal{F}$ limit is a subset of the GH limit (cf. Theorem 9.4.3). There are a number of theorems which apply Gromov's Compactness theorem combined with this theorem, and then prove the GH and $\mathcal{F}$ limits agree under additional hypotheses including noncollapsing, $\mathbf{M}(M_j) \geq V_0 > 0$. We call these $\mathcal{F}$ =GH compactness theorems.

The author and Wenger prove in [61] a $\mathcal{F}$ =GH compactness theorem for sequences of manifolds without boundary that either have uniform linear contractibility functions or are noncollapsing with Ric ≥ 0. Perales has extended this to allow boundaries with various conditions on the boundary in [46]. Matveev-Portegies have extended the result without boundary to uniform negative lower bounds on Ricci curvature in [42]. The author, Huang and Lee have proven a $\mathcal{F}$ =GH compactness theorem for sequences of integral current spaces, (X, d_j, T), with varying bounded distance functions d_j in the Appendix to [28]. Li and Perales have proven a $\mathcal{F}$ =GH compactness theorem for noncollapsing integral current spaces (X_j, d_j, T_j) with nonnegative Alexandrov curvature (including manifolds with nonnegative sectional curvature) in [41]. It is unknown whether integral current spaces satisfying various generalized notions of Ricci curvature have $\mathcal{F}$ =GH compactness theorems.

In the setting with Scal ≥ 0, we do not in general have GH limits. Thus we need compactness theorems with weaker hypotheses that do not imply GH convergence of subsequences. Wenger's Compactness Theorem was proven in [63] and stated in the following form in [61]:

Theorem 9.4.4. Wenger Compactness
If M_j are integral current space of dimension m satisfying the following

$$\mathbf{M}(M_j) \leq V_0 \qquad \mathbf{M}(\partial M_j) \leq A_0 \qquad \mathrm{diam}\,(\mathrm{M_j}) \leq \mathrm{D_0} \tag{9.125}$$

then there exists a subsequence $M_{j_k} \xrightarrow{\mathcal{F}} M_\infty$ where M_∞ is an integral current spaces of dimension m possibly **0**.

Perales has applied this theorem in [47] to prove two $\mathcal{F}$ compactness theorems. One assumes the given sequence of oriented manifolds satisfies

$$\mathrm{Ric}_j \geq 0 \qquad \mathrm{vol}(\partial M_j) \leq A_0 \qquad H_{\partial M_j} \geq H_0 \qquad \mathrm{diam}\,(\mathrm{M_j}) \leq \mathrm{D_0} \tag{9.126}$$

and the other assumes the given sequence satisfies

$$\mathrm{Ric}_j \geq 0 \qquad \mathrm{vol}(\partial M_j) \leq A_0 \qquad H_{\partial M_j} \geq H_0 > 0 \qquad \mathrm{diam}\,(\partial \mathrm{M_j}) \leq \mathrm{D_0}. \tag{9.127}$$

Here H is the mean curvature with respect to the outward pointing normal. Note that in (9.127) the only condition on the interior of the manifold is Ric ≥ 0.

Observe that in both of these theorems, we could renormalize the manifolds to have $\mathrm{vol}(\partial M_j) = A_0$. When $H_0 \geq 0$, these sequences have Hawking mass as in (9.8) uniformly bounded above:

$$m_H(\partial M_j) \leq m_0 = \sqrt{\frac{A_0}{16\pi}}\left(1 - \frac{1}{16\pi}A_0 H_0^2\right). \tag{9.128}$$

This leads naturally to the following conjecture which could be a step towards proving almost rigidity of the Positive Mass Theorem or Bartnik's Conjecture [4]:

Conjecture 9.4.5. Hawking Mass Compactness
Given a sequence of three dimensional oriented manifolds M_j^3 satisfying

$$\mathrm{vol}(M_j) \leq V_0 \qquad \mathrm{vol}(\partial M_j) = A_0 \qquad \mathrm{diam}\,(\mathrm{M_j}) \leq \mathrm{D_0} \tag{9.129}$$

$$\mathrm{Scal}_j \geq 0 \qquad H_{\partial M_j} \geq 0 \qquad m_H(\partial M_j) \leq m_0 \tag{9.130}$$

and either no closed interior minimal surfaces or $\mathrm{MinA}(M_j) \geq A_1 > 0$, *then a subsequence converges in the intrinsic flat sense*

$$M_{j_k} \xrightarrow{\mathcal{F}} M_\infty \quad \text{and} \quad \mathbf{M}(M_{j_k}) \to \mathbf{M}(M_\infty) \tag{9.131}$$

and M_∞ *satisfies (9.130) in some generalized sense (cf. Section 9.7.5). One might replace Hawking mass with another quasilocal mass in this conjecture.*

LeFloch and the author have proven this Hawking Mass Compactness Conjecture in the rotationally symmetric setting assuming that there are no closed interior minimal surfaces in [40]. This is shown by proving H^1_{loc} convergence of a subsequence of the manifolds with a well chosen gauge, and then proving the H^1_{loc} limit is a $\mathcal{F}$ limit using Theorem 9.5.2. In general it is unknown whether H^1_{loc} convergence implies $\mathcal{F}$ convergence, but here there is also monotonicity of the Hawking mass to help. Since the limit space is a rotationally symmetric manifold with a metric tensor $g \in H^1_{loc}$, it is possible to define generalized notions of nonnegative scalar curvature using (9.3) in a weak sense. It is also possible to define Hawking mass and show (9.130) hold on the limit spaces as well.

Gromov has conjectured vaguely that intrinsic flat convergence may preserve some notion of $\mathrm{Scal} \geq 0$ in [24] and [23]. Considering the examples and the above conjecture, we propose the following Scalar Compactness Theorem which requires a uniform lower bound on the area of a closed minimal surface:

Conjecture 9.4.6. Scalar Compactness
Given a sequence of oriented manifolds M^3_j *with* $\partial M^3_j = \mathbf{0}$ *satisfying*

$$\mathrm{vol}(M_j) \leq V_0 \quad \mathrm{diam}\,(\mathrm{M_j}) \leq \mathrm{D_0} \quad \mathrm{Scal_j} \geq 0 \quad \mathrm{MinA(M_j)} \geq \mathrm{A_1} > 0 \tag{9.132}$$

then a subsequence converges in the intrinsic flat sense

$$M_{j_k} \xrightarrow{\mathcal{F}} M_\infty \quad \text{and} \quad \mathbf{M}(M_{j_k}) \to \mathbf{M}(M_\infty) \tag{9.133}$$

and M_∞ *has* $\mathrm{Scal}_\infty \geq 0$ *in some generalized sense (cf. Section 9.7.5).*

A proof of this Scalar Compactness Theorem in the rotationally symmetric case might imitate the proof of the Hawking Compactness Theorem of the author with LeFloch, however the work in [40] very strongly relies on the existence of a boundary to choose a gauge. Nevertheless a very similar proof should work and would make a nice problem for a doctoral student. Quite a different technique would be needed to handle other settings. In the graph case as in [28] there is no H^1_{loc} convergence.

See Section 9.7.5 for more about how generalized $\mathrm{Scal}_\infty \geq 0$ might be defined.

9.5 Theorems which imply Intrinsic Flat Convergence

In this section we present theorems which have been applied to prove sequences of spaces converge in the intrinsic flat sense. In the previous section we have already presented compactness theorems which imply intrinsic flat convergence of subsequences. Here we present theorems where geometric constraints and relationships between a pair of spaces are used to bound the intrinsic flat distance between them.

Before we begin, note that in Section 5 of [61], the author and Wenger proved that if the Gromov Lipschitz distance between two Riemannian manifolds is small, then the intrinsic flat distance is small. In particular if the manifolds are close in the C_0 sense then they are close in the intrinsic flat sense. This theorem may now be viewed as a special case of Theorem 9.5.2 included below. More general statements about pairs of integral current spaces with such bounds also appear in [61].

9.5.1 Using Riemannian Embeddings to estimate $d_{\mathcal{F}}$

Recall that in the definition of intrinsic flat convergence, one must find distance preserving maps of the pair of manifolds into a common complete metric space, Z, before estimating the flat distance between the images. If, however, one only has Riemannian isometric embeddings of the manifolds into a common Riemannian manifold, then one may apply the following theorem proven by Lee and the author in [39] to estimate the intrinsic flat distance between the spaces.

Theorem 9.5.1. *If $\phi_i : M_i^m \to N^{m+1}$ are Riemannian isometric embeddings with embedding constants C_{M_i} where*

$$C_M := \sup_{p,q\in M} |d_M(p,q) - d_N(\phi(p),\phi(q))|, \tag{9.134}$$

and if they are disjoint and lie in the boundary of a region $W \subset N$ then

$$\begin{aligned} d_{\mathcal{F}}(M_1, M_2) \le\ & S_{M_1}\left(\mathrm{vol}_m(M_1) + \mathrm{vol}_{m-1}(\partial M_1)\right) && (9.135)\\ & + S_{M_2}\left(\mathrm{vol}_m(M_2) + \mathrm{vol}_{m-1}(\partial M_2)\right) && (9.136)\\ & + \mathrm{vol}_{m+1}(W) + \mathrm{vol}_m(V) && (9.137)\end{aligned}$$

where $V = \partial W \setminus \left(\phi_1(M_1) \cup \phi_2(M_2)\right)$ and $S_{M_i} = \sqrt{C_{M_i}(\mathrm{diam}\,(\mathrm{M_i}) + \mathrm{C_{M_i}})}$.

This theorem is proven in [39] by explicitly constructing a geodesic metric space,

$$Z = W_0 \cup W_1 \cup W_2 \subset N \times [0, S_M], \tag{9.138}$$

where $W_0 = \{(x,0) : x \in N\}$ and

$$W_i = \{(x,s) : x \in \phi_i(M_i),\ s \in [0, S_{M_i}]\}, \tag{9.139}$$

and by proving $\psi_i(x) = (\phi_i(x), S_{M_i})$ are distance preserving maps into Z. Then taking $B = [W_0] + [W_1] - [W_2]$ and $A = V_0 + V_1 - V_2$ where $V_0 = [\{(x, 0) : x \in V\}]$ and

$$V_i = [\{(x, s) : x \in \phi_i(\partial M_i),\ s \in [0, S_{M_i}]\}]. \tag{9.140}$$

With the appropriate orientations, we get $\psi_{1\#}[M_1] - \psi_{2\#}[M_2] = A + \partial B$. The estimate then follows because $d_{\mathcal{F}}(M_1, M_2) \leq \mathbf{M}(A) + \mathbf{M}(B)$.

9.5.2 Smooth Convergence Away from Singular Sets

Suppose one has a pair of Riemannian manifolds containing subregions that are close in the C_0 sense. Then one can estimate the intrinsic flat distance between these manifolds, using estimates on the volumes of the regions where they are different and additional information as proven by Lakzian and the author in [36]:

Theorem 9.5.2. *Suppose $M_1 = (M, g_1)$ and $M_2 = (M, g_2)$ are oriented precompact Riemannian manifolds with diffeomorphic subregions $U_i \subset M_i$ and diffeomorphisms $\psi_i : U \to U_i$ such that the following hold*

$$(1+\epsilon)^{-2}\psi_2^* g_2(V, V) < \psi_1^* g_1(V, V) < (1+\epsilon)^2 \psi_2^* g_2(V, V) \qquad \forall V \in TU, \tag{9.141}$$

$$D_{U_i} = \sup\{\mathrm{diam}_{M_i}(W) : W \text{ is a connected component of } U_i\}, \tag{9.142}$$

$$\lambda = \sup_{x,y \in U} |d_{M_1}(\psi_1(x), \psi_1(y)) - d_{M_2}(\psi_2(x), \psi_2(y))|. \tag{9.143}$$

Then the Gromov-Hausdorff distance between the metric completions is bounded,

$$d_{GH}(\bar{M}_1, \bar{M}_2) \leq a + 2\bar{h} + \max\left\{d_H^{M_1}(U_1, M_1), d_H^{M_2}(U_2, M_2)\right\} \tag{9.144}$$

and the intrinsic flat distance between the settled completions is bounded,

$$\begin{aligned} d_{\mathcal{F}}(M_1', M_2') \leq\ & \mathrm{vol}_m(M_1 \setminus U_1) + \mathrm{vol}_m(M_2 \setminus U_2) && (9.145)\\ & + (2\bar{h} + a)\left(\mathrm{vol}_{m-1}(\partial U_1) + \mathrm{vol}_{m-1}(\partial U_2)\right) && (9.146)\\ & + (2\bar{h} + a)\left(\mathrm{vol}_m(U_1) + \mathrm{vol}_m(U_2))\right). && (9.147) \end{aligned}$$

where $a = a(\epsilon, D_{U_1}, D_{U_2})$ converges to 0 as $\epsilon \to 0$ for fixed values of D_{U_i}, and where $h = h(\epsilon, \lambda, D_{U_1}, D_{U_2})$ converges to 0 as both $\epsilon \to 0$ and $\lambda \to 0$ for fixed D_{U_i}. Explicit formulas for $a(\epsilon, D_{U_1}, D_{U_2})$ and $h(\epsilon, \lambda, D_{U_1}, D_{U_2})$ are given in [36].

Theorem 9.5.2 is proven in [36] by explicitly constructing a common metric space

$$Z = M_1 \sqcup_{U_1} (U \times [0, 2\bar{h} + a]) \sqcup_{U_2} M_2 \tag{9.148}$$

where $U \times [0, 2\bar{h} + a]$ is a product manifold with a precisely given metric g', and M_1 is glued to it along $U_1 \subset M_1$ which is isometric to $U \times \{0\}$ and M_2 is glued to it along

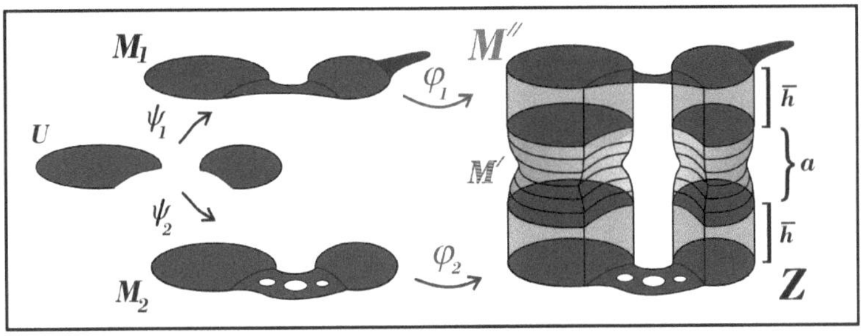

Fig. 9.4. Estimating the intrinsic flat distance. Image owned by the author and Sajjad Lakzian.

$U_2 \subset M_2$ which is isometric to $U \times \{2\bar{h} + a\}$ as in Figure 9.4. The metric g' is chosen so that $\phi_i : M_i \to Z$ are distance preserving. In particular $g' = dt^2 + g_1$ on $U \times [0, \bar{h}]$ where $\bar{h}$ is chosen as in the theorem statement to guarantee that there are no short paths between points in M_1 that run through $U \times [0, 2\bar{h} + a]$. Similarly, $g' = dt^2 + g_2$ on $U \times [a + \bar{h}, a + 2\bar{h}]$. On the middle, $U \times [\bar{h}, a + \bar{h}]$, there is a hemispheric warping between the metrics g_1 and g_2. Once an explicit Z has been found, then an explicit $A \in \mathbf{I}_m(Z)$ is found where $\mathbf{M}(A)$ is the sum of the terms in (9.145)-(9.146) and an explicit $B \in \mathbf{I}_{m+1}(Z)$ is found where $\mathbf{M}(B)$ is the term in (9.147). The details are easy to follow in [36].

Theorem 9.5.2 has been applied in work of the author and Lakzian to prove the rotationally symmetric Hawking Mass Compactness Theorem [40]. It has been applied by Lakzian to study Ricci flow through neck pinch singularities in [33]. It has been applied by Lakzian in [35] to prove $\mathcal{F}$ convergence of sequences of manifolds which converge smoothly away from singular sets. In particular, Lakzian proves the following. Let $M_j = (M, g_j)$ be a sequence of compact oriented Riemannian manifolds with a set $S \subset M$ such that $\mathcal{H}^{n-1}(S) = 0$, and a connected precompact exhaustion, W_k, of $M \setminus S$ satisfying

$$\bar{W}_k \subset W_{k+1} \text{ with } \bigcup_{k=1}^{\infty} W_k = M \setminus S \tag{9.149}$$

with g_j converge smoothly to g_∞ on each W_k,

$$\mathrm{diam}_{M_j}(W_k) \le D_0 \qquad \forall i \ge j, \tag{9.150}$$

$$\mathrm{vol}_{g_j}(\partial W_k) \le A_0, \tag{9.151}$$

and

$$\mathrm{vol}_{g_j}(M \setminus W_k) \le V_k \text{ where } \lim_{k\to\infty} V_k = 0. \tag{9.152}$$

Then

$$\lim_{j\to\infty} d_{\mathcal{F}}(M_j', N') = 0 \tag{9.153}$$

where N' is the settled completion of $N = (M \setminus S, g_\infty)$ as in (9.97). If one also assumes $\mathrm{Ric}_j \geq H$, then

$$\lim_{j\to\infty} d_{GH}(M_j, \bar{N}) = 0. \tag{9.154}$$

Other theorems about smooth convergence away from singular sets are proven in [35] as well.

9.5.3 Pairs of Integral Current Spaces

The following theorem concerns pairs of integral current spaces which share the same space, X, and the same current structure, T, but have different distance functions. This happens, for example, when one has a pair of oriented Riemannian manifolds with an orientation preserving diffeomorphism between them. It can also be applied to pairs of integral current spaces with a biLipschitz current preserving map between them. The Gromov-Hausdorff part of this theorem was proven by Gromov in [21] and the intrinsic flat part was proven by the author with Huang and Lee in the Appendix to [28].

Theorem 9.5.3. *Fix a precompact m-dimensional integral current space (X, d_0, T) with $\partial T = 0$ and fix $\lambda > 0$. Suppose that d_1 is another distance on X such that*

$$\epsilon = \sup\left\{|d_1(p,q) - d_0(p,q)| \,:\, p, q \in X\right\}. \tag{9.155}$$

Then we have the following:

$$d_{GH}\left((X, d_1), (X, d_0)\right) \leq 2\epsilon \tag{9.156}$$

$$d_{\mathcal{F}}\left((X, d_1, T), (X, d_0, T)\right) \leq 2^{(m+1)/2}\lambda^{m+1}(2\epsilon)\mathbf{M}(T). \tag{9.157}$$

Note that the hypothesis of this theorem is satisfied when an integral current space, (X, d, T), has an almost distance preserving map into another metric space, $F : X \to Y$, and one defines a new distance d_1 on X as $d_1(p,q) = d_Y(F(p), F(q))$. It was applied by the author with Huang and Lee in [28] as one of the final steps towards proving the almost rigidity for the Positive Mass Theorem in the graph setting.

Problem 9.5.4. *It seems natural that one could apply Theorem 9.5.3 to prove a more general theorem about pairs of integral current spaces, (X_i, d_i, T_i), which contain a pair of regions $U_i \subset X_i$ with a current preserving isometry between the regions. If one simply uses the restricted distances then there is no need for an embedding constant or a λ as in the prior two sections. However, one needs to extend the Theorem 9.5.3 to consider the setting with boundary.*

9.6 Theorems about Intrinsic Flat Limits

In this section we assume we have a sequence, $M_j \xrightarrow{\text{GH}} M_\infty$ or $M_j \xrightarrow{\mathcal{F}} M_\infty$ and present theorems about limits of points in these spaces, limits of functions on these spaces, continuity and semicontinuity of various quantities on these spaces. Recall that we have already mentioned a number of such results in Subsection 9.4.2 and we will be refering to those results here as well. We close this section with a discussion of the setting where one has intrinsic flat convergence with volume continuity.

9.6.1 Limits of Points and Points with no Limits

When studying sequences of converging Riemannian manifolds, M_j, one often wishes to understand what happens to points, $x_j \in M_j$: *where do they converge and when do they disappear?* One is also interested in understanding limit points, $x_\infty \in M_\infty$, by considering $x_j \in M_j$ that converge to these points. This can easily be understood when the M_j are a sequence of converging submanifolds lying in a given space, and it is clear that under flat convergence of submanifolds some sequences of points will disappear in the limit. Defining converging sequences of points when M_j are distinct Riemannian manifolds is much more difficult.

Under Gromov-Hausdorff convergence

$$(X_j, d_j) \xrightarrow{\text{GH}} (X_\infty, d_\infty), \tag{9.158}$$

one can apply Gromov's Embedding Theorem (9.117) to define what it means to say $x_j \to x_\infty$ for $x_j \in X_j$ as follows: there exists a compact metric space, Z, and distance preserving maps, $\varphi_j : X_j \to Z$, such that

$$d_H^Z(\varphi_j(X_j), \varphi_\infty(X_\infty)) \to 0 \text{ and } d_Z(\varphi_j(x_j), \varphi_\infty(x_\infty)) \to 0. \tag{9.159}$$

Note that x_∞ is not unique: if $F : X_\infty \to X_\infty$ is an isometry then $F(x_\infty)$ is also a limit of x_j. This is a natural consequence of the fact that the GH distance is between isometry classes of metric spaces. Gromov shows that

$$\forall x_\infty \in X_\infty \; \exists x_j \in X_j \text{ such that } x_j \to x_\infty. \tag{9.160}$$

In fact, there exist functions, $H_j : X_\infty \to X_j$, and distance preserving maps, $\varphi_j : X_j \to Z$, satisfying (9.159) such that if $x_j = H_j(x_\infty)$ then $x_j \to x_\infty$ and

$$d_j(H_j(x), H_j(y)) \to d_\infty(x, y). \tag{9.161}$$

Furthermore, for any $r > 0$, we have converging closed balls

$$x_j \to x_\infty \;\Rightarrow\; \left(\bar{B}(x_j, r), d_j\right) \xrightarrow{\text{GH}} \left(\bar{B}(x_\infty, r), d_\infty\right). \tag{9.162}$$

By the compactness of Z, there is a Bolzano-Weierstass Theorem:

$$x_j \in X_j \quad \Rightarrow \quad \exists j_k \text{ s.t. } x_{j_k} \to x_\infty \in X_\infty. \tag{9.163}$$

Combining (9.161) with (9.163) we have diameter continuity:

$$\lim_{j\to\infty} \operatorname{diam}(\mathrm{X_j}) = \operatorname{diam}(\mathrm{X}_\infty). \tag{9.164}$$

See for example [10] and [56] for more details.

Now suppose we have intrinsic flat convergence,

$$M_j = (X_j, d_j, T_j) \xrightarrow{\mathcal{F}} M_\infty = (X_\infty, d_\infty, T_\infty). \tag{9.165}$$

In [56], the author applied the Intrinsic Flat Embedding Theorem as in (9.119) to say a sequence $x_j \in X_j$ is *Cauchy* if there exists a complete metric space, Z, a point, $z_\infty \in Z$, and distance preserving maps, $\phi_j : X_j \to Z$, such that

$$d_F^Z\left(\phi_{j\#}T_j, \phi_{\infty\#}T_\infty\right) \to 0 \tag{9.166}$$

and $\phi_j(x_j) \to z_\infty$. One says the sequence has *no limit* in $\bar{X}_\infty$ if

$$z_\infty \notin \phi_\infty(\bar{X}_\infty). \tag{9.167}$$

One says the points *converge* $x_j \to x_\infty$ in X_∞ (or respectively in $\bar{X}_\infty$) if there exists x_∞ in X_∞ (or respectively in $\bar{X}_\infty$) such that $z_\infty = \phi_\infty(x_\infty)$:

$$z_\infty \in \phi_\infty(X_\infty) = \operatorname{Set}(\phi_{\infty\#}T_\infty) \quad \left(\text{or respectively } z_\infty \in \phi_\infty(\bar{X}_\infty)\right). \tag{9.168}$$

Note that x_∞ is not unique: if $F : X_\infty \to X_\infty$ is a current preserving isometry then $F(x_\infty)$ is also a limit of x_j. This is a natural consequence of the fact that the $\mathcal{F}$ distance is between current preserving isometry classes of integral current spaces. Below we will provide additional conditions (9.176) and (9.177) which demonstrate that Cauchy sequences of points which disappear with respect to one sequence of ϕ_j satisfying (9.166) cannot be Cauchy sequences of points that converge with respect to another sequence of ϕ_j satisfying (9.166) and vice versa.

In [56] the author proves that if (9.165) then

$$\forall x_\infty \in \bar{X}_\infty \ \textit{ exists } x_j \in X_j \text{ such that } x_j \to x_\infty. \tag{9.169}$$

In fact there exist *convergence functions*, $H_j : X_\infty \to X_j$, and distance preserving maps, $\phi_j : X_j \to Z$, satisfying (9.166) such that $H_j(x_\infty) \to x_j$ as in (9.159) with

$$d_j(H_j(x), H_j(y)) \to d_\infty(x, y) \text{ and } H_j(\partial M_\infty) \subset \partial M_j. \tag{9.170}$$

In fact the author proves in [56][Theorem 5.1] the following:

$$\text{if } M_j \xrightarrow{\mathcal{F}} M_\infty \text{ where } M_\infty \text{ is nonzero and precompact,} \tag{9.171}$$

$$\text{then there exist } N_j \subset M_j \text{ such that } N_j \xrightarrow{\text{GH}} \bar{M}_\infty. \tag{9.172}$$

Additional properties of the N_j are described there. In general, intrinsic flat limits need not be precompact and need not have finite diameter [61].

Since Z is only complete, we do not have a simple Bolzano-Weierstrass Theorem. In fact points may disappear under flat convergence even in a compact Z. So we do not have continuity of diameter. However, we have semicontinuity of diameter

$$\liminf_{j\to\infty} \operatorname{diam}(\mathrm{M_j}) \geq \operatorname{diam}(\mathrm{M_\infty}) \tag{9.173}$$

and depth

$$\liminf_{j\to\infty} \operatorname{Depth}(M_j) \geq \operatorname{Depth}(M_\infty) \tag{9.174}$$

where

$$\operatorname{Depth}(M) = \sup\left\{d_X(x,y):\; x\in X,\; y\in \operatorname{Set}(\partial T)\right\} \in [0,\infty]. \tag{9.175}$$

Recall that for almost every $r > 0$ one may view a ball in an integral current spaces as an integral current space itself $S(x,r) = \left(\operatorname{Set}\left(T_j \llcorner B(x,r)\right),\, d,\, T \llcorner B(x,r)\right)$. One can also examine the convergence of balls when $M_j \xrightarrow{\mathcal{F}} M_\infty$. In [56] the author proves that if $x_j \to x_\infty$ as in (9.169), then there is a subsequence j_k such that

$$S(x_{j_k}, r) \xrightarrow{\mathcal{F}} S(x_\infty, r) \neq \mathbf{0} \quad \textit{for almost every } r > 0. \tag{9.176}$$

If a Cauchy sequence of points, $x_j \in M_j$, has no limit in $\bar{X}_\infty$ as in (9.167) then

$$\exists \delta > 0 \text{ s.t. } S(x_{j_k}, r) \xrightarrow{\mathcal{F}} 0 \quad \textit{for almost every } r \in (0,\delta). \tag{9.177}$$

Since (9.176) and (9.177) are intrinsic notions which do not depend on a choice of distance preserving maps, one concludes that a Cauchy sequence of points which has no limit in $\bar{X}_\infty$ with respect to one sequence of ϕ_j satisfying (9.166) cannot be a Cauchy sequence of points that converges with respect to another sequence of ϕ_j satisfying (9.166), and vice versa. Far more subtle is determining when exactly a sequence of points converges to a point in $\bar{X}_\infty \setminus X_\infty$.

Let us further consider converging sequences of points, $x_j \to x_\infty$. As a consequence of mass semicontinuity as in (9.60), we have for almost every $r > 0$

$$\liminf_{k\to\infty} \mathbf{M}(S(x_{j_k}, r)) \geq \mathbf{M}(S(x_\infty, r)). \tag{9.178}$$

As a consequence of (9.59), we have convergence of spheres for almost every $r > 0$

$$\partial S(x_{j_k}, r) \xrightarrow{\mathcal{F}} \partial S(x_\infty, r) \tag{9.179}$$

Combining sphere convergence as in (9.179) with filling volume continuity as in (9.122), Portegies and the author have proven that for almost every $r > 0$

$$\operatorname{FillVol}(\partial S(x_{j_k}, r)) \xrightarrow{\mathcal{F}} \operatorname{FillVol}(\partial S(x_\infty, r)) \tag{9.180}$$

as well as continuity of another notion called the sliced filling volume in [49]. In Theorem 4.27 of that paper, this continuity is applied to determine when a sequence converges in $X_\infty = \mathrm{Set}(T_\infty)$. The filling volume case of this theorem observes that

$$\mathbf{M}(S(x_\infty, r)) \geq \mathrm{FillVol}(\partial S(x_\infty, r)) = \lim_{k\to\infty} \mathrm{FillVol}(\partial S(x_{j_k}, r)), \tag{9.181}$$

which implies that if there is a uniform lower bound $C > 0$ such that

$$\mathrm{FillVol}(\partial S(x_j, r)) \geq Cr^m \qquad \forall j \in \mathbb{N} \tag{9.182}$$

then by (9.48), $x_\infty \in \mathrm{Set}(T_\infty) = X_\infty$. An idea similar to this one was applied earlier in an extrinsic way by the author and Wenger in [60]. The goal was to show that when a sequence of manifolds has nonnegative Ricci curvature or has a linear contractibility function, then the GH and $\mathcal{F}$ limits agree. This method has been applied to prove intrinsic flat and GH limits agree under a variety of different conditions by the author with Portegies, by Munn, by Perales, by Perales-Li, and by Matveev-Portegies [41, 42, 45, 46, 49].

With only lower scalar curvature bounds on the sequence one does not in general have GH and intrinsic flat limits that agree (see examples below). In fact one may not have any GH limit even for a subsequence. Nevertheless this method might be applied to determine which points in the sequence of manifolds are disappearing and which remain.

The author and Portegies prove a Bolzano-Weierstrass Theorem for sequences of points with bounds on their filling or sliced filling volume [49][Theorem 4.30]. One consequence of this theorem is that if a sequence satisfies (9.181) then a subsequence is Cauchy (using an argument involving the fact that the limit space has finite mass) and then by the above method, the sequence converges to a point in X_∞. There is also a Bolzano-Weierstrass Theorem for sequences of points which does not require a lower bound on the filling volumes of spheres, but instead requires

$$\exists r_0 > 0 \;\; s.t. \;\; \liminf_{j\to\infty} d_{\mathcal{F}}(S(x_j, r), \mathbf{0}) \geq h(r) > 0 \;\; \textit{for a.e. } r \in (0, r_0] \tag{9.183}$$

to obtain a subsequence which is Cauchy and converges in $\bar{X}_\infty$ [56][Theorem 7.1]. Note that the theorems and proofs in [56] are very easy to read. The more technically difficult results in [49] involving filling volumes and sliced filling volumes are better in that they provide more precise controls under weaker hypotheses.

9.6.2 Limits of Functions

Recall that when using the compactness and weak rigidity method to prove an almost rigidity theorem as in (9.19), one proves that $M_j \to M_0$, where M_0 is a specific given rigid space, by first using a compactness theorem to show a subsequence, $M_{j_k} \to M_\infty$

and then proving $M_\infty = M_0$. One way to prove that $M_\infty = M_0$ is to construct an isometry between these spaces as a limit of functions from M_{j_k} to M_0. Theorems which produce limits of subsequences of functions are called Arzela-Ascoli Theorems.

First let us describe what we mean by a limit function. Suppose $F_j : X_j \to Y_j$ are functions, and $X_j \to X_\infty$ and $Y_j \to Y_\infty$ in the GH or $\mathcal{F}$ sense. Then we say $F_\infty : X_\infty \to Y_\infty$ is their limit, denoted $F_j \to F_\infty$, if

$$F_j(x_j) \to F_\infty(x_\infty) \text{ whenever } x_j \to x_\infty. \tag{9.184}$$

More precisely, $F_j \to F_\infty$ if there exist convergence functions, $H_j : X_\infty \to X_j$ and $H_j' : Y_\infty \to Y_j$, as in (9.170) such that

$$F_j \circ H_j(x) = H_j' \circ F_\infty(x) \qquad \forall x \in X_\infty. \tag{9.185}$$

The Gromov-Hausdorff Arzela-Ascoli Theorem states that if one has compact metric spaces, $X_j \xrightarrow{\text{GH}} X_\infty$ and $Y_j \xrightarrow{\text{GH}} Y_\infty$, and if $F_j : X_j \to Y_j$ are equicontinuous

$$\forall \epsilon > 0\ \exists \delta_\epsilon > 0 \text{ such that } d_{X_j}(x, x') < \delta_\epsilon \Rightarrow d_{Y_j}(F_j(x), F_j(x')) \le \epsilon \tag{9.186}$$

then $F_j \to F_\infty$ where $F_\infty : X_\infty \to Y_\infty$ satisfies (9.186). This theorem is a direct consequence of Gromov's Embedding Theorem combined with the standard proof of the classical Arzela-Ascoli Theorem (cf. [56]). Furthermore, if F_j are surjective then the limit F_∞ is surjective. If the F_j are isometries on balls of radius $r_0 > 0$, then the limit is an isometry on balls of radius $r_0 > 0$. This was applied by the author in joint work with Wei to prove that the GH limits of manifolds with Ricci ≥ 0 have universal covering spaces [58] and that the covering spectrum is continuous with respect to Gromov-Hausdorff convergence in [59].

It is not absolutely necessary that the sequence of functions be equicontinuous. Gromov proves the same result for ϵ_j almost isometries $F_j : X_j \to Y_j$,

$$|d_{Y_j}(F_j(p), F_j(q)) - d_{X_j}(p, q)| < \varepsilon_j \text{ and } Y_j \subset T_{\varepsilon_j}(X_j) \tag{9.187}$$

with $\epsilon_j \to 0$, producing a limit $F_\infty : X_\infty \to Y_\infty$ which is an isometry [21]. See also the Burago-Burago-Ivanov text [10]. Almost isometries are applied to prove the GH almost rigidity theorems of Colding in [15] and of Cheeger-Colding in [13] by constructing almost isometries from the M_j to the rigid space, M_0. In [55], a GH Arzela-Ascoli Theorem which only requires almost equicontinuity is proven by the author in order to prove another GH almost rigidity theorem.

One cannot hope for an $\mathcal{F}$ Arzela-Ascoli Theorem which is as powerful as the GH Arzela-Ascoli Theorem. In Example 9.2.5 one has no limit for the geodesics $\gamma_j : [0, 1] \to M_j$ running through the increasingly thin tunnels. Nevertheless two useful Arzela-Ascoli Theorems with additional hypotheses were proven in [56].

Suppose $F_j : M_j \to M_j'$ are (surjective) current preserving isometries on balls of radius $r_0 > 0$ and $M_j \xrightarrow{\mathcal{F}} M_\infty \neq \mathbf{0}$ and $M_j' \xrightarrow{\mathcal{F}} M_\infty' \neq \mathbf{0}$. Then a subsequence of F_j

converges to $F_\infty : M_\infty \to M'_\infty$ which is also a (surjective) current preserving isometry on balls of radius $r_0 > 0$ [56]. This theorem has been applied in joint work of the author with Sinaei to study the intrinsic flat convergence of covering spaces and the covering spectrum [54].

Suppose $F_j : M_j \to Y_j$ are equicontinuous maps as in (9.186) where M_j are integral current space and Y_j are compact metric spaces such that $M_j \xrightarrow{\mathcal{F}} M_\infty \neq \mathbf{0}$ and $Y_j \xrightarrow{\text{GH}} Y_\infty$. Then F_j converge to $F_\infty : M_\infty \to Y_\infty$ which also satisfies (9.186). Furthermore, if F_j are surjective then the limit F_∞ is surjective [56]. Keep in mind that this includes equicontinuous functions, $F_j : M_j \to [a, b]$, and embeddings in compact regions in Euclidean space, $F_j : M_j \to \mathbb{E}^N$. This theorem has been applied jointly with Huang and Lee to prove Almost Rigidity of the Positive Mass Theorem in the graph setting [28].

Conjectured related Arzela-Ascoli theorems are suggested in [56] and the proofs there are not difficult to read. These theorems are particularly useful when proving almost rigidity theorems to try to construct isometries from the limit M_∞ of a subsequence M_j to the desired rigid space, M_0.

Problem 9.6.1. *Suppose $F_j : M_j \to Y$ has $\mathrm{Lip}(F_j) \le 1$ with Y compact (including Riemannian embeddings and graphs), and M_j satisfy the hypothesis of Wenger's Compactness Theorem as in (9.125). Then we know a subsequence $M_j \xrightarrow{\mathcal{F}} M_\infty = (X_\infty, d_\infty, T_\infty)$. We can see $F_{j\#}[M_j]$ satisfies the hypothesis of the Ambrosio-Kirchheim Compactness as in (9.64), so we know a subsequence $F_{j\#}[M_j]$ converges in the flat sense to some $S_\infty \in \mathbf{I}_m(Y)$. If $M_\infty \neq \mathbf{0}$ we know by the Arzela-Ascoli Theorem above that a subsequence, also denoted F_j, converges to $F_\infty : X_\infty \to Y$. We conjecture that*

$$M_\infty = (X_\infty, d_\infty, T_\infty) \text{ with } F_{\infty\#} T_\infty = S_\infty. \tag{9.188}$$

This conjecture captures the key steps used in joint work of the author with Huang and Lee to prove the Almost Rigidity of the Positive Mass Theorem in the graph setting in [28]. In that proof we have M_j^m which satisfy our almost rigidity conditions, show they satisfy the hypothesis of Wenger's Compactness Theorem, and then study their images as graphs in $Y = \mathbb{E}^{m+1}$ and obtain (9.188). Some of the techniques there may be useful towards proving this conjecture in general.

9.6.3 Intrinsic Flat with Volume Convergence

Recall that intrinsic flat convergence does not imply volume convergence. One does have semicontinuity, $M_j \xrightarrow{\mathcal{F}} M_\infty$ implies $\liminf_{j\to\infty} \mathbf{M}(M_j) \ge \mathbf{M}(M_\infty)$, but even when M_j and M_∞ are Riemannian manifolds the volumes need not converge (as seen in the examples with cancellation). Note that all the examples presented above with wells, bubbling and sewing do have volume convergence. We introduce the following:

Definition 9.6.2. *The intrinsic flat volume distance between two integral current spaces, $M_j = (X_j, d_j, T_j)$ is*

$$d_{V\mathcal{F}}(M_1, M_2) = d_{\mathcal{F}}(M_1, M_2) + |\mathbf{M}(M_1) - \mathbf{M}(M_2)| \tag{9.189}$$

So $M_j \xrightarrow{V\mathcal{F}} M_\infty$ iff $\mathbf{M}(M_j) \to \mathbf{M}(M_\infty)$ and $M_j \xrightarrow{\mathcal{F}} M_\infty$. Note that points may still disappear as in the Ilmanen Example. However there is no cancellation.

Portegies studied the properties of $M_j = (X_j, d_j, T_j)$ such that $M_j \xrightarrow{V\mathcal{F}} M_\infty$. Applying his Lemma 2.7 in [48], we see that the metric measure spaces, $(X_j, d_j, ||T_j||)$ converge in a measured sense. That is, there are distance preserving maps, $\phi_j : X_j \to Z$ such that

$$\phi_j||T_j|| \to \phi_j||T_\infty|| \text{ weakly as measures in } Z. \tag{9.190}$$

Here Z is only complete and one need not have GH convergence (as in Ilmanen's Example). Portegies then applied this to prove that the Laplace eigenvalues of converging sequences of manifolds are upper semicontinuous,

$$\limsup_{j\to\infty} \lambda_k(M_j) \le \lambda_k(M_\infty). \tag{9.191}$$

He presents examples showing this is false without $\mathbf{M}(M_j) \to \mathbf{M}(M_\infty)$.

Note that the Hawking Mass Compactness Conjecture and Scalar Compactness Conjecture both propose that a subsequence $M_{j_k} \xrightarrow{V\mathcal{F}} M_\infty$ as is shown in the compactness theorem proven jointly with LeFloch [40]. Also note that in the work of the author with Lee, Huang, and Stavrov proving various special cases towards almost rigidity of the Positive Mass Theorem, it is proven that $M_j \xrightarrow{V\mathcal{F}} M_0$ [39][28][57]. Matveev and Portegies prove $M_{j_k} \xrightarrow{V\mathcal{F}} M_\infty$ when Ricci curvature is uniformly bounded below [42][Theorem 4.1].

Keep in mind that $M_{j_k} \xrightarrow{V\mathcal{F}} M_\infty$ alone will not suffice to achieve generalized Scal > 0 properties on M_∞ as seen in the problematic examples involving sewing of a single curve to a point will have volume convergence and yet the limit in (9.5) will fail to be nonnegative for a ball about that limit point. Note however that if one assumes (9.6) holds at every point p in every M_j with a uniform lower bound on $r_p \ge r_0 > 0$, then (9.6) can be shown to hold with inequalities on M_∞.

9.7 Results and Conjectures about Limits of Manifolds with Nonnegative Scalar Curvature

In this final section of the paper we state our almost rigidity conjectures precisely and survey known results towards proving these conjectures. We suggest special cases which might be proven more easily. Note that completely proving any almost rigidity theorem is significantly more difficult than proving the corresponding rigidity theorem. One must either reprove the rigidity theorem in a quantitative way obtaining

(9.18) using the *explicit control method* as described in the introduction, or one must prove the Rigidity Theorem on a generalized class of spaces obtaining (9.19) using the *compactness and weak rigidity method*. Recall that we have already proposed two compactness conjectures in Section 9.4.3. In the final two subsections of this paper we discuss generalized notions of nonnegative scalar curvature as proposed by Gromov and regularity theory.

9.7.1 Almost Rigidity of the Positive Mass Theorem and the Bartnik Conjecture

Consider the class, $\mathcal{M}$, of asymptotically flat three dimensional Riemannian manifolds with nonnegative scalar curvature and no interior closed minimal surfaces and either no boundary or the boundary is an outermost minimizing surface. This is the physically natural class of spaces used to prove the Penrose Inequality as discussed in the introduction. By the Positive Mass Theorem, if $M \in \mathcal{M}$ has $m_{ADM}(M) = 0$ then M is isometric to Euclidean space. In [39], Lee and the author proposed that almost rigidity for the Positive Mass Theorem should be provable using intrinsic flat convergence, and demonstrated that it is false for GH convergence.

Conjecture 9.7.1. *Fix $D > 0$, $r_0 > 0$. Let $M_j^3 \in \mathcal{M}$ and let $\Sigma_j \subset M_j^3$ be special surfaces with* $\operatorname{Area}(\Sigma_j) = 4\pi r_0^2$ *and $\Sigma_\infty = \partial B(0, r_0) \subset \mathbb{E}^3$. We conjecture that*

$$m_{ADM}(M_j) \to 0 \quad \Rightarrow \quad d_{Vol\mathcal{F}}\left(T_D(\Sigma_j), T_D(\Sigma_\infty)\right) \to 0, \tag{9.192}$$

where $T_D(\Sigma)$ is the tubular neighborhood of radius D around Σ, or alternatively

$$m_{ADM}(M_j) \to 0 \quad \Rightarrow \quad d_{Vol\mathcal{F}}\left(\Omega_j, \Omega_\infty\right) \to 0, \tag{9.193}$$

where Ω_j is the interior of Σ_j with $\operatorname{Depth}(\Omega_j) \le D$.

Note that Lee and the author were being deliberately vague as to what a special surface, Σ_j, should be in this conjecture. A number of more precisely stated special cases of this conjecture were provided in the final section of [39] along with brief ideas as to how one might approach the proof of the conjecture in those cases. Most of these special cases are still open.

Lee and the author proved that (9.192) holds when M_j^3 have metric tensors of the form $g_j = dr^2 + f_j(r)^2 g_{\mathbb{S}^2}$ and $\Sigma_j = r^{-1}(r_0)$. The proof uses Geroch monotonicity and Theorem 9.5.1 to obtain explicit controls on $T_D(\Sigma_j)$[39]. LeFloch and the author examined these explicit controls in more detail in [40]. The author and Stavrov proved this conjecture when M_j^3 are Brill-Lindquist geometrostatic manifolds and Σ_j are large spheres in [57]. That proof is also completed using explicit controls: bounds on the metric tensor are found on carefully selected regions within the manifolds followed by an application of Theorem 9.5.2.

Huang, Lee and the author proved (9.193) when $\Omega_j^3 \subset M_j$ are graph manifolds, which have Riemannian embeddings, $\Psi_j : M_j^3 \to \mathbb{E}^4$ as graphs. We assumed controls on the Σ_j and other technical properties on the M_j^3 [28]. Using the properties of graph manifolds with Scal ≥ 0 and $m_{ADM}(M_j) \to 0$ we first proved that $\mathrm{vol}(\Omega_j) \to \mathrm{vol}(B(0, r_0))$. We applied Wenger's Compactness Theorem and an Arzela Ascoli Theorem to prove a subsequence of the $\Omega_j \xrightarrow{\mathcal{F}} \Omega_\infty$ and Ψ_j converge to $\Psi_\infty : \Omega_\infty \to \mathbb{E}^4$ with $\mathrm{Lip}(\Psi_\infty) \leq 1$. By the lower semicontinuity of mass we showed

$$\mathbf{M}(\Psi_{\infty\#}[\Omega_\infty]) \leq \mathbf{M}([\Omega_\infty]) \leq \mathrm{vol}(B(0, r_0)). \tag{9.194}$$

We used the controls on Σ_j to prove they Lipschitz converge to $\partial\Omega_\infty$ and that $\Psi_\infty : \partial\Omega_\infty \to \partial B(0, r_0) \times \{0\}$ is biLipschitz. Thus $\partial\Psi_{\infty\#}[\Omega_\infty] = [\partial B(0, r_0) \times \{0\}]$. Combining this with (9.194) implies $Psi_{\infty\#}[\Omega_\infty]$ solves the Plateau Problem. So

$$\Psi_{\infty\#}[\Omega_\infty] = [B(0, r_0) \times \{0\}] \text{ and } \mathbf{M}(\Psi_{\infty\#}[\Omega_\infty]) = \mathrm{vol}(B(0, r_0))(4/3). \tag{9.195}$$

This implies equality in (9.194) so Ψ_∞ must be an isometry: $\Omega_\infty = B(0, r_0)$ [28].

Remark 9.7.2. *One might consider applying the Huisken isoperimetric mass as in (9.11) to prove Conjecture 9.7.1 with Σ_j chosen to be uniformly asymptotically spherical in M_j with $m_{ADM}(M_j) \to 0$ so that*

$$m_{ISO}(\Omega_j) = \frac{2}{\mathrm{Area}(\partial\Omega_j)} \left(\mathrm{vol}(\Omega_j) - \frac{\mathrm{Area}(\partial\Omega_j)^{3/2}}{6\sqrt{\pi}} \right) \to 0. \tag{9.196}$$

Observe that (9.196) immediately implies

$$\mathrm{vol}(\Omega_j) \to A_0^{3/2}/(6\sqrt{\pi}) = (4/3)\pi r_0^3 = \mathrm{vol}(B(0, r_0)). \tag{9.197}$$

So by Wenger's Compactness Theorem a subsequence converges $\Omega_j \xrightarrow{\mathcal{F}} \Omega_\infty$. By lower semicontinuity of mass we have $\mathbf{M}(\Omega_\infty) \leq \mathrm{vol}(B(0, r_0))$.

Suppose we also assume that there exist Riemannian isometric embeddings $\Psi_j : \Omega_j \to \mathbb{E}^N$ with the property that Ψ_j restricted to $\partial\Omega_j$ is uniformly biLipschitz to $\partial B(0, r_0) \times \{0, \ldots, 0\}$. Then exactly as in the above description of the proof in [28] we have Ω_∞ isometric to $B(0, r_0)$. Here the only place we used Scal ≥ 0 was when we replaced $m_{ADM}(M_j) \to 0$ by (9.196) applying Miao's proof that $m_{ISO}(\Omega)$ is close to $m_{ADM}(M)$ for large round $\partial\Omega$ in [18]. Note that $m_{ISO}(\Omega)$ for $M^3 \in \mathcal{M}$ does not imply $\Omega = B(0, r_0)$; one needs to impose some asymptotic roundness on the $\partial\Omega$ as well as Scal ≥ 0 even to obtain rigidity.

Problem 9.7.3. *Shi and Tam proved in [53] that the Brown-York Mass, $m_{BY}(\partial\Omega)$, is nonnegative if* Scal ≥ 0 *on Ω and $\partial\Omega$ has positive Gauss curvature, and*

$$m_{BY}(\partial\Omega) = 0 \;\Rightarrow\; \Omega \subset \mathbb{E}^3. \tag{9.198}$$

This mass (which agrees with the Liu-Yau mass in this setting) is defined using a Riemannian isometric embedding $\Psi : \partial\Omega \to \mathbb{E}^3$, the mean curvature, H, of $\partial\Omega \subset M^3$ and the mean curvature, H_0, of $\Psi(\partial\Omega) \subset \mathbb{E}^3$ as follows:

$$m_{BY}(\partial\Omega) = \frac{1}{8\pi} \int_{\partial\Omega} H_0 - H \, d\sigma. \tag{9.199}$$

The Arzela-Ascoli Theorems proven above might be helpful towards proving Conjecture 9.4.5 for the Brown-York Mass, including semicontinuity of the Brown-York mass under $\mathcal{F}$ convergence, and almost rigidity of the Shi-Tam Rigidity Theorem. This might also be applied to prove the almost rigidity of the Positive Mass Theorem, or the Bartnik Conjecture. One of the biggest difficulties here is that mean curvature must be defined in a generalized way and controlled under intrinsic flat convergence.

Problem 9.7.4. *Huisken and Ilmanen defined a weak notion of mean curvature in their proof of the Penrose Inequality (which can also be applied to prove the Positive Mass Theorem) [30]. Perhaps one might try to use their method to prove Almost Rigidity of the Positive Mass Theorem. One might consider a limit space M_∞ and attempt to define Huisken-Ilmanen's weak inverse mean curvature flow and prove Geroch monotonicity on M_∞. What regularity is needed on M_∞? What notion of nonnegative scalar curvature is required?*

Recall that in [20], Geroch proved that if $N_t : \mathbb{S}^2 \to M^3$ evolves by smooth inverse mean curvature flow,

$$\frac{d}{dt}x = \frac{\nu}{H} \quad \text{where} \quad x = N_t(p), \ \nu \text{ is the normal to } N_t \text{ at } x, \tag{9.200}$$

$$\text{and } H \text{ is the mean curvature of } N_t \text{ at } x, \tag{9.201}$$

and M^3 has Scal ≥ 0 then then the Hawking mass, $m_H(N_t)$, is nondecreasing:

$$t_2 > t_1 \ \Rightarrow \ m_H(N_{t_2}) \geq m_H(N_{t_1}). \tag{9.202}$$

In [30], Huisken-Ilmanen introduced the weak inverse mean curvature flow,

$$N_t = \partial\{x : \ u(x) < t\} \ \text{ where } \ \mathrm{div}_M\left(\frac{\nabla u}{|\nabla u|}\right) = |\nabla u|, \tag{9.203}$$

proving it also satisfies Geroch monotonicity as in (9.202), and with the right boundary conditions $\lim_{t\to\infty} m_H(N_t) = m_{ADM}(M)$. They defined a weak mean curvature for the level sets of u to be $H = \nabla u$ almost everywhere. One may naturally ask what regularity was needed on the limit space to prove these results.

Alternatively one might apply Huisken-Ilmanen's method on the sequence M_j rather than on M_∞. One might consider sequences of $u = u_j$ satisfying (9.203) on M_j and consider limits $u_j \to u_\infty$, if it is difficult to define (9.203) on M_∞ itself. On each M_j one can define (9.203) and then one has $m_H(N_t)$ nearly constant. This was a key step in the proof of the Almost Rigidity of the Positive Mass Theorem in the rotationally symmetric case [39]. It would be interesting to investigate this even in the setting with smooth inverse mean curvature flow. In the setting where one only has weak inverse mean curvature flow, then the N_t may skip over entire regions. In private communication with the author, Huisken has suggested that it might be possible to bound the volume of the skipped regions using his isoperimetric masses. These same techniques might also be applied to prove the Almost Rigidity of the Penrose Inequality.

Problem 9.7.5. *Recall that the Penrose Inequality for $M^3 \in \mathcal{M}$:*

$$m_{ADM}(M^3) \geq m_H(\partial M^3) = \sqrt{\frac{\text{Area}(\Sigma)}{16\pi}} \tag{9.204}$$

and Penrose Rigidity:

$$m_{ADM}(M^3) = m_H(\partial M^3) \quad \Rightarrow \quad M^3 \text{ is isometric to } M_{Sch,m} \tag{9.205}$$

where $M_{Sch,m}$ is the Riemannian Schwarschild space with mass $m = m_{ADM}(M^3)$:

$$M_{Sch,m} = \left(\mathbb{R}^3 \setminus \{0\}, (1 + \tfrac{m}{2r})^2 \delta\right). \tag{9.206}$$

Almost rigidity for the Penrose Inequality was conjectured jointly with Lee in [38] where it was proven in the rotationally symmetric setting. This has not yet been explored for graph manifolds nor for Brill-Lindquist Geometrostatic manifolds. These are perhaps easy enough to assign as a first project to a doctoral student as the techniques in [28] and [57] should directly apply. Proving it in general would involve all the same difficultes as proving the almost rigidity for the Positive Mass Theorem and more.

9.7.2 The Bartnik Conjecture

Bartnik's quasilocal mass of a region $\Omega_0 \subset M^3$ where $M_0^3 \in \mathcal{M}$ and $\partial M_0^3 \subset \Omega_0$ was defined in [3] as an infimum of the ADM masses of extensions, M, of Ω_0:

$$m_B(\Omega_0) = \inf\{m_{ADM}(M) | \, M \in \mathcal{PM}(\Omega_0)\} \tag{9.207}$$

where $M \in \mathcal{PM} \subset \mathcal{M}$ if it contains an isometric image of Ω: $\Omega \subset M$. Bartnik conjectured that this infimum is achieved by what he called the *minimal mass extension* and that this minimal mass extension is scalar flat and static. Significant research in this area has been completed by Corvino in [16] in which the properties of a minimal mass extension are proven assuming it exists. Miao has searched for static extensions using perturbative methods in [44]. To prove the minimal mass extension exists in general one may consider the following approach:

Conjecture 9.7.6 (Bartnik Conjecture). *There exists a sequence*

$$M_j \in \mathcal{PM}(\Omega_0) \text{ such that } m_{ADM}(M_j) \to m_B(\Omega_0). \tag{9.208}$$

with a limit $M_j \to M_\infty$ such that $m_{ADM}(M_\infty) = m_B(\Omega_0) = \lim_{j\to\infty} m_{ADM}(M_j)$, where M_∞ is asymptotically flat, smooth, scalar flat and static.

By the definition of Bartnik Mass we know there is a sequence satisfying (9.208). We propose that one may be able to prove an intrinsic flat compactness theorem for $\Omega_{R,j} \subset M_j$ where $\Omega_{R,j} = T_R(\partial\Omega)$ or perhaps $\Omega_{R,j} \supset \Omega_0$ with well chosen $\partial\Omega_{R,j} = \Sigma_{R,j} \subset M_j$. Note that by Wenger's Compactness Theorem we only need

$$\mathrm{vol}(\Omega_{R,j}) \le V_R \text{ and } \mathrm{vol}(\partial\Omega_{R,j}) \le A_R \tag{9.209}$$

to obtain a subsequence $\Omega_{R,j} \to \Omega_{R,\infty}$ since $\mathrm{diam}\,(\Omega_{R,j}) \le 2R + \mathrm{diam}\,(\Omega_0)$. Then we could glue together an M_∞ from these limit regions $\Omega_{R,\infty}$ and $\Omega_0 \subset M_\infty$. Our conjectured Hawking mass compactness theorem would imply that M_∞ has generalized Scal ≥ 0 and a well controlled Hawking mass.

However, one must be warned that an M_∞ obtained in this manner need not be asymptotically flat. In [38], Lee and the author proved that a sequence M_j approaching equality in the Penrose Inequality may develop a longer and longer neck. The recent examples of Mantoulidis and Schoen which satisfy (9.208) also develop increasingly long necks so that M_j smoothly converge on regions around Ω_0 to an M_∞ which is not asymptotically flat and has no ADM mass [43]. For the regions near infinity these M_j are simply Schwarzschild space. Thus a new construction is needed which shortens these necks so that the M_j are in some sense uniformly asymptotically flat. Then one could try to prove a subsequence converges in the intrinsic flat sense to an aymptotically flat limit.

Once one has proven $M_j \to M_\infty$ where M_∞ is asymptotically flat and has generalized Scal ≥ 0, then one must prove the semicontinuity of the ADM masses. In [31] Jauregui has proven this semicontinuity in a variety of settings including intrinsic flat convergence in the rotationally symmetric case, by applying the Hawking mass Compactness Theorem proven by the author with LeFloch in [40]. Jauregui-Lee have proven semicontinuity of the ADM mass under C_0 convergence using the Huisken isoperimetric mass in [32]. These techniques might apply more generally.

9.7.3 Almost Rigidity of the Scalar Torus Theorem

The Scalar Torus Theorem states any M^n diffeomorphic to $\mathbb{T}^n$ with Scal ≥ 0 is isometric to a flat torus. It was proven using minimal surfaces for $n \le 7$ by Schoen and Yau in [50]. It was proven by Gromov and Lawson in all dimensions using spinors and the Lichnerowetz formula in [25]. Gromov proposed the following almost rigidity conjecture vaguely in [23]. To avoid collapsing and expanding we have normalized the manifolds with the volume and diameter bounds. To avoid bubbling, we have added the uniform lower bound on MinA of (9.13). This conjecture is false for GH convergence as seen in Example 9.2.3.

Conjecture 9.7.7. *Let M_j^3 be diffeomorphic to $\mathbb{T}^3$ with $\mathrm{Scal}(M_j^3) \geq -1/j$ and*

$$\mathrm{vol}(M_j^3) = V_0 \quad \mathrm{diam}\,(\mathrm{M}_j^3) \leq \mathrm{D}_0 \quad \mathrm{MinA}(\mathrm{M}_j^3) \geq \mathrm{A}_0 > 0. \tag{9.210}$$

Then a subsequence $M_{j_k} \xrightarrow{\mathcal{F}} M_0$ where M_0 is a flat torus. Possibly $M_{j_k} \xrightarrow{\mathcal{VF}} M_0$.

Note that by Wenger's Compactness Theorem, we know $M_{j_k} \xrightarrow{\mathcal{F}} M_\infty$, and by semi-continuity we have $\mathrm{diam}\,(\mathrm{M}_\infty) \leq \mathrm{D}_0$ and $\mathbf{M}(M_\infty) \leq V_0$. We do not know if M_∞ is the **0** integral current space. So even proving that much would be interesting. Some points will disappear as seen in Example 9.2.3, so one must find a sequence of points which doesn't disappear.

Even assuming that $M_\infty \neq \mathbf{0}$ one needs to prove that it has some sort of generalized Scal ≥ 0 which is strong enough to prove torus rigidity. Gromov suggests a few such notions which should be strong enough in [23] and prove C^0 limits of M_j^3 as above satisfy these conditions. Examining his proofs and considering filling volumes and sliced filling as defined in joint work of the author with Portegies [49] might be helpful. Note that Bamler has proven Scal ≥ 0 persists under C^0 convergence using Ricci flow in [2].

Perhaps a more approachable problem would be to consider first M_j^3 which are graphs over the standard $\mathbb{T}^3$ and try to prove $M_j^3 \xrightarrow{\mathcal{F}} \mathbb{T}^3$ using methods similar to those used in joint work of the author with Huang and Lee [28]. Another possibility is to consider M_j^3 with metric tensors $g_j = dx^2 + dy^2 + f_j(x,y)dz^2$ or $g_j = a_j(z)^2 dx^2 + b_j^2(z)dy^2 + dz^2$ and first try to prove the warping functions a_j, b_j and f_j converge in the H^1_{loc} sense to a metric with generalized Scal ≥ 0 as in joint work of the author with LeFloch [40]. It is possible that spinors and the Lichnerowetz formula might be formulated weakly for an H^1_{loc} metric tensor. More likely one can find a partial differential inequality on the warping functions which implies the rigidity theorem and can be shown to persist weakly under H^1_{loc}. Finally one can use an Arzela–Ascoli Theorem to relate the limit space obtained under H^1_{loc} convergence and the intrinsic flat limit.

9.7.4 Almost Rigidity Theorems and Ricci Flow

Bray, Brendle and Neves proved the Cover Splitting Rigidity Theorem as in (9.15) and Bray, Brendle, Eichmair and Neves have proven the $\mathbb{RP}^3$ Rigidity Theorem as in (9.16) using Ricci flow [8] [7]. Here we propose the following:

Conjecture 9.7.8 (Almost Rigidity of the $\mathbb{RP}^3$ Rigidity Theorem).
Given a sequence of M_j^3 that are diffeomorphic to $\mathbb{RP}^3$ with

$$\mathrm{Scal}_j \geq \tfrac{6j}{j-1} \quad \mathrm{MinA}(M^3) = 2\pi \quad \mathrm{vol}(M_j^3) \leq V_0 \quad \mathrm{diam}\,(\mathrm{M}_j^3) \leq \mathrm{D}_0 \tag{9.211}$$

then $M_j^3 \xrightarrow{\mathcal{F}} M_0 = \mathbb{RP}^3$ or possibly even $M_j^3 \xrightarrow{\mathcal{VF}} M_0 = \mathbb{RP}^3$.

One may construct examples with wells demonstrating that $M_j^3 \xrightarrow{\text{GH}} \mathbb{RP}^3$ may fail.

Conjecture 9.7.9 (Almost Rigidity of Cover Splitting Theorem).
Given a sequence of M_j^3 that contain noncontractible $\mathbb{S}^2$ and have

$$\mathrm{Scal}_j \geq \tfrac{2j}{j-1} \quad \mathrm{MinA}(M^3) = 4\pi \quad \mathrm{vol}(M_j^3) \leq V_0 \quad \mathrm{diam}\,(\mathrm{M}_\mathrm{j}^3) \leq \mathrm{D}_0 \tag{9.212}$$

then a subsequence $M_{j_k}^3 \xrightarrow{\mathcal{F}} M_0$ where $\tilde{M}_0$ is isometric to $\mathbb{S}^2 \times \mathbb{R}$.

Problem 9.7.10. *Can one construct M_j with $Scalar \geq 2j/(j-1)$ that are diffeomorphic to $\mathbb{S}^2 \times \mathbb{S}^1$ and contain balls of radius $1/4$ that are isometric to balls in rescaled standard spheres? If so then one can attach wells and have no GH limit.*

One approach to proving these conjectures would be to use volume renormalized Ricci flow, $M_{j,t}$, of M_j and show $M_{j,t}$ flows as $t \to \infty$ to an $M_{j,\infty}$ which is isometric to M_0. One may simplify things by assuming a smooth Ricci flow exists for all time, or try to prove this, or attempt to deal with Ricci flow through singularities. If

$$d_{Vol\mathcal{F}}(M_{j,t}, M_{j,s}) < \epsilon\left(\tfrac{1}{t} - \tfrac{1}{s}\right) \text{ where } \lim_{\delta\to 0} \epsilon(\delta) = 0 \tag{9.213}$$

where ϵ is independant of j, then we have the conjecture. Continuity of Ricci flow with respect to the intrinsic flat distance has been studied by Lakzian in [34] but only to analyze the Ricci flow through a neck pinch singularity. The author believes one might be able to obtain (9.213) constructing an explicit

$$Z = \{(x, r) | \, x \in M_{j,(1/r)},\ r \in [(1/t), (1/s)]\} \text{ with } g = dr^2 + f(r)^2 g_{j,(1/r)}. \tag{9.214}$$

Again one might consider special cases where M_j are known to be warped products.

9.7.5 Gromov's Prisms and Generalized Nonnegative Scalar Curvature

Gromov suggested in [24] that if a sequence of $M_j \xrightarrow{\mathcal{F}} M_\infty$ has Scal ≥ 0 then M_∞ might have Scal ≥ 0 in some generalized sense. In [23], Gromov proved the Gauss Bonnet Prism Inequality for prisms in manifolds with Scal ≥ 0:

$$\sum_{i=1}^{3} \alpha_i \geq \pi \text{ and } \sum_{i=1}^{3} \alpha_i = \pi \quad \Rightarrow \quad P \text{ is a prism in } \mathbb{E}^3 \tag{9.215}$$

where α_i are bounds on the dihedral angles between the sides of the prism which are minimal surfaces. He suggests that the Prism Inequality could be used to define generalized Scal ≥ 0, and then be applied to prove the torus rigidity theorem on such limit spaces.

Before we can define such a generalized notion of nonnegative scalar curvature on integral current spaces, we would need to prove more regularity on the $\mathcal{F}$ limit spaces.

Ordinarily there is no notion of angle on an integral current space, nor is an integral current space a geodesic space. In fact it might even be the zero space. It should be noted that Ilmanen, the author and Wenger conjectured in [61][Conjecture 4.18] that "a converging sequence of three dimensional Riemannian manifolds with positive scalar curvature, a uniform lower bound on volume, and no interior closed minimal surfaces converges without cancellation to a nonzero integral current space." Combining Gromov's ideas with this one we introduce the following conjecture:

Conjecture 9.7.11. *Suppose* $M_j^3 \xrightarrow{\mathcal{F}} M_\infty$ *or* $M_j^3 \xrightarrow{V\mathcal{F}} M_\infty$ *where* $\mathrm{Scal}_j \geq 0$ *and*

$$\mathrm{MinA}(M_j) \geq A_0 > 0 \quad \mathrm{vol}(M_j) \in [V_0, V_1] \subset (0, \infty) \quad \mathrm{diam}\,(\mathrm{M_j}) \leq \mathrm{D_0}. \tag{9.216}$$

Then we have the following:

(a) M_∞ *is a nonzero integral current space. In fact there is no cancellation without collapse: if* $p_j \in M_j$ *has no limit in* $\bar{M}_\infty$ *then* $\exists \delta > 0$ *s.t.* $\mathrm{vol}(B(p_j, \delta)) \to 0$.

(b) M_∞ *is geodesic: if* $p, q \in M_\infty$ *there exist* $p_j \to p$ *and* $q_j \to q$ *with midpoints* x_j *that converge to* x_∞ *which is a midpoint between* p *and* q *in* M_∞.

(c) There is a notion of angle between geodesics emanating from a point.

(d) There is a notion of dihedral angle between two surfaces at $p \in \Sigma \cap \Sigma' \subset M_\infty$.

(e) Gromov's Gauss-Bonnet Prism Inequality as in (9.215) holds on M_∞.

(f) $\forall p \in M_\infty\ \exists r_p > 0\ s.t.\ \forall r < r_p \quad V_p(r) = \mathbf{M}(B(p, r))/(4\pi r^3/3) \leq 1.$

Note that (a)-(d) are regularity properties for our limit spaces. As seen in examples above, they do not hold on limits of manifolds with $\mathrm{Scal}_j \geq 0$ unless the $\mathrm{MinA}(M_j) \geq A_0 > 0$. The notion of sliced filling volume developed in joint work with Portegies in [49] might be helpful towards proving (a) and (b). It is quite possible that (c) is false but that one can still prove (d) using the limit process to define the dihedral angle. In the special case where there exist Riemannian embeddings $\Psi_j : M_j \to \mathbb{E}^N$, one can use an Arzela-Ascoli Theorem to obtain $\Psi_\infty : M_j \to \mathbb{E}^N$. Then one can use $\mathbb{E}^N$ to define angles as needed in (c) or (d) and examine the semicontinuity of such angles. Note that any Riemannian manifold satisfies (a)-(d) so they do not capture a generalized notion of nonnegative scalar curvature.

Now (e)-(f) are properties which capture $\mathrm{Scal} \geq 0$. Property (e) proposed by Gromov needs a notion of angle so one needs to prove (d) first or use embeddings into some large $\mathbb{E}^N$ to even define what this means. Then one need only prove semicontinuity of the angles. The power of property (e) is described in [23] including ideas towards the possibility that (e) implies (f). Recall that (f) on a Riemannian manifold is equivalent to $\mathrm{Scal} \geq 0$ but was not powerful enough to prove any global properties of such spaces. Nor is (f) continuous with respect to $Vol\mathcal{F}$ convergence unless one assumes a uniform lower bound $r_p \geq r_0 > 0$ for all $p \in M_j$ for all $j \in \mathbb{N}$. Nevertheless any natural notion of Scal ≥ 0 on an integral current space ought to imply (f). Note that Gromov also proves hyperbolic and spherical prism inequalities on spaces with lower bounds on scalar curvature which are negative or positive, respectively. He suggests that such such inequalities might be used to generalize these lower bounds on scalar curvature. One might prove they imply the appropriate limit as in (9.5) on M_∞.

Remark 9.7.12. *In Remark 9.7.3 we proposed that the Brown York mass might be semi-continuous with respect to $\mathcal{F}$ convergence. One might devise a way to define the Brown-York mass, $m_{BY}(\partial P)$, where P is a prism. These definitions are likely only to involve integrals of dihedral angles. It is quite possible that it would be easier to study the limits of Brown-York masses of prisms than to even define dihedral angles on the limit spaces. If one can do this, then might try to replace (e) with*

(e_{BY}) *Shi-Tam Nonnegativity of the Brown-York Mass of Prisms: $m_{BY}(\partial P) \geq 0$.*

Perhaps the consequences Gromov devises using (e) might be concluded from (e_{BY}).

9.7.6 Almost Rigidity of the Positive Mass Theorem and Regularity of $\mathcal{F}$ Limits

It should be noted that in the famous work of Cheeger-Colding on the properties of metric measure limits of Riemannian manifolds with lower bounds on Ricci curvature, a key step in proving regularity and the existence of Euclidean tangent spaces at regular points, was the proof of their Almost Splitting Theorem [14]. In their work a point p in a limit space M_∞ has a Euclidean tangent space if the sequence of rescaled balls

$$(B(p, r_j), d_j/r_j) \text{ where } r_j \to 0 \tag{9.217}$$

converges in the GH sense to a ball $B(0, 1)$ in Euclidean space. The Toponogov Splitting Theorem was similarly used to prove regularity results for Alexandrov Spaces by Burago-Gromov-Perelman [11]. However, there is no splitting theorem for manifolds with nonnegative scalar curvature.

Remark 9.7.13. *Let us consider $M_j \xrightarrow{\mathcal{F}} M_\infty$ where the M_j satisfy conditions like those in the proposed Scalar Compactness Conjecture,*

$$\mathrm{Scal}_j \geq 0 \text{ and } \mathrm{MinA}(M_j) \geq A_0 > 0 \tag{9.218}$$

and $p \in M_\infty$. If one has proven a local version of the almost rigidity of the Positive Mass Theorem then one can determine settings in which

$$(B(p, r_j), d_j/r_j, [B(p, r_j)]) \xrightarrow{\mathcal{F}} (B(0, 1), d_{\mathbb{E}^3}, [B(0, 1)]). \tag{9.219}$$

This may possibly then be applied in the place of a splitting theorem to prove some sort of regularity on the limit space, M_∞.

Note that without the assumptions in (9.218), an intrinsic flat limit may have no regular points in the sense described in (9.219). In the original paper defining intrinsic flat convergence by the author with Wenger [61], an example of a sequence of M_j is given which converges in the intrinsic flat sense to taxicab space,

$$M^2_{taxi} = ([0,1] \times [0,1], d_{taxi}, [[0,1] \times [0,1]]) \tag{9.220}$$

where

$$d_{taxi}\left((x_1, x_2), (y_1, y_2)\right) = |x_1 - y_1| + |x_2 - y_2|. \tag{9.221}$$

Such a space has no notion of angles and no prism properties. It is possible to imagine how sewing methods could be used to construct a sequence of M_j^3 with $\mathrm{Scal}_j \geq -1/j$ that converges to M^3_{taxi}. However, the $\mathrm{MinA}(M_j) \geq A_0 > 0$ condition fails on such examples. The taxi limit example in [61] satisfies $\mathrm{MinA}(M_j) \geq A_0 > 0$ but contains points for which the scalar curvature decreases to negative infinity.

Given M_∞ as in Remark 9.7.13 one may even be able to prove that almost every point $p \in M_\infty$ satisfies (9.219). Thus it may well be worthwhile to examine to what extent one may use this kind of regularity to define dihedral angles, mean curvatures, quasilocal masses and a generalized notion of nonnegative scalar curvature on M_∞.

Acknowledgment: The author would like to thank the following mathematicians for discussions over the past few years related to this paper and for suggested references: Bartnik, Brendle, Carlotto, Cederbaum, Eichmair, Gigli, Huisken, Ilmanen, Khuri, Gromov, Huang, Jauregui, Lawson, Lee, LeFloch, Miao, Morgan, Sakovich, Schoen, Shi, Stavrov, Wang and Yau. The author's research was funded by an individual research grant NSF-DMS-1309360 and a PSC-CUNY Research Grant. This survey was written as notes for a course at the 2016 School in Geometric Analysis at the Lake Como School for Advanced Studies sponsored by IISS, ERC, GNAMPA, and DISAT; organized by Besson, Pigola, Impera, Rimoldi, Setti, Troyanov, Valtorta, and Veronelli.

Bibliography

[1] L. Ambrosio and B. Kirchheim: Currents in metric spaces. Acta Math., 185(1):1–80, (2000).

[2] R. Bamler: A Ricci flow proof of a result by gromov on lower bounds for scalar curvature. Mathematical Research Letters, 23(2):325–337, 2016.

[3] R. Bartnik: The mass of an asymptotically flat manifold. Comm. Pure Appl. Math., 39(5):661–693, (1986).

[4] R. Bartnik: New definition of quasilocal mass. Phys. Rev. Lett., 62(20):2346–2348, (1989).

[5] J. Basilio, J. Dodziuk, and C. Sormani: Sewing Riemannian Manifolds with Positive Scalar Curvature, Preprint: arXiv:1703.00984 (2017)

[6] J. Basilio and C. Sormani: Sequences of three-dimensional manifolds with positive scalar curvature. preprint to appear, (2017).

[7] H. Bray, S. Brendle, M. Eichmair, and A. Neves: Area-minimizing projective planes in 3-manifolds. Comm. Pure Appl. Math., 63(9):1237–1247, (2010).

[8] H. Bray, S. Brendle, and A. Neves: Rigidity of area-minimizing two-spheres in three-manifolds. Comm. Anal. Geom., 18(4):821–830, (2010).
[9] H. L. Bray: Proof of the Riemannian Penrose inequality using the positive mass theorem. J. Differential Geom., 59(2):177–267, (2001).
[10] Burago, D., Burago, Y., and Ivanov, S. (2001). A course in metric geometry, volume 33 of Graduate Studies in Mathematics. American Mathematical Society, Providence, RI.
[11] Y. Burago, M. Gromov, and G. Perel′man: A. D. Aleksandrov spaces with curvatures bounded below. Uspekhi Mat. Nauk, 47(2(284)):3–51, 222, (1992).
[12] A. Carlotto, O. Chodosh and M. Eichmair: Effective versions of the positive mass theorem. Inventiones mathematicae, (2016).
[13] J. Cheeger and T. H. Colding: Lower bounds on Ricci curvature and the almost rigidity of warped products. Ann. of Math. (2), 144(1):189–237, (1996).
[14] J. Cheeger and T. H. Colding: On the structure of spaces with Ricci curvature bounded below. I. J. Differential Geom., 46(3):406–480, (1997).
[15] T. H. Colding: Ricci curvature and volume convergence. Ann. of Math. (2), 145(3):477–501, (1997).
[16] J. Corvino: Scalar curvature deformation and a gluing construction for the Einstein constraint equations. Comm. Math. Phys., 214(1):137–189, (2000).
[17] E. DeGiorgi: Problema di plateau generale e funzionali geodetici. Atti Sem. Mat. Fis. Univ. Modena, 43:285–292, (1995).
[18] X.-Q. Fan, Y. Shi and L.-F. Tam: Large-sphere and small-sphere limits of the Brown-York mass. Comm. Anal. Geom., 17(1):37–72, (2009).
[19] H. Federer and W. H. Fleming: Normal and integral currents. Ann. of Math. (2), 72:458–520, (1960).
[20] R. Geroch: Energy extraction. Ann. New York Acad. Sci., 224:108–117, (1973).
[21] M. Gromov: Structures Métriques pour les Variétés Riemanniennes, volume 1 of Textes Mathématiques [Mathematical Texts]. CEDIC, Paris. Edited by J. Lafontaine and P. Pansu, (1981).
[22] M. Gromov: Filling Riemannian manifolds. J. Differential Geom., 18(1):1–147, (1983).
[23] M. Gromov: Dirac and Plateau billiards in domains with corners. Cent. Eur. J. Math., 12(8):1109–1156, (2014).
[24] M. Gromov: Plateau-Stein manifolds. Cent. Eur. J. Math., 12(7):923–951, (2014).
[25] M. Gromov and H. B. Jr. Lawson: Spin and scalar curvature in the presence of a fundamental group. I. Ann. of Math. (2), 111(2):209–230, (1980).
[26] M. Gromov and H. B. Jr. Lawson: Spin and scalar curvature in the presence of a fundamental group. I. Ann. of Math. (2), 111(2):209–230, (1980).
[27] R. S. Hamilton: The formation of singularities in the Ricci flow. In Surveys in differential geometry, Vol. II (Cambridge, MA, 1993), pages 7–136. Int. Press, Cambridge, MA, (1995).
[28] L.-H. Huang, D. Lee and C. Sormani: Stability of the positive mass theorem for graphical hypersurfaces of Euclidean space. to appear Journal fur die Riene und Angewandte Mathematik, (2016).
[29] G. Huisken: An isoperimetric concept for the mass in general relativity. Presentation at IAS, (2009).
[30] G. Huisken and T. Ilmanen: The inverse mean curvature flow and the Riemannian Penrose inequality. J. Differential Geom., 59(3):353–437, (2001).
[31] J. Jauregui: On the lower semicontinuity of the adm mass. arXiv:1411.3699, (2016).
[32] J. Jauregui and D. A. Lee: Lower semicontinuity of mass under C^0 convergence and Huisken's isoperimetric mass. arXiv:1602.00732, (2016).
[33] S. Lakzian: Intrinsic flat continuity of Ricci flow through neckpinch singularities. Geom. Dedicata, 179:69–89, (2015).

[34] S. Lakzian: Intrinsic flat continuity of Ricci flow through neckpinch singularities. Geom. Dedicata, 179:69–89, (2015).
[35] S. Lakzian: On diameter controls and smooth convergence away from singularities. Differential Geom. Appl., 47:99–129, (2016).
[36] S. Lakzian and C. Sormani: Smooth convergence away from singular sets. Comm. Anal. Geom., 21(1):39–104, (2013).
[37] H. B. Jr. Lawson and M.-L. Michelsohn: Spin geometry, volume 38 of Princeton Mathematical Series. Princeton University Press, Princeton, NJ, (1989).
[38] D. A. Lee and C. Sormani: Near-equality of the Penrose inequality for rotationally symmetric Riemannian manifolds. Ann. Henri Poincaré, 13(7):1537–1556, (2012).
[39] D. A. Lee and C. Sormani: Stability of the positive mass theorem for rotationally symmetric riemannian manifolds. Journal fur die Riene und Angewandte Mathematik (Crelle's Journal), 686, (2014).
[40] P. G. LeFloch and C. Sormani: The nonlinear stability of rotationally symmetric spaces with low regularity. J. Funct. Anal., 268(7):2005–2065, (2015).
[41] N. Li and R. Perales: On the Sormani-Wenger intrinsic flat convergence of Alexandrov spaces. arXiv:1411.6854, (2015).
[42] R. Matveev and J. Portegies: Intrinsic flat and Gromov-Hausdorff convergence of manifolds with Ricci curvature bounded below, to appear in the Journal of Geometric Analysis, arXiv:1510.07547, (2015).
[43] C. Mantoulidis and R. Schoen: On the Bartnik mass of apparent horizons. Classical and Quantum Gravity, 32(20):205002, 2015.
[44] P. Miao: On existence of static metric extensions in general relativity. Comm. Math. Phys., 241(1):27–46, (2003).
[45] M. Munn: Intrinsic flat convergence with bounded ricci curvature. arXiv:1405.3313, (2014).
[46] R. Perales: Convergence of manifolds and metric spaces with boundary. arXiv:math/1505.01792, (2015).
[47] R. Perales: Volumes and limits of manifolds with ricci curvature and mean curvature bounds, to appear in Journal of Differential Geometry and Applications, arXiv:math/1404.0560, (2015).
[48] J. W. Portegies: Semicontinuity of eigenvalues under intrinsic flat convergence. Calc. Var. Partial Differential Equations, 54(2):1725–1766, (2015).
[49] J. Portegies and C. Sormani: Properties of the intrinsic flat distance. arXiv:1210.3895, 2014.
[50] R. Schoen and S. T. Yau: Existence of incompressible minimal surfaces and the topology of three-dimensional manifolds with nonnegative scalar curvature. Ann. of Math. (2), 110(1):127–142, (1979).
[51] R. Schoen and S. T. Yau: On the proof of the positive mass conjecture in general relativity. Comm. Math. Phys., 65(1):45–76, (1979).
[52] R. Schoen and S. T. Yau: On the structure of manifolds with positive scalar curvature. Manuscripta Math., 28(1-3):159–183, (1979).
[53] Y. Shi and L.-F. Tam: Positive mass theorem and the boundary behaviors of compact manifolds with nonnegative scalar curvature. J. Differential Geom., 62(1):79–125, (2002).
[54] Z. Sinaei and C. Sormani: Limits of covering spaces under intrinsic flat convergence. Geom Dedicata, 184, 83–114, (2016).
[55] C. Sormani: Friedmann cosmology and almost isotropy. Geom. Funct. Anal., 14(4):853–912, (2004).
[56] C. Sormani: Intrinsic flat Arzela-Ascoli theorems, to appear in Communications in Analysis and Geometry, preprint on arXiv: 1402.6066, (2014).
[57] C. Sormani and I. Stavrov: Geometrostatic manifolds of small adm mass. to appear, (2017).

[58] C. Sormani and G. Wei: Hausdorff convergence and universal covers. Trans. Amer. Math. Soc., 353(9):3585–3602 (electronic), (2001).

[59] C. Sormani and G. Wei: The covering spectrum of a compact length space. J. Differential Geom., 67(1):35–77, (2004).

[60] C. Sormani and S. Wenger: Weak convergence and cancellation, appendix by Raanan Schul and Stefan Wenger. Calculus of Variations and Partial Differential Equations, 38(1-2), (2010).

[61] C. Sormani and S. Wenger: Intrinsic flat convergence of manifolds and other integral current spaces. Journal of Differential Geometry, 87, (2011).

[62] S. Wenger: Flat convergence for integral currents in metric spaces. Calc. Var. Partial Differential Equations, 28(2):139–160, (2007).

[63] S. Wenger: Compactness for manifolds and integral currents with bounded diameter and volume. Calculus of Variations and Partial Differential Equations, 40(3-4):423–448, (2011).

[64] H. Whitney: Geometric integration theory. Princeton University Press, Princeton, N. J., (1957).

[65] E. Witten: A new proof of the positive energy theorem. Comm. Math. Phys., 80(3):381–402, (1981).

Kolibri 360 GmbH
Ferdinand-Jühlke-Straße 7
99095 Erfurt, Deutschland
produktsicherheit@kolibri360.de